NOTATION

ELEMENTARY LINEAR ALGEBRA

THE PRINDLE, WEBER, AND SCHMIDT SERIES IN MATHEMATICS

Althoen and Bumcrot, *Introduction to Discrete Mathematics*

Brown and Sherbert, *Introductory Linear Algebra with Applications*

Buchthal and Cameron, *Modern Abstract Algebra*

Burden and Faires, *Numerical Analysis*, Fourth Edition

Cullen, *Linear Algebra and Differential Equations*

Cullen, *Mathematics for the Biosciences*

Eves, *In Mathematical Circles*

Eves, *Mathematical Circles Adieu*

Eves, *Mathematical Circles Revisited*

Eves, *Mathematical Circles Squared*

Eves, *Return to Mathematical Circles*

Fletcher and Patty, *Foundations of Higher Mathematics*

Geltner and Peterson, *Geometry for College Students*

Gilbert and Gilbert, *Elements of Modern Algebra*, Second Edition

Gobran, *Beginning Algebra*, Fourth Edition

Gobran, *College Algebra*

Gobran, *Intermediate Algebra*, Fourth Edition

Gordon, *Calculus and the Computer*

Hall, *Algebra for College Students*

Hall and Bennett, *College Algebra with Applications*, Second Edition

Hartfiel and Hobbs, *Elementary Linear Algebra*

Hunkins and Mugridge, *Applied Finite Mathematics*, Second Edition

Kaufmann, *Algebra for College Students*, Third Edition

Kaufmann, *Algebra with Trigonometry for College Students*, Second Edition

Kaufmann, *College Algebra*

Kaufmann, *College Algebra and Trigonometry*

Kaufmann, *Elementary Algebra for College Students*, Third Edition

Kaufmann, *Intermediate Algebra for College Students*, Third Edition

Kaufmann, *Precalculus*

Kaufmann, *Trigonometry*

Keisler, *Elementary Calculus: An Infinitesimal Approach*, Second Edition

Kirkwood, *Introduction to Real Analysis*

Laufer, *Discrete Mathematics and Applied Modern Algebra*

Nicholson, *Linear Algebra with Applications*

Pasahow, *Mathematics for Electronics*

ELEMENTARY LINEAR ALGEBRA
THIRD EDITION

Stewart Venit and Wayne Bishop
California State University at Los Angeles

PWS–KENT PUBLISHING COMPANY
Boston

PWS–KENT
Publishing Company

20 Park Plaza
Boston, Massachusetts 02116

PWS–KENT Publishing Company is a division of Wadsworth, Inc.

Library of Congress Cataloging-in-Publication Data

Venit, Stewart.
    Elementary linear algebra/Stewart Venit & Wayne Bishop.—3rd ed.

        p.   cm.
    Includes index.
    ISBN 0-534-91689-9
    1. Algebras, Linear.   I. Bishop, Wayne   II. Title.
QA184.V46 1989
512.9--dc 19   88-23345
                                                                CIP

Printed in the United States of America
  90 91 92 93—10 9 8 7 6 5 4 3

Sponsoring Editor   *David Geggis*
Production Coordinator   *Elise Kaiser*
Composition   *Universities Press Ltd.*
Interior/Cover Design   *Elise Kaiser*
Manufacturing Coordinator   *Margaret Sullivan Higgins*
Cover Printer   *New England Book Components*
Text Printer/Binder   *Maple-Vail Book Manufacturing Group*

Cover etching "Zapateado" by Caroline Stone. Used with permission of the artist.

PREFACE

This text is designed primarily for a first course in linear algebra for majors in mathematics, computer science, engineering, or the sciences. Knowledge of calculus is not required, except for some clearly marked optional material in the second half of the text.

FEATURES

EMPHASIS

For the most part, we have emphasized the computational aspects of linear algebra. One of our basic goals has been to provide a highly readable text that develops students' abilities to apply the material as well as to recall it. We have tried to avoid the pitfall of stating and proving elegant theorems without showing how they can be used to solve concrete problems. Throughout the text, algorithmic procedures are carefully described to guide students in their thinking; these procedures are listed for easy reference on the inside front cover.

ARRANGEMENT OF MATERIAL

Since a course in linear algebra usually serves as a transition between the heavily computational calculus and the more theoretical upper division courses in many disciplines, we have tried to maintain a reasonable balance between algorithms and theory. To as great an extent as possible, the easier, more computational material appears at the beginning of each chapter. Proofs are often delayed until the end of a section or at least until an illustrative example has been worked.

We have also tried to ease a difficult transition within the course itself. In most other texts, after the first few chapters have dealt with the relatively straightforward material on linear systems, matrices, and determinants, a quantum leap in difficulty takes place. The topics of abstract vector space, subspace, span, linear dependence, basis, and dimension appear in rapid, bewildering succession. Chapter 4 of this text introduces the notions of linear independence, subspace, basis, and dimension within the friendly confines of R^m. When these topics are then studied in the context of abstract vector spaces in Chapter 5, students will be less intimidated by them than is normally the case.

CHAPTER REVIEWS

Review material is provided at the end of each of the first seven chapters. This material consists of a very brief synopsis of the subject matter, a list of the computational skills that the student should have acquired, "key words" (terms introduced in the chapter), "key theorems," a set of review exercises, a self-test, and answers to the self-test.

WORKING WITH THE COMPUTER

The computer is playing an ever-increasing role in mathematics in general and in linear algebra in particular. Acknowledging this fact, we provide Computational Notes and Computational Exercises throughout the text, which are denoted by a computer logo. The former relate the topic under discussion to the computer, while the latter are meant to be solved by means of "canned" programs. The publisher offers adopters of this text two software packages containing such programs:

1. A minimal package, free of charge, consisting of a diskette (formatted for one of several microcomputers) that contains the BASIC programs listed in the Appendix.
2. Adopters of this edition may request a free site license for MAX—The MAtriX Algebra Calculator, Version 3.1. An associated problem book for students is also available. A demonstration package may be requested from the publisher.

APPLICATIONS

There are two types of applications presented in this text. Sprinkled throughout the first seven chapters are "mini-applications" labeled Spotlight. These are about one page in length and help to answer, in a relatively painless fashion, the perennial student question, "What is this stuff good for anyway?"

The last two chapters contain ten much more detailed applications of linear algebra to various disciplines. The prerequisite material necessary for each application is given on the next page, as well as the relevant section. Adhering to these prerequisites, it would be easy to intersperse these applications throughout the course rather than leaving them for the end.

Application (section)	Prerequisite
The interpolating polynomial (8.1)	Chapter 2
The least square polynomial (8.2)	Chapter 2
Electrical circuits and pipe networks (8.3)	Section 2.2
Leontief models (8.4)	Chapter 2
Markov chains (8.5)	Section 2.3 (Section 7.1, if desired)
Quadric surfaces (8.6)	Sections 7.1, 7.2
Continuous least squares (8.7)	Sections 5.1, 5.2, 5.4
Fourier series (8.7)	Sections 5.1, 5.2, 5.4
Linear differential equations (8.8)	Sections 5.1, 5.2, 6.1–6.3
Linear programming (9.1–9.4)	Chapter 2

POSSIBLE COURSE OUTLINES

This text can be used for both quarter and semester-length courses and by both minimally prepared and mathematically advanced students. A few representative course outlines are given below.

1. Minimally prepared students—quarter course

Chapters 1 and 2

Chapter 3 (computational aspects)

Chapter 4, Sections 4.1–4.3

Chapter 5, Sections 5.1, 5.2 (if desired—not necessary)

Chapter 6, Section 6.1 (and 6.2, 6.3 if time permits)

Chapter 7, Section 7.1

2. Average class—semester course

Chapter 1 (quickly)

Chapters 2 and 3

Chapter 4, Sections 4.1–4.3

Chapter 5, Sections 5.1–5.3

Chapter 6, Sections 6.1–6.4

Chapter 7, Sections 7.1, 7.2

Chapters 8 and 9 as desired and as time permits

3. Advanced class, semester course

Chapters 1–7, all sections (the easier material as assigned reading)

Chapters 8 and 9 as desired and as time permits

SUPPLEMENTARY MATERIAL

The publisher can provide the following items to interested instructors:

Instructor's Manual, keyed to the text. For each chapter the manual contains an overview, remarks concerning the material, answers to even-numbered exercises (answers to the odd-numbered ones are at the back of this text), and solutions to selected exercises.

Software Packages are described earlier in this preface (see "Working with the Computer").

ACKNOWLEDGMENTS

We are grateful to the many people who have helped us create the three editions of this text. Of special importance were our colleagues in colleges and universities, whose reviews of the various manuscripts had a profound effect on the resulting books. The reviewers who have helped shape this third edition are: Benny Evans, *Texas A & M University*; L. C. Grove, *University of Arizona*; Joel K. Haack, *Oklahoma State University*; Norman K. Lee, *Ball State University*; James Rowell, *University of Minnesota, Duluth*; and David Sprows, *Villanova University*.

The reviewers of previous editions are: Alfonso Azpeitia, *University of Massachusetts, Boston*; Jean Bevis, *Georgia State University*; Clark Benson, *University of Arizona*; Gordon Brown, *University of Colorado*; Judith Cederberg, *St. Olaf College*; Richard Chamberlain, *California State University, Los Angeles*; Sam Councilman, *California State University, Long Beach*; Burton Fein, *Orgeon State University*; Leonard Fuller, *Kansas State University*; Michael Greeley, *University of Massachusetts, Boston*; David Kullman, *Miami University (Ohio)*; Ancel Mewborn, *University of North Carolina, Chapel Hill*; Edgar Stout, *University of Washington*; Robert Devaney, *Tufts University*; Richard Porter, *Northeastern University*; Harris Shultz, *California State University, Fullerton*; Ray Spring, *Ohio University*; and Harlan Stevens, *Pennsylvania State University*.

We have been fortunate to be associated with people at PWS–KENT who are as interested as we in creating a textbook of the highest quality. These include our editor David Geggis, production editor Elise Kaiser, and copyeditor David Hoyt (of Hoyt Publishing Services), all of whom were instrumental in producing the third edition. Finally, we would like to thank our wives, Corinne and Judi, and our children, Tamara, Jeffrey, and Anthony, for their encouragement and unflagging support.

Stewart Venit
Wayne Bishop

CONTENTS

9 LINEAR PROGRAMMING 383

SPOTLIGHTS

VELOCITY 7
Basic vector arithmetic is an aid to navigation.

TORQUE 21
Cross product can be used to help change a flat tire.

DISTANCE FROM A POINT TO A PLANE 31
It's easy to compute if you use linear algebra.

APPLICATIONS OF LINEAR SYSTEMS 54
They are an indispensable tool for the applied mathematician's toolbox.

REVENUE AS MATRIX MULTIPLICATION 80
Matrix multiplication really does make sense.

ELEMENTARY LINEAR ALGEBRA

1

GEOMETRY OF R^m

In the physical sciences some quantities, such as mass and pressure, can be adequately described by simply giving their magnitude (size). Others, such as force and velocity, require not only magnitude but direction as well. The latter quantities are usually represented by *vectors*.

In Sections 1.1 through 1.3 we will develop the properties of vectors in the plane and 3-space. Here it is easy to draw pictures of the situation and let intuition be our guide. In Section 1.4 we will then generalize our discussion to *m*-space, R^m. The terminology and results given in this section will be used throughout the rest of the text.

1.1   VECTORS IN R^2 AND R^3

Each point in a plane can be identified with a unique ordered pair of real numbers by using a pair of perpendicular lines called *coordinate axes* (Figure 1.1). In a similar manner each point in 3-space can be represented by a unique ordered triple of real numbers (Figure 1.2). Throughout this text we will denote the set of real numbers by **R**, the set of ordered pairs of real numbers by **R**^{2}, and the set of ordered triples by **R**^{3}.

In addition to identifying a point in 2-space or 3-space, an ordered pair or triple will be used to identify the *directed line segment* from the origin to the point indicated (Figures 1.3 and 1.4). This interpretation of an ordered pair or triple is called a **vector** in **R**^{2} or **R**^{3}. Vectors will be identified by boldfaced, lowercase, English letters. For example, $\mathbf{u} = (u_1, u_2, u_3)$ represents the vector in **R**^{3} with **components** u_1, u_2, and u_3. When we wish to speak of a *point* in **R**^{2} or **R**^{3}, we will use uppercase English letters to denote it, as in "the point P" or "the point $P(x_1, x_2, x_3)$." By convention, a point denoted by U will have coordinates (u_1, u_2) in **R**^{2} or (u_1, u_2, u_3) in **R**^{3}. Thus the point P and the vector **p** will be represented by the same ordered pair or triple. In other words, the *endpoint* of the vector **p** is the point P.

FIGURE 1.1

FIGURE 1.2

FIGURE 1.3

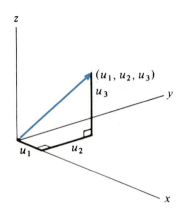

FIGURE 1.4

NOTE   Sometimes vectors are defined as directed line segments between *any* two points. We will reserve *vector* for the special case of a directed line segment that emanates from (begins at) the origin. More general directed line segments are important in geometric and physical applications. They will be discussed later in this section and used throughout this chapter.

···

Two vectors **u** and **v** are **equal** if their *corresponding components* are equal; that is, $u_1 = v_1$ and $u_2 = v_2$ (and in $\mathbf{R}^3$ $u_3 = v_3$). The **zero vector** is the origin of the coordinate system, $\mathbf{0} = (0, 0)$ (or $(0, 0, 0)$). Thus the statement $\mathbf{u} \neq \mathbf{0}$ is equivalent to saying that at least one component of **u** is nonzero.

Two nonzero vectors **u** and **v** are **collinear** if the lines they determine are one and the same; in other words, if the *points U, V,* and 0 are collinear points (Figure 1.5).

FIGURE 1.5

LENGTH OF A VECTOR

Recall that the distance between any two points in $\mathbf{R}^2$ or $\mathbf{R}^3$ can be given by the *distance formula*. The **distance** between two vectors $\mathbf{u}$ and $\mathbf{v}$ is defined to be the distance between their endpoints. Thus in $\mathbf{R}^2$,

$$d(\mathbf{u}, \mathbf{v}) = \sqrt{(v_1 - u_1)^2 + (v_2 - u_2)^2}.$$

Similarly, for $\mathbf{u}$ and $\mathbf{v}$ in $\mathbf{R}^3$,

$$d(\mathbf{u}, \mathbf{v}) = \sqrt{(v_1 - u_1)^2 + (v_2 - u_2)^2 + (v_3 - u_3)^2}.$$

By considering the right triangles indicated in Figures 1.6 and 1.7, we see that the distance formula is an application of the Pythagorean theorem. For example, in Figure 1.6 the length of the hypotenuse of the right triangle is $d(\mathbf{u}, \mathbf{v})$, and the legs have lengths $v_1 - u_1$ and $v_2 - u_2$ (or, more generally, $|v_1 - u_1|$ and $|v_2 - u_2|$). Thus

$$(d(\mathbf{u}, \mathbf{v}))^2 = (v_1 - u_1)^2 + (v_2 - u_2)^2,$$

and, by taking square roots, we obtain the first of the distance formulas. (For the second formula this process is repeated twice.)

FIGURE 1.6

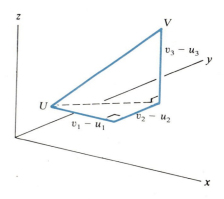

FIGURE 1.7

DEFINITION

The **length (norm, magnitude)** of a vector **u** is the distance from the origin to the endpoint of **u** and is denoted $\|\mathbf{u}\|$:

$$\|\mathbf{u}\| = d(\mathbf{0}, \mathbf{u}) = \begin{cases} \sqrt{u_1^2 + u_2^2}, & \text{if } \mathbf{u} \text{ is in } \mathbf{R}^2, \\ \sqrt{u_1^2 + u_2^2 + u_3^2}, & \text{if } \mathbf{u} \text{ is in } \mathbf{R}^3. \end{cases}$$

NOTE   If $\|\mathbf{u}\| = 0$, then $\mathbf{u} = \mathbf{0}$ (and conversely). If $\|\mathbf{u}\| = 1$, the vector **u** is called a **unit vector**.

··

EXAMPLE 1.1

Find $\|\mathbf{u}\|$, where $\mathbf{u} = (2, -1, 3)$.

Solution

$$\|\mathbf{u}\| = \sqrt{2^2 + (-1)^2 + 3^2} = \sqrt{4 + 1 + 9} = \sqrt{14}.$$

SCALAR MULTIPLICATION

An element of the set of real numbers **R** is called a **scalar**.

DEFINITION

Let c be a scalar and **u** a vector in $\mathbf{R}^2$ or $\mathbf{R}^3$. The **scalar multiple** of **u** by c is the vector

$$c\mathbf{u} = \begin{cases} (cu_1, cu_2), & \text{if } \mathbf{u} \text{ is in } \mathbf{R}^2, \\ (cu_1, cu_2, cu_3), & \text{if } \mathbf{u} \text{ is in } \mathbf{R}^3. \end{cases}$$

In other words, $c\mathbf{u}$ is the vector obtained by multiplying each component of **u** by the scalar c.

EXAMPLE 1.2

Find $c\mathbf{u}$ for each scalar c and vector **u**.

a. $c = 3, \mathbf{u} = (1, -4, 4)$ **b.** $c = 0, \mathbf{u} = (2, 4)$

Solution   **a.** $c\mathbf{u} = (3, -12, 12)$ **b.** $c\mathbf{u} = (0, 0) = \mathbf{0}$

Examples of the geometric interpretation of $c\mathbf{u}$ are given in Figure 1.8 and will be justified by the next theorem.

FIGURE 1.8

THEOREM I   Let $\mathbf{u}$ be a nonzero vector in $\mathbf{R}^2$ or $\mathbf{R}^3$, and let c be a scalar. Then $\mathbf{u}$ and $c\mathbf{u}$ are collinear, and

    a. if $c > 0$, then $\mathbf{u}$ and $c\mathbf{u}$ have the same direction,

    b. if $c < 0$, then $\mathbf{u}$ and $c\mathbf{u}$ have opposite directions,

    c. $\|c\mathbf{u}\| = |c|\,\|\mathbf{u}\|$.

Proof   We will prove this theorem for vectors in $\mathbf{R}^2$. (The proof is similar in $\mathbf{R}^3$.) Figures 1.9 and 1.10 illustrate the situation when $c > 0$ (part (a)) and $c < 0$ (part (b)), respectively. In each figure, the two right triangles are *similar*, and hence their corresponding angles are equal, because their corresponding sides are in proportion; that is, the slopes of the lines determined by $\mathbf{u} = (u_1, u_2)$ and $c\mathbf{u} = (cu_1, cu_2)$,

$$\frac{u_2}{u_1} \quad \text{and} \quad \frac{cu_2}{cu_1},$$

are equal. Moreover, both lines have a point in common (the origin) and thus are identical. This shows that $\mathbf{u}$ and $c\mathbf{u}$ are collinear.

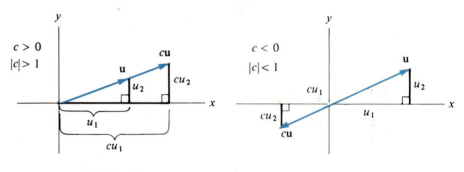

FIGURE 1.9   FIGURE 1.10

If $c > 0$ (part (a)), then the signs of the corresponding components of $\mathbf{u}$ and $c\mathbf{u}$ are the same, and the vectors have the same direction. If $c < 0$ (part (b)), the signs are reversed, so the vectors have opposite directions.

To verify part (c), we apply the distance formula:

$$\|c\mathbf{u}\| = \|(cu_1, cu_2)\| = \sqrt{(cu_1)^2 + (cu_2)^2}$$
$$= \sqrt{c^2(u_1^2 + u_2^2)} = \sqrt{c^2}\sqrt{u_1^2 + u_2^2} = |c|\,\|\mathbf{u}\|.$$

EXAMPLE 1.3   Find the *midpoint* of the vector $\mathbf{u} = (2, 1, -4)$.

Solution   Since $(\tfrac{1}{2})\mathbf{u}$ has the same direction as $\mathbf{u}$ and has half its length, the point corresponding to

$(\frac{1}{2})\mathbf{u}$ is the midpoint of $\mathbf{u}$. Thus the midpoint of the given vector is

$$\left(\frac{1}{2}\right)(2, 1, -4) = \left(1, \frac{1}{2}, -2\right).$$

EXAMPLE 1.4   For a nonzero vector $\mathbf{u}$, find a unit vector with the same direction.

Solution   We will show that the desired vector is $\dfrac{1}{\|\mathbf{u}\|}\mathbf{u}$. It has the same direction as $\mathbf{u}$ by Theorem 1a, and it has length 1 (it is a unit vector) because $\left\|\dfrac{1}{\|\mathbf{u}\|}\mathbf{u}\right\| = \dfrac{1}{\|\mathbf{u}\|}\|\mathbf{u}\|$ by Theorem 1c.

NOTE   The scalar multiplier result (Theorem 1) holds in reverse as well: if $\mathbf{u}$ and $\mathbf{v}$ are nonzero collinear vectors, then $\mathbf{v} = c\mathbf{u}$ for some scalar c. The argument can be based on similar triangles, where c is taken as the constant of proportionality (or its negative if $\mathbf{u}$ and $\mathbf{v}$ have opposite directions).

···

VECTOR ADDITION

DEFINITION   Let $\mathbf{u}$ and $\mathbf{v}$ be vectors in $\mathbf{R}^2$ or $\mathbf{R}^3$. The **sum** of $\mathbf{u}$ and $\mathbf{v}$ is the vector

$$\mathbf{u} + \mathbf{v} = \begin{cases} (u_1 + v_1, u_2 + v_2), & \text{if } \mathbf{u}, \mathbf{v} \text{ are in } \mathbf{R}^2, \\ (u_1 + v_1, u_2 + v_2, u_3 + v_3), & \text{if } \mathbf{u}, \mathbf{v} \text{ are in } \mathbf{R}^3. \end{cases}$$

In other words, $\mathbf{u} + \mathbf{v}$ is the vector obtained by adding the corresponding components.

EXAMPLE 1.5   Find the sum of each pair of vectors.

a. $\mathbf{u} = (2, 1, -1), \mathbf{v} = (1, 2, 3)$ **b.** $\mathbf{u} = (3, -1), \mathbf{v} = (1, 3)$

Solution   **a.** $\mathbf{u} + \mathbf{v} = (3, 3, 2)$ **b.** $\mathbf{u} + \mathbf{v} = (4, 2)$

We now look at the *geometric interpretation* of the sum of two vectors. The endpoint of $\mathbf{u} + \mathbf{v}$ is located by the endpoint of the directed line segment with the same direction and length as $\mathbf{v}$, but originating at the endpoint of $\mathbf{u}$ (Figure 1.11). Another way to view $\mathbf{u} + \mathbf{v}$ (for noncollinear vectors) is as the diagonal from $\mathbf{0}$ of the parallelogram determined by $\mathbf{u}$ and $\mathbf{v}$ (Figure 1.12).

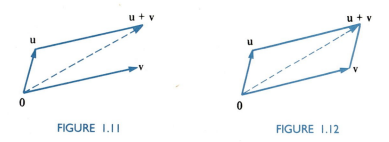

FIGURE 1.11   FIGURE 1.12

The preceding remarks are justified by Theorem 2 on the next page.

S P O T L I G H T

VELOCITY

As an airplane flies through still air, its engines propel it at a certain speed in a certain direction; it has a given *velocity*. Its velocity can be represented by a vector whose length is proportional to the speed of the plane and which points in the direction of flight. Now suppose that the plane encounters a wind blowing at a certain speed in a certain direction. We can represent *its* velocity by another vector. In this situation, we can ask the question, "At what speed, and in what direction, will the plane fly?" In other words, what will be the plane's *resultant velocity*? To find this vector, we simply add (by the "parallelogram method") the two given velocities. For example, if a plane is flying due East at 400 mph when it is struck by a 50-mph gust of wind blowing toward the Northeast, the resultant velocity will be as indicated in the figure below.

THEOREM 2   For nonzero vectors **u** and **v** the directed segment from the endpoint of **u** to the endpoint of **u** + **v** is parallel and equal in length to the vector **v**.

Proof   We will show that the figure determined by 0, U, P, and V, where P is the endpoint of **u** + **v**, is a parallelogram. In a plane it is sufficient to show that the two pairs of opposite sides are equal in length. By the distance formula,

$$
\begin{aligned}
d(\mathbf{u}, \mathbf{u} + \mathbf{v}) &= d((u_1, u_2), (u_1 + v_1, u_2 + v_2)) \\
&= \sqrt{((u_1 + v_1) - u_1)^2 + ((u_2 + v_2) - u_2)^2} \\
&= \sqrt{v_1^2 + v_2^2} = \|\mathbf{v}\| = d(\mathbf{0}, \mathbf{v}).
\end{aligned}
$$

Similarly, since $d(\mathbf{v}, \mathbf{u} + \mathbf{v}) = d(\mathbf{0}, \mathbf{u})$, the figure is a parallelogram. In **R**^{3} these equations are as easy to check, but this is not quite enough to prove the assertion. It must also be confirmed that **u** + **v** is *coplanar* with **0**, **u**, and **v**. (See Exercise 42 of Section 1.4.)

For each vector **u**, the vector that has the same length as **u** but points in the opposite direction is called the **negative** of **u** and will be denoted −**u**. By Theorem 1, $(-1)\mathbf{u}$ has these properties, so $(-1)\mathbf{u} = -\mathbf{u}$. It follows from the definition of scalar multiplication that

$$
-\mathbf{u} = \begin{cases} (-u_1, -u_2), & \text{if } \mathbf{u} \text{ is in } \mathbf{R}^2, \\ (-u_1, -u_2, -u_3), & \text{if } \mathbf{u} \text{ is in } \mathbf{R}^3. \end{cases}
$$

Now we can define subtraction of vectors in terms of addition by "adding the negative." That is, we define the **difference** of vectors **u** and **v** to be

$$
\mathbf{u} - \mathbf{v} = \mathbf{u} + (-\mathbf{v}).
$$

For example, if $\mathbf{u} = (0, -1, 2)$ and $\mathbf{v} = (-3, 0, 5)$, then $-\mathbf{u} = (0, 1, -2)$ and $\mathbf{u} - \mathbf{v} = (3, -1, -3)$.

The properties of vector addition and subtraction and scalar multiplication are given in Theorem 3 and follow directly from the definitions of these operations. To demonstrate how the arguments can be constructed, the proofs of 3a, 3b, and 3f are included, and the others are left as exercises. In the proofs we assume that the vectors are in **R**^{2}. (For **R**^{3} just add a third component.)

THEOREM 3   Let **u**, **v**, and **w** be vectors in **R**^{2} or **R**^{3}, and let c and d be scalars. Then

    a. $\mathbf{u} + \mathbf{v} = \mathbf{v} + \mathbf{u}$ (vector addition is commutative),

    b. $(\mathbf{u} + \mathbf{v}) + \mathbf{w} = \mathbf{u} + (\mathbf{v} + \mathbf{w})$ (vector addition is associative),

c. $\mathbf{u} + \mathbf{0} = \mathbf{u}$,

d. $\mathbf{u} + (-\mathbf{u}) = \mathbf{0}$,

e. $(cd)\mathbf{u} = c(d\mathbf{u})$,

f. $(c + d)\mathbf{u} = c\mathbf{u} + d\mathbf{u}$

g. $c(\mathbf{u} + \mathbf{v}) = c\mathbf{u} + c\mathbf{v}$ } (distributive laws),

h. $1\mathbf{u} = \mathbf{u}$,

i. $(-1)\mathbf{u} = -\mathbf{u}$,

j. $0\mathbf{u} = \mathbf{0}$.

Proof in part

3a.
$$\begin{aligned}
\mathbf{u} + \mathbf{v} &= (u_1 + v_1, u_2 + v_2)\\
&= (v_1 + u_1, v_2 + u_2)\\
&= \mathbf{v} + \mathbf{u}.
\end{aligned}$$

3b.
$$\begin{aligned}
(\mathbf{u} + \mathbf{v}) + \mathbf{w} &= (u_1 + v_1, u_2 + v_2) + (w_1, w_2)\\
&= ((u_1 + v_1) + w_1, (u_2 + v_2) + w_2)\\
&= (u_1 + (v_1 + w_1), u_2 + (v_2 + w_2))\\
&= (u_1, u_2) + (v_1 + w_1, v_2 + w_2)\\
&= \mathbf{u} + (\mathbf{v} + \mathbf{w}).
\end{aligned}$$

3f.
$$\begin{aligned}
(c + d)\mathbf{u} &= ((c + d)u_1, (c + d)u_2)\\
&= (cu_1 + du_1, cu_2 + du_2)\\
&= (cu_1, cu_2) + (du_1, du_2)\\
&= c\mathbf{u} + d\mathbf{u}.
\end{aligned}$$

TRANSLATION

Subtraction of vectors is a particularly useful way of discussing directed line segments that do not emanate from the origin. For points P and Q in $\mathbf{R}^2$ or $\mathbf{R}^3$, we will denote the directed line segment from P to Q by the symbol $\overrightarrow{PQ}$. Two directed segments $\overrightarrow{PQ}$ and $\overrightarrow{RS}$ are said to be **equivalent** if they have the same direction and length (Figure 1.13). The next theorem relates equivalent directed segments to vector subtraction.

FIGURE 1.13

THEOREM 4   Let U and V be distinct points in $\mathbf{R}^2$ or $\mathbf{R}^3$. Then the vector $\mathbf{v} - \mathbf{u}$ is equivalent to the directed line segment from U to V. In other words, $\overrightarrow{UV}$ is parallel to the vector $\mathbf{v} - \mathbf{u}$, and $d(\mathbf{u}, \mathbf{v}) = \|\mathbf{v} - \mathbf{u}\|$.

Proof   We add $\mathbf{u}$ and $\mathbf{v} - \mathbf{u}$:

$$\begin{aligned}
\mathbf{u} + (\mathbf{v} - \mathbf{u}) &= \mathbf{u} + (\mathbf{v} + (-\mathbf{u})) && \text{(definition of subtraction)} \\
&= (\mathbf{u} + \mathbf{v}) + (-\mathbf{u}) && \text{(Theorem 3b)} \\
&= (\mathbf{v} + \mathbf{u}) + (-\mathbf{u}) && \text{(Theorem 3a)} \\
&= \mathbf{v} + (\mathbf{u} + (-\mathbf{u})) && \text{(Theorem 3b)} \\
&= \mathbf{v} + \mathbf{0} && \text{(Theorem 3d)} \\
&= \mathbf{v} && \text{(Theorem 3c).}
\end{aligned}$$

Since $\mathbf{v}$ is the sum of $\mathbf{u}$ and $\mathbf{v} - \mathbf{u}$, by Theorem 2, the vector $\mathbf{v} - \mathbf{u}$ is parallel and equal in length (hence equivalent) to the directed line segment from U to V (Figure 1.14).

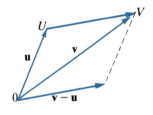

FIGURE 1.14

NOTE   The process of replacing the directed line segment $\overrightarrow{UV}$ with the vector $\mathbf{v} - \mathbf{u}$ is often called *translating U to the origin*. In reverse, the process is called *translating to U*.

··

EXAMPLE 1.6   Let l_1 and l_2 be lines in $\mathbf{R}^3$. Suppose that l_1 is determined by the points $P(2, 2, 1)$ and $Q(-1, 0, 2)$ and that l_2 is determined by the points $R(2, 1, 1)$ and $S(8, 5, -1)$. Are the lines parallel?

Solution   We translate the directed segments $\overrightarrow{PQ}$ and $\overrightarrow{RS}$ to the origin and compare them:

$$\mathbf{q} - \mathbf{p} = (-3, -2, 1),$$
$$\mathbf{s} - \mathbf{r} = (6, 4, -2).$$

Since one of these is a scalar multiple of the other, $\mathbf{s} - \mathbf{r} = -2(\mathbf{q} - \mathbf{p})$, they determine the same line through the origin. But by Theorem 4, $\mathbf{q} - \mathbf{p}$ is parallel to l_1 and $\mathbf{s} - \mathbf{r}$ is parallel to l_2. Thus l_1 and l_2 are parallel.

ALTERNATE NOTATION

Notice that a vector (a, b) in $\mathbf{R}^2$ may be written

$$(a, b) = a(1, 0) + b(0, 1).$$

Similarly, (a, b, c) in $\mathbf{R}^3$ may be written

$$(a, b, c) = a(1, 0, 0) + b(0, 1, 0) + c(0, 0, 1).$$

In many engineering and physics books the symbols reserved for these special vectors are $\mathbf{i} = (1, 0, 0)$, $\mathbf{j} = (0, 1, 0)$, and $\mathbf{k} = (0, 0, 1)$. Then the vector (a, b, c) is written as $a\mathbf{i} + b\mathbf{j} + c\mathbf{k}$. In $\mathbf{R}^2$ we have only $\mathbf{i}$ and $\mathbf{j}$ and drop the last component.

EXAMPLE 1.7

Express $4\mathbf{i} - 3\mathbf{k}$ as an ordered triple.

Solution

Since $\mathbf{j}$ is missing, its coefficient is 0, and the result is $4\mathbf{i} - 3\mathbf{k} = (4, 0, -3)$.

1.1 EXERCISES

In Exercises 1–16 let the vectors be given as follows:

$$\mathbf{r} = (-1, 1), \qquad \mathbf{u} = (3, 1, 0),$$
$$\mathbf{s} = (2, 0), \qquad \mathbf{v} = (2, 0, 1),$$
$$\mathbf{t} = (3, -2), \qquad \mathbf{w} = (-1, -2, 3).$$

Perform each indicated operation.

1. $4\mathbf{w}$
2. $-3\mathbf{v}$
3. $-\mathbf{r}$
4. $\mathbf{s} + \mathbf{t}$
5. $\mathbf{u} + \mathbf{v}$
6. $\mathbf{t} - \mathbf{r}$
7. $\mathbf{w} - \mathbf{v}$
8. $3\mathbf{r} + 2\mathbf{s} - \mathbf{t}$
9. $\mathbf{u} - 5\mathbf{v} + 2\mathbf{w}$
10. $\|\mathbf{t}\|$
11. $\|\mathbf{w}\|$
12. $d(\mathbf{s}, \mathbf{t})$
13. $d(\mathbf{w}, \mathbf{v})$
14. $\|\mathbf{w} - \mathbf{v}\|$
15. $\|\mathbf{w}\| - \|\mathbf{v}\|$
16. $\|5(\mathbf{w} - \mathbf{v})\|$

In Exercises 17 and 18 find the unit vector with the same direction as the given one.

17. $(3, -4)$ **18.** $(2, 1, -3)$

In Exercises 19 and 20 find the unit vector having the opposite direction of the given one.

19. $(2, -1, 2)$ **20.** $(-1, 2)$

In Exercises 21–24 find the vector that is equivalent to the directed line segment from the first point to the second.

21. $(-1, 2), (3, 4)$ **22.** $(2, 1, 0), (-1, 3, 4)$

23. $(1, 5, 7), (2, 1, 0)$ **24.** $(1, -2), (-2, 3)$

25. Express $(2, 0, -3)$ in $\mathbf{i}, \mathbf{j}, \mathbf{k}$ form.

26. Express $3\mathbf{i} + 2\mathbf{j} - \mathbf{k}$ as an ordered triple.

27. Find the length of $2\mathbf{i} + 3\mathbf{j} - \mathbf{k}$.

28. Simplify $3(2\mathbf{i} + 7\mathbf{j} - \mathbf{k}) - 2(\mathbf{i} - 4\mathbf{j})$.

29. Prove the properties of Theorem 3 that remain unproved. Assume the vectors are in $\mathbf{R}^3$.

30. Prove that the midpoint of the line segment determined by the points P and Q is $(\frac{1}{2})\mathbf{p} + (\frac{1}{2})\mathbf{q}$. (*Hint:* This can be done directly by the distance formula, but using Theorem 4 is easier.)

31. Prove directly from the distance formula that $d(\mathbf{p}, \mathbf{q}) = \|\mathbf{p} - \mathbf{q}\|$.

32. Sketch a picture of the set of endpoints of all unit vectors in $\mathbf{R}^2$. What is the figure? What about in $\mathbf{R}^3$?

33. Describe the geometric figure consisting of the set of all solutions $\mathbf{x}$ in $\mathbf{R}^2$ to the equation $\|\mathbf{x} - \mathbf{p}\| = r$ for $\mathbf{p}$ in $\mathbf{R}^2$ and r in $\mathbf{R}$. What about in $\mathbf{R}^3$?

34. Let P and Q be distinct points and t a real number. Let $\mathbf{x}(t) = (1 - t)\mathbf{p} + t\mathbf{q}$. Use Theorem 4 to prove that

$$\frac{d(\mathbf{p}, \mathbf{x}(t))}{d(\mathbf{p}, \mathbf{q})} = |t|.$$

35. Use Exercise 34 to find a point one-third of the distance from P to Q.

36. Let l_1 be the line determined by $(2, 1, 3)$ and $(1, 2, -1)$, and l_2 be the line determined by $(0, 2, 3)$ and $(-1, 1, 2)$. Is l_1 parallel to l_2?

37. Let l_1 be the line determined by $(3, 1, 2)$ and $(4, 3, 1)$ and similarly l_2 be the line determined by $(1, 3, -3)$ and $(-1, -1, -1)$. Is l_1 parallel to l_2?

38. Let P, Q, and R be the three vertices of a triangle. Using Theorem 4, prove that the line segment that joins the midpoints of two of the sides is parallel to and half the length of the third side.

1.2 DOT AND CROSS PRODUCTS

In this section we will define and develop the properties of two types of vector products: the *dot product,* which associates a scalar with every pair of vectors and is of general interest, and the *cross product,* which yields another vector but is defined only in $\mathbf{R}^3$.

DOT PRODUCT

Since two nonzero vectors are assumed to have a common initial point (the origin), they form an angle. The following operation will be useful in describing this angle.

DEFINITION

Let **u** and **v** be vectors in $\mathbf{R}^2$ or $\mathbf{R}^3$. The **dot product** (or **standard inner product**) of **u** and **v** is denoted by $\mathbf{u} \cdot \mathbf{v}$ and is defined as follows:

$$\mathbf{u} \cdot \mathbf{v} = \begin{cases} u_1 v_1 + u_2 v_2, & \text{if } \mathbf{u}, \mathbf{v} \text{ are in } \mathbf{R}^2, \\ u_1 v_1 + u_2 v_2 + u_3 v_3, & \text{if } \mathbf{u}, \mathbf{v} \text{ are in } \mathbf{R}^3. \end{cases}$$

In words, $\mathbf{u} \cdot \mathbf{v}$ is the sum of the products of corresponding components of **u** and **v**.

EXAMPLE 1.8

Compute the dot product of each pair of vectors.

a. $\mathbf{u} = (1, 2, 3), \mathbf{v} = (-2, 4, 1)$

b. $\mathbf{u} = (1, 2), \mathbf{v} = (3, -4)$

c. $\mathbf{u} = (2, 1, -3), \mathbf{v} = (1, 1, 1)$

Solution

a. $\mathbf{u} \cdot \mathbf{v} = (1, 2, 3) \cdot (-2, 4, 1) = (1)(-2) + (2)(4) + (3)(1) = 9.$

b. $\mathbf{u} \cdot \mathbf{v} = (1, 2) \cdot (3, -4) = (1)(3) + (2)(-4) = -5.$

c. $\mathbf{u} \cdot \mathbf{v} = (2, 1, -3) \cdot (1, 1, 1) = (2)(1) + (1)(1) + (-3)(1) = 0.$

Note that $\mathbf{u} \cdot \mathbf{v}$ is a *scalar* rather than a vector. For this reason, dot product is often called **scalar product.** The algebraic properties of dot product are included in the next theorem.

THEOREM 1   Let $\mathbf{u}, \mathbf{v}$, and $\mathbf{w}$ be vectors in $\mathbf{R}^2$ or $\mathbf{R}^3$, and let c be a scalar. Then

a. $\mathbf{u} \cdot \mathbf{v} = \mathbf{v} \cdot \mathbf{u}$ (dot product is commutative),

b. $c(\mathbf{u} \cdot \mathbf{v}) = (c\mathbf{u}) \cdot \mathbf{v} = \mathbf{u} \cdot (c\mathbf{v})$ (scalars factor out),

c. $\mathbf{u} \cdot (\mathbf{v} + \mathbf{w}) = \mathbf{u} \cdot \mathbf{v} + \mathbf{u} \cdot \mathbf{w}$ (distributive law),

d. $\mathbf{u} \cdot \mathbf{0} = 0,$

e. $\mathbf{u} \cdot \mathbf{u} = \|\mathbf{u}\|^2.$

Proof in part   Assume that the vectors are in $\mathbf{R}^2$. The case of $\mathbf{R}^3$ is entirely analogous; just add a third component.

1a. $\mathbf{u} \cdot \mathbf{v} = u_1 v_1 + u_2 v_2$

$\qquad\qquad = v_1 u_1 + v_2 u_2$

$\qquad\qquad = \mathbf{v} \cdot \mathbf{u}.$

1b. $c(\mathbf{u} \cdot \mathbf{v}) = c(u_1 v_1 + u_2 v_2)$

$\qquad\qquad = c u_1 v_1 + c u_2 v_2 = \begin{cases} (c u_1) v_1 + (c u_2) v_2 = (c\mathbf{u}) \cdot \mathbf{v}, \\ u_1 (c v_1) + u_2 (c v_2) = \mathbf{u} \cdot (c\mathbf{v}). \end{cases}$

1e. $\mathbf{u} \cdot \mathbf{u} = u_1 u_1 + u_2 u_2$

$\qquad\qquad = u_1^2 + u_2^2$

$\qquad\qquad = \|\mathbf{u}\|^2.$

The proofs of 1c and 1d are left as exercises.

The geometric interpretation of $\mathbf{u} \cdot \mathbf{v}$ is particularly interesting. If $\mathbf{u}$ and $\mathbf{v}$ are nonzero vectors, they form an *angle*, since they both originate at the origin **0** (Figure 1.15). We will use *radian* measure for the angle (although degree measure could be used as well), and we will use the smallest nonnegative number possible. In other words, if θ is the measure of the angle, we have $0 \le \theta \le \pi$. The next theorem describes the relationship between this angle and $\mathbf{u} \cdot \mathbf{v}$.

FIGURE 1.15

THEOREM 2   Let $\mathbf{u}$ and $\mathbf{v}$ be vectors in $\mathbf{R}^2$ or $\mathbf{R}^3$, and let θ be the angle between $\mathbf{u}$ and $\mathbf{v}$. Then

$$\mathbf{u} \cdot \mathbf{v} = \|\mathbf{u}\| \, \|\mathbf{v}\| \cos \theta,$$

or, equivalently for nonzero $\mathbf{u}$ and $\mathbf{v}$,

$$\cos \theta = \frac{\mathbf{u} \cdot \mathbf{v}}{\|\mathbf{u}\| \, \|\mathbf{v}\|}.$$

Proof   Consider the triangle determined by $\mathbf{u}$ and $\mathbf{v}$ (Figure 1.16). Let

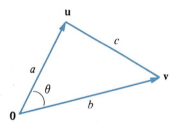

FIGURE 1.16

$a = \|\mathbf{u}\|$, $b = \|\mathbf{v}\|$, and $c = \|\mathbf{v} - \mathbf{u}\|$. Recall from Section 1.1 that $\mathbf{v} - \mathbf{u}$ is equivalent to the directed line segment $\overrightarrow{UV}$, so c is the length of the third side of the triangle. To this triangle we apply the law of cosines, $c^2 = a^2 + b^2 - 2ab \cos \theta$. Substituting for a, b, and c, we have

$$\|\mathbf{v} - \mathbf{u}\|^2 = \|\mathbf{u}\|^2 + \|\mathbf{v}\|^2 - 2 \|\mathbf{u}\| \|\mathbf{v}\| \cos \theta. \tag{1}$$

We need a more convenient form for $\|\mathbf{v} - \mathbf{u}\|^2$.

$$
\begin{aligned}
\|\mathbf{v} - \mathbf{u}\|^2 &= (\mathbf{v} - \mathbf{u}) \cdot (\mathbf{v} - \mathbf{u}) & \text{(Theorem 1e)} \\
&= (\mathbf{v} - \mathbf{u}) \cdot \mathbf{v} - (\mathbf{v} - \mathbf{u}) \cdot \mathbf{u} & \text{(Theorem 1c)} \\
&= \mathbf{v} \cdot \mathbf{v} - \mathbf{u} \cdot \mathbf{v} - \mathbf{v} \cdot \mathbf{u} + \mathbf{u} \cdot \mathbf{u} & \text{(Exercise 35)} \\
&= \mathbf{v} \cdot \mathbf{v} - \mathbf{u} \cdot \mathbf{v} - \mathbf{u} \cdot \mathbf{v} + \mathbf{u} \cdot \mathbf{u} & \text{(Theorem 1a)} \\
&= \|\mathbf{v}\|^2 - 2(\mathbf{u} \cdot \mathbf{v}) + \|\mathbf{u}\|^2 & \text{(Theorem 1e)}.
\end{aligned}
$$

Returning now to Equation (1),

$$\|\mathbf{v}\|^2 - 2(\mathbf{u} \cdot \mathbf{v}) + \|\mathbf{u}\|^2 = \|\mathbf{u}\|^2 + \|\mathbf{v}\|^2 - 2 \|\mathbf{u}\| \|\mathbf{v}\| \cos \theta,$$

$$-2(\mathbf{u} \cdot \mathbf{v}) = -2 \|\mathbf{u}\| \|\mathbf{v}\| \cos \theta,$$

$$\mathbf{u} \cdot \mathbf{v} = \|\mathbf{u}\| \|\mathbf{v}\| \cos \theta.$$

EXAMPLE 1.9

Find the cosine of the angle between each pair of vectors.

a. $\mathbf{u} = (1, 2, 3)$, $\mathbf{v} = (-2, 4, 1)$

b. $\mathbf{u} = (2, 1, -3)$, $\mathbf{v} = (1, 1, 1)$

Solution

a.

$$
\begin{aligned}
\cos \theta &= \frac{\mathbf{u} \cdot \mathbf{v}}{\|\mathbf{u}\| \|\mathbf{v}\|} \\
&= \frac{(1)(-2) + (2)(4) + (3)(1)}{\sqrt{1^2 + 2^2 + 3^2} \sqrt{(-2)^2 + 4^2 + 1^2}} \\
&= \frac{9}{\sqrt{14} \sqrt{21}} \\
&= \frac{3\sqrt{6}}{14}.
\end{aligned}
$$

(Using tables or a hand calculator, θ is approximately 1.02 radians.)

b.
$$\cos \theta = \frac{\mathbf{u} \cdot \mathbf{v}}{\|\mathbf{u}\| \, \|\mathbf{v}\|}$$

$$= \frac{0}{\|\mathbf{u}\| \, \|\mathbf{v}\|}$$

$$= 0.$$

(Here the angle θ is $\pi/2$ radians, so θ is a right angle.)

EXAMPLE 1.10

Find the cosine of the angle at P determined by the directed line segments $\overrightarrow{PQ}$ and $\overrightarrow{PR}$, where $P(1, 2)$, $Q(1 + \sqrt{3}, -1)$ and $R(1 - \sqrt{3}, 3)$.

Solution

We need vectors to apply Theorem 2, so we translate $\overrightarrow{PQ}$ and $\overrightarrow{PR}$ to the origin by subtracting $\mathbf{p}$ from $\mathbf{q}$ and $\mathbf{r}$. This results in an angle at the origin determined by the vectors $\mathbf{q} - \mathbf{p}$ and $\mathbf{r} - \mathbf{p}$ (Figure 1.17). Since these vectors are respectively equivalent (and hence parallel) to $\overrightarrow{PQ}$ and $\overrightarrow{PR}$, the angle they form is the same, and we may use Theorem 2,

$$\cos \theta = \frac{(\mathbf{q} - \mathbf{p}) \cdot (\mathbf{r} - \mathbf{p})}{\|\mathbf{q} - \mathbf{p}\| \, \|\mathbf{r} - \mathbf{p}\|}$$

$$= \frac{(\sqrt{3}, -3) \cdot (-\sqrt{3}, 1)}{\|(\sqrt{3}, -3)\| \, \|(-\sqrt{3}, 1)\|}$$

$$= \frac{-6}{\sqrt{12} \, \sqrt{4}}$$

$$= -\frac{\sqrt{3}}{2}.$$

(This time the angle θ is $5\pi/6$ radians.)

FIGURE 1.17

We will investigate further the relationship between dot product and angle after introducing the following concept.

DEFINITION

> Vectors **u** and **v** in $\mathbf{R}^2$ or $\mathbf{R}^3$ are **orthogonal** if $\mathbf{u} \cdot \mathbf{v} = 0$.

For example, the vectors $\mathbf{u} = (2, 1, -3)$ and $\mathbf{v} = (1, 1, 1)$ of Example 1.8 are orthogonal. In Example 1.9 we saw that these same vectors form a right angle. The words *orthogonal* (from Greek), *perpendicular* (from Latin), and *normal* (also Latin) mean essentially the same thing, "meet at right angles." However, usage sometimes makes one word preferable to another. As examples, two vectors are usually said to be *orthogonal*, two lines or planes are *perpendicular*, and a vector is *normal* to a plane.

THEOREM 3   Let **u** and **v** be nonzero vectors in $\mathbf{R}^2$ or $\mathbf{R}^3$, and let θ be the angle they form. Then θ is

   i. an acute angle if $\mathbf{u} \cdot \mathbf{v} > 0$,

   ii. a right angle if $\mathbf{u} \cdot \mathbf{v} = 0$,

   iii. an obtuse angle if $\mathbf{u} \cdot \mathbf{v} < 0$.

Proof   From Theorem 2 we have the equation

$$\mathbf{u} \cdot \mathbf{v} = \|\mathbf{u}\| \, \|\mathbf{v}\| \cos \theta.$$

Since $\|\mathbf{u}\|$ and $\|\mathbf{v}\|$ are positive, it follows that $\mathbf{u} \cdot \mathbf{v}$ is zero, positive, or negative as $\cos \theta$ is zero, positive, or negative. However, $\cos \theta > 0$ implies that θ is an acute angle. Similarly, $\cos \theta = 0$ implies that θ is a right angle, and $\cos \theta < 0$ implies that θ is obtuse.

In Example 1.9a we calculated $\cos \theta$ to be $3\sqrt{6}/14$ and gave the approximate measure of the angle. Without tables or a calculator, θ is seen to be an acute angle by Theorem 3. Similarly, the angle in Example 1.10 must be obtuse, since $(\mathbf{q} - \mathbf{p}) \cdot (\mathbf{r} - \mathbf{p}) = -6$.

CROSS PRODUCT

It may have seemed surprising that the dot product of two vectors is a scalar instead of a vector. There is, in $\mathbf{R}^3$, another type of product of two vectors that produces a vector.

DEFINITION

> Let $\mathbf{u} = (u_1, u_2, u_3)$ and $\mathbf{v} = (v_1, v_2, v_3)$ be vectors in $\mathbf{R}^3$. The **cross product** of **u** and **v** is the vector
>
> $$\mathbf{u} \times \mathbf{v} = (u_2 v_3 - u_3 v_2, \; u_3 v_1 - u_1 v_3, \; u_1 v_2 - u_2 v_1).$$

Fortunately, there is a convenient way of remembering this formula. Given an array of numbers

$$\begin{bmatrix} a & b \\ c & d \end{bmatrix},$$

we define the **determinant** of the array as the difference of the diagonal products,

$$\det \begin{bmatrix} a & b \\ c & d \end{bmatrix} = ad - bc.$$

With this notation the expression for $\mathbf{u} \times \mathbf{v}$ becomes

$$\mathbf{u} \times \mathbf{v} = \left(\det \begin{bmatrix} u_2 & u_3 \\ v_2 & v_3 \end{bmatrix}, -\det \begin{bmatrix} u_1 & u_3 \\ v_1 & v_3 \end{bmatrix}, \det \begin{bmatrix} u_1 & u_2 \\ v_1 & v_2 \end{bmatrix} \right).$$

The arrays that appear in this expression are easy to remember by use of the following device: write the components of $\mathbf{u}$ above those of $\mathbf{v}$, forming the rectangular array

$$\begin{bmatrix} u_1 & u_2 & u_3 \\ v_1 & v_2 & v_3 \end{bmatrix}.$$

Then the square arrays that appear in the expression for $\mathbf{u} \times \mathbf{v}$ are obtained by successively deleting the first, second, and third columns of this rectangular array.

EXAMPLE 1.11   Find $\mathbf{u} \times \mathbf{v}$ where $\mathbf{u} = (1, 2, -1)$ and $\mathbf{v} = (0, 2, 3)$.

Solution   Construct the rectangular array $\begin{bmatrix} 1 & 2 & -1 \\ 0 & 2 & 3 \end{bmatrix}$. Then

$$\mathbf{u} \times \mathbf{v} = \left(\det \begin{bmatrix} 2 & -1 \\ 2 & 3 \end{bmatrix}, -\det \begin{bmatrix} 1 & -1 \\ 0 & 3 \end{bmatrix}, \det \begin{bmatrix} 1 & 2 \\ 0 & 2 \end{bmatrix} \right)$$

$$= ((2)(3) - (-1)(2), -(1)(3) + (-1)(0), (1)(2) - (2)(0))$$

$$= (6 + 2, -3 - 0, 2 - 0)$$

$$= (8, -3, 2).$$

COMPUTATIONAL NOTE   In the course of your study of linear algebra, you will encounter many tedious computational problems such as the calculation of cross products. The Computational Notes that appear from time to time throughout the text will describe how the computer can be used to solve the problem under discussion and, occasionally, the pitfalls of doing so. Corresponding Computational Exercises will appear at the end of every section that contains these notes. In the Computational Exercises, you will be asked to use a computer program (possibly supplied by your instructor or computer lab) to help you solve these problems.

....

For our purposes the most important property of cross product is given by the following theorem.

THEOREM 4 The **vector u × v is orthogonal to both u and v.**

Proof We compute $\mathbf{u} \cdot (\mathbf{u} \times \mathbf{v})$:

$$(u_1, u_2, u_3) \cdot (u_2v_3 - u_3v_2, u_3v_1 - u_1v_3, u_1v_2 - u_2v_1)$$

$$= u_1u_2v_3 - u_1u_3v_2 + u_2u_3v_1 - u_2u_1v_3 + u_3u_1v_2 - u_3u_2v_1$$

$$= 0.$$

Thus, by definition, **u** is orthogonal to **u** × **v**. Similarly, $\mathbf{v} \cdot (\mathbf{u} \times \mathbf{v}) = 0$, so **v** is orthogonal to **u** × **v**.

EXAMPLE 1.12

Find a vector that is orthogonal to both $\mathbf{u} = (1, 2, -1)$ and $\mathbf{v} = (0, 2, 3)$.

Solution

Taking the cross product of these vectors as in Example 1.11, we get $\mathbf{u} \times \mathbf{v} = (8, -3, 2)$. By Theorem 4, this vector is orthogonal to both **u** and **v**.

Other properties of cross product are given in the following theorem.

THEOREM 5 Let **u**, **v** be vectors in $\mathbf{R}^3$ and c a scalar. Then

a. $\mathbf{u} \times \mathbf{v} = -(\mathbf{v} \times \mathbf{u})$,

b. $\mathbf{u} \times (\mathbf{v} + \mathbf{w}) = (\mathbf{u} \times \mathbf{v}) + (\mathbf{u} \times \mathbf{w})$ ⎫

c. $(\mathbf{u} + \mathbf{v}) \times \mathbf{w} = (\mathbf{u} \times \mathbf{w}) + (\mathbf{v} \times \mathbf{w})$ ⎬ (distributive laws),

d. $c(\mathbf{u} \times \mathbf{v}) = (c\mathbf{u}) \times \mathbf{v} = \mathbf{u} \times (c\mathbf{v})$ (scalars factor out),

e. $\mathbf{u} \times \mathbf{0} = \mathbf{0} \times \mathbf{u} = \mathbf{0}$,

f. $\mathbf{u} \times \mathbf{u} = \mathbf{0}$,

g. $\|\mathbf{u} \times \mathbf{v}\| = \|\mathbf{u}\| \|\mathbf{v}\| \sin \theta = \sqrt{\|\mathbf{u}\|^2 \|\mathbf{v}\|^2 - (\mathbf{u} \cdot \mathbf{v})^2}$,

where θ is the angle determined by **u** and **v**.

Proof in part

5a. Let $\mathbf{u} = (u_1, u_2, u_3)$ and $\mathbf{v} = (v_1, v_2, v_3)$, and simply compute both $\mathbf{u} \times \mathbf{v}$ and $\mathbf{v} \times \mathbf{u}$ and notice that one vector is the negative of the other. (The important fact here is the geometrical interpretation: $\mathbf{u} \times \mathbf{v}$ and $\mathbf{v} \times \mathbf{u}$ have the same length but point in opposite directions.)

5b. Let $\mathbf{u}$ and $\mathbf{v}$ be given as above, and $\mathbf{w} = (w_1, w_2, w_3)$. Then

$$
\begin{aligned}
\mathbf{u} \times (\mathbf{v} + \mathbf{w}) &= (u_1, u_2, u_3) \times (v_1 + w_1, v_2 + w_2, v_3 + w_3) \\
&= (u_2(v_3 + w_3) - u_3(v_2 + w_2), u_3(v_1 + w_1) - u_1(v_3 + w_3), \\
&\quad\ u_1(v_2 + w_2) - u_2(v_1 + w_1)) \\
&= (u_2 v_3 + u_2 w_3 - u_3 v_2 - u_3 w_2, u_3 v_1 + u_3 w_1 - u_1 v_3 - u_1 w_3, \\
&\quad\ u_1 v_2 + u_1 w_2 - u_2 v_1 - u_2 w_1) \\
&= (u_2 v_3 - u_3 v_2, u_3 v_1 - u_1 v_3, u_1 v_2 - u_2 v_1) \\
&\quad + (u_2 w_3 - u_3 w_2, u_3 w_1 - u_1 w_3, u_1 w_2 - u_2 w_1) \\
&= \mathbf{u} \times \mathbf{v} + \mathbf{u} \times \mathbf{w}.
\end{aligned}
$$

Proofs of 5c through 5f are straightforward computations and are left as exercises.

5g. Note the similarity of the statement 5g with the rule relating dot product and the cosine of the angle, $\|\mathbf{u}\| \|\mathbf{v}\| \cos \theta = \mathbf{u} \cdot \mathbf{v}$. Starting with this equation, we square both sides to obtain

$$\|\mathbf{u}\|^2 \|\mathbf{v}\|^2 \cos^2 \theta = (\mathbf{u} \cdot \mathbf{v})^2.$$

Since $\cos^2 \theta = 1 - \sin^2 \theta$,

$$\|\mathbf{u}\|^2 \|\mathbf{v}\|^2 (1 - \sin^2 \theta) = (\mathbf{u} \cdot \mathbf{v})^2,$$

$$\|\mathbf{u}\|^2 \|\mathbf{v}\|^2 \sin^2 \theta = \|\mathbf{u}\|^2 \|\mathbf{v}\|^2 - (\mathbf{u} \cdot \mathbf{v})^2.$$

Taking square roots,

$$\|\mathbf{u}\| \|\mathbf{v}\| \sin \theta = \sqrt{\|\mathbf{u}\|^2 \|\mathbf{v}\|^2 - (\mathbf{u} \cdot \mathbf{v})^2}.$$

It is a straightforward but tedious computation to verify that $\|\mathbf{u} \times \mathbf{v}\| = \sqrt{\|\mathbf{u}\|^2 \|\mathbf{v}\|^2 - (\mathbf{u} \cdot \mathbf{v})^2}$. Simply expand each expression and verify that they are the same. The details are left as an exercise.

NOTE   Although cross product is distributive over addition (Theorem 5b and 5c), it is *not* commutative (Theorem 5a). Moreover, it is easily shown that the cross product is not associative either (see Exercise 25).

Geometrically, Theorem 5g tells us that the length of $\mathbf{u} \times \mathbf{v}$ is the area of the parallelogram determined by $\mathbf{u}$ and $\mathbf{v}$ (Figure 1.18). To see why this is so, let h be the height of the parallelogram from the endpoint of $\mathbf{v}$ to the side

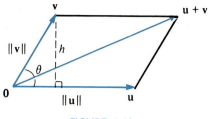

FIGURE 1.18

determined by **u**. Then the area A of the parallelogram is given by $A = \|\mathbf{u}\|\, h$. However, $\sin\theta = h/\|\mathbf{v}\|$, or $h = \|\mathbf{v}\| \sin\theta$. Making the substitution, we conclude that $A = \|\mathbf{u}\|\, \|\mathbf{v}\| \sin\theta$.

SPOTLIGHT

TORQUE

If you have to remove a stubborn nut from a bolt (for example, a lug nut on an automobile wheel), then the longer your wrench, the easier the job will be. This is because the *torque* applied to the nut depends not only on the force you exert on the wrench, but also on the distance from its point of application to the nut. This force and (directed) distance can be represented by vectors, since direction as well as magnitude is involved. Letting **F** be the force applied and **r** the directed distance, the torque, **T**, is the vector given by the cross product

$$\mathbf{T} = \mathbf{F} \times \mathbf{r}.$$

Now, by Theorem 5g, the magnitude of the torque is given by

$$\|\mathbf{T}\| = \|\mathbf{F}\|\, \|\mathbf{r}\|\, |\sin\theta|.$$

Thus, to generate the most torque, you should push as hard as you can (to maximize $\|\mathbf{F}\|$), on the *end* of the wrench (to maximize $\|\mathbf{r}\|$), in a direction perpendicular to the wrench (since $\sin 90° = 1$, this makes $\sin\theta$ as large as possible).

EXAMPLE 1.13

Find the area of the triangle determined by the three points $P(1, 2, 3)$, $Q(-3, 2, 1)$, and $R(2, 4, 5)$.

Solution

We translate $\overrightarrow{PQ}$ and $\overrightarrow{PR}$ to the origin to obtain the vectors $\mathbf{u} = \mathbf{q} - \mathbf{p}$ and $\mathbf{v} = \mathbf{r} - \mathbf{p}$. By the preceding remarks, $\|\mathbf{u} \times \mathbf{v}\|$ is the area of the parallelogram determined by $\mathbf{u}$ and $\mathbf{v}$. Thus the area of the triangle determined by $\mathbf{u}$ and $\mathbf{v}$ is $\frac{1}{2}\|\mathbf{u} \times \mathbf{v}\|$ (Figure 1.19).

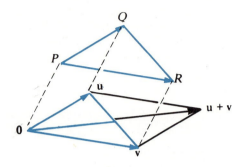

FIGURE 1.19

Since $\mathbf{u} = (-4, 0, -2)$, and $\mathbf{v} = (1, 2, 2)$, we have

$$\frac{1}{2}\|\mathbf{u} \times \mathbf{v}\| = \frac{1}{2}\left\|\left(\det\begin{bmatrix} 0 & -2 \\ 2 & 2 \end{bmatrix}, -\det\begin{bmatrix} -4 & -2 \\ 1 & 2 \end{bmatrix}, \det\begin{bmatrix} -4 & 0 \\ 1 & 2 \end{bmatrix}\right)\right\|$$

$$= \frac{1}{2}\|(4, 6, -8)\| = \|(2, 3, -4)\| = \sqrt{29}.$$

1.2 EXERCISES

In Exercises 1–10 let the vectors be given as follows:

$$\mathbf{s} = (-1, 2), \qquad \mathbf{u} = (3, -1, 4),$$
$$\mathbf{t} = (2, 3), \qquad \mathbf{v} = (-1, -3, 1),$$
$$\mathbf{w} = (-1, 1, 2).$$

Perform each indicated operation.

1. $\mathbf{s} \cdot \mathbf{t}$ **2.** $\mathbf{s} \cdot (3\mathbf{t})$

3. $\mathbf{u} \cdot \mathbf{v}$ **4.** $(3\mathbf{v}) \cdot (-2\mathbf{u})$

5. $\mathbf{u} \times \mathbf{v}$ **6.** $(\mathbf{v} \times \mathbf{u}) \cdot (-\mathbf{w})$

7. $\mathbf{v} \times (2\mathbf{u})$ **8.** $(2\mathbf{u} + \mathbf{v}) \times (3\mathbf{w})$

9. $(\mathbf{u} \times \mathbf{v}) \cdot \mathbf{w}$ **10.** $\mathbf{u} \cdot (\mathbf{w} \times \mathbf{v})$

In Exercises 11 and 12 determine if the angle formed by the two vectors is acute, right, or obtuse.

11. $(-2, 1), (5, 2)$

12. $(-1, 2, 3), (2, 0, 4)$

In Exercises 13–16 determine the cosine of the angle formed by the two vectors.

13. $(-2, 1), (5, 2)$

14. $(-1, 2, 3), (2, 0, 4)$

15. $(1, 0, 1), (-1, -1, 0)$

16. $(1, -1), (2, 0)$

In Exercises 17 and 18 find a vector that is orthogonal to both vectors.

17. $(1, 1, 3), (2, 1, -1)$

18. $(-1, 2, 3), (2, 0, 4)$

19. Find the area of the triangle with vertices at $(2, 1, 3)$, $(1, 0, 2)$, and $(-1, 1, 2)$.

20. Find the area of the parallelogram determined by the vectors $\mathbf{u} = (-1, 1, 0)$ and $\mathbf{v} = (2, 3, -1)$.

21. Prove that for any vector $\mathbf{u} = (a, b)$ in $\mathbf{R}^2$ the vector $\mathbf{r} = (-b, a)$ is orthogonal to $\mathbf{u}$.

The expressions in Exercises 22–24 are undefined. Explain why.

22. $\mathbf{u} \times (\mathbf{v} \cdot \mathbf{w})$

23. $\mathbf{u} \cdot (\mathbf{v} \cdot \mathbf{w})$

24. $\|\mathbf{u} \cdot \mathbf{v}\|$

25. Show that cross product is not associative. That is, it is not always true that $(\mathbf{u} \times \mathbf{v}) \times \mathbf{w} = \mathbf{u} \times (\mathbf{v} \times \mathbf{w})$. (*Hint:* Try $(\mathbf{i} \times \mathbf{i}) \times \mathbf{j}$ and $\mathbf{i} \times (\mathbf{i} \times \mathbf{j})$, where $\mathbf{i} = (1, 0, 0)$ and $\mathbf{j} = (0, 1, 0)$.)

26. Use Theorem 5g to prove that $\mathbf{u}$ and $\mathbf{v}$ are collinear if and only if $\mathbf{u} \times \mathbf{v} = \mathbf{0}$.

27. Prove the properties in Theorem 1 that remain unproved.

28. Use Theorem 1e to prove the *parallelogram law,*

$$\|\mathbf{u} + \mathbf{v}\|^2 + \|\mathbf{u} - \mathbf{v}\|^2 = 2\|\mathbf{u}\|^2 + 2\|\mathbf{v}\|^2.$$

Interpret this result geometrically to conclude "The sum of the squares of the sides of a parallelogram is equal to the sum of the squares of the diagonals."

29. Use Theorem 2 and the trigonometric identity, $\cos(\pi - \theta) = -\cos\theta$, to conclude that the angles formed by $\mathbf{u}$ and $\mathbf{v}$ and by $-\mathbf{u}$ and $\mathbf{v}$ are supplementary (add to π radians).

30. Prove the properties of Theorem 5 that remain unproved.

31. Let $\mathbf{u} = (u_1, u_2, u_3)$ be a unit vector in $\mathbf{R}^3$ in the direction of $\mathbf{v}$, that is $\mathbf{u} = (1/\|\mathbf{v}\|)\mathbf{v}$. Use Theorem 2 to prove that u_1, u_2, and u_3 are the cosines of the angles formed by $\mathbf{v}$ with $(1, 0, 0)$, $(0, 1, 0)$, and $(0, 0, 1)$, respectively. The components of $\mathbf{u}$ are called the *direction cosines* of $\mathbf{v}$.

32. Use the result of Exercise 31 to find the direction cosines of the vector $(1, 2, -2)$.

33. Use Figure 1.20 to show that the *volume of the parallelepiped* determined by the vectors $\mathbf{u}$, $\mathbf{v}$, and $\mathbf{w}$ in $\mathbf{R}^3$ is given by

$$V = |(\mathbf{u} \times \mathbf{v}) \cdot \mathbf{w}|.$$

(Recall that $V = Ah,$ the area of the base times the perpendicular height.)

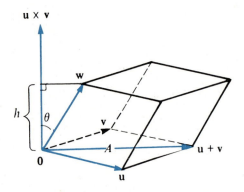

FIGURE I.20

34. Use the result of Exercise 33 to find the volume of the parallelepiped determined by the vectors $\mathbf{u} = (2, 1, -1)$, $\mathbf{v} = (3, -1, 0)$, and $\mathbf{w} = (1, 1, -1)$.

35. Show that for all vectors $\mathbf{u}, \mathbf{v}$, and $\mathbf{w}$ in $\mathbf{R}^3$,

$$\mathbf{u} \cdot (\mathbf{v} - \mathbf{w}) = \mathbf{u} \cdot \mathbf{v} - \mathbf{u} \cdot \mathbf{w}.$$

In Exercises 36 and 37 use a computer program* to find the cross product of the given vectors.

36. $\mathbf{u} = (0.32, -1.41, 0.13)$, $\mathbf{v} = (5.19, 0.01, -3.68)$

37. $\mathbf{u} = (\frac{1}{2}, \frac{1}{3}, \frac{1}{4})$, $\mathbf{v} = (\frac{1}{5}, \frac{1}{6}, \frac{1}{7})$

1.3 LINES AND PLANES

In this section we will show how vector equations and functions can be used to describe lines and planes in $\mathbf{R}^3$.

POINT-NORMAL FORM FOR A PLANE

Intuition tells us that the set of all directed line segments in $\mathbf{R}^3$ from a given point and perpendicular to a given directed line segment is a plane (Figure 1.21). We will use this idea to develop an algebraic description of a plane.

Let P be a point in $\mathbf{R}^3$, let $\mathbf{n}$ be a nonzero vector, and let Π denote the plane that contains the point P and is perpendicular to the line determined by $\mathbf{n}$ (Figure 1.22). Let X be an arbitrary point in Π, and consider the directed segment $\overrightarrow{PX}$. The vector $\mathbf{x} - \mathbf{p}$ is equivalent to $\overrightarrow{PX}$ (Theorem 4 of Section

*The programs mentioned in the Computational Exercises are available directly from the publisher for instructors using this text.

FIGURE 1.21

1.1) and therefore is orthogonal to **n**. In terms of dot product, we have

$$\mathbf{n} \cdot (\mathbf{x} - \mathbf{p}) = 0. \tag{1}$$

In other words, any point X in the plane Π satisfies Equation (1). Furthermore, if a point Y is not in Π, then $\mathbf{p} - \mathbf{y}$ cannot be orthogonal to **n**. Thus Equation (1) completely describes the plane Π. Equation (1) is called an equation in **point-normal form** for the plane, and **n** is called a **normal** for the plane. Note that any nonzero scalar multiple of **n** and any other point in the plane yield an equally valid equation, so the point-normal form is by no means unique.

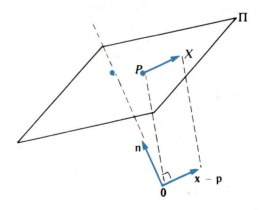

FIGURE 1.22

EXAMPLE 1.14   Find a point-normal form for the plane that contains the point $P(2, 1, 3)$ with normal vector $\mathbf{n} = (1, 2, -2)$.

Solution   From Equation (1) we have

$$\mathbf{n} \cdot (\mathbf{x} - \mathbf{p}) = 0, \quad \text{or} \quad (1, 2, -2) \cdot (\mathbf{x} - (2, 1, 3)) = 0.$$

Sometimes an expanded form of Equation (1) is taken as the point-normal form. Specifically, let $\mathbf{n} = (n_1, n_2, n_3)$, $\mathbf{p} = (p_1, p_2, p_3)$, and $\mathbf{x} = (x, y, z)$. Then $\mathbf{n} \cdot (\mathbf{x} - \mathbf{p}) = 0$ becomes

$$(n_1, n_2, n_3) \cdot ((x, y, z) - (p_1, p_2, p_3)) = 0$$
$$(n_1, n_2, n_3) \cdot (x - p_1, y - p_2, z - p_3) = 0$$
$$n_1(x - p_1) + n_2(y - p_2) + n_3(z - p_3) = 0. \tag{2}$$

Equation (2) is equivalent to Equation (1), and it often goes by the same name. For instance, by letting $\mathbf{x} = (x, y, z)$ and performing the subtraction and dot product operations, the equation of Example 1.14 may be written

$$(x - 2) + 2(y - 1) - 2(z - 3) = 0.$$

The most efficient procedure for writing a "clean" equation for a plane through P with normal vector $\mathbf{n}$ results from expanding (1) in a slightly different manner,

$$\mathbf{n} \cdot (\mathbf{x} - \mathbf{p}) = 0$$
$$\mathbf{n} \cdot \mathbf{x} - \mathbf{n} \cdot \mathbf{p} = 0$$
$$\mathbf{n} \cdot \mathbf{x} = \mathbf{n} \cdot \mathbf{p}.$$

Letting $\mathbf{n} = (a, b, c)$, $\mathbf{x} = (x, y, z)$, and $\mathbf{n} \cdot \mathbf{p} = d$, we obtain an equation for the plane in **standard form:**

$$ax + by + cz = d. \tag{3}$$

EXAMPLE 1.15   Find an equation in standard form for the plane through the given point with normal $\mathbf{n}$.

a. $(-1, 1, 2)$, $\mathbf{n} = (2, 3, 4)$ **b.** the origin, $\mathbf{n} = (2, -1, 3)$

Solution   **a.** Since

$$(2, 3, 4) \cdot (-1, 1, 2) = 9,$$

Equation (3) yields

$$2x + 3y + 4z = 9.$$

b. Here $\mathbf{p} = \mathbf{0}$ and $\mathbf{n} \cdot \mathbf{p} = 0$, so we have simply

$$2x - y + 3z = 0.$$

Suppose we are given an equation for a plane in standard form

$$ax + by + cz = d, \qquad (a, b, c) \neq \mathbf{0},$$

and we wish to transform it into point-normal form. The plane in question must be orthogonal to the vector $\mathbf{n} = (a, b, c)$, but we still need a point in the plane to get a point-normal form for it. One such point is easily found. If $a \neq 0$, then $(d/a, 0, 0)$ satisfies the equation. If $b \neq 0$, we can use $(0, d/b, 0)$, and similarly, if $c \neq 0$, we can use $(0, 0, d/c)$. Letting P be any of these, it is a simple matter to show that $\mathbf{n} \cdot (\mathbf{x} - \mathbf{p}) = 0$ reduces to the original equation $ax + by + cz = d$.

EXAMPLE 1.16

Find an equation in point-normal form for the plane described by the equation $3x - y + 2z = 5$.

Solution

A normal vector is $(3, -1, 2)$, and the point $(0, -5, 0)$ satisfies the equation. Thus

$$(3, -1, 2) \cdot (\mathbf{x} - (0, -5, 0)) = 0$$

is one such equation.

PLANE DETERMINED BY THREE POINTS

A plane Π in $\mathbf{R}^3$ is determined by three noncollinear points P, Q, and R. If three such points are specified, the cross product of two vectors can be used to find a normal vector. Specifically, the vector $\mathbf{q} - \mathbf{p}$ is equivalent to $\overrightarrow{PQ}$, and $\mathbf{r} - \mathbf{p}$ is equivalent to $\overrightarrow{PR}$ (Figure 1.23). Since P, Q, and R are not collinear, $\mathbf{q} - \mathbf{p}$ and $\mathbf{r} - \mathbf{p}$ are noncollinear vectors, and by Theorem 4 of Section 1.2,

$$\mathbf{n} = (\mathbf{q} - \mathbf{p}) \times (\mathbf{r} - \mathbf{p})$$

is a nonzero vector orthogonal to both $\mathbf{q} - \mathbf{p}$ and $\mathbf{r} - \mathbf{p}$. Thus $\mathbf{n}$ is normal to the plane determined by $\mathbf{q} - \mathbf{p}$ and $\mathbf{r} - \mathbf{p}$, and hence to Π as well. Using this vector, $\mathbf{n}$, the equation

$$\mathbf{n} \cdot (\mathbf{x} - \mathbf{p}) = 0$$

describes the plane Π.

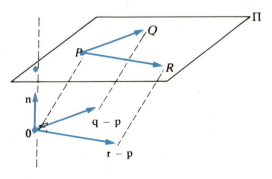

FIGURE 1.23

EXAMPLE 1.17   Find an equation for the plane determined by the points $P(2, 1, 3)$, $Q(4, 2, -1)$, and $R(6, 2, 2)$.

Solution   For a normal vector we have

$$\mathbf{n} = (\mathbf{q} - \mathbf{p}) \times (\mathbf{r} - \mathbf{p}) = (2, 1, -4) \times (4, 1, -1) = (3, -14, -2).$$

Thus a point-normal form for the plane is

$$(3, -14, -2) \cdot (\mathbf{x} - (2, 1, 3)) = 0,$$

or in standard form,

$$3x - 14y - 2z = -14.$$

POINT-PARALLEL FORM FOR A LINE

Since the usual form for the equation of a line in $\mathbf{R}^2$, $ax + by = c$, becomes an equation of a plane in $\mathbf{R}^3$, it is natural to ask how we can describe a line in $\mathbf{R}^3$. Let $\mathbf{v}$ be a nonzero vector and P a point in $\mathbf{R}^3$. We wish to describe the line through P and parallel to $\mathbf{v}$ (Figure 1.24).

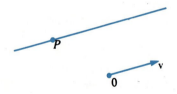

FIGURE 1.24

From Section 1.1 we know that a point is on the line determined by the vector $\mathbf{v}$ if and only if it has the form $t\mathbf{v}$ for some scalar t. From the properties of vector addition, the directed segment from P to the endpoint of $\mathbf{p} + t\mathbf{v}$ is parallel and equal in length to the vector $t\mathbf{v}$. But then the endpoint of $\mathbf{p} + t\mathbf{v}$ must lie on the line determined by P and the endpoint of $\mathbf{p} + \mathbf{v}$ (Figure 1.25).

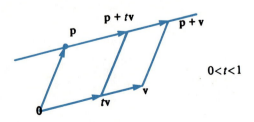

FIGURE 1.25

Thus any point X on the line through P and parallel to $\mathbf{v}$ is the endpoint of a vector of the form $\mathbf{p} + t\mathbf{v}$. We write this as $\mathbf{x} = \mathbf{p} + t\mathbf{v}$ or, since $\mathbf{x}$ depends upon the value of t, as

$$\mathbf{x}(t) = \mathbf{p} + t\mathbf{v}. \tag{4}$$

Equation (4) is called the **point-parallel form** for the line through the point P and parallel to the vector $\mathbf{v}$. Note that this equation is not an equation in the same sense as Equations (1) through (3). It is instead a *function*, where different vectors $\mathbf{x}(t)$ are obtained for different values of the variable t. For instance, $\mathbf{x}(0) = \mathbf{p}$ and $\mathbf{x}(1) = \mathbf{p} + \mathbf{v}$. Similarly, the endpoint of $\mathbf{x}(\frac{1}{2})$ is the midpoint of the segment from P to the endpoint of $\mathbf{p} + \mathbf{v}$, and $\mathbf{x}(2) = \mathbf{p} + 2\mathbf{v}$ and $\mathbf{x}(-1) = \mathbf{p} - \mathbf{v}$ are as pictured in Figure 1.26.

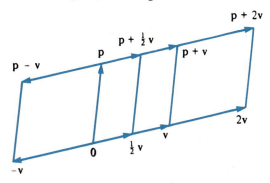

FIGURE 1.26

It is also important to note that, although we were discussing lines in $\mathbf{R}^3$, no special use of $\mathbf{R}^3$ was made in the discussion leading to Equation (4). Therefore the interpretation of the equation holds in both $\mathbf{R}^2$ and $\mathbf{R}^3$.

EXAMPLE 1.18

Give a point-parallel form for the line through the point $P(2, 2, 0)$ and parallel to the vector $\mathbf{v} = (-3, -1, 2)$.

Solution

From Equation (4) we have

$$\mathbf{x}(t) = (2, 2, 0) + t(-3, -1, 2).$$

EXAMPLE 1.19

Give a point-parallel form for the line in $\mathbf{R}^2$ given by the equation $y = 2x - 1$.

Solution

We need to find a point on the given line and a vector parallel to it. The given line has the y-intercept $(0, -1)$, so let $\mathbf{p} = (0, -1)$. Moreover the line has slope $2 = 2/1$, so the vector $\mathbf{v} = (1, 2)$ is parallel to it. Thus we have

$$\mathbf{x}(t) = (0, -1) + t(1, 2).$$

It is sometimes convenient to write Equation (4) in terms of its coordinates. To do this, let $P(p_1, p_2, p_3)$, $V(v_1, v_2, v_3)$, and $X(x, y, z)$ be points in $\mathbf{R}^3$. Then Equation (4) becomes

$$(x, y, z) = (p_1, p_2, p_3) + t(v_1, v_2, v_3)$$
$$= (p_1 + tv_1, p_2 + tv_2, p_3 + tv_3).$$

Equating coordinates, we have

$$x = p_1 + tv_1,$$
$$y = p_2 + tv_2, \qquad (5)$$
$$z = p_3 + tv_3.$$

These equations are called **parametric equations** for the line. Taken together, they are equivalent to the original Equation (4). Of course, there will only be equations for x and y in the case of $\mathbf{R}^2$.

EXAMPLE 1.20

Find parametric equations for the line through the point $(2, 2, 0)$ and parallel to the vector $(-3, -1, 2)$.

Solution

From Example 1.18 we have the point-parallel form for the line

$$\mathbf{x}(t) = (2, 2, 0) + t(-3, -1, 2).$$

Equating coordinates, we have the parametric equations

$$x = 2 - 3t,$$
$$y = 2 - t,$$
$$z = 2t.$$

EXAMPLE 1.21

Find the point of intersection of the line $\mathbf{x}(t) = (2, 1, 3) + t(2, -2, 1)$ with the plane $x + 2y - z = 7$.

Solution

We substitute the parametric equations for the line,

$$x = 2 + 2t,$$
$$y = 1 - 2t,$$
$$z = 3 + t,$$

into the equation for the plane,

$$(2 + 2t) + 2(1 - 2t) - (3 + t) = 7.$$

Solving for t, we obtain $t = -2$. Thus the point is

$$\mathbf{x}(-2) = (2, 1, 3) + (-2)(2, -2, 1)$$
$$= (-2, 5, 1).$$

DISTANCE FROM A POINT TO A PLANE

Determining the (perpendicular) distance from a given point P to a given plane Π may seem like a difficult problem, but using the tools of linear algebra we can easily derive a formula to accomplish this. Let Π be given in point-normal form by $\mathbf{n} \cdot (\mathbf{x} - \mathbf{q}) = 0$. Since $\mathbf{n}$ is normal to Π, it provides a parallel vector needed to describe the line through P perpendicular to Π. Therefore the point-parallel form for this line is $\mathbf{x}(t) = \mathbf{p} + t\mathbf{n}$.

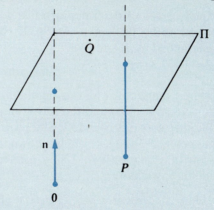

To find the point of intersection of this line with Π, we substitute into the equation for Π:

$$\mathbf{n} \cdot ((\mathbf{p} + t\mathbf{n}) - \mathbf{q}) = 0,$$

$$\mathbf{n} \cdot (t\mathbf{n} - (\mathbf{q} - \mathbf{p})) = 0,$$

$$\mathbf{n} \cdot (t\mathbf{n}) - \mathbf{n} \cdot (\mathbf{q} - \mathbf{p}) = 0,$$

$$t(\mathbf{n} \cdot \mathbf{n}) = \mathbf{n} \cdot (\mathbf{q} - \mathbf{p}),$$

$$t = \frac{\mathbf{n} \cdot (\mathbf{q} - \mathbf{p})}{\mathbf{n} \cdot \mathbf{n}} = \frac{\mathbf{n} \cdot (\mathbf{q} - \mathbf{p})}{\|\mathbf{n}\|^2}.$$

Thus the point of intersection is the one corresponding to

$$\mathbf{x}(t) = \mathbf{p} + \left(\frac{\mathbf{n} \cdot (\mathbf{q} - \mathbf{p})}{\|\mathbf{n}\|^2} \right) \mathbf{n}.$$

Finally, the distance from $\mathbf{p}$ to $\mathbf{x}(t)$ is

$$d(\mathbf{p}, \mathbf{x}(t)) = \|\mathbf{x}(t) - \mathbf{p}\| = \left\| \left(\frac{\mathbf{n} \cdot (\mathbf{q} - \mathbf{p})}{\|\mathbf{n}\|^2} \right) \mathbf{n} \right\|$$

$$= \frac{|\mathbf{n} \cdot (\mathbf{q} - \mathbf{p})|}{\|\mathbf{n}\|^2} \|\mathbf{n}\| = \frac{|\mathbf{n} \cdot (\mathbf{q} - \mathbf{p})|}{\|\mathbf{n}\|}.$$

For example, to find the distance from the point $P(2, 1, 3)$ to the plane $2x - 3y - z = 1$, we first need to express $2x - 3y - z = 1$ in point-normal form. One such form is $(2, -3, -1) \cdot (\mathbf{x} - (0, 0, -1)) = 0$. Thus we can take $\mathbf{n} = (2, -3, -1)$ and $\mathbf{q} = (0, 0, -1)$. Substituting into the equation derived above, we have

$$\frac{|\mathbf{n} \cdot (\mathbf{q} - \mathbf{p})|}{\|\mathbf{n}\|} = \frac{|(2, -3, -1) \cdot (-2, -1, -4)|}{\|(2, -3, -1)\|} = \frac{3}{\sqrt{14}}.$$

TWO-POINT FORM FOR A LINE

Another way of determining a line geometrically is by two distinct points. Let P and Q be distinct points in $\mathbf{R}^2$ or $\mathbf{R}^3$. The **two-point form** for the line determined by P and Q is

$$\mathbf{x}(t) = (1 - t)\mathbf{p} + t\mathbf{q}, \tag{6}$$

where t ranges over all real numbers.

We now show that the set of all points determined by Equation (6) is the line that contains P and Q. Since $\mathbf{x}(0) = \mathbf{p}$ and $\mathbf{x}(1) = \mathbf{q}$, the set does contain the points P and Q. It can be shown that the set is a line by converting it to the form of Equation (4). Using properties of scalar multiplication and vector addition, we have

$$\mathbf{x}(t) = (1 - t)\mathbf{p} + t\mathbf{q}$$
$$= \mathbf{p} - t\mathbf{p} + t\mathbf{q}$$
$$= \mathbf{p} + t(\mathbf{q} - \mathbf{p}).$$

Thus the set of endpoints of all such $\mathbf{x}(t)$ is the line that contains the point P and is parallel to $\mathbf{q} - \mathbf{p}$ (Figure 1.27).

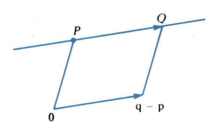

FIGURE 1.27

EXAMPLE 1.22   Describe the line through the points $(1, 2, 3)$ and $(-1, 1, 2)$ in both the two-point form and parametrically.

Solution

From Equation (6) we have the two-point form,

$$\mathbf{x}(t) = (1 - t)(1, 2, 3) + t(-1, 1, 2).$$

To find the parametric equations, let $\mathbf{x} = (x, y, z)$ and perform the indicated operations on the right:

$$(x, y, z) = (1 - t, 2 - 2t, 3 - 3t) + (-t, t, 2t)$$
$$= (1 - 2t, 2 - t, 3 - t).$$

Equating coordinates, we have the parametric equations,

$$x = 1 - 2t,$$
$$y = 2 - t,$$
$$z = 3 - t.$$

1.3 EXERCISES

In Exercises 1–5 give equations in both point-normal and standard form of the plane described.

1. Through $P(2, -1, -4)$ with normal $\mathbf{n} = (3, 1, 2)$

2. Through $P(1, 2, 3)$ with normal $\mathbf{n} = (-3, 0, 1)$

3. Through the origin with normal $\mathbf{n} = (2, 1, 3)$

4. Through $P(-1, 2, 3)$, $Q(3, 2, 5)$, and $R(-1, 1, 2)$

5. Through $P(-1, 2, 3)$, $Q(0, 2, 0)$, and $R(1, 1, 3)$

In Exercises 6–8 convert the equation in standard form to point-normal form for a plane in $\mathbf{R}^3$.

6. $2x - y + z = 5$ **7.** $2x + 3y = 1$ **8.** $z = 0$

If, in $\mathbf{R}^2$, P is a point and $\mathbf{n}$ is a nonzero vector, then $\mathbf{n} \cdot (\mathbf{x} - \mathbf{p}) = 0$ is an equation of the line through P perpendicular to the direction $\mathbf{n}$. Borrowing terminology from $\mathbf{R}^3$, we call this the *point-normal* form for the given line. Give the point-normal form for each $\mathbf{R}^2$ line described in Exercises 9–12.

9. Through $P(-1, 2)$ with normal $\mathbf{n} = (2, 1)$

10. Through $P(2, -1)$ with normal $\mathbf{n} = (0, 2)$

11. Through $P(-2, 5)$ and perpendicular to the line $x + 2y = 5$

12. Through $P(1, -3)$ and perpendicular to the line $3x - 4y = 1$

In Exercises 13–21 give a point-parallel or two-point form, and parametric equations for the $\mathbf{R}^2$ or $\mathbf{R}^3$ line described.

13. Through $P(2, 1, -3)$ and parallel to $\mathbf{v} = (1, 2, 2)$

14. Through $P(3, -1)$ and parallel to $\mathbf{v} = (2, 3)$

15. Through $P(2, -3, 1)$ and parallel to the x-axis

16. Through $P(1, 2, -1)$ and $Q(2, -1, 3)$

17. Through $P(2, 0, -2)$ and $Q(1, 4, 2)$

18. Through $P(1, 2, 3)$ and perpendicular to the plane $2x - y - 2z = 4$

19. Through $P(2, 4, 5)$ and perpendicular to the plane $5x - 5y - 10z = 2$

20. Through $P(3, -2)$ and perpendicular to the line $3x - 5y = 1$

21. Through $P(1, -1)$ and perpendicular to the line $x - 3y = 4$

22. Let $\mathbf{p} = (p_1, p_2)$, $\mathbf{u} = (u_1, u_2)$, and $\mathbf{x}(t) = \mathbf{x} = (x, y)$. Show that the point-parallel form $\mathbf{x}(t) = \mathbf{p} + t\mathbf{u}$ describes the same line in $\mathbf{R}^2$ as the point-normal form $\mathbf{n} \cdot (\mathbf{x} - \mathbf{p}) = 0$, where $\mathbf{n} = (u_2, -u_1)$.

23. Convert the line $\mathbf{x}(t) = (2, 3) + t(1, -3)$ into point-normal form. (See Exercise 22.)

24. Convert the line $2x - y = 5$ into point-parallel form.

25. Find the distance from the point $(3, 2, 3)$ to the plane $x - 2y + z = 5$. (See Spotlight.)

26. Show that this section's Spotlight formula for the distance from a point to a plane in $\mathbf{R}^3$ yields the distance from a point to a line in $\mathbf{R}^2$.

27. Find the distance from the point $(2, -3)$ to the line $x + 2y = 5$. (See Exercise 26.)

28. Find the point of intersection of the line $\mathbf{x}(t) = (3, 0, 1) + t(-2, 1, 0)$ with the plane $3x + 2y - 3z = 3$.

29. Find the point of intersection of the line $\mathbf{x}(t) = (2, 1, 1) + t(-1, 0, 4)$ with the plane $x - 3y - z = 1$.

30. Let Π be the plane described by $\mathbf{n} \cdot (\mathbf{x} - \mathbf{q}) = 0$ and l the line $\mathbf{x}(t) = \mathbf{p} + t\mathbf{u}$ $(\mathbf{n}, \mathbf{u} \neq \mathbf{0})$. Show the following:

    (a) If $\mathbf{n} \cdot \mathbf{u} = 0$, then l is contained in Π or *parallel* to it (they have no points in common).

    (b) If $\mathbf{n} \cdot \mathbf{u} \neq 0$, then l and Π intersect in the point with coordinates

$$\mathbf{p} + \left(\frac{\mathbf{n} \cdot (\mathbf{q} - \mathbf{p})}{\mathbf{n} \cdot \mathbf{u}} \right) \mathbf{u},$$

which is at a distance of

$$\left| \frac{\mathbf{n} \cdot (\mathbf{q} - \mathbf{p})}{\mathbf{n} \cdot \mathbf{u}} \right| \|\mathbf{u}\|$$

from P.

31. Use Exercise 30 to show that the line $\mathbf{x}(t) = (2, 4, 5) + t(1, 0, 1)$ and the plane $2x - y - 2z = 5$ are parallel.

32. Two planes are *parallel* if they have no points in common. Show that the planes $2x - 3y + z = 1$ and $4x - 6y + 2z = 5$ are parallel. (*Hint:* First show that, if they have one point in common, they are one and the same plane.)

33. Show that the planes $3x - 2y + z = 1$ and $x - y - 2z = 1$ are not parallel *without* finding any point of intersection.

34. Find the distance from the point $P(1, -1, 3)$ to the line l given by $\mathbf{x}(t) = (-1, 0, 2) + t(2, 1, 4)$. (*Hint:* Find the point of intersection of l with the plane through P and perpendicular to l.)

35. The lines $\mathbf{x}(t) = (2, 1, 3) + t(0, -2, -1)$ and $\mathbf{y}(s) = (3, 1, 5) + s(1, 0, 2)$ intersect at $(2, 1, 3)$ (let $t = 0$ and $s = -1$). Find the cosine of the angle of intersection. (*Hint:* Just use the parallel vectors. The angle θ is restricted to $0 \leq \theta \leq \pi/2$ radians, so use Exercise 29 of Section 1.2.)

36. By considering complementary angles (ones whose sum is $\pi/2$ radians), it can be shown that the angle between two intersecting lines has the same measure as the angle formed by normal vectors to the lines (or its supplement). Find the cosine of the angle between the $\mathbf{R}^2$ lines $2x - y = 3$ and $x + 5y = 9$.

37. The *angle between two intersecting planes* is that of the lines formed by intersecting the planes with a third plane perpendicular to the line of intersection of the planes. As in Exercise 36, it suffices to use normal vectors to the planes. Find the cosine of the angle between the planes $3x - y + z = 3$ and $2x + 4y - 2z = 1$.

1.4 EUCLIDEAN m-SPACE

Just about everything we have studied in the first three sections can be generalized to Euclidean m-space, $\mathbf{R}^m$. In this section we introduce $\mathbf{R}^m$ and consider many of these generalizations. The terminology and operations presented here will be used throughout the text.

VECTORS IN $\mathbf{R}^m$

For $m \geq 1$ let $\mathbf{R}^m$ denote the set of all ordered m-tuples,

$$\mathbf{u} = (u_1, u_2, \ldots, u_m),$$

where the u_i are in $\mathbf{R}$, the set of real numbers. An element of $\mathbf{R}^m$ is called a **vector** or **m-vector**. The numbers $u_1, \ldots, u_m$ are called the **components** of $\mathbf{u}$ and, when unspecified, are denoted by the same letter as the vector. In other words, "m-vector $\mathbf{v}$" means the vector $\mathbf{v} = (v_1, v_2, \ldots v_m)$. However, in the case of $m = 2$ or $m = 3$ we will sometimes continue to use $\mathbf{x} = (x, y)$ or $\mathbf{x} = (x, y, z)$.

Two vectors $\mathbf{u}$ and $\mathbf{v}$ are **equal** if their *corresponding components* are equal—if $u_i = v_i$ for each i. The **zero vector** in $\mathbf{R}^m$ is the m-vector $\mathbf{0} = (0, 0, \ldots, 0)$. For a vector $\mathbf{u}$ in $\mathbf{R}^m$, the **negative** of $\mathbf{u}$ is the vector $-\mathbf{u} = (-u_1, -u_2, \ldots, -u_m)$.

LENGTH OF A VECTOR

DEFINITION

The **distance** between vectors $\mathbf{u}$ and $\mathbf{v}$ in $\mathbf{R}^m$ is given by the **distance formula**,

$$d(\mathbf{u}, \mathbf{v}) = \sqrt{(v_1 - u_1)^2 + (v_2 - u_2)^2 + \cdots + (v_m - u_m)^2}.$$

The **length** (**norm, magnitude**) of a vector **u** is its distance from **0**,

$$\|\mathbf{u}\| = d(\mathbf{0}, \mathbf{u}) = \sqrt{u_1^2 + u_2^2 + \cdots + u_m^2}.$$

NOTE   If $\|\mathbf{u}\| = 0$, then $\mathbf{u} = \mathbf{0}$ (and conversely). A vector **u** is called a **unit vector** if $\|\mathbf{u}\| = 1$.

EXAMPLE 1.23   Find $\|\mathbf{u}\|$ where $\mathbf{u} = (1, 2, 3, -2)$.

Solution

$$\|\mathbf{u}\| = \sqrt{1^2 + 2^2 + 3^2 + (-2)^2} = \sqrt{18} = 3\sqrt{2}.$$

We will now extend the definition of the operations on vectors given in Section 1.1 to vectors in $\mathbf{R}^m$.

SCALAR MULTIPLICATION

DEFINITION

Let c be a scalar and **u** a vector in $\mathbf{R}^m$. The **scalar multiple** of **u** by c is the vector

$$c\mathbf{u} = (cu_1, cu_2, \ldots, cu_m).$$

That is, $c\mathbf{u}$ is the vector obtained by multiplying each component of **u** by the scalar c.

If **u** and **v** are nonzero vectors, and there exists a scalar such that $\mathbf{v} = c\mathbf{u}$, we say **u** and **v** are **collinear**. If this is the case, we say **u** and **v** *have the same direction* if c is positive and *have opposite directions* if c is negative.

THEOREM 1   Let **u** be a vector in $\mathbf{R}^m$ and c a scalar. Then

$$\|c\mathbf{u}\| = |c|\, \|\mathbf{u}\|.$$

Proof $\|c\mathbf{u}\| = \|(cu_1, \ldots, cu_m)\| = \sqrt{(cu_1)^2 + \cdots + (cu_m)^2}$
$$= \sqrt{c^2(u_1^2 + \cdots + u_m^2)} = |c|\, \|\mathbf{u}\|.$$

Note that this proof is exactly the same (with more components) as was given for Theorem 1 of Section 1.1. Furthermore, the interpretation of the result is exactly the same: Multiplication of a vector by a scalar changes the length by the same factor (in absolute value).

VECTOR ADDITION

DEFINITION

Let **u** and **v** be vectors in $\mathbf{R}^m$. The **sum** of **u** and **v** is the vector

$$\mathbf{u} + \mathbf{v} = (u_1 + v_1, u_2 + v_2, \ldots, u_m + v_m).$$

In words, **u** + **v** is the vector obtained by adding the corresponding components.

The **negative** of **u** is the vector

$$-\mathbf{u} = (-u_1, -u_2, \ldots, -u_m),$$

and the **difference** of **u** and **v** is

$$\mathbf{u} - \mathbf{v} = \mathbf{u} + (-\mathbf{v}) = (u_1 - v_1, u_2 - v_2, \ldots, u_m - v_m).$$

EXAMPLE 1.24 Letting $c = 3$, $\mathbf{u} = (1, 3, 5, -1)$, and $\mathbf{v} = (1, 0, 2, 1)$, find $3\mathbf{u} - \mathbf{v}$.

Solution

$$\begin{aligned} 3\mathbf{u} - \mathbf{v} &= 3(1, 3, 5, -1) - (1, 0, 2, 1) \\ &= (3, 9, 15, -3) - (1, 0, 2, 1) \\ &= (2, 9, 13, -4). \end{aligned}$$

The following properties of vector addition and scalar multiplication follow directly from the definitions of these operations. The proofs are exactly the same as those given for Theorem 3 of Section 1.1 (with more components) and are left as exercises.

THEOREM 2 Let **u**, **v**, and **w** be vectors in $\mathbf{R}^m$, and let c and d be scalars. Then

a. $\mathbf{u} + \mathbf{v} = \mathbf{v} + \mathbf{u}$ (vector addition is commutative),

b. $(\mathbf{u} + \mathbf{v}) + \mathbf{w} = \mathbf{u} + (\mathbf{v} + \mathbf{w})$ (vector addition is associative),

c. $\mathbf{u} + \mathbf{0} = \mathbf{u}$,

d. $\mathbf{u} + (-\mathbf{u}) = \mathbf{0}$,

e. $(cd)\mathbf{u} = c(d\mathbf{u})$,

f. $(c + d)\mathbf{u} = c\mathbf{u} + d\mathbf{u}$

g. $c(\mathbf{u} + \mathbf{v}) = c\mathbf{u} + c\mathbf{v}$ (distributive laws),

h. $1\mathbf{u} = \mathbf{u}$,

i. $(-1)\mathbf{u} = -\mathbf{u}$,

j. $0\mathbf{u} = \mathbf{0}$.

DOT PRODUCT

DEFINITION

> Let $\mathbf{u}$ and $\mathbf{v}$ be vectors in $\mathbf{R}^m$. The **dot product** (or **standard inner product**) of $\mathbf{u}$ and $\mathbf{v}$ is
>
> $$\mathbf{u} \cdot \mathbf{v} = u_1 v_1 + u_2 v_2 + \cdots + u_m v_m.$$
>
> In words, $\mathbf{u} \cdot \mathbf{v}$ is the sum of the products of the corresponding components of $\mathbf{u}$ and $\mathbf{v}$.

NOTE   The definition of $\mathbf{u} \cdot \mathbf{v}$ can be given more concisely using *summation notation*. Given a set of numbers $a_1, a_2, \ldots, a_m$, the **summation notation** $\sum_{k=1}^{m} a_k$ represents their sum. In other words,

$$\sum_{k=1}^{m} a_k = a_1 + a_2 + \cdots + a_m.$$

Here k is called the *index of summation*.

Using summation notation, the dot product of $\mathbf{u}$ and $\mathbf{v}$ in $\mathbf{R}^m$ can be written as

$$\mathbf{u} \cdot \mathbf{v} = \sum_{k=1}^{m} u_k v_k.$$

Variables other than k can be used for the index of summation, so $\sum_{i=1}^{m} u_i v_i$ and $\sum_{j=1}^{m} u_j v_j$ also represent $\mathbf{u} \cdot \mathbf{v}$. Throughout the rest of this text, whenever appropriate we will use summation notation in addition to the longer expanded form of the sum.

· ·

Because the dot product of two vectors is a scalar, it is often called the **scalar product.** The algebraic properties of dot product are included in the next theorem. The proofs are exactly the same as those given in Theorem 1 of Section 1.2 (with more components) and are left as exercises.

THEOREM 3   Let $\mathbf{u}$, $\mathbf{v}$, and $\mathbf{w}$ be vectors in $\mathbf{R}^m$ and let c be a scalar. Then

  a. $\mathbf{u} \cdot \mathbf{v} = \mathbf{v} \cdot \mathbf{u}$ (dot product is commutative),

  b. $c(\mathbf{u} \cdot \mathbf{v}) = (c\mathbf{u}) \cdot \mathbf{v} = \mathbf{u} \cdot (c\mathbf{v})$ (scalars factor out),

  c. $\mathbf{u} \cdot (\mathbf{v} + \mathbf{w}) = \mathbf{u} \cdot \mathbf{v} + \mathbf{u} \cdot \mathbf{w}$ (distributive law),

  d. $\mathbf{u} \cdot \mathbf{0} = 0$,

  e. $\mathbf{u} \cdot \mathbf{u} = \|\mathbf{u}\|^2.$

DEFINITION

> Vectors **u** and **v** in $\mathbf{R}^m$ are **orthogonal** if $\mathbf{u} \cdot \mathbf{v} = 0$.

Everything so far has been so analogous to Sections 1.1 and 1.2 that you probably now expect to see *cross product* defined for vectors in $\mathbf{R}^m$. There *is* an analog for the cross product in $\mathbf{R}^m$, but it is complicated and of little value, so we will omit it.

LINES IN $\mathbf{R}^m$

The geometry of $\mathbf{R}^m$ for $m \geq 4$ is even more interesting than it is in the case of $m = 2$ and $m = 3$. The problem, of course, is that we cannot draw pictures of all of it at once. We must focus on subregions for pictures, and often even this device will be of little help.

Generalizing the $\mathbf{R}^2$ and $\mathbf{R}^3$ cases, we define *points* in $\mathbf{R}^m$ to be *m*-tuples of real numbers. We will denote a point by, for example, $P(u_1, u_2, \ldots, u_m)$ to distinguish it from the vector $\mathbf{p} = (u_1, u_2, \ldots, u_m)$ represented by the same *m*-tuple. Although we have no intuitive feel for what a *line* might be in $\mathbf{R}^m$ when $m \geq 4$, we can again generalize the $\mathbf{R}^2$ and $\mathbf{R}^3$ cases to make a reasonable definition. To preserve the idea that two points determine a line, we therefore follow the *two-point form* for a line in $\mathbf{R}^3$, Equation (6) of Section 1.3, to define the geometric terms that follow.

DEFINITION

> Let P and Q be distinct points in $\mathbf{R}^m$, and let
>
> $$\mathbf{x}(t) = (1 - t)\mathbf{p} + t\mathbf{q}. \tag{1}$$
>
> **a.** The set of all $\mathbf{x}(t)$ for t in the real numbers is the **line** determined by P and Q.
>
> **b.** The set of all $\mathbf{x}(t)$ for $0 \leq t \leq 1$ is the **line segment** determined by P and Q. By taking the $\mathbf{x}(t)$ in their natural order (start at $\mathbf{x}(0)$ and finish at $\mathbf{x}(1)$), we have the **directed line segment** from P to Q.

We can now speak of a nonzero vector **u** in $\mathbf{R}^m$ as a directed line segment $\overrightarrow{PQ}$ from the origin by letting $\mathbf{p} = \mathbf{0}$ and $\mathbf{q} = \mathbf{u}$. By letting t range over all real numbers, we have the *line determined* by **u**. Two nonzero vectors **u** and **v** are collinear if the lines they determine are one and the same.

Since we can view a vector **u** as a directed segment from **0**, we can generalize the *point-parallel form* for a line. The set of all vectors

$$\mathbf{x}(t) = \mathbf{p} + t\mathbf{v} \tag{2}$$

is the line that contains P and is parallel to **v**, where t is a real number and $\mathbf{v} \neq \mathbf{0}$.

EXAMPLE 1.25

Give a point-parallel form for the line through $P(2, 1, -1, 3)$ and parallel to $\mathbf{v} = (1, 0, -2, 1)$.

Solution

$$\mathbf{x}(t) = (2, 1, -1, 3) + t(1, 0, -2, 1).$$

We can also view this function parametrically as before. Let

$$\mathbf{x}(t) = (x, y, z, w).$$

Equating coordinates, we have the *parametric equations*,

$$x = 2 + t,$$
$$y = 1,$$
$$z = -1 - 2t,$$
$$w = 3 + t.$$

Thus we can deal with lines just as easily in $\mathbf{R}^m$ with $m \geq 4$ as in $\mathbf{R}^3$.

LINEAR EQUATIONS

Let P be a point, and let $\mathbf{n}$ be a nonzero vector in $\mathbf{R}^m$. The **point-normal form** for P and $\mathbf{n}$ is the equation

$$\mathbf{n} \cdot (\mathbf{x} - \mathbf{p}) = 0. \tag{3}$$

Recall that in $\mathbf{R}^3$ this is the equation of the *plane* through P with normal $\mathbf{n}$. In the general case $\mathbf{R}^m$, the set of all solutions to Equation (3) is called the **hyperplane** through P with normal $\mathbf{n}$.

EXAMPLE 1.26

Find a point-normal form for the hyperplane that contains the point $P(2, 1, 0, 3, 1)$ with normal $\mathbf{n} = (2, 1, -1, 3, 0)$.

Solution

From Equation (3), we have the point-normal form,

$$(2, 1, -1, 3, 0) \cdot (\mathbf{x} - (2, 1, 0, 3, 1)) = 0.$$

Other forms of the equation of a plane are obtained by letting $\mathbf{n} = (a_1, a_2, \ldots, a_m)$, $\mathbf{p} = (p_1, p_2, \ldots, p_m)$, $\mathbf{x} = (x_1, x_2, \ldots, x_m)$, $b = \mathbf{n} \cdot \mathbf{p}$, and expanding Equation (3). The alternate point-normal form is

$$a_1(x_1 - p_1) + a_2(x_2 - p_2) + \cdots + a_m(x_m - p_m) = 0. \tag{4}$$

The **standard form** is obtained from Equation (3) by distributing

$\mathbf{n} \cdot \mathbf{x} - \mathbf{n} \cdot \mathbf{p} = 0$ and transposing to obtain

$$a_1 x_1 + a_2 x_2 + \cdots + a_m x_m = b. \qquad (5)$$

An equation that can be put into the form of Equation (5) is called a **linear equation** in the variables $x_1, x_2, \ldots, x_m$. This is a generalization of the equation $ax_1 + bx_2 = c$ (which describes a line in $\mathbf{R}^2$) and $ax_1 + bx_2 + cx_3 = d$ (which describes a plane in $\mathbf{R}^3$). (See Section 1.3.)

Note that a linear equation contains no products of variables (such as x_1^2 or $x_1 x_2$) or functions of variables (such as $\sin x_1$, e^{x_2}, $\sqrt{x_4}$) other than multiplication by a constant scalar. For example,

$$3x_1 - 5x_1 x_2 + 2x_2 = 1 \qquad \text{and} \qquad 3x - y^2 = 7$$

are *not* linear equations.

Given a linear equation in standard form (Equation 5), it is an easy matter to put it in point-normal form. Let $\mathbf{n} = (a_1, \ldots, a_m)$ be the vector of coefficients of the variables. For any i from 1 to m with $n_i \neq 0$, let $\mathbf{p} = (0, \ldots, 0, b/a_i, 0, \ldots, 0)$, where the coordinate b/a_i is in the ith position. Then $\mathbf{n} \cdot (\mathbf{x} - \mathbf{p}) = 0$ is equivalent to the original equation, as can easily be confirmed by expansion of the dot product.

EXAMPLE 1.27

Express the linear equation

$$2x_1 + 3x_2 - x_3 - x_4 = 5$$

in point-normal form for a hyperplane in $\mathbf{R}^4$.

Solution

Letting $\mathbf{n} = (2, 3, -1, -1)$, we have $n_3 = -1 \neq 0$, so the point $P(0, 0, 5/(-1), 0) = (0, 0, -5, 0)$ is in the hyperplane. The desired equation is

$$(2, 3, -1, -1) \cdot (\mathbf{x} - (0, 0, -5, 0)) = 0.$$

PLANE DETERMINED BY TWO VECTORS

We can discuss lines in $\mathbf{R}^m$ (point-parallel form) and hyperplanes in $\mathbf{R}^m$ (point-normal form), but so far we have no means of discussing other types of regions. This can be accomplished in a manner analogous to the point-parallel form by adding more terms—one more for a plane, two more for a three-dimensional region, and so on. As an example, we describe a plane in $\mathbf{R}^m$ determined by two noncollinear vectors.

Let $\mathbf{u}$ and $\mathbf{v}$ be noncollinear vectors; that is, one is not a scalar multiple of the other. Let X be any point in the plane determined by $\mathbf{u}$ and $\mathbf{v}$. The line through X parallel to the vector $\mathbf{v}$ must intersect the line determined by $\mathbf{u}$. Since any point on the latter line is a scalar multiple of $\mathbf{u}$, the point of intersection is $s\mathbf{u}$ for some real number s (Figure 1.28). Similarly, the line

FIGURE 1.28

through X parallel to **u** must intersect the line determined by **v** in some point $t\mathbf{v}$. Then the quadrilateral **0**, $s\mathbf{u}$, **x**, $t\mathbf{v}$ is a parallelogram, and by Theorem 2 of Section 1.1,

$$\mathbf{x} = s\mathbf{u} + t\mathbf{v}.$$

Thus every point in the plane determined by **u** and **v** is representable as the sum of a multiple of **u** and a multiple of **v**. A representation for a vector **x** of this type is called a **linear combination** of **u** and **v**. Conversely, every linear combination $s\mathbf{u} + t\mathbf{v}$ lies in the plane of **u** and **v**. Therefore we may view the plane of **u** and **v** as the set of all points given by the function $\mathbf{x} = \mathbf{x}(s, t)$, where

$$\mathbf{x} = s\mathbf{u} + t\mathbf{v},$$

and s and t range independently over all real numbers.

If we start with three noncoplanar vectors **u**, **v**, and **w**, we have an analogous situation. The set of all linear combinations $\mathbf{x} = \mathbf{x}(r, s, t)$ of **u**, **v**, and **w** given by

$$\mathbf{x} = r\mathbf{u} + s\mathbf{v} + t\mathbf{w}$$

is the 3-space determined by **u**, **v**, and **w**.

The plane and 3-space just described contain the origin. For the general case of a plane determined by three noncollinear points or 3-space determined by four noncoplanar points, translate to the origin as in Section 1.1, use the procedure described, and then translate back. We demonstrate with an example.

EXAMPLE 1.28   Describe the plane determined by $P(1, 2, 1, 1)$, $Q(3, -1, 4, 0)$, and $R(1, -1, 0, 2)$ in $\mathbf{R}^4$.

Solution   We translate P to the origin. The vectors $\mathbf{q} - \mathbf{p}$ and $\mathbf{r} - \mathbf{p}$ then determine a parallel plane consisting of all linear combinations $s(\mathbf{r} - \mathbf{p}) + t(\mathbf{q} - \mathbf{p})$. Adding **p** to each such point, we translate back, and the desired plane is given by the set of all points of the form

$$\mathbf{x} = \mathbf{p} + s(\mathbf{r} - \mathbf{p}) + t(\mathbf{q} - \mathbf{p}),$$

where s and t are real numbers. That is,

$$\mathbf{x} = (1, 2, 1, 1) + s(0, -3, -1, 1) + t(2, -3, 3, -1)$$

for all real numbers s and t.

1.4 EXERCISES

In Exercises 1–12 let the vectors be given as follows:

$$\mathbf{u} = (2, 1, 3, 0), \qquad \mathbf{v} = (-1, 1, 2, 1), \qquad \mathbf{w} = (2, -2, 0, 6).$$

Perform each indicated operation.

1. $\mathbf{u} + \mathbf{v}$

2. $3\mathbf{u} - \mathbf{v} + 2\mathbf{w}$

3. $2(\mathbf{u} + 2\mathbf{v} + 3\mathbf{w})$

4. $d(\mathbf{v}, \mathbf{w})$

5. $\mathbf{v} \cdot \mathbf{w}$

6. $\|\mathbf{w}\|$

7. $\|3\mathbf{w}\|$

8. $\dfrac{1}{\|\mathbf{w}\|} \mathbf{w}$

9. $d(\mathbf{v}, -\mathbf{u})$

10. $(\mathbf{u} \cdot \mathbf{v})(-\mathbf{w})$

11. $\|2(\mathbf{v} - \mathbf{w})\|$

12. $(3\mathbf{v}) \cdot (5\mathbf{w})$

13. Which pairs of the following vectors are orthogonal?

$$\mathbf{u}_1 = (2, 1, 3, -1), \mathbf{u}_2 = (3, 1, 0, 2), \mathbf{u}_3 = (0, -2, 1, 1), \mathbf{u}_4 = (2, 0, 3, -3).$$

14. Prove that, for $\mathbf{u}, \mathbf{v}$ in $\mathbf{R}^m$, $d(\mathbf{u}, \mathbf{v}) = \|\mathbf{u} - \mathbf{v}\|$.

15. Prove that, for a nonzero vector $\mathbf{u}$ in $\mathbf{R}^m$, $(1/\|\mathbf{u}\|)\mathbf{u}$ is the unit vector in the direction of $\mathbf{u}$.

In Exercises 16 and 17 find the unit vector with the same direction as the one given.

16. $(1, 1, 3, 0, 5)$

17. $(2, 1, -1, 0, 3, 4)$

In Exercises 18 and 19 find all hyperplanes orthogonal to the given vector.

18. $(1, 0, -1, 0, 1)$

19. $(-2, -1, 0, 1)$

20. Give two-point and point-parallel forms and parametric equations for the line in $\mathbf{R}^5$ determined by $P(2, 1, 0, 3, 1)$ and $Q(1, -1, 3, 0, 5)$.

21. Give two-point and point-parallel forms and parametric equations for the line in $\mathbf{R}^4$ determined by $P(-1, 0, 3, 2)$ and $Q(-1, 0, 4, 5)$.

22. Give point-normal and standard forms for the hyperplane through $(-2, 1, 4, 0)$ with normal $(1, 2, -1, 3)$.

23. Give point-normal and standard forms for the hyperplane through $(3, 4, 5, 6, 7)$ with normal $(1, -1, 1, -1, 1)$.

24. Find a point-normal form for the linear equation $2x_1 - 3x_2 + x_4 - x_5 = 2$ in $\mathbf{R}^5$.

25. Repeat Exercise 24 viewing the equation in $\mathbf{R}^6$.

In Exercises 26–29 explain why each equation fails to be linear.

26. $x(2y + 3z) = 5$

27. $x_1 + 2x_2 + x_3 - x_4x_5 = 3$

28. $x + 2 \sin y + 3z = 0$

29. $x^2 + y^2 = 0$

30. Find the midpoint of the line segment that joins $(2, 1, 3, 5)$ with $(6, 3, 2, 1)$.

31. Prove the properties of Theorem 2.

32. The *angle between two nonzero vectors* **u** and **v** in $\mathbf{R}^m$ is defined to be the unique number θ, $0 \le \theta \le \pi$ radians such that $\cos \theta = \mathbf{u} \cdot \mathbf{v}/(\|\mathbf{u}\| \, \|\mathbf{v}\|)$. Find the cosine of the angle between the vectors $(3, 1, 1, 2, 1)$ and $(0, 2, 1, -2, 0)$.

33. Prove the properties of Theorem 3.

34. Describe the plane determined by the points $(3, 1, 0, 2, 1)$, $(2, 1, 4, 2, 0)$, and $(-1, 2, 1, 3, 1)$.

35. Describe the 3-space determined by the points $(3, 1, 0, 2, 1)$, $(2, 1, 4, 2, 0)$, $(-1, 2, 1, 3, 1)$, and $(0, 2, 0, 1, 0)$.

36. Prove the *parallelogram law* in $\mathbf{R}^m$, $\|\mathbf{u} + \mathbf{v}\|^2 + \|\mathbf{u} - \mathbf{v}\|^2 = 2 \|\mathbf{u}\|^2 + 2 \|\mathbf{v}\|^2$.

37. In metric geometry a point B is *between* A and C if and only if $d(A, B) + d(B, C) = d(A, C)$. For distinct **p** and **q** in $\mathbf{R}^m$ prove that $\mathbf{x}(t) = (1 - t)\mathbf{p} + t\mathbf{q}$ is between **p** and **q** if $0 \le t \le 1$.

38. Use Exercise 14 to show that $d(\mathbf{p}, \mathbf{p} + t\mathbf{u}) = |t| \, \|\mathbf{u}\|$.

39. Find a point one-third of the distance from $(2, 1, 3, -4)$ to $(2, -2, 0, 2)$.

40. Let P, Q, and R be the vertices of a triangle in $\mathbf{R}^m$. Prove that the segment that joins the midpoints of two of the sides is parallel to and half the length of the third side.

41. Let **u**, $\mathbf{v}_1$, and $\mathbf{v}_2$ be vectors in $\mathbf{R}^m$ and c_1 and c_2 scalars. If **u** is orthogonal to both $\mathbf{v}_1$ and $\mathbf{v}_2$, prove that **u** is orthogonal to $c_1\mathbf{v}_1 + c_2\mathbf{v}_2$.

42. Interpret the result of Exercise 41 geometrically to show that in $\mathbf{R}^3$ the sum of two vectors is coplanar with the two vectors.

43. Extend the result of Exercise 41 to any number of vectors $\mathbf{v}_1, \mathbf{v}_2, \ldots, \mathbf{v}_n$.

CHAPTER I REVIEW

This chapter introduced the concept of *vectors* in $\mathbf{R}^2$ (the plane), $\mathbf{R}^3$ (3-space) and $\mathbf{R}^m$ with $m > 3$. Operations on vectors (such as *addition* and *dot product*) were defined, and geometric notions (such as *length, line,* and *plane*) were discussed. You should now be able to determine the following:

1. the length of a vector and the distance between vectors;

2. the sum, difference and dot product of two vectors;

3. scalar multiples of vectors;

4. a vector equivalent to a given directed line segment;

5. the cross product of two vectors in $\mathbf{R}^3$;

6. the equations of lines in $\mathbf{R}^2$, lines and planes in $\mathbf{R}^3$, and lines and hyperplanes in $\mathbf{R}^m$.

KEY WORDS

vector	components of a vector
equality of vectors	zero vector
collinear vectors	distance between vectors
length (or norm) of a vector	scalar
multiplication by a scalar	sum and difference of vectors
negative of a vector	equivalent vectors
dot product (scalar product)	orthogonal vectors
cross product	point-normal form for a plane in $\mathbf{R}^3$, a hyperplane in $\mathbf{R}^m$
standard form (of a linear equation)	point-parallel and two-point form for a line
parametric equations of a line	line segment in $\mathbf{R}^m$
directed line segment in $\mathbf{R}^m$	linear combination of vectors

SELF-TEST

Answers appear at the end of this chapter.

1. Perform each operation on the indicated vectors:

$$\mathbf{w} = (2, 1, 0, -1)$$
$$\mathbf{x} = (-3, 2, -1, 0)$$
$$\mathbf{y} = (3, 2, 1)$$
$$\mathbf{z} = (-2, 3, -1).$$

(a) $2\mathbf{x} \cdot \mathbf{w}$

(b) $\mathbf{y} \times \mathbf{z}$

(c) $\|3\mathbf{x}\|$

(d) $d(3\mathbf{y}, 6\mathbf{z})$ [Section 1.2, 1.4]

2. Give the two-point form and the point-parallel form for a line in $\mathbf{R}^4$ that contains the points $P(1, -1, 1, -1)$ and $Q(4, 3, 2, 1)$.   [Section 1.4]

3. Determine the point (if any) of intersection of the line $\mathbf{x}(t) = t(1, 2, 4) + (-1, 0, 1)$ with the plane in $\mathbf{R}^3$ given by $2x - y + z = 5$.   [Section 1.3]

4. Find the line of all vectors in $\mathbf{R}^3$ that are perpendicular to the plane $2x - 3y + z = 6$.   [Section 1.3]

5. Find the line in $\mathbf{R}^3$ that is perpendicular to the plane $2x - 3y + z = 6$ and passes through the point $(2, 1, -3)$.   [Section 1.3]

6. Express the hyperplane $x - y + 2z - w = 6$ in $\mathbf{R}^4$ in point-normal form.   [Section 1.4]

7. Determine whether or not the following lines in $\mathbf{R}^4$ are parallel:

$$\mathbf{x}(t) = t(2, 1, 0, 3) + (1, -1, 1, -1),$$
$$\mathbf{x}(t) = t(1, -1, 1, -1) + (2, 1, 0, 4). \qquad \text{[Section 1.4]}$$

8. In plane geometry it is common to use the fact that if two lines are each perpendicular to the same line, they are parallel. Show by a *specific* example that this need not be the case in $\mathbf{R}^3$.   [Section 1.3]

9. Find a standard-form equation for the plane that contains the three points $(1, 0, 1)$, $(1, 1, 2)$, $(3, 2, 0)$.   [Section 1.3]

10. Being careful to explain each step in your argument, show that for any vectors $\mathbf{u}$ and $\mathbf{v}$ in $\mathbf{R}^n$ it is always true that

$$(\mathbf{u} + \mathbf{v}) \cdot (\mathbf{u} - \mathbf{v}) = \mathbf{u} \cdot \mathbf{u} - \mathbf{v} \cdot \mathbf{v}. \qquad \text{[Section 1.4]}$$

REVIEW EXERCISES

In Exercises 1–12 let the vectors be given as follows:

$$\mathbf{u} = (1, 2, 0) \qquad \mathbf{x} = (1, 0, 3, 0, 2)$$
$$\mathbf{v} = (-1, 0, 3) \qquad \mathbf{y} = (-1, 1, 0, 1, 4),$$

and perform the indicated operations.

1. $\mathbf{x} - 3\mathbf{y}$

2. $2\mathbf{u} + 4\mathbf{v}$

3. $d(2\mathbf{u}, \mathbf{v})$

4. $d(\mathbf{x} + \mathbf{y}, \mathbf{x})$

5. $\|-2\mathbf{x}\|$

6. $\|\mathbf{u} - \mathbf{v}\|$

7. $\mathbf{u} \times \mathbf{v}$

8. $\left(\dfrac{1}{\|\mathbf{y}\|}\right)\mathbf{y}$

9. $\mathbf{x} \cdot \mathbf{y}$

10. $(\mathbf{u} + \mathbf{v}) \cdot (\mathbf{u} - \mathbf{v})$

11. Determine the unit vector that has a direction opposite to that of $\mathbf{v}$.

12. Find a vector that has the same direction as that of $\mathbf{x}$ but whose length is 2.

13. Find an equation of the plane (in $\mathbf{R}^3$) that is perpendicular to the z-axis and that passes through the point $(1, 2, 3)$.

14. Find an equation of the plane (in $\mathbf{R}^3$) that contains the points $(1, 0, 0)$, $(0, 1, 0)$, and $(0, 0, 1)$.

15. The equation of a hyperplane in $\mathbf{R}^4$ is $3x + y - 2z + w = 1$. Give a point-normal form for this equation.

16. A plane (in $\mathbf{R}^3$) has equation $(2, -1, 3) \cdot ((x, y, z) - (-1, 3, 0)) = 0$. Give this equation in standard form.

17. Find an equation of the line in $\mathbf{R}^3$ that completely contains the vector $\mathbf{v} = (1, 0, 2)$.

18. Determine the point (in $\mathbf{R}^4$) of intersection of the line $\mathbf{x}(t) = (1, 0, 1, 0) + t(0, 1, 0, 1)$ and the hyperplane $x + y + z + w = 0$.

19. Determine all planes (in $\mathbf{R}^3$) orthogonal to the vector $(1, 1, 1)$.

20. Let $\mathbf{u}$ and $\mathbf{v}$ be vectors in $\mathbf{R}^5$ and let c be a scalar. Which of the following operations are *not* defined?

(a) $3\mathbf{u} - c\mathbf{v}$ (b) $5c - \mathbf{u}$ (c) $\mathbf{u} \times \mathbf{v}$

(d) $\mathbf{u} \cdot (5\mathbf{v})$ (e) $\|c\mathbf{v}\|$ (f) $\|\mathbf{u} \cdot \mathbf{v}\|$

ANSWERS TO SELF-TEST

1. (a) -8 (b) $(-5, 1, 13)$ (c) $3\sqrt{14}$ (d) $3\sqrt{74}$

2. $\mathbf{x}(t) = (1 - t)(1, -1, 1, -1) + t(4, 3, 2, 1) = t(3, 4, 1, 2) + (1, -1, 1, -1)$.

3. Replacing $x = t - 1$, $y = 2t$, $z = 4t + 1$, solve $2(t - 1) - 2t + 4t + 1 = 5$ for t. Thus $t = \frac{3}{2}$, and the point of intersection is $\mathbf{x}(\frac{3}{2}) = (\frac{1}{2}, 3, 7)$.

4. $\mathbf{x}(t) = t(2, -3, 1)$.

5. $\mathbf{x}(t) = t(2, -3, 1) + (2, 1, -3)$.

6. $(1, -1, 2, -1) \cdot (\mathbf{x} - (6, 0, 0, 0)) = 0$.

7. No.

8. Answers vary. For example,

$$l: \mathbf{x}(t) = t(0, 0, 1), \qquad l_1: \mathbf{x}(t) = t(1, 0, 0), \qquad l_2: \mathbf{x}(t) = t(1, 1, 0).$$

9. $-3x + 2y - 2z = -5$

10. $(\mathbf{u} + \mathbf{v}) \cdot (\mathbf{u} - \mathbf{v}) = (\mathbf{u} + \mathbf{v}) \cdot \mathbf{u} - (\mathbf{u} + \mathbf{v}) \cdot \mathbf{v}$

$$= \mathbf{u} \cdot \mathbf{u} + \mathbf{v} \cdot \mathbf{u} - \mathbf{u} \cdot \mathbf{v} - \mathbf{v} \cdot \mathbf{v}$$

$$= \mathbf{u} \cdot \mathbf{u} + \mathbf{u} \cdot \mathbf{v} - \mathbf{u} \cdot \mathbf{v} - \mathbf{v} \cdot \mathbf{v}$$

$$= \mathbf{u} \cdot \mathbf{u} - \mathbf{v} \cdot \mathbf{v}$$

2

LINEAR EQUATIONS AND MATRICES

Linear algebra originated with the study of linear equations, and this topic is still of major importance in virtually all disciplines that make use of mathematics. We will begin this chapter by presenting a technique for solving systems of linear equations with the aid of *matrices*. We will then investigate the properties of matrices in some detail and use these properties to develop the theory of linear systems.

2.1 SYSTEMS OF LINEAR EQUATIONS

Linear equations were introduced in Section 1.4, with an emphasis on their geometrical interpretation in Euclidean m-space. In this section we will recall some terminology associated with systems of linear equations and present a technique for solving them.

A **linear equation** in the variables $x_1, x_2, \ldots, x_n$ is one that can be put in the form

$$a_1 x_1 + a_2 x_2 + \cdots + a_n x_n = b,$$

where b and the coefficients a_i are constants and not all a_i equal zero. A **system of linear equations** (or **linear system**) is simply a finite set of linear equations. An n-vector $(s_1, s_2, \ldots, s_n)$ is a **solution** to a linear system (in n variables) if it satisfies every equation in the system. The variables in a linear system are also referred to as *unknowns*; of course, they may be denoted by other symbols such as x, y, z, or w.

Recall from Section 1.4 that the set of solutions to a linear equation is a line in $\mathbf{R}^2$, a plane in $\mathbf{R}^3$, and, generally speaking, a hyperplane in $\mathbf{R}^n$. From a geometric point of view, a solution to a system of linear equations represents a point that lies in the *intersection* of the regions described by the individual equations.

EXAMPLE 2.1   The linear system

$$2x = y - 4z$$
$$y = 2z$$

is a system of two equations in the three unknowns x, y, and z. One solution of this system is $(-1, 2, 1)$, since $x = -1$, $y = 2$, $z = 1$ satisfies *both* equations. However, this is not the only solution. For example, $(0, 0, 0)$ and $(\frac{1}{2}, -1, -\frac{1}{2})$ are also solutions. In fact, any vector of the form $(-t, 2t, t)$ is a solution, where t represents an arbitrary real number. Consequently there are infinitely many solutions to this system—one for each value of t. For example, the solutions $(-1, 2, 1)$, $(0, 0, 0)$, and $(\frac{1}{2}, -1, -\frac{1}{2})$ are obtained by setting t equal to 1, 0, and $-\frac{1}{2}$, respectively. A solution set of this type is called a *one-parameter family* of solutions; here t is the *parameter*.

To check that all vectors of the form $(-t, 2t, t)$ are solutions, simply substitute the components for the corresponding variables in each equation. This substitution results in an identity in both cases, namely, $2(-t) = (2t) - 4(t)$ and $(2t) = 2(t)$. Consequently all such vectors are indeed solutions.

Note that the solution in Example 2.1 may be written as $(x, y, z) = t(-1, 2, 1)$, the point-parallel form of the line through the origin in $\mathbf{R}^3$ determined by the vector $(-1, 2, 1)$. Since each of the two given equations represents a plane in $\mathbf{R}^3$, we see that their intersection is this line (Figure 2.1).

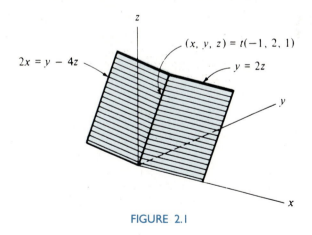

FIGURE 2.1

EXAMPLE 2.2   The linear system

$$x + y = 0$$
$$x + y = 1$$

is a system of two equations in two unknowns that does not have a solution. This can be seen by the following argument. Suppose (s_1, s_2) were a solution. Then the first equation states that the two components s_1 and s_2 add up to zero, whereas the second states that they add up to 1. Both statements cannot be true at the same time, and consequently no pair of numbers (s_1, s_2) can satisfy both equations simultaneously.

Geometrically speaking, the given equations represent lines in $\mathbf{R}^2$. We have thus shown that these lines do not intersect—they must be parallel (Figure 2.2).

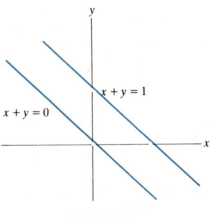

FIGURE 2.2

NOTE If a linear system has no solution, we say that it is **inconsistent.** If it has at least one solution, it is **consistent.** For example, the system of Example 2.1 is consistent, whereas that of Example 2.2 is inconsistent.

EXAMPLE 2.3   The system of equations

$$
\begin{aligned}
x_1 + x_2 \;\; + x_3 + \;\; x_4 &= 4 \\
x_1 - x_2 \qquad\quad + 2x_4 &= 2 \\
x_1 + x_1 x_2 \qquad\;\; - \;\; x_4 &= 1
\end{aligned}
$$

is *not* a linear system, because the third equation is not linear. (Remember from Section 1.4 that, if a product of variables occurs in an equation, the equation is not linear.) Although the first and second equations are linear, a set of equations forms a linear system if and only if *all* equations in the set are linear.

In solving a system of linear equations, it is convenient to begin by putting the system into **standard form,** which means writing it in such a way that

  i. each equation is in *standard form* (Section 1.4), and

  ii. like variables are aligned in vertical columns.

Note that the system of Example 2.2 is in standard form, but that of Example 2.1 is not. To put the latter into standard form, we simply transpose the variables to the left-hand side, obtaining

$$2x - y + 4z = 0$$
$$y - 2z = 0.$$

We now proceed to describe a technique known as an *elimination procedure* for solving linear systems of equations. The basic idea of this technique is to transform the original linear system into simpler and simpler systems that have the same solutions as the one given. The final system should be so simple that its solution set can be easily determined.

Each step of this transformation process is accomplished by applying one of the three following **elementary operations:**

i. Multiply an equation in the system by a nonzero scalar.

ii. Interchange the positions of two equations in the system.

iii. Replace an equation with the sum of itself and a multiple of another equation of the system. *ADD MULTIPLE OF ONE EQN TO ANOTHER.*

As we will see at the end of this section, performing any finite sequence of these elementary operations on a linear system of equations results in an **equivalent system,** one with the same set of solutions. Although a description of a *systematic* solution procedure based on these elementary operations will be deferred until Section 2.2, we will use this procedure to solve the examples that follow. In reading through these examples, you may begin to see the structure of this systematic procedure. However, in working the exercises for this section, feel free to use the elementary operations in any order you wish so as to arrive at a system with an "obvious" solution. It is often better to be clever than to be systematic!

EXAMPLE 2.4

Solve

$$-y + z = 3 \tag{A1}$$
$$x - y - z = 0 \tag{B1}$$
$$-x \quad\quad - z = -3. \tag{C1}$$

Solution

Interchange Equations (A1) and (B1):

$$x - y - z = 0 \tag{A2}$$
$$-y + z = 3 \tag{B2}$$
$$-x \quad\quad - z = -3. \tag{C2}$$

Replace (C2) with the sum of (C2) and 1 times (A2):

$$x - y - \quad z = 0 \tag{A3}$$
$$-y + \quad z = 3 \tag{B3}$$
$$-y - 2z = -3. \tag{C3}$$

Multiply (B3) by -1:

$$x - y - z = 0 \qquad \text{(A4)}$$
$$y - z = -3 \qquad \text{(B4)}$$
$$-y - 2z = -3. \qquad \text{(C4)}$$

Replace (A4) with the sum of (A4) and 1 times (B4). Then replace (C4) with the sum of (C4) and 1 times (B4):

$$x \quad - 2z = -3 \qquad \text{(A5)}$$
$$y - z = -3 \qquad \text{(B5)}$$
$$- 3z = -6. \qquad \text{(C5)}$$

Multiply (C5) by $-\frac{1}{3}$:

$$x \quad - 2z = -3 \qquad \text{(A6)}$$
$$y - z = -3 \qquad \text{(B6)}$$
$$z = 2. \qquad \text{(C6)}$$

Replace (A6) with the sum of (A6) and 2 times (C6). Then replace (B6) with the sum of (B6) and 1 times (C6):

$$x \qquad = 1 \qquad \text{(A7)}$$
$$y \qquad = -1 \qquad \text{(B7)}$$
$$z = 2. \qquad \text{(C7)}$$

The only solution of this final system, and hence of the original one, is $(1, -1, 2)$.

Geometrically, we have shown that the intersection of the three planes given by Equations (A1), (B1), and (C1) is the single point $(1, -1, 2)$ (Figure 2.3).

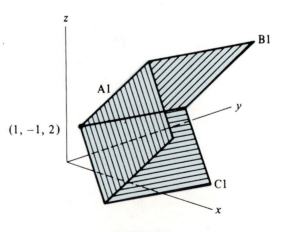

FIGURE 2.3

EXAMPLE 2.5   Solve

$$x_1 \qquad\quad + 3x_3 + x_4 = 0 \qquad\qquad (A1)$$
$$-x_1 + 2x_2 + x_3 + x_4 = 0 \qquad\qquad (B1)$$
$$-x_1 + x_2 - x_3 \qquad\quad = 0. \qquad\qquad (C1)$$

Solution   To solve the given system, replace (B1) with the sum of (B1) and 1 times (A1). Then replace (C1) with the sum of (C1) and 1 times (A1):

$$x_1 \qquad\quad + 3x_3 + x_4 = 0 \qquad\qquad (A2)$$
$$2x_2 + 4x_3 + 2x_4 = 0 \qquad\qquad (B2)$$
$$x_2 + 2x_3 + x_4 = 0. \qquad\qquad (C2)$$

Multiply (B2) by $\frac{1}{2}$:

$$x_1 \qquad\quad + 3x_3 + x_4 = 0 \qquad\qquad (A3)$$
$$x_2 + 2x_3 + x_4 = 0 \qquad\qquad (B3)$$
$$x_2 + 2x_3 + x_4 = 0. \qquad\qquad (C3)$$

Replace (C3) with the sum of (C3) and -1 times (B3):

$$x_1 \qquad\quad + 3x_3 + x_4 = 0 \qquad\qquad (A4)$$
$$x_2 + 2x_3 + x_4 = 0 \qquad\qquad (B4)$$
$$0 = 0. \qquad\qquad (C4)$$

Equation (C4) is an identity. It places no restriction on the values of the unknowns, and we discard it:

$$x_1 \qquad\quad + 3x_3 + x_4 = 0 \qquad\qquad (A4)$$
$$x_2 + 2x_3 + x_4 = 0. \qquad\qquad (B4)$$

Now solve (A4) for x_1 and (B4) for x_2:

$$x_1 = -3x_3 - x_4$$
$$x_2 = -2x_3 - x_4.$$

Finally, let $x_3 = s$ and $x_4 = t$, obtaining the solution

$$x_1 = -3s - t$$
$$x_2 = -2s - t$$
$$x_3 = s$$
$$x_4 = t,$$

or in vector form, $(-3s - t, -2s - t, s, t)$, where the parameters s and t range independently over the real numbers. Once again we get a linear system with infinitely many solutions, but here they form a *two-parameter family*.

Among the solutions are

$$(0, 0, 0, 0) \qquad \text{when } s = 0, t = 0,$$

$$(-1, -1, 0, 1) \qquad \text{when } s = 0, t = 1,$$

$$\left(1, \frac{1}{2}, -\frac{1}{2}, \frac{1}{2}\right) \qquad \text{when } s = -\frac{1}{2}, t = \frac{1}{2}.$$

Geometrically, the solution set will be the common points of intersection of the given *hyperplanes* in $\mathbf{R}^4$: $x_1 + 3x_3 + x_4 = 0$, $-x_1 + 2x_2 + x_3 + x_4 = 0$, and $-x_1 + x_2 - x_3 = 0$. Since any solution may be expressed as

$$\mathbf{x} = s(-3, -2, 1, 0) + t(-1, -1, 0, 1),$$

where $\mathbf{x} = (x_1, x_2, x_3, x_4)$, the set of all such points is the plane in $\mathbf{R}^4$ determined by the vectors $(-3, -2, 1, 0)$ and $(-1, -1, 0, 1)$.

SPOTLIGHT

APPLICATIONS OF LINEAR SYSTEMS

Systems of linear equations arise in a myriad of applications. A few of these are described in detail in Chapters 8 and 9.

- Linear systems arise in determining the curve that best fits a set of data (see Sections 8.1 and 8.2).
- Linear systems can be used to determine the magnitude of the current in various parts of an electric circuit (see Section 8.3).
- Linear systems describe the economic interplay in certain situations involving producers and consumers (see Section 8.4).
- Linear systems arise in the process of maximizing or minimizing certain functions subject to certain restrictions (see Chapter 9).

In fact, the majority of problems that occur in applications of mathematics, when analyzed, lead to a system of linear equations. The ability to solve linear systems is certainly one of the most valuable tools in the applied mathematician's toolbox.

We will close this section with a theorem that justifies the elimination procedure used in Examples 2.4 and 2.5.

THEOREM 1 If any one of the three elementary operations is performed on a given system of linear equations, an equivalent linear system is obtained.

Proof We will show this to be true for elementary operation (iii) only. The statement is certainly true for elementary operation (ii), since this operation does not alter the system, and we will leave the part of the proof concerning operation (i) as an exercise.

Let two equations of the given linear system be

$$a_1x_1 + a_2x_2 + \cdots + a_nx_n = b, \tag{1}$$

$$c_1x_1 + c_2x_2 + \cdots + c_nx_n = d. \tag{2}$$

Suppose we replace Equation (2) with the sum of itself and k times Equation (1), and leave all other equations in the system as they stand. We need to verify that $(s_1, s_2, \ldots, s_n)$ is a solution of the given system if and only if it is a solution of the transformed system. Since all the other equations remain unchanged, we need only verify this for Equation (2), which has been transformed into

$$(c_1 + ka_1)x_1 + (c_2 + ka_2)x_2 + \cdots + (c_n + ka_n)x_n = d + kb. \tag{3}$$

First, assume $(s_1, s_2, \ldots, s_n)$ satisfies the original system. Then, in particular, it satisfies (1) and (2), so that

$$a_1s_1 + a_2s_2 + \cdots + a_ns_n = b,$$

$$c_1s_1 + c_2s_2 + \cdots + c_ns_n = d.$$

Multiplying the first of these by k, we have

$$k(a_1s_1 + a_2s_2 + \cdots + a_ns_n) = kb,$$

which when added to the second yields

$$k(a_1s_1 + a_2s_2 + \cdots + a_ns_n) + (c_1s_1 + c_2s_2 + \cdots + c_ns_n) = kb + d.$$

Finally, rearranging terms, we obtain

$$(c_1 + ka_1)s_1 + (c_2 + ka_2)s_2 + \cdots + (c_n + ka_n)s_n = d + kb, \tag{4}$$

which says that $(s_1, s_2, \ldots, s_n)$ satisfies Equation (3).

Now assume that $(s_1, s_2, \ldots, s_n)$ satisfies the transformed system. This implies that Equation (4) holds. We also know that Equation (1) is satisfied, so that $a_1s_1 + a_2s_2 + \cdots + a_ns_n = b$. This in turn implies that $k(a_1s_1 + a_2s_2 + \cdots + a_ns_n) = kb$. Subtracting this last equation from Equation (4) yields

$$c_1s_1 + c_2s_2 + \cdots + c_ns_n = d,$$

as desired.

2.1 EXERCISES

In Exercises 1–6 determine whether or not the given system of equations is linear.

1. $x_1 - 3x_2 = x_3 - 4$
$x_4 = 1 - x_1$
$x_1 + x_4 + x_3 - 2 = 0$

2. $2x - \sqrt{y} + 3z = -1$
$x + 2y - z = 2$
$4x - y = -1$

3. $3x - xy = 1$
$x + 2xy - y = 0$

4. $y = 2x - 1$
$y = -x$

5. $2x_1 - \sin x_2 = 3$
$x_2 = x_1 + x_3$
$-x_1 + x_2 - 3x_3 = 0$

6. $x_1 + x_2 + x_3 = 1$
$-2x_1^2 - 2x_3 = -1$
$3x_2 - x_3 = 2$

In Exercises 7–10 put the given system of equations into standard form.

7. $z = 6 - y$
$z = x + y$
$y + z - 3 = x$

8. $x_2 - x_1 = 3 - x_3 + x_4$
$x_2 = 0$
$x_3 = 1 - x_1 - x_2$

9. $x_2 + x_1 = x_4 - x_3$
$x_1 - 1 = 0$
$x_4 - 3 = x_2$

10. $x + y = 0$
$x - z = 1$
$x - 1 = y$

In Exercises 11–14 show that the given linear system has the solution indicated.

11. $x - 3y + z = 0$
$x + y - 3z = 0$
$x - y - z = 0$
Solution: $(2s, s, s)$

12. $x_1 - 3x_2 - x_3 + x_4 = 1$
$-x_1 + 2x_2 + x_3 + x_4 = 0$
Solution: $(-2 + s + 5t, -1 + 2t, s, t)$

13. $v - 2w + z = 1$
$2u - v - z = 0$
$4u + v - 6w + z = 3$
Solution: $(\frac{1}{2} + s, 1 + 2s - t, s, t)$

14. $2x - 3y + z = 1$
$-x + 2z = 0$
$3x - 3y - z = 1$
Solution: $(2s, -\frac{1}{3} + \frac{5}{3}s, s)$

In Exercises 15–18 solve the given linear system.

15. $x - 4y = 1$
$-2x + 8y = -2$

16. $x_1 + x_2 - x_3 + x_4 = 2$
$-2x_1 + x_3 = -4$
$x_1 - x_2 - 2x_3 + 2x_4 = -1$

17. $x_1 + x_2 + x_3 + x_4 = 1$
$2x_1 + 3x_2 + 3x_3 = 1$
$-x_1 - 2x_2 - 2x_3 + x_4 = 0$
$- x_2 - x_3 + 2x_4 = 1$

18. $2u - v = 0$
$-3u + 2v = 0$
$3u - v = 0$

In Exercises 19–22 the given equations represent planes in $\mathbf{R}^3$. Describe the region (point, line, or plane), if any, that is the intersection of these planes.

19. $x + y + z = -4$
$x + 2y = 1$
$2y + 3z = -2$

20. $y + z = 1$
$x + y + 2z = -1$
$x + z = 1$

21. $x - y + z = 0$
$y - 2z = 1$
$2x - y = 1$

22. $-x + 2y - z = 1$
$2x - 4y + 2z = -2$
$x - 2y + z = -1$

23. Find the value of c such that the system

$$
\begin{aligned}
cx \quad\quad + z &= 0 \\
2y - 4z &= 0 \\
2x - y \quad\quad &= 0
\end{aligned}
$$

has a solution other than $(0, 0, 0)$.

24. Find all values of k such that the system

$$
\begin{aligned}
x_1 - x_2 + 2x_3 &= 0 \\
x_2 - x_3 &= k \\
-x_1 + 2x_2 - 3x_3 &= 1
\end{aligned}
$$

has no solution.

25. Prove that, if we multiply any equation of an arbitrary linear system by a nonzero scalar, the resultant linear system is equivalent to the original one.

26. Show that the linear system

$$
\begin{aligned}
ax + by &= e \\
cx + dy &= f,
\end{aligned}
$$

where a, b, c, d, e, f, are constants, has a unique solution if $ad - bc \neq 0$. Express this solution in terms of a, b, c, d, e, and f.

2.2   MATRICES AND ROW REDUCTION OF LINEAR SYSTEMS

In Section 2.1 systems of linear equations were solved by means of an elimination procedure. In this section we will introduce the notion of a *matrix* and then use *matrices* (the plural of matrix) to write each step of this solution technique in a more concise form.

A **matrix** is a rectangular array of numbers, each of which is called an **entry** of the matrix. We will enclose the rectangular array in brackets. For example,

$$
\begin{bmatrix} 2 & 1 \\ -1 & 2 \\ 0 & 5 \end{bmatrix}, \tag{1}
$$

$$
\begin{bmatrix} 1 & -1 & 0 \\ 0 & 5 & 0 \end{bmatrix}, \tag{2}
$$

and

$$
\begin{bmatrix} 1 & \sqrt{2} & \pi \end{bmatrix} \tag{3}
$$

are all matrices. Each horizontal line of numbers is called a **row** of the matrix; each vertical one is called a **column.** For example, in matrix (1) the third row consists of the entries 0 and 5, whereas the first column is 2, −1, 0. In other books you may see the rectangular array enclosed in parentheses. For example, matrix (2) might be written

$$\begin{pmatrix} 1 & -1 & 0 \\ 0 & 5 & 0 \end{pmatrix}.$$

Matrices arise naturally in the study of systems of linear equations. If we take a given linear system, put it in standard form, and then delete all unknowns and plus and equals signs, we obtain a rectangular array of numbers. This array is called the **augmented matrix** of the system. (When writing an augmented matrix, it is customary to draw a vertical line separating its last column—the one that corresponds to the right side constants of the system—from the rest of the matrix.)

EXAMPLE 2.6

Find the augmented matrices for the following linear systems:

a. $2x - 3y + z = 1$
$\quad x \quad\quad + 2z = 0$
$\quad\quad - y - z = -5$

b. $x_1 - 3x_2 - 1 = x_4 - 4x_3$
$\quad\quad - x_1 + 2x_3 = 0.$

Solution

a. The augmented matrix is

$$\begin{bmatrix} 2 & -3 & 1 & | & 1 \\ 1 & 0 & 2 & | & 0 \\ 0 & -1 & -1 & | & -5 \end{bmatrix}.$$

b. We first rewrite the system in standard form. This yields

$$x_1 - 3x_2 + 4x_3 - x_4 = 1$$
$$- x_1 \quad\quad + 2x_3 \quad\quad = 0.$$

Thus the augmented matrix is

$$\begin{bmatrix} 1 & -3 & 4 & -1 & | & 1 \\ -1 & 0 & 2 & 0 & | & 0 \end{bmatrix}.$$

NOTE When an unknown is "missing" in the given system (when its coefficient is zero), put a zero in the appropriate place in the augmented matrix corresponding to this system.

In solving the linear systems of Examples 2.4 and 2.5 (Section 2.1), we made use of the *elementary operations*:

i. Multiply an equation by a nonzero scalar.

ii. Interchange the positions of two equations.

iii. Replace an equation by the sum of itself and a multiple of another equation.

In an analogous manner, we define for matrices the **elementary row operations:**

    i. Multiply a row by a nonzero scalar.

    ii. Interchange the position of two rows.

    iii. Replace a row by the sum of itself and a multiple of another row.

ADD MULTIPLE OF ONE EQN. TO ANOTHER.

NOTE   When we say "multiply a row by," we mean more precisely, "multiply each entry of a row by"; and when we say "the sum of two rows," we mean "the sum of the corresponding entries in the two rows."

EXAMPLE 2.7

Given the matrix

$$\begin{bmatrix} 1 & -1 & 0 & 4 \\ 2 & 0 & -3 & 1 \\ 5 & -2 & 3 & 0 \end{bmatrix},$$

if we multiply the third row by 3 and replace the second row by the sum of itself and -2 times the first, we obtain the matrix

$$\begin{bmatrix} 1 & -1 & 0 & 4 \\ 0 & 2 & -3 & -7 \\ 15 & -6 & 9 & 0 \end{bmatrix}.$$

Performing elementary row operations on the augmented matrix of a given linear system has the same effect as performing the analogous elementary operations on the system itself. In other words, the matrix resulting from the elementary row operations is the same as the augmented matrix of the linear system resulting from the analogous elementary operations.

EXAMPLE 2.8

Consider the linear system that follows, with augmented matrix at the right:

$$\begin{array}{rcl} x_1 - 2x_2 & = & 1 \\ 2x_2 + x_3 & = & -4 \\ 3x_1 - 4x_2 + x_3 & = & 0 \end{array} \qquad \begin{bmatrix} 1 & -2 & 0 & | & 1 \\ 0 & 2 & 1 & | & -4 \\ 3 & -4 & 1 & | & 0 \end{bmatrix}.$$

Let us perform the following elementary operations on the system. Replace the third equation with the sum of itself and -3 times the first, and multiply the second equation by $\frac{1}{2}$. Now perform the analogous elementary row operations on the matrix—replace the third row with the sum of itself and -3 times the first, and multiply the second row by $\frac{1}{2}$. Then the resulting system and matrix are

$$
\begin{aligned}
x_1 - 2x_2 \phantom{+ \tfrac{1}{2}x_3} &= 1 \\
x_2 + \tfrac{1}{2}x_3 &= -2 \\
2x_2 + x_3 &= -3
\end{aligned}
\qquad
\begin{bmatrix}
1 & -2 & 0 & \vline & 1 \\
0 & 1 & \tfrac{1}{2} & \vline & -2 \\
0 & 2 & 1 & \vline & -3
\end{bmatrix}.
$$

Note that the resulting matrix is the augmented matrix for the resulting system.

Consequently, in implementing the elimination procedure, we can perform elementary row operations on the augmented matrices instead of elementary operations on the system itself. This will reduce the amount of writing necessary, as well as the possibilities for error, and lends itself nicely to calculator and computer solution of linear systems. As an illustration of this process, we will solve Example 2.4 of Section 2.1, using the same sequence of operations, but this time performing them on matrices.

EXAMPLE 2.9

Use elementary row operations on matrices to solve the linear system (see Example 2.4):

$$
\begin{aligned}
-y + z &= 3 \\
x - y - z &= 0 \\
-x - z &= -3.
\end{aligned}
$$

Solution

We first write the augmented matrix for the system:

$$
\begin{bmatrix}
0 & -1 & 1 & \vline & 3 \\
1 & -1 & -1 & \vline & 0 \\
-1 & 0 & -1 & \vline & -3
\end{bmatrix}.
$$

Now we proceed as in Example 2.4. Interchange the first and second rows:

$$
\begin{bmatrix}
1 & -1 & -1 & \vline & 0 \\
0 & -1 & 1 & \vline & 3 \\
-1 & 0 & -1 & \vline & -3
\end{bmatrix}.
$$

Replace the third row with the sum of itself and 1 times the first row:

$$
\begin{bmatrix}
1 & -1 & -1 & \vline & 0 \\
0 & -1 & 1 & \vline & 3 \\
0 & -1 & -2 & \vline & -3
\end{bmatrix}.
$$

Multiply the second row by -1:

$$
\begin{bmatrix}
1 & -1 & -1 & \vline & 0 \\
0 & 1 & -1 & \vline & -3 \\
0 & -1 & -2 & \vline & -3
\end{bmatrix}.
$$

Replace the first row with the sum of itself and 1 times the second row. Then replace the third row with the sum of itself and 1 times the second row:

$$\left[\begin{array}{ccc|c} 1 & 0 & -2 & -3 \\ 0 & 1 & -1 & -3 \\ 0 & 0 & -3 & -6 \end{array}\right].$$

Multiply the third row by $-\frac{1}{3}$:

$$\left[\begin{array}{ccc|c} 1 & 0 & -2 & -3 \\ 0 & 1 & -1 & -3 \\ 0 & 0 & 1 & 2 \end{array}\right].$$

Replace the first row with the sum of itself and 2 times the third row. Then replace the second row with the sum of itself and 1 times the third row:

$$\left[\begin{array}{ccc|c} 1 & 0 & 0 & 1 \\ 0 & 1 & 0 & -1 \\ 0 & 0 & 1 & 2 \end{array}\right].$$

Change back to equation form (write the linear system for which this is the augmented matrix):

$$\begin{aligned} x \qquad\qquad &= 1 \\ y \qquad &= -1 \\ z &= 2. \end{aligned}$$

Thus the solution is $(1, -1, 2)$.

NOTE   When one matrix can be obtained from another by means of a finite sequence of elementary row operations, the two matrices are said to be **row equivalent.** For example, any two of the matrices in Example 2.9 are row equivalent.

··

The following theorem is a restatement of Theorem 1 of Section 2.1. It tells us that although the augmented matrices are changing as we perform elementary row operations on them, the underlying linear systems all have the same solutions.

THEOREM 1   If two augmented matrices are row equivalent, the corresponding linear systems are equivalent—they have the same solutions.

The final matrix of Example 2.9 is an especially simple one. It is said to be in *row-reduced echelon form.* The basic goal of the elimination procedure is to transform the original augmented matrix (by means of elementary row operations) into one that has this special form.

DEFINITION

A matrix is in **row-reduced echelon form** if it satisfies the following:

  i. In each row that does not consist entirely of zeros, the first nonzero entry is a 1 (we call such an element a **leading 1**).

  ii. In each column that contains a leading 1 of some row, all other entries are zero.

  iii. In any two rows with some nonzero entries, the leading 1 of the higher row is farther to the left.

  iv. Any row that contains only zeros is lower than any row with some nonzero entries.

EXAMPLE 2.10

The augmented matrices corresponding to the final linear systems in Examples 2.4 and 2.5 (Section 2.1) are, respectively,

$$
\left[\begin{array}{ccc|c}
1 & 0 & 0 & 1 \\
0 & 1 & 0 & -1 \\
0 & 0 & 1 & 2
\end{array}\right]
\quad \text{and} \quad
\left[\begin{array}{cccc|c}
1 & 0 & 3 & 1 & 0 \\
0 & 1 & 2 & 1 & 0 \\
0 & 0 & 0 & 0 & 0
\end{array}\right].
$$

Each is in row-reduced echelon form.

EXAMPLE 2.11

The matrix

$$
\left[\begin{array}{ccccc}
1 & 2 & 0 & -1 & 0 \\
0 & 0 & 1 & 2 & 1 \\
0 & 0 & 0 & 0 & 0 \\
0 & 0 & 0 & 0 & 0
\end{array}\right]
$$

is in row-reduced echelon form.

EXAMPLE 2.12

The matrices

$$
\left[\begin{array}{ccc}
1 & 1 & 0 \\
0 & 1 & 0 \\
0 & 0 & 1
\end{array}\right], \quad
\left[\begin{array}{ccc}
1 & 0 & 0 \\
0 & 0 & 1 \\
0 & 1 & 0
\end{array}\right], \quad
\left[\begin{array}{cc}
1 & 0 \\
0 & 2
\end{array}\right], \quad \text{and} \quad
\left[\begin{array}{ccc}
0 & 0 & 0 \\
1 & 0 & 0
\end{array}\right]
$$

are *not* in row-reduced echelon form. (They violate conditions ii, iii, i, and iv, respectively, of the definition.)

As we have said, the basic goal of the matrix elimination procedure is to transform, by means of a systematic sequence of elementary row operations, the original augmented matrix into one that is in row-reduced echelon form. We then change back to the "equation form" of the linear system and complete the solution process as in Example 2.9. This procedure is often called **Gauss–Jordan elimination** and will be described in the next two examples.

EXAMPLE 2.13

Solve the following system by Gauss–Jordan elimination:

$$
\begin{aligned}
x_3 + 2x_4 &= 3 \\
2x_1 + 4x_2 - 2x_3 \quad\quad\ &= 4 \\
2x_1 + 4x_2 - x_3 + 2x_4 &= 7.
\end{aligned}
$$

Solution

Step 0. Write the augmented matrix for the system:

$$
\left[\begin{array}{cccc|c}
0 & 0 & 1 & 2 & 3 \\
2 & 4 & -2 & 0 & 4 \\
2 & 4 & -1 & 2 & 7
\end{array}\right].
$$

Step 1a. Obtain a leading 1 in the first row, first column. (If there is a 0 in this position, interchange the first row with a row below it so that a nonzero entry appears there.) Interchange the first and second rows:

$$
\left[\begin{array}{cccc|c}
2 & 4 & -2 & 0 & 4 \\
0 & 0 & 1 & 2 & 3 \\
2 & 4 & -1 & 2 & 7
\end{array}\right].
$$

(If the nonzero entry in the first row, first column is not a 1, multiply the first row by the reciprocal of this entry.) Multiply the first row by $\frac{1}{2}$:

$$
\left[\begin{array}{cccc|c}
1 & 2 & -1 & 0 & 2 \\
0 & 0 & 1 & 2 & 3 \\
2 & 4 & -1 & 2 & 7
\end{array}\right].
$$

Step 1b. Obtain zeros in other positions in the first column by adding appropriate multiples of the first row to other rows. Replace the third row with the sum of itself and -2 times the first:

$$
\left[\begin{array}{cccc|c}
1 & 2 & -1 & 0 & 2 \\
0 & 0 & 1 & 2 & 3 \\
0 & 0 & 1 & 2 & 3
\end{array}\right].
$$

Step 2. Obtain a leading 1 in the second row, second column. (If there is a 0 in this position, interchange the second row with a row below it so that a nonzero entry appears there. If this is not possible, go to the next column.) It is not possible, since all entries in the second column below the first row are zero.

Step 3a. Obtain a leading 1 in the second row, third column. There is one there already.

Step 3b. Obtain zeros in the other positions in the third column by adding appropriate multiples of the second row to the other rows. Replace the first row with the sum of itself and 1 times the second row. Then replace the third row with the sum of itself and -1 times the second row:

$$\begin{bmatrix} 1 & 2 & 0 & 2 & | & 5 \\ 0 & 0 & 1 & 2 & | & 3 \\ 0 & 0 & 0 & 0 & | & 0 \end{bmatrix}.$$

Step 3c. Place any newly created zero rows at the bottom of the matrix. It is already there. (The matrix is now in row-reduced echelon form.)

Step 4. Change back to a system of equations, ignoring any rows that contain only zeros:

$$x_1 + 2x_2 \qquad + 2x_4 = 5$$
$$x_3 + 2x_4 = 3.$$

Step 5. Solve each equation for the unknown whose coefficient is a leading 1 in the final augmented matrix

$$x_1 = 5 - 2x_2 - 2x_4$$
$$x_3 = 3 - 2x_4.$$

The unknowns appearing on the right-hand side are taken to be parameters. Set $x_2 = s$ and $x_4 = t$. We have here a "two-parameter family of solutions" (see Example 2.5). All solutions are of the form $(5 - 2s - 2t, s, 3 - 2t, t)$ or, equivalently, $(5, 0, 3, 0) + s(-2, 1, 0, 0) + t(-2, 0, -2, 1)$.

EXAMPLE 2.14   Solve the following system by Gauss–Jordan elimination:

$$x_1 + 2x_2 \qquad\qquad = 1$$
$$x_1 + 2x_2 + 3x_3 + x_4 = 0$$
$$-x_1 - x_2 + x_3 + x_4 = -2$$
$$x_2 + x_3 + x_4 = -1$$
$$-x_2 + 2x_3 \qquad = 0.$$

Solution   **Step 0.** Write the augmented matrix for the system:

$$\begin{bmatrix} 1 & 2 & 0 & 0 & | & 1 \\ 1 & 2 & 3 & 1 & | & 0 \\ -1 & -1 & 1 & 1 & | & -2 \\ 0 & 1 & 1 & 1 & | & -1 \\ 0 & -1 & 2 & 0 & | & 0 \end{bmatrix}.$$

Step 1a. Obtain a leading 1 in the first row, first column. There is one there already.

Step 1b. Obtain zeros in the other positions in the first column. Replace the second row with the sum of itself and −1 times the first row. Then replace the third row with the sum of itself and 1 times the first row:

$$\begin{bmatrix} 1 & 2 & 0 & 0 & | & 1 \\ 0 & 0 & 3 & 1 & | & -1 \\ 0 & 1 & 1 & 1 & | & -1 \\ 0 & 1 & 1 & 1 & | & -1 \\ 0 & -1 & 2 & 0 & | & 0 \end{bmatrix}.$$

Step 2a. Obtain a leading 1 in the second row, second column. (If there is a 0 in this position, interchange the second row with a row below it to create a nonzero entry there.) Interchange the second and third rows:

$$\begin{bmatrix} 1 & 2 & 0 & 0 & | & 1 \\ 0 & 1 & 1 & 1 & | & -1 \\ 0 & 0 & 3 & 1 & | & -1 \\ 0 & 1 & 1 & 1 & | & -1 \\ 0 & -1 & 2 & 0 & | & 0 \end{bmatrix}.$$

We now have a 1 in the required position.

Step 2b. Obtain zeros in the other positions in the second column. Replace the first row with the sum of itself and −2 times the second row. Next replace the fourth row with the sum of itself and −1 times the second row. Then replace the fifth row with the sum of itself and 1 times the second row:

$$\begin{bmatrix} 1 & 0 & -2 & -2 & | & 3 \\ 0 & 1 & 1 & 1 & | & -1 \\ 0 & 0 & 3 & 1 & | & -1 \\ 0 & 0 & 0 & 0 & | & 0 \\ 0 & 0 & 3 & 1 & | & -1 \end{bmatrix}.$$

Step 2c. Place any newly created zero rows at the bottom of the matrix. Interchange the fourth and fifth rows:

$$\begin{bmatrix} 1 & 0 & -2 & -2 & | & 3 \\ 0 & 1 & 1 & 1 & | & -1 \\ 0 & 0 & 3 & 1 & | & -1 \\ 0 & 0 & 3 & 1 & | & -1 \\ 0 & 0 & 0 & 0 & | & 0 \end{bmatrix}.$$

Step 3a. Obtain a leading 1 in the third row, third column. (If there is a nonzero entry there and it is not a 1, multiply the third row by the reciprocal of this entry.) Multiply the third row by $\frac{1}{3}$:

$$\begin{bmatrix} 1 & 0 & -2 & -2 & \bigm| & 3 \\ 0 & 1 & 1 & 1 & \bigm| & -1 \\ 0 & 0 & 1 & \frac{1}{3} & \bigm| & -\frac{1}{3} \\ 0 & 0 & 3 & 1 & \bigm| & -1 \\ 0 & 0 & 0 & 0 & \bigm| & 0 \end{bmatrix}.$$

Step 3b. Obtain zeros in the other positions of the third column. Replace the first row with the sum of itself and 2 times the third row. Next replace the second row with the sum of itself and -1 times the third row. Then replace the fourth row with the sum of itself and -3 times the third row:

$$\begin{bmatrix} 1 & 0 & 0 & -\frac{4}{3} & \bigm| & \frac{7}{3} \\ 0 & 1 & 0 & \frac{2}{3} & \bigm| & -\frac{2}{3} \\ 0 & 0 & 1 & \frac{1}{3} & \bigm| & -\frac{1}{3} \\ 0 & 0 & 0 & 0 & \bigm| & 0 \\ 0 & 0 & 0 & 0 & \bigm| & 0 \end{bmatrix}.$$

Step 3c. Place any newly created zero rows at the bottom of the matrix. It is already there. (The matrix is now in row-reduced echelon form.)

Step 4. Change back to a system of equations ignoring any zero rows:

$$x_1 \qquad\qquad -\frac{4}{3}x_4 = \frac{7}{3}$$

$$x_2 \qquad +\frac{2}{3}x_4 = -\frac{2}{3}$$

$$x_3 + \frac{1}{3}x_4 = -\frac{1}{3}.$$

Step 5. Solve each equation for the unknown whose coefficient is a leading 1:

$$x_1 = \frac{7}{3} + \frac{4}{3}x_4$$

$$x_2 = -\frac{2}{3} - \frac{2}{3}x_4$$

$$x_3 = -\frac{1}{3} - \frac{1}{3}x_4.$$

We set $x_4 = s$ and obtain the "one-parameter family of solutions":

$$\left(\frac{7}{3} + \frac{4}{3}s, -\frac{2}{3} - \frac{2}{3}s, -\frac{1}{3} - \frac{1}{3}s, s\right)$$

or, equivalently,

$$\left(\frac{7}{3}, -\frac{2}{3}, -\frac{1}{3}, 0\right) + s\left(\frac{4}{3}, -\frac{2}{3}, -\frac{1}{3}, 1\right).$$

Before continuing, we will summarize the steps involved in the heart of the Gauss–Jordan elimination procedure: the transformation of the original augmented matrix to row-reduced echelon form.

PROCEDURE TO ROW-REDUCE A MATRIX

i. In the leftmost column whose entries are not all zero, obtain a *leading 1* at the top of this column by using elementary row operations (i) and (ii).

ii. Obtain zeros in all other positions in this column by using elementary operation (iii).

iii. Place any newly created zero rows at the bottom of the matrix by using elementary operation (ii).

iv. Repeat step (i) for the *submatrix* obtained by deleting all rows containing previously obtained leading 1's, and then do steps (ii) and (iii) on the full matrix. Continue until row-reduced echelon form is obtained.

Using the Gauss–Jordan procedure, we can transform any matrix into row-reduced echelon form by means of a sequence of elementary row operations. We could also obtain this form by performing elementary row operations in a different order (see, for example, the note following Theorem 3). However, we would still arrive at the same row-reduced echelon form. These facts are stated without proof in the following two theorems. The first of these theorems could be proved by giving a precise description of the Gauss–Jordan procedure, which has been informally described in the preceding pages. A proof of the second can be constructed with the aid of the tools introduced in Chapter 4.

THEOREM 2 Every matrix can be transformed by a finite sequence of elementary row operations into one that is in row-reduced echelon form.

THEOREM 3 The row-reduced echelon form of a matrix is unique.

NOTE In doing computations by hand, it is often more convenient to alter the sequence of elementary row operations to avoid fractions. This can be accomplished by delaying the placement of a leading 1 in certain rows until the end, together with some additional row multiplications. For example, at the end of Step 2 of Example 2.14, the augmented matrix has only integral entries:

$$\begin{bmatrix} 1 & 0 & -2 & -2 & 3 \\ 0 & 1 & 1 & 1 & -1 \\ 0 & 0 & 3 & 1 & -1 \\ 0 & 0 & 3 & 1 & -1 \\ 0 & 0 & 0 & 0 & 0 \end{bmatrix}.$$

To avoid fractions until the last step, we proceed as follows:

1. Multiply the first row by 3.
2. Multiply the second row by 3:

$$\begin{bmatrix} 3 & 0 & -6 & -6 & 9 \\ 0 & 3 & 3 & 3 & -3 \\ 0 & 0 & 3 & 1 & -1 \\ 0 & 0 & 3 & 1 & -1 \\ 0 & 0 & 0 & 0 & 0 \end{bmatrix}.$$

3. Replace the first row with the sum of itself and 2 times the third row.
4. Replace the second row with the sum of itself and -1 times the third row.
5. Replace the fourth row with the sum of itself and -1 times the third row:

$$\begin{bmatrix} 3 & 0 & 0 & -4 & 7 \\ 0 & 3 & 0 & 2 & -2 \\ 0 & 0 & 3 & 1 & -1 \\ 0 & 0 & 0 & 0 & 0 \\ 0 & 0 & 0 & 0 & 0 \end{bmatrix}.$$

6. Multiply the first, second, and third rows by $\frac{1}{3}$:

$$\begin{bmatrix} 1 & 0 & 0 & -\frac{4}{3} & \frac{7}{3} \\ 0 & 1 & 0 & \frac{2}{3} & -\frac{2}{3} \\ 0 & 0 & 1 & \frac{1}{3} & -\frac{1}{3} \\ 0 & 0 & 0 & 0 & 0 \\ 0 & 0 & 0 & 0 & 0 \end{bmatrix}.$$

The matrix is now in row-reduced echelon form. Complete the solution process exactly as in Example 2.14.

In Example 2.9 the given linear system had a unique solution, whereas in Examples 2.13 and 2.14 there were infinitely many solutions. As we know from Section 2.1, a linear system may also have no solution. The following example illustrates what happens in the Gauss–Jordan procedure in such a case.

EXAMPLE 2.15   Solve

$$x_1 \qquad + x_3 = \quad 1$$
$$x_2 - x_3 = -1$$
$$2x_1 + x_2 + x_3 = \quad 2.$$

Solution   The augmented matrix for this system is

$$\begin{bmatrix} 1 & 0 & 1 & | & 1 \\ 0 & 1 & -1 & | & -1 \\ 2 & 1 & 1 & | & 2 \end{bmatrix}.$$

To transform it into row-reduced echelon form, first replace the third row with the sum of itself and -2 times the first row:

$$\begin{bmatrix} 1 & 0 & 1 & | & 1 \\ 0 & 1 & -1 & | & -1 \\ 0 & 1 & -1 & | & 0 \end{bmatrix}.$$

Then replace the third row with the sum of itself and -1 times the second, obtaining

$$\begin{bmatrix} 1 & 0 & 1 & | & 1 \\ 0 & 1 & -1 & | & -1 \\ 0 & 0 & 0 & | & 1 \end{bmatrix}.$$

Although this matrix is not in row-reduced echelon form, we see that the third row represents the equation

$$0x_1 + 0x_2 + 0x_3 = 1,$$

which cannot be satisfied by any choice of x_1, x_2, and x_3. Thus this equation, and hence the system, has no solution.

Geometrically, we have shown here that the planes in $\mathbf{R}^3$ whose equations are the given ones have no common points of intersection (see Figure 2.4).

FIGURE 2.4

COMPUTATIONAL NOTE Gauss–Jordan elimination is not the most efficient matrix elimination procedure for use on computers. This honor belongs to a method known as **Gaussian elimination.** In the latter, elementary row operations are used to transform the original augmented matrix into **echelon form,** defined as follows:

i. The leftmost nonzero element in each row is called a **leading entry.**

ii. In each column that contains a leading entry of some row, all elements *below* the leading entry are zero.

iii. In any two rows with leading entries, the leading entry of the higher row is farther to the left.

iv. Any row that contains only zeros is lower than any row with some nonzero entries.

Note that compared to *row-reduced* echelon form, echelon form does not require the leading entry in a row to be a *one,* nor does it require *all* entries in a column with a leading entry to be zero. For example, the matrix

$$\begin{bmatrix} 2 & 1 & -1 & 3 & 5 \\ 0 & 0 & -3 & 6 & 1 \\ 0 & 0 & 0 & 2 & 4 \\ 0 & 0 & 0 & 0 & 0 \end{bmatrix}$$

is in echelon form. Once the original augmented matrix is transformed into echelon form, the linear system is solved by a process of *back-substitution,* as illustrated in Example 2.16.

··

EXAMPLE 2.16

Solve the system of Example 2.9 by Gaussian elimination:

$$\begin{aligned} -y + z &= 3 \\ x - y - z &= 0 \\ -x \qquad - z &= -3. \end{aligned}$$

Solution

The augmented matrix is

$$\left[\begin{array}{ccc|c} 0 & -1 & 1 & 3 \\ 1 & -1 & -1 & 0 \\ -1 & 0 & -1 & -3 \end{array}\right].$$

Interchange the first and second rows:

$$\left[\begin{array}{ccc|c} 1 & -1 & -1 & 0 \\ 0 & -1 & 1 & 3 \\ -1 & 0 & -1 & -3 \end{array}\right].$$

Replace the third row with the sum of itself and the first row:

$$\begin{bmatrix} 1 & -1 & -1 & | & 0 \\ 0 & -1 & 1 & | & 3 \\ 0 & -1 & -2 & | & -3 \end{bmatrix}.$$

Replace the third row with the sum of itself and -1 times the second row:

$$\begin{bmatrix} 1 & -1 & -1 & | & 0 \\ 0 & -1 & 1 & | & 3 \\ 0 & 0 & -3 & | & -6 \end{bmatrix}.$$

The matrix is now in echelon form. Convert back to equational form:

$$x - y - z = 0$$
$$-y + z = 3$$
$$-3z = -6.$$

We now solve this system by **back-substitution.** First, solve the last equation for z, obtaining $z = 2$. Next, substitute this value of z into the second equation, and solve it for y, obtaining $y = -1$. Finally, substitute $z = 2$, $y = -1$ into the first equation to get $x = 1$. As before, the solution is $x = 1$, $y = -1$, $z = 2$, or $(1, -1, 2)$.

2.2 EXERCISES

In Exercises 1–4 find the augmented matrix for each of the systems.

1. $2x - 3y + z = 0$
$\quad x \quad\quad - 2z = 1$
$\quad\quad -4y + z = -1$

2. $x_1 = 3 - x_2 + x_3$
$\quad x_2 = x_4 - x_3$
$\quad x_3 = x_1 + x_4 + 1$

3. $u = v$
$\quad v = w$

4. $x + w = 0$
$\quad y + z = 0$

In Exercises 5–10 determine which matrices are in row-reduced echelon form.

5. $\begin{bmatrix} 0 & 1 \\ 1 & 0 \end{bmatrix}$

6. $\begin{bmatrix} 1 & 2 & 0 \\ 0 & 0 & 1 \end{bmatrix}$

7. $\begin{bmatrix} 1 & 0 & 2 \\ 0 & 1 & 1 \end{bmatrix}$

8. $\begin{bmatrix} 1 & 2 & 0 & 3 \\ 0 & 1 & 0 & 2 \\ 0 & 0 & 1 & 0 \end{bmatrix}$

9. $\begin{bmatrix} 1 & 0 & 0 \\ 0 & 0 & 0 \\ 0 & 1 & 0 \end{bmatrix}$

10. $\begin{bmatrix} 2 & 0 & 0 \\ 0 & 2 & 0 \\ 0 & 0 & 2 \end{bmatrix}$

In Exercises 11–16 the given matrix is the augmented matrix for a linear system in the variables x_1, x_2, and x_3. Solve the system.

11. $\begin{bmatrix} 1 & 0 & 0 & | & 0 \\ 0 & 1 & 0 & | & 2 \\ 0 & 0 & 1 & | & -1 \end{bmatrix}$

12. $\begin{bmatrix} 1 & 0 & 1 & | & 0 \\ 0 & 1 & -1 & | & 0 \\ 0 & 0 & 0 & | & 0 \end{bmatrix}$

13. $\begin{bmatrix} 1 & 0 & 0 & | & 0 \\ 0 & 1 & 0 & | & 0 \\ 0 & 0 & 0 & | & 1 \end{bmatrix}$

14. $\begin{bmatrix} 1 & 0 & 0 & | & 0 \\ 0 & 1 & 0 & | & 0 \\ 0 & 0 & 0 & | & 0 \end{bmatrix}$

15. $\begin{bmatrix} 1 & 2 & 3 & | & 4 \\ 0 & 0 & 0 & | & 0 \end{bmatrix}$

16. $\begin{bmatrix} 1 & 1 & 1 & | & 0 \\ 0 & 0 & 0 & | & 1 \\ 0 & 0 & 0 & | & 0 \end{bmatrix}$

In Exercises 17–26 solve the linear systems by the Gauss–Jordan matrix elimination method.

17. $\begin{aligned} x_1 - 2x_2 &= 1 \\ -2x_1 + 4x_2 &= -2 \end{aligned}$

18. $\begin{aligned} -x + y &= 0 \\ 2x - 2y &= 0 \end{aligned}$

19. $\begin{aligned} x - 2y + z &= 5 \\ -2x + 3y + z &= 1 \\ x + 3y + 2z &= 2 \end{aligned}$

20. $\begin{aligned} -x_1 + x_2 - 2x_3 &= 1 \\ x_1 + x_2 + 2x_3 &= -1 \\ x_1 + 3x_2 + 2x_3 &= -1 \end{aligned}$

21. $\begin{aligned} 2x_1 - 3x_2 + x_3 &= 1 \\ -x_1 \quad\quad + 2x_3 &= 0 \\ 3x_1 - 3x_2 - x_3 &= 1 \end{aligned}$

22. $\begin{aligned} x + y &= 0 \\ x + y &= -z \\ -x + 1 &= y \\ 1 + x + 2z &= 0 \end{aligned}$

23. $\begin{aligned} w &= x + y + z \\ w &= 2x - 3y + z - 1 \\ w &= -x + y - 2z + 2 \\ w &= 4x - 3y + 4z \end{aligned}$

24. $\begin{aligned} x_1 + 2x_2 \quad\quad - x_4 &= 0 \\ x_1 - 2x_2 + x_3 + x_4 &= 0 \\ 2x_1 - 3x_2 \quad\quad - x_4 &= 0 \end{aligned}$

25. $\begin{aligned} v - 2w + z &= 1 \\ 2u - v \quad - z &= 0 \\ 4u + v - 6w + z &= 3 \end{aligned}$

26. $\begin{aligned} -x_1 + 3x_2 \quad\quad + x_4 &= 0 \\ 2x_1 - 5x_2 + x_3 - x_4 &= 1 \\ x_2 + x_3 + x_4 &= 1 \\ x_1 + x_2 + x_3 \quad\quad &= 0 \end{aligned}$

In Exercises 27 and 28 solve the given linear system without introducing fractions until the last step (if necessary at all).

27. $\begin{aligned} 2x_1 - x_2 + 3x_3 &= 4 \\ -x_1 + 2x_2 - x_3 &= 0 \\ 2x_1 + x_2 + x_3 &= 4 \end{aligned}$

28. $\begin{aligned} 2x_2 - x_3 &= 1 \\ 2x_1 - x_2 + x_3 &= 1 \\ 2x_1 + x_2 \quad &= 2 \end{aligned}$

29.–38. Solve Exercises 17–26 by the Gaussian elimination procedure.

39. Given a system of linear equations in three variables with the augmented matrix

$$\begin{bmatrix} 1 & 0 & 1 & | & 1 \\ 0 & 1 & 1 & | & 2 \\ 0 & 2 & k & | & k \end{bmatrix},$$

for what value of k is there no solution for the system? Is there any value of k for which there are infinitely many solutions?

40. The possible row-reduced echelon forms of the matrix $\begin{bmatrix} a & b \\ c & d \end{bmatrix}$

are $\begin{bmatrix} 1 & 0 \\ 0 & 1 \end{bmatrix}, \begin{bmatrix} 1 & k \\ 0 & 0 \end{bmatrix}, \begin{bmatrix} 0 & 1 \\ 0 & 0 \end{bmatrix},$ and $\begin{bmatrix} 0 & 0 \\ 0 & 0 \end{bmatrix}.$

List the possible row-reduced echelon forms of the matrix

$$\begin{bmatrix} a & b & c \\ d & e & f \\ g & h & i \end{bmatrix}.$$

 COMPUTATIONAL NOTE A digital computer rounds certain fractions (such as $\frac{1}{3}$) to a finite (relatively small) number of decimal places. Consequently "round-off errors" may be introduced at each step of an algorithm such as Gauss–Jordan or Gaussian elimination. In "small" problems, like those in this exercise set, it is unlikely that these errors will have a significant effect on the accuracy of the result. However, it is possible. The cumulative effect of round-off errors is a function not only of the size of a linear system but also of its coefficients. Exercises 41 and 42 explore this phenomenon.

···

 41. Solve Exercises 17–26 by using a computer program for Gauss–Jordan or Gaussian elimination. How do the answers compare with the ones you obtained by hand?

 42. Solve the linear system whose augmented matrix is given as follows by using a computer program for Gauss–Jordan or Gaussian elimination:

$$\left[\begin{array}{cccccc|c} 1 & \frac{1}{2} & \frac{1}{3} & \frac{1}{4} & \frac{1}{5} & \frac{1}{6} & 1 \\ \frac{1}{2} & \frac{1}{3} & \frac{1}{4} & \frac{1}{5} & \frac{1}{6} & \frac{1}{7} & 0 \\ \frac{1}{3} & \frac{1}{4} & \frac{1}{5} & \frac{1}{6} & \frac{1}{7} & \frac{1}{8} & 0 \\ \frac{1}{4} & \frac{1}{5} & \frac{1}{6} & \frac{1}{7} & \frac{1}{8} & \frac{1}{9} & 0 \\ \frac{1}{5} & \frac{1}{6} & \frac{1}{7} & \frac{1}{8} & \frac{1}{9} & \frac{1}{10} & 0 \\ \frac{1}{6} & \frac{1}{7} & \frac{1}{8} & \frac{1}{9} & \frac{1}{10} & \frac{1}{11} & 0 \end{array}\right]$$

Compare your result with the "exact" solution $(36, -630, 3360, -7560, 7560, -2772)$.

2.3 OPERATIONS ON MATRICES

In the previous section we used matrices to provide a convenient notation for performing the elimination procedure. Later in this chapter we will use them in another way to provide a second technique for solving certain systems of linear equations. Matrices will also appear frequently throughout the remainder of this text, in several different settings. Consequently it will be useful at this point to establish some of the notation and definitions concerning this important mathematical tool.

BASIC TERMINOLOGY

Recall from Section 2.2 that a **matrix** is simply a rectangular array of **entries,** which are usually real or complex numbers. If a given matrix has m rows and n columns, it is called an **m × n matrix** (read, "m by n matrix"), and the numbers m and n are the **dimensions** of the matrix. The matrices

$$\begin{bmatrix} 1 & 3 \\ -1 & 0 \\ 2 & 4 \end{bmatrix},\tag{1}$$

$$\begin{bmatrix} -1 & 0 & 1 \\ 4 & 2 & -3 \end{bmatrix},\tag{2}$$

and

$$\begin{bmatrix} \frac{1}{2} \\ -6 \\ 4 \end{bmatrix}\tag{3}$$

are, respectively, 3×2, 2×3, and 3×1 matrices. We sometimes refer to an $m \times 1$ matrix (with $m > 1$) as a **column vector** and a $1 \times n$ matrix ($n > 1$) as a **row vector.** For example, the matrix (3) is called a column vector.

Capital (uppercase) letters, such as A, B, and C, will be used to denote matrices, whereas the entries of a matrix will be denoted by the corresponding lowercase letter with "double subscripts." For example, to describe a general 2×4 matrix, we could write

$$A = \begin{bmatrix} a_{11} & a_{12} & a_{13} & a_{14} \\ a_{21} & a_{22} & a_{23} & a_{24} \end{bmatrix}.$$

Note that the first subscript gives the row number, whereas the second gives the column number. As another example, if

$$B = \begin{bmatrix} 3 & 6 & -1 \\ 4 & 2 & 0 \\ 1 & -3 & -5 \end{bmatrix},$$

then $b_{12} = 6$, $b_{21} = 4$, and $b_{33} = -5$.

Occasionally, when we want to make explicit this uppercase/lowercase convention, we will write something like, "Let $A = [a_{ij}]$ and $B = [b_{ij}]$." All this means is, "Let A be a matrix with entries denoted a_{ij} and let B be a matrix with entries denoted b_{ij}." Moreover, if C is an $m \times n$ matrix, then the subscripts i and j of its entries take on the values $1, 2, \ldots, m$ and $1, 2, \ldots, n$, respectively. However, we will rarely state this explicitly.

Before proceeding to a description of matrix operations, we will give one more definition. Given matrices A and B, the entries a_{ij} and b_{kp} are **corresponding entries** if and only if $i = k$ and $j = p$. In other words, an entry of A *corresponds* to an entry of B if and only if they occupy the same position in their respective matrices. For example, if

$$A = \begin{bmatrix} 1 & 2 & 3 \\ 4 & 5 & 6 \end{bmatrix} \quad \text{and} \quad B = \begin{bmatrix} 7 & 8 & 9 \\ -1 & -2 & -3 \end{bmatrix},$$

then 1 and 7, 2 and 8, and 4 and −1 are all pairs of corresponding entries. We shall speak of corresponding entries only when the two matrices in question have the same dimensions.

DEFINITION

Equality of matrices. Two matrices **A** and **B** are **equal**, written $A = B$, if they have the same dimensions and their corresponding entries are equal.

Formally, an $m \times n$ matrix A is equal to a $p \times q$ matrix B if and only if $m = p$, $n = q$, and $a_{ij} = b_{ij}$ for all i and j.

EXAMPLE 2.17

Which of the following matrices are equal:

$$A = \begin{bmatrix} 1 & 2 & -1 \\ 4 & 0 & 1+2 \end{bmatrix}, \quad B = \begin{bmatrix} 1 & 2 \\ 4 & 0 \end{bmatrix},$$

$$C = \begin{bmatrix} 1 & \frac{4}{2} & -1 \\ 4 & 0 & 3 \end{bmatrix}, \quad D = \begin{bmatrix} 1 & 2 & -1 \\ 4 & 0 & 1 \end{bmatrix}?$$

Solution

We have $A = C$, but this is the only pair of equal matrices.

SUM, DIFFERENCE, AND SCALAR MULTIPLICATION

DEFINITION

Sum of matrices. Let A and B be matrices with the same dimensions. The **sum** of A and B, written $A + B$, is the matrix obtained by adding corresponding entries of A and B.

Formally, we have $C = A + B$ if and only if $c_{ij} = a_{ij} + b_{ij}$ for all i and j.

EXAMPLE 2.18

Find the matrix C that is the sum of

$$A = \begin{bmatrix} 3 & 6 \\ -1 & 5 \\ 0 & 2 \end{bmatrix} \quad \text{and} \quad B = \begin{bmatrix} \frac{1}{2} & -4 \\ -2 & 0 \\ -3 & -2 \end{bmatrix}.$$

Solution

$$C = \begin{bmatrix} 3+\frac{1}{2} & 6+(-4) \\ -1+(-2) & 5+0 \\ 0+(-3) & 2+(-2) \end{bmatrix} = \begin{bmatrix} \frac{7}{2} & 2 \\ -3 & 5 \\ -3 & 0 \end{bmatrix}.$$

DEFINITION   **Multiplication by a scalar.**   Let A be a matrix and c be a scalar. Then the **scalar multiple** cA is the matrix obtained by multiplying each entry of A by the scalar c.

Formally, we have $B = cA$ if and only if $b_{ij} = ca_{ij}$ for all i and j.

EXAMPLE 2.19   Let

$$A = \begin{bmatrix} 1 & -1 & 0 & 3 \\ 2 & 6 & -4 & 8 \end{bmatrix}.$$

Find $3A$ and $(-1)A$.

Solution   We have

$$3A = \begin{bmatrix} 3 & -3 & 0 & 9 \\ 6 & 18 & -12 & 24 \end{bmatrix}, \quad \text{and} \quad (-1)A = \begin{bmatrix} -1 & 1 & 0 & -3 \\ -2 & -6 & 4 & -8 \end{bmatrix}.$$

DEFINITION   The **negative** of a matrix B, written $-B$, is the matrix $(-1)B$. It is obtained from B by simply reversing the sign of every entry.

DEFINITION   **Difference of matrices.**   Let A and B be matrices with the same dimensions. The **difference** of A and B, written $A - B$, is the matrix C given by $C = A + (-B)$.

NOTE   The matrix $A - B$ can be obtained by simply subtracting each entry of B from the corresponding entry of A.

EXAMPLE 2.20   Find $C = 2A - B$, where

$$A = \begin{bmatrix} 1 & 2 \\ -1 & 0 \end{bmatrix}, \quad B = \begin{bmatrix} 0 & 4 \\ -1 & 0 \end{bmatrix}.$$

Solution   We have

$$2\begin{bmatrix} 1 & 2 \\ -1 & 0 \end{bmatrix} - \begin{bmatrix} 0 & 4 \\ -1 & 0 \end{bmatrix} = \begin{bmatrix} 2 & 4 \\ -2 & 0 \end{bmatrix} - \begin{bmatrix} 0 & 4 \\ -1 & 0 \end{bmatrix} = \begin{bmatrix} 2 & 0 \\ -1 & 0 \end{bmatrix}.$$

DEFINITION   The matrix analog of the number 0 is called a **zero matrix**. A zero matrix, denoted by 0, is a matrix of any dimensions consisting of only zero entries.

EXAMPLE 2.21   The matrices

$$\begin{bmatrix} 0 & 0 & 0 \\ 0 & 0 & 0 \\ 0 & 0 & 0 \end{bmatrix}, \quad \begin{bmatrix} 0 & 0 & 0 \\ 0 & 0 & 0 \end{bmatrix}, \quad \begin{bmatrix} 0 \\ 0 \\ 0 \end{bmatrix}, \quad \text{and} \quad [0 \ 0 \ 0 \ 0]$$

are all zero matrices.

A zero matrix has the property that $A + 0 = A$ and $0 + A = A$ for any matrix A for which the sums are defined.

MATRIX MULTIPLICATION

We will now define the product of two matrices. From the way the other matrix operations have been defined, you might guess that we will obtain the product of two matrices by simply multiplying corresponding entries. The definition of *product* given below is much more complicated than this but also considerably more useful in applications. This section's Spotlight provides a simple example of how our definition arises in a natural way. Another example is given in Section 2.4, and still others appear throughout the text.

Before getting to the definition, we will introduce some notation. Given an $m \times n$ matrix A, the vector consisting of the entries of the ith row of A will be denoted $\mathbf{a}_i$, whereas the vector consisting of the entries of the jth column of A will be denoted $\mathbf{a}^j$. For example, if

$$A = \begin{bmatrix} 4 & 3 & -1 \\ 1 & 2 & 5 \\ 0 & 1 & 0 \end{bmatrix},$$

then $\mathbf{a}^1 = (4, 1, 0)$, $\mathbf{a}_1 = (4, 3, -1)$, and $\mathbf{a}_3 = (0, 1, 0)$. Also recall from Section 1.4 that the dot product of the vectors $\mathbf{x} = (x_1, x_2, \ldots, x_r)$ and $\mathbf{y} = (y_1, y_2, \ldots, y_r)$, written $\mathbf{x} \cdot \mathbf{y}$, is given by the number $x_1 y_1 + x_2 y_2 + \cdots + x_r y_r$. For example, for the given matrix A, we have $\mathbf{a}^1 \cdot \mathbf{a}_1 = 4 \cdot 4 + 1 \cdot 3 + 0(-1) = 19$ and $\mathbf{a}_1 \cdot \mathbf{a}^3 = 11$.

DEFINITION

Matrix multiplication.   Let A and B be matrices such that the number of columns of A is equal to the number of rows of B. Suppose that the number of rows of A is m and the number of columns of B is q. The **product** AB is the $m \times q$ matrix C, defined by

$$c_{ij} = \mathbf{a}_i \cdot \mathbf{b}^j.$$

In words, to obtain the entry in the ith row and jth column of C, we take the dot product of the ith row of A with the jth column of B.

NOTE If A is an $m \times n$ matrix, and B is a $p \times q$ matrix, then the product AB is defined only if $n = p$ and, if so, AB is an $m \times q$ matrix. The entry of AB in the ith row and jth column is obtained by multiplying each entry in the ith row of A by the corresponding entry in the jth column of B, and adding up all such products. In other words,

$$c_{ij} = a_{i1}b_{1j} + a_{i2}b_{2j} + \cdots + a_{in}b_{nj},$$

or, using summation notation,

$$c_{ij} = \sum_{k=1}^{n} a_{ik}b_{kj}.$$

EXAMPLE 2.22

Given the pairs of matrices A and B, find the products AB and BA, if defined:

a.
$$A = \begin{bmatrix} 1 & 2 & 3 \\ -3 & -2 & -1 \end{bmatrix}, \quad B = \begin{bmatrix} -4 & 5 \\ 0 & 4 \\ -5 & 0 \end{bmatrix}.$$

b.
$$A = \begin{bmatrix} 1 & 2 & 3 \\ 4 & 5 & 6 \\ 7 & 8 & 9 \end{bmatrix}, \quad B = \begin{bmatrix} x \\ y \\ z \end{bmatrix}.$$

Solution

a. Here A is a 2×3 matrix, and B is a 3×2 matrix. Since the number of columns of A is equal to the number of rows of B, the product AB is defined. It has 2 rows (the same number as A) and 2 columns (the same number as B). Let $C = AB$. Then

$$c_{11} = \mathbf{a}_1 \cdot \mathbf{b}^1 = (1, 2, 3) \cdot (-4, 0, -5)$$
$$= 1(-4) + 2(0) + 3(-5) = -19,$$

$$c_{12} = \mathbf{a}_1 \cdot \mathbf{b}^2 = (1, 2, 3) \cdot (5, 4, 0)$$
$$= 1(5) + 2(4) + 3(0) = 13,$$

$$c_{21} = \mathbf{a}_2 \cdot \mathbf{b}^1 = (-3, -2, -1) \cdot (-4, 0, -5)$$
$$= (-3)(-4) + 0 + (-1)(-5) = 17,$$

$$c_{22} = \mathbf{a}_2 \cdot \mathbf{b}^2 = (-3, -2, -1) \cdot (5, 4, 0)$$
$$= (-3)(5) + (-2)(4) + 0 = -23.$$

Thus

$$C = \begin{bmatrix} -19 & 13 \\ 17 & -23 \end{bmatrix}.$$

In terms of remembering the definition and actually doing the computations to get the ijth entry of AB, it is probably better to think, "go across the ith row of A and down the jth column of B," forming products of corresponding entries and adding the results.

We illustrate this device in finding BA. Now BA is defined, since B (now the "first matrix") has 2 columns, which is the same as the number of rows of A (the "second matrix"). Also, BA will have 3 rows (as does B) and 3 columns (as does A). Letting $D = BA$, we have

$$D = \begin{bmatrix} -4 & 5 \\ 0 & 4 \\ -5 & 0 \end{bmatrix} \begin{bmatrix} 1 & 2 & 3 \\ -3 & -2 & -1 \end{bmatrix}.$$

We find

d_{11} by "going across first row of B and down first column of A," so $d_{11} = (-4)(1) + 5(-3) = -19$;

d_{12} by "going across first row of B and down second column of A," so $d_{12} = (-4)(2) + 5(-2) = -18$;

d_{21} by "going across second row of B and down first column of A," so $d_{21} = 0(1) + 4(-3) = -12$.

Similarly,

$$d_{13} = (-4)(3) + (5)(-1) = -17,$$
$$d_{22} = (0)(2) + (4)(-2) = -8,$$
$$d_{23} = (0)(3) + (4)(-1) = -4,$$
$$d_{31} = (-5)(1) + (0)(-3) = -5,$$
$$d_{32} = (-5)(2) + (0)(-2) = -10,$$
$$d_{33} = (-5)(3) + (0)(-1) = -15.$$

Thus

$$D = BA = \begin{bmatrix} -19 & -18 & -17 \\ -12 & -8 & -4 \\ -5 & -10 & -15 \end{bmatrix}.$$

Note that $AB \neq BA$ in this example. In fact, they do not even have the same dimensions.

b. The product AB is defined, since A has 3 columns and B has 3 rows, and it is a 3×1 matrix. We have

$$AB = \begin{bmatrix} 1 & 2 & 3 \\ 4 & 5 & 6 \\ 7 & 8 & 9 \end{bmatrix} \begin{bmatrix} x \\ y \\ z \end{bmatrix} = \begin{bmatrix} x + 2y + 3z \\ 4x + 5y + 6z \\ 7x + 8y + 9z \end{bmatrix}.$$

However, the product BA is not defined, since the number of columns of B (one) is not equal to the number of rows of A (three).

SPOTLIGHT

REVENUE AS MATRIX MULTIPLICATION

The revenue (income) that a company receives for a certain product is calculated by multiplying the product's price by the number sold. Suppose that the Three's Company markets three products: widgets, xylogs, and yoodels. The price of each is listed in Table 1, and the annual sales figures for four years are given in Table 2.

	Widgets	Xylogs	Yoodels
Price	$7.29	$8.39	$9.49

TABLE 1

Sales	1985	1986	1987	1988
Widgets	5690	6530	3270	2310
Xylogs	4355	4765	3105	2135
Yoodels	8230	9450	5400	4110

TABLE 2

To determine the company's total revenue for 1985, for example, we multiply the price of each product by the corresponding sales figure and add the results:

$$(7.29)(5690) + (8.39)(4355) + (9.49)(8230) = \$156{,}121.25.$$

Notice that if we view Table 1 as a 1×3 matrix A and Table 2 as a 3×4 matrix B, then this result is just the dot product of the only row of A and the first column of B. Similarly, the total revenue for the other three years can be obtained by multiplying the row of A by the appropriate column of B. Thus, if we form the matrix product AB, we will get a 1×4 matrix whose entries are the total revenue for the four years listed:

$$AB = [156{,}121.25 \quad 177{,}262.55 \quad 101{,}135.25 \quad 73{,}756.45]$$

It should be apparent from Example 2.22 that multiplication of matrices "acts differently" from multiplication of numbers. The latter has the property of *commutativity* (for all numbers a and b, $ab = ba$), but matrix multiplication does not.

One situation in which the products AB and BA are both defined (although they still need not be equal) is when A and B have the same dimensions and are *square*. A matrix is **square** if it has the same number of rows as columns. In this case, the number of rows (or columns) is called the **order** of the matrix. For example, the matrix A of Example 2.22b is a square matrix of order 3.

DEFINITION

Powers of a matrix. Let A be a square matrix. Then,

$$A^1 = A$$
$$A^n = \underbrace{AA \cdots A}_{n}$$

EXAMPLE 2.23   Let

$$A = \begin{bmatrix} 1 & 0 \\ -1 & 2 \end{bmatrix}.$$

Find the matrices A^1, A^2, and A^3.

Solution

$$A^1 = A = \begin{bmatrix} 1 & 0 \\ -1 & 2 \end{bmatrix},$$

$$A^2 = AA = \begin{bmatrix} 1 & 0 \\ -1 & 2 \end{bmatrix}\begin{bmatrix} 1 & 0 \\ -1 & 2 \end{bmatrix} = \begin{bmatrix} 1 & 0 \\ -3 & 4 \end{bmatrix},$$

$$A^3 = A^2A = \begin{bmatrix} 1 & 0 \\ -3 & 4 \end{bmatrix}\begin{bmatrix} 1 & 0 \\ -1 & 2 \end{bmatrix} = \begin{bmatrix} 1 & 0 \\ -7 & 8 \end{bmatrix}.$$

NOTE   We found A^3 in Example 2.23 by forming the product A^2A^1, the exponents adding just as they do when the base is a number. In general, if A is a square matrix, and m and n are positive integers, we have $A^mA^n = A^{m+n}$. It is also true that under the same conditions, $(A^m)^n = A^{mn}$.

The entries a_{ij} of a square matrix A for which $i = j$ form the **main diagonal** of A. For example, the main diagonal of

$$B = \begin{bmatrix} 3 & -1 & 4 \\ 0 & 1 & 2 \\ -2 & 5 & -3 \end{bmatrix}$$

consists of 3, 1, and −3. A square matrix in which every element *not* on the main diagonal (every "off diagonal" element) is zero is called a **diagonal matrix.** A special type of diagonal matrix is the **identity matrix,** denoted by *I*, in which each entry on the main diagonal is 1.

EXAMPLE 2.24

The matrices

$$\begin{bmatrix} 1 & 0 \\ 0 & 1 \end{bmatrix}, \qquad \begin{bmatrix} 1 & 0 & 0 \\ 0 & 1 & 0 \\ 0 & 0 & 1 \end{bmatrix}, \qquad \text{and} \qquad \begin{bmatrix} 2 & 0 \\ 0 & -3 \end{bmatrix}$$

are all diagonal matrices. The first two are identity matrices of orders 2 and 3, respectively. The last is not an identity matrix.

The identity matrix is so named because it has the property that $AI = A$ and $IA = A$ for any matrix A for which the products are defined. Thus it is the matrix analog of the number 1, the "multiplicative identity" for numbers. Also following the convention for numbers, we define

$$A^0 = I$$

for any square matrix A.

TRANSPOSE OF A MATRIX

It is sometimes necessary to interchange the rows and columns of a matrix. This operation is called *transposing* the matrix.

DEFINITION

Transpose of a matrix. The **transpose** of the $m \times n$ matrix A, written A^T, is the $n \times m$ matrix B whose *j*th column is the *j*th row of A. Equivalently, $B = A^T$ if and only if $b_{ij} = a_{ji}$ for each *i* and *j*.

EXAMPLE 2.25

Find the transpose of

a. $A = \begin{bmatrix} 1 & 2 \\ -1 & 3 \end{bmatrix}$ **b.** $B = \begin{bmatrix} 1 & -1 \\ 2 & 0 \\ 3 & 4 \end{bmatrix}$

Solution

a. $A^T = \begin{bmatrix} 1 & -1 \\ 2 & 3 \end{bmatrix}$ **b.** $B^T = \begin{bmatrix} 1 & 2 & 3 \\ -1 & 0 & 4 \end{bmatrix}$

RULES GOVERNING OPERATIONS ON MATRICES

We close this section with a theorem that states some of the rules governing the matrix operations already discussed.

THEOREM 1 (Properties of matrices and rules of operation)   Let A, B, and C be matrices and a and b scalars. Assume that the dimensions of the matrices are such that each operation is defined. Then

- **a.** $A + B = B + A$ (commutative law for addition),
- **b.** $A + (B + C) = (A + B) + C$ (associative law for addition),
- **c.** $A + 0 = A$,
- **d.** $A + (-A) = 0$,
- **e.** $A(BC) = (AB)C$ (associative law for multiplication),
- **f.** $AI = A, IB = B$,
- **g.** $A(B + C) = AB + AC$
  $(B + C)A = BA + CA$ (distributive laws),
- **h.** $a(B + C) = aB + aC$,
- **i.** $(a + b)C = aC + bC$,
- **j.** $(ab)C = a(bC)$,
- **k.** $1A = A$,
- **l.** $A0 = 0, 0B = 0$,
- **m.** $a0 = 0$,
- **n.** $a(AB) = (aA)B = A(aB)$,
- **o.** $(A + B)^T = A^T + B^T$,
- **p.** $(AB)^T = B^T A^T$.

Before we prove some of these assertions, we will give a few illustrative examples.

EXAMPLE 2.26   Let

$$A = \begin{bmatrix} 1 & 0 & -1 \\ 2 & 1 & -1 \\ 3 & -2 & 0 \end{bmatrix}, \quad B = \begin{bmatrix} -1 & 2 & 3 \\ 0 & 1 & 0 \\ -2 & -1 & 0 \end{bmatrix}, \quad C = \begin{bmatrix} -2 & 1 & 1 \\ 1 & 1 & 0 \\ 3 & 0 & -3 \end{bmatrix}.$$

Find $A(BC)$, $(AB)C$, $A(B + C)$, and $AB + AC$.

Solution

$$A(BC) = \begin{bmatrix} 1 & 0 & -1 \\ 2 & 1 & -1 \\ 3 & -2 & 0 \end{bmatrix} \begin{bmatrix} 13 & 1 & -10 \\ 1 & 1 & 0 \\ 3 & -3 & -2 \end{bmatrix} = \begin{bmatrix} 10 & 4 & -8 \\ 24 & 6 & -18 \\ 37 & 1 & -30 \end{bmatrix},$$

$$(AB)C = \begin{bmatrix} 1 & 3 & 3 \\ 0 & 6 & 6 \\ -3 & 4 & 9 \end{bmatrix} \begin{bmatrix} -2 & 1 & 1 \\ 1 & 1 & 0 \\ 3 & 0 & -3 \end{bmatrix} = \begin{bmatrix} 10 & 4 & -8 \\ 24 & 6 & -18 \\ 37 & 1 & -30 \end{bmatrix}.$$

(Note that $A(BC) = (AB)C$ for these choices of A, B, and C.)

$$A(B + C) = \begin{bmatrix} 1 & 0 & -1 \\ 2 & 1 & -1 \\ 3 & -2 & 0 \end{bmatrix} \begin{bmatrix} -3 & 3 & 4 \\ 1 & 2 & 0 \\ 1 & -1 & -3 \end{bmatrix} = \begin{bmatrix} -4 & 4 & 7 \\ -6 & 9 & 11 \\ -11 & 5 & 12 \end{bmatrix},$$

$$AB + AC = \begin{bmatrix} 1 & 3 & 3 \\ 0 & 6 & 6 \\ -3 & 4 & 9 \end{bmatrix} + \begin{bmatrix} -5 & 1 & 4 \\ -6 & 3 & 5 \\ -8 & 1 & 3 \end{bmatrix} = \begin{bmatrix} -4 & 4 & 7 \\ -6 & 9 & 11 \\ -11 & 5 & 12 \end{bmatrix}.$$

(Note that $A(B + C) = AB + AC$ for these A, B, and C.)

Of course, Example 2.26 does not constitute a *proof* of the fact that $A(BC) = (AB)C$ or that $A(B + C) = AB + AC$ for *all* matrices A, B, and C.

Proof of Theorem I (in part)

1b. Let $M = A + (B + C)$ and $M' = (A + B) + C$. Then $m_{ij} = a_{ij} + (b_{ij} + c_{ij}) = (a_{ij} + b_{ij}) + c_{ij} = m'_{ij}$ (since addition of numbers is associative).

1e. This part of the proof illustrates the value of summation notation. Assume that A is $m \times n$, B is $n \times p$ and C is $p \times q$. Since

$$(BC)_{kj} = \sum_{s=1}^{p} b_{ks} c_{sj},$$

we have

$$(A(BC))_{ij} = \sum_{k=1}^{n} a_{ik}(BC)_{kj} = \sum_{k=1}^{n} a_{ik}\left(\sum_{s=1}^{p} b_{ks} c_{sj}\right) = \sum_{k=1}^{n} \sum_{s=1}^{p} a_{ik} b_{ks} c_{sj}$$

$$= \sum_{s=1}^{p}\left(\sum_{k=1}^{n} a_{ik} b_{ks}\right) c_{sj} = \sum_{s=1}^{p} (AB)_{is} c_{sj}$$

$$= ((AB)C)_{ij},$$

which proves the assertion.

1n. Let $C = a(AB)$, $D = (aA)B$, and $E = A(aB)$. Then $c_{ij} = a(\mathbf{a}_i \cdot \mathbf{b}^j) = (a\mathbf{a}_i) \cdot \mathbf{b}^j = d_{ij}$ (by Theorem 3b of Section 1.4). Moreover $d_{ij} = (a\mathbf{a}_i) \cdot \mathbf{b}^j = \mathbf{a}_i \cdot (a\mathbf{b}^j) = e_{ij}$ (same theorem). Thus $c_{ij} = d_{ij} = e_{ij}$, as desired.

2.3 EXERCISES

In Exercises 1–12 let

$$A = \begin{bmatrix} -1 & 0 & 1 \\ 2 & -1 & 3 \\ 0 & 1 & -2 \end{bmatrix}, \quad B = \begin{bmatrix} 0 & 4 & -2 \\ 3 & 1 & 2 \\ -1 & 0 & 1 \end{bmatrix},$$

and let I be the identity matrix of order 3. Find each as indicated.

1. $A + B$
2. $A - B$
3. $2A - 3B$
4. $A + I$
5. $A - \lambda I$ (λ, a scalar)
6. $A + \lambda B$ (λ, a scalar)
7. A^T
8. A^0
9. AB
10. $(AB)^T$
11. $B^T A^T$
12. $A^T B^T$

In Exercises 13–16 find each as indicated, where

$$A = \begin{bmatrix} 2 & -1 & \frac{1}{2} \\ 1 & 0 & -2 \end{bmatrix}.$$

13. A^T
14. $(A^T)^T$
15. The columns $\mathbf{a}^1$ and $\mathbf{a}^3$ and the row $\mathbf{a}_2$
16. $\mathbf{a}^1 \cdot \mathbf{a}^3$ and $\mathbf{a}_1 \cdot \mathbf{a}_2$

In Exercises 17–22 find AB and BA when these products are defined.

17. $A = \begin{bmatrix} 1 & 2 & 0 \\ -2 & 0 & 1 \\ 0 & 0 & 3 \end{bmatrix}, \quad B = \begin{bmatrix} 1 & 1 & 1 \\ 3 & -1 & 0 \\ 2 & -2 & 0 \end{bmatrix}$

18. $A = \begin{bmatrix} 1 \\ 0 \\ 0 \end{bmatrix}, \quad B = [-3 \quad 2 \quad 1]$

19. $A = \begin{bmatrix} -1 & 1 & 0 \\ 2 & 3 & 5 \end{bmatrix}, \quad B = \begin{bmatrix} 0 & 4 & -2 \\ 3 & 6 & -1 \end{bmatrix}$

20. $A = \begin{bmatrix} -1 & 1 & 0 \\ 2 & 3 & 5 \end{bmatrix}, \quad B = \begin{bmatrix} 0 & 4 & -2 \\ 3 & 6 & -1 \\ 1 & 2 & -3 \end{bmatrix}$

21. $A = \begin{bmatrix} 2 & 1 & 0 \\ -1 & 4 & 3 \end{bmatrix}, \quad B = \begin{bmatrix} 1 & 0 & 0 \\ 0 & 1 & 0 \\ 0 & 0 & 1 \end{bmatrix}$

22. $A = \begin{bmatrix} 1 & 2 & 1 \\ -2 & 1 & -2 \end{bmatrix}, \quad B = \begin{bmatrix} 0 & 0 \\ 0 & 0 \\ 0 & 0 \\ 0 & 0 \end{bmatrix}$

23. Find

$$\begin{bmatrix} 1 & 3 \\ -1 & 1 \end{bmatrix}\begin{bmatrix} 0 & 1 & -1 \\ 1 & 0 & 1 \end{bmatrix} - 3\begin{bmatrix} 4 & 0 & -2 \\ 1 & \frac{2}{3} & 4 \end{bmatrix}.$$

24. Find $3A - 0B + 6I$ if

$$A = \begin{bmatrix} \frac{1}{2} & 2 \\ 5 & \frac{1}{4} \end{bmatrix}, \qquad B = \begin{bmatrix} \frac{1}{3} & \frac{1}{4} \\ \frac{1}{5} & \frac{1}{6} \end{bmatrix},$$

and I is the identity matrix of order 2.

In Exercises 25–28 find the powers A^2 and A^3 for the given matrix A.

25. $A = \begin{bmatrix} 1 & -1 \\ -1 & 2 \end{bmatrix}$

26. $A = \begin{bmatrix} 1 & 2 \\ 3 & 4 \end{bmatrix}$

27. $A = \begin{bmatrix} 1 & 0 & 0 \\ 2 & 1 & 0 \\ 3 & 2 & 1 \end{bmatrix}$

28. $A = \begin{bmatrix} 1 & 0 & 2 \\ -2 & -1 & 0 \\ 0 & 0 & 2 \end{bmatrix}$

In Exercises 29–32 let

$$A = \begin{bmatrix} 3 & 0 & -1 \\ 0 & 0 & 2 \\ 1 & 0 & -2 \end{bmatrix}, \qquad B = \begin{bmatrix} 1 & 2 & 1 \\ 0 & 0 & 0 \\ -1 & 1 & 0 \end{bmatrix}, \qquad C = \begin{bmatrix} 4 & 0 & -3 \\ -2 & -1 & 0 \\ 1 & 1 & 1 \end{bmatrix},$$

and verify each equation.

29. $(A + B)C = AC + BC$

30. $(2A)B = 2(AB)$

31. $A + C = C + A$

32. $(A + B) + C = (C + A) + B$

33. Prove the remaining parts of Theorem 1.

34. Use Theorem 1 to show that $(A + B) + C = (C + A) + B$ for all matrices A, B, and C for which the sums are defined.

For each matrix in Exercises 35 and 36, find a general formula for A^n.

35. $A = \begin{bmatrix} 1 & 1 \\ 0 & 1 \end{bmatrix}$

36. $A = \begin{bmatrix} 1 & -1 \\ -1 & 1 \end{bmatrix}$

37. A matrix A is called an *involution* if $A^2 = I$. Give an example of a 2×2 matrix (not equal to I) that is an involution.

38. A matrix A is *idempotent* if $A^2 = A$. Give an example of a 2×2 matrix (not equal to I or 0) that is idempotent.

39. Prove that $(A + B)^T = A^T + B^T$ for all matrices A and B of the same dimensions.

40. Prove that $(A^T)^T = A$ for all matrices A.

In Exercises 41 and 42 let A and B be square matrices of order 10 with

$$A = \begin{bmatrix} 1 & 2 & 3 & \cdots & 10 \\ 2 & 3 & 4 & \cdots & 11 \\ 3 & 4 & 5 & \cdots & 12 \\ \vdots & \vdots & \vdots & & \vdots \\ 10 & 11 & 12 & \cdots & 19 \end{bmatrix}, \qquad B = \begin{bmatrix} 1 & -1 & 1 & -1 & \cdots & -1 \\ 0 & 1 & -1 & 1 & \cdots & 1 \\ 0 & 0 & 1 & -1 & \cdots & -1 \\ \vdots & \vdots & \vdots & \vdots & & \vdots \\ 0 & 0 & 0 & 0 & \cdots & 1 \end{bmatrix}.$$

Using a computer program for matrix multiplication, find the products.

41. AB and BA

42. $(AB)^T$, $A^T B^T$, and $B^T A^T$

2.4 MATRIX EQUATIONS AND INVERSES

In this section we will develop a very concise way of writing a system of linear equations. This concise form will suggest another solution procedure that in turn will lead to the definition and investigation of the notion of the *inverse* of a square matrix.

MATRIX FORM OF A LINEAR SYSTEM

A system of m linear equations in the n unknowns $x_1, x_2, \ldots, x_n$ can be written in the general form:

$$
\begin{aligned}
a_{11}x_1 + a_{12}x_2 + \cdots + a_{1n}x_n &= b_1 \\
a_{21}x_1 + a_{22}x_2 + \cdots + a_{2n}x_n &= b_2 \\
&\;\;\vdots \\
a_{m1}x_1 + a_{m2}x_2 + \cdots + a_{mn}x_n &= b_m.
\end{aligned}
$$

The a_{ij} and the b_i are, respectively, the **coefficients** and **right-side constants** of the system. The "double subscript" notation is particularly convenient for the coefficients. Note that in the ith equation, the coefficient of the jth unknown is a_{ij}. For example, in the second equation the coefficient of the first unknown is denoted a_{21}.

Now recall from Section 2.3 that two matrices are equal if and only if all pairs of corresponding entries are equal. For this reason we can write the linear system just given as an equation between two $m \times 1$ matrices:

$$
\begin{bmatrix}
a_{11}x_1 + a_{12}x_2 + \cdots + a_{1n}x_n \\
a_{21}x_1 + a_{22}x_2 + \cdots + a_{2n}x_n \\
\vdots \\
a_{m1}x_1 + a_{m2}x_2 + \cdots + a_{mn}x_n
\end{bmatrix}
=
\begin{bmatrix}
b_1 \\
b_2 \\
\vdots \\
b_m
\end{bmatrix}.
$$

However, by the definition of matrix multiplication, the matrix on the left can be written as the product of an $m \times n$ matrix and an $n \times 1$ matrix, so

$$
\begin{bmatrix}
a_{11} & a_{12} & \cdots & a_{1n} \\
a_{21} & a_{22} & \cdots & a_{2n} \\
\vdots & \vdots & & \vdots \\
a_{m1} & a_{m2} & \cdots & a_{mn}
\end{bmatrix}
\begin{bmatrix}
x_1 \\
x_2 \\
x_3 \\
\vdots \\
x_n
\end{bmatrix}
=
\begin{bmatrix}
b_1 \\
b_2 \\
\vdots \\
b_m
\end{bmatrix}.
$$

Finally, denoting the $m \times n$ matrix with entries a_{ij} by A, the $n \times 1$ matrix with entries x_i by X, and the $m \times 1$ matrix with entries b_i by B, we can write the given system of linear equations in the concise matrix form

$$AX = B.$$

Here A is called the **coefficient matrix** of the system; X is the matrix, or (column) vector, of unknowns; and B is the matrix, or (column) vector, of constants. Since X and B are column vectors, we can write them using vector notation as $\mathbf{x}$ and $\mathbf{b}$. This leads to the conventional way of writing a linear system in **matrix form:**

$$A\mathbf{x} = \mathbf{b}.$$

EXAMPLE 2.27   Write in matrix form the linear systems

a.
$$\begin{aligned} x \quad\quad + 2z &= 3 \\ 2x - 3y - 4z &= -1 \\ -y + z &= 0 \end{aligned}$$

b.
$$\begin{aligned} x_1 - 3x_2 \quad\quad + x_4 &= 0 \\ - x_2 + x_3 - x_4 &= 0 \end{aligned}$$

Solution

a.
$$\begin{bmatrix} 1 & 0 & 2 \\ 2 & -3 & -4 \\ 0 & -1 & 1 \end{bmatrix} \begin{bmatrix} x \\ y \\ z \end{bmatrix} = \begin{bmatrix} 3 \\ -1 \\ 0 \end{bmatrix}$$

b.
$$\begin{bmatrix} 1 & -3 & 0 & 1 \\ 0 & -1 & 1 & -1 \end{bmatrix} \begin{bmatrix} x_1 \\ x_2 \\ x_3 \\ x_4 \end{bmatrix} = \begin{bmatrix} 0 \\ 0 \end{bmatrix}$$

Both are written as $A\mathbf{x} = \mathbf{b}$, where in part (a),

$$A = \begin{bmatrix} 1 & 0 & 2 \\ 2 & -3 & -4 \\ 0 & -1 & 1 \end{bmatrix}, \quad \mathbf{x} = \begin{bmatrix} x \\ y \\ z \end{bmatrix}, \quad \mathbf{b} = \begin{bmatrix} 3 \\ -1 \\ 0 \end{bmatrix};$$

and in part (b),

$$A = \begin{bmatrix} 1 & -3 & 0 & 1 \\ 0 & -1 & 1 & -1 \end{bmatrix}, \quad \mathbf{x} = \begin{bmatrix} x_1 \\ x_2 \\ x_3 \\ x_4 \end{bmatrix}, \quad \mathbf{b} = \begin{bmatrix} 0 \\ 0 \end{bmatrix}.$$

NOTE   It is often convenient to write a column vector on a horizontal line, so we establish the convention that $(c_1, c_2, \ldots, c_k)$ will denote the $k \times 1$ column vector

$$\begin{bmatrix} c_1 \\ c_2 \\ \vdots \\ c_k \end{bmatrix}.$$

Note that parentheses and commas distinguish our column vector written horizontally from the $1 \times k$ row vector $[c_1 \quad c_2 \quad \cdots \quad c_k]$. This new convention also conforms with previous usage. Earlier in this chapter we wrote solutions of a linear system with n unknowns as n-tuples $(x_1, x_2, \ldots, x_n)$. We may still write solutions in this manner. For example, the solution of the system

$$\begin{matrix} 3x_1 - x_2 = 0 \\ 2x_1 + x_2 = 5 \end{matrix} \quad \text{or} \quad \begin{bmatrix} 3 & -1 \\ 2 & 1 \end{bmatrix} \begin{bmatrix} x_1 \\ x_2 \end{bmatrix} = \begin{bmatrix} 0 \\ 5 \end{bmatrix}$$

may be written as either

$$\begin{bmatrix} 1 \\ 3 \end{bmatrix} \quad \text{or} \quad (1, 3).$$

THE INVERSE OF A MATRIX

Writing a linear system in the form $A\mathbf{x} = \mathbf{b}$ suggests a deceptively simple means of solution. Were this a single (scalar) equation, $ax = b$ ($a \neq 0$), with unknown x, we could quickly solve it by multiplying both sides by the *multiplicative inverse* (the *reciprocal*) of a. The solution is $x = a^{-1}b$ ($= (1/a)b = b/a$). Taking our cue from this simpler situation, it is natural to ask if we can solve the matrix equation $A\mathbf{x} = \mathbf{b}$ in the same way. The answer is, "Yes, sometimes," but the process, which is developed next, takes considerably more work.

We must first define the (multiplicative) inverse of a matrix and then determine a procedure for finding it when it exists. The multiplicative inverse of a number a is a number b, such that $ab = 1$. (Since multiplication of numbers is commutative, we also have $ba = 1$.) Consequently a natural candidate for the inverse of a matrix A would be a matrix B, such that $AB = I$. However, matrix multiplication is not commutative, so we also require that $BA = I$. Note that if both products AB and BA are defined and equal to the same matrix I, then A and B must both be square and of the same order as I.

DEFINITION

Let A be a square matrix. If there exists a matrix B such that $AB = BA = I$, we say that A is **invertible** (or **nonsingular**) and that B is the **inverse** of A. If A has no inverse, it is said to be **noninvertible** (or **singular**).

EXAMPLE 2.28

Show that the inverse of the matrix A is B, where

$$A = \begin{bmatrix} 2 & -1 \\ 0 & 1 \end{bmatrix} \quad \text{and} \quad B = \begin{bmatrix} \frac{1}{2} & \frac{1}{2} \\ 0 & 1 \end{bmatrix}.$$

Solution   We have

$$AB = \begin{bmatrix} 2 & -1 \\ 0 & 1 \end{bmatrix}\begin{bmatrix} \frac{1}{2} & \frac{1}{2} \\ 0 & 1 \end{bmatrix} = \begin{bmatrix} 1 & 0 \\ 0 & 1 \end{bmatrix} = I,$$

and

$$BA = \begin{bmatrix} \frac{1}{2} & \frac{1}{2} \\ 0 & 1 \end{bmatrix}\begin{bmatrix} 2 & -1 \\ 0 & 1 \end{bmatrix} = \begin{bmatrix} 1 & 0 \\ 0 & 1 \end{bmatrix} = I,$$

as desired.

In the definition and in Example 2.28 we have spoken of *the* inverse of a matrix. This terminology implies that a given matrix has at most one inverse. The following theorem justifies the use of this word.

THEOREM 1   If a matrix A is invertible, then the inverse is unique.

Proof   Suppose that A has inverses B and C. We show that B and C must be equal as follows: $B = BI = B(AC) = (BA)C = IC = C$.

NOTE   We will denote the inverse of an invertible matrix A by the symbol A^{-1}. From the definition it follows that if $A^{-1} = B$, then B is invertible, and $B^{-1} = A$ as well.

....

We will now develop a technique for finding the inverse of a square matrix. First we state a theorem that will simplify this task somewhat. Its proof is given in Theorem 9 of Section 3.2.

THEOREM 2   Let A be a square matrix. If a square matrix B exists such that $AB = I$, then $BA = I$ as well, and thus $B = A^{-1}$.

Theorem 2 tells us that the inverse of an $n \times n$ matrix A is the $n \times n$ matrix X that solves the equation $AX = I$, where I is the identity matrix of order n. Before establishing a procedure to find this solution, we will consider a particular example that illustrates the basic idea.

EXAMPLE 2.29   Find the inverse of

$$A = \begin{bmatrix} 1 & 1 & 1 \\ 0 & 2 & 1 \\ 1 & 0 & 1 \end{bmatrix}.$$

Solution

We seek a 3×3 matrix X such that $AX = I$, that is,

$$\begin{bmatrix} 1 & 1 & 1 \\ 0 & 2 & 1 \\ 1 & 0 & 1 \end{bmatrix} \begin{bmatrix} x_{11} & x_{12} & x_{13} \\ x_{21} & x_{22} & x_{23} \\ x_{31} & x_{32} & x_{33} \end{bmatrix} = \begin{bmatrix} 1 & 0 & 0 \\ 0 & 1 & 0 \\ 0 & 0 & 1 \end{bmatrix}.$$

Now the first, second, and third columns of the matrix product on the left must be equal to, respectively, the first, second, and third columns of the identity matrix. Symbolically, we write this as $(AX)^1 = \mathbf{i}^1, (AX)^2 = \mathbf{i}^2,$ and $(AX)^3 = \mathbf{i}^3$. By the definition of matrix multiplication,

$$(AX)^1 = \begin{bmatrix} \mathbf{a}_1 \cdot \mathbf{x}^1 \\ \mathbf{a}_2 \cdot \mathbf{x}^1 \\ \mathbf{a}_3 \cdot \mathbf{x}^1 \end{bmatrix} = \begin{bmatrix} (1,1,1) \cdot (x_{11}, x_{21}, x_{31}) \\ (0,2,1) \cdot (x_{11}, x_{21}, x_{31}) \\ (1,0,1) \cdot (x_{11}, x_{21}, x_{31}) \end{bmatrix} = \begin{bmatrix} x_{11} + x_{21} + x_{31} \\ 2x_{21} + x_{31} \\ x_{11} \qquad + x_{31} \end{bmatrix}.$$

This is exactly the result we get if we multiply the matrix A by the first column of X; that is,

$$A\mathbf{x}^1 = \begin{bmatrix} 1 & 1 & 1 \\ 0 & 2 & 1 \\ 1 & 0 & 1 \end{bmatrix} \begin{bmatrix} x_{11} \\ x_{21} \\ x_{31} \end{bmatrix} = \begin{bmatrix} x_{11} + x_{21} + x_{31} \\ 2x_{21} + x_{31} \\ x_{11} \qquad + x_{31} \end{bmatrix}.$$

Thus $(AX)^1 = A\mathbf{x}^1$. Similarly, $(AX)^2 = A\mathbf{x}^2$ and $(AX)^3 = A\mathbf{x}^3$. Therefore, to find X, we can solve the three linear systems

$$A\mathbf{x}^1 = \mathbf{i}^1, \qquad A\mathbf{x}^2 = \mathbf{i}^2, \qquad \text{and} \qquad A\mathbf{x}^3 = \mathbf{i}^3.$$

We do this by Gauss–Jordan elimination (Section 2.2), forming the augmented matrices

$$\left[\begin{array}{ccc|c} 1 & 1 & 1 & 1 \\ 0 & 2 & 1 & 0 \\ 1 & 0 & 1 & 0 \end{array}\right], \quad \left[\begin{array}{ccc|c} 1 & 1 & 1 & 0 \\ 0 & 2 & 1 & 1 \\ 1 & 0 & 1 & 0 \end{array}\right], \quad \left[\begin{array}{ccc|c} 1 & 1 & 1 & 0 \\ 0 & 2 & 1 & 0 \\ 1 & 0 & 1 & 1 \end{array}\right],$$

and transforming each to its row-reduced echelon form

$$\left[\begin{array}{ccc|c} 1 & 0 & 0 & 2 \\ 0 & 1 & 0 & 1 \\ 0 & 0 & 1 & -2 \end{array}\right], \quad \left[\begin{array}{ccc|c} 1 & 0 & 0 & -1 \\ 0 & 1 & 0 & 0 \\ 0 & 0 & 1 & 1 \end{array}\right], \quad \left[\begin{array}{ccc|c} 1 & 0 & 0 & -1 \\ 0 & 1 & 0 & -1 \\ 0 & 0 & 1 & 2 \end{array}\right].$$

Thus $\mathbf{x}^1 = (2, 1, -2), \mathbf{x}^2 = (-1, 0, 1),$ and $\mathbf{x}^3 = (-1, -1, 2),$ and we have

$$A^{-1} = \begin{bmatrix} 2 & -1 & -1 \\ 1 & 0 & -1 \\ -2 & 1 & 2 \end{bmatrix}.$$

The solution procedure of Example 2.29 contains a gross inefficiency. Since the coefficient matrices of the three linear systems $A\mathbf{x}^1 = \mathbf{i}^1, A\mathbf{x}^2 = \mathbf{i}^2,$ and $A\mathbf{x}^3 = \mathbf{i}^3$ are identical, all three augmented matrices are transformed using the same row operations. Therefore we can save a lot of work by doing all the row reductions at the same time. To do this, we form the 3×6 augmented matrix whose first three columns are those of A and whose last three columns are $\mathbf{i}^1, \mathbf{i}^2$, and $\mathbf{i}^3$ (that is, the last three columns are those of I):

$$\left[\begin{array}{ccc|ccc} 1 & 1 & 1 & 1 & 0 & 0 \\ 0 & 2 & 1 & 0 & 1 & 0 \\ 1 & 0 & 1 & 0 & 0 & 1 \end{array}\right].$$

We now transform this matrix to the row-reduced echelon form

$$\left[\begin{array}{ccc|ccc} 1 & 0 & 0 & 2 & -1 & -1 \\ 0 & 1 & 0 & 1 & 0 & -1 \\ 0 & 0 & 1 & -2 & 1 & 2 \end{array}\right].$$

Then the fourth, fifth, and sixth columns of the last matrix are $\mathbf{x}^1$, $\mathbf{x}^2$, and $\mathbf{x}^3$, the columns of A^{-1}. Thus (as before)

$$A^{-1} = \left[\begin{array}{ccc} 2 & -1 & -1 \\ 1 & 0 & -1 \\ -2 & 1 & 2 \end{array}\right].$$

NOTE   In effect, to form the original 3×6 augmented matrix, we adjoined the identity matrix of order 3 to the coefficient matrix A. We will denote this matrix by $[A \mid I]$. Similarly, the augmented matrix for the system $A\mathbf{x} = \mathbf{b}$ can be denoted $[A \mid \mathbf{b}]$.

....

We now turn to the general case of attempting to find the inverse of an $n \times n$ matrix A. This can be done by solving the matrix equation $AX = I$ (Theorem 2). As in Example 2.29 the latter is equivalent to solving the n linear systems

$$A\mathbf{x}^1 = \mathbf{i}^1, A\mathbf{x}^2 = \mathbf{i}^2, \ldots, A\mathbf{x}^n = \mathbf{i}^n$$

for the columns $\mathbf{x}^i$ of A^{-1}. To solve these systems simultaneously, we transform the augmented matrix $[A \mid I]$ into row-reduced echelon form. The last n columns of the latter form the inverse of A *provided that each of the systems $A\mathbf{x}^i = \mathbf{i}^i$ has a solution*. If one of them does not, the matrix equation $AX = I$ has no solution, and hence the inverse of A does not exist. We summarize this discussion in the form of a procedure.

PROCEDURE FOR FINDING A MATRIX INVERSE

Let A be a square matrix, and let I be the identity matrix of the same order. Form the augmented matrix $[A \mid I]$ and transform it into row-reduced echelon form $[C \mid D]$. Then

i.   If C is the identity matrix, $D = A^{-1}$,

ii.   if C is not the identity matrix, A is not invertible.

....

SPOTLIGHT

SOLVING MATRIX EQUATIONS

In some applications it becomes necessary to solve several linear systems (say, k of them), all of which have the same coefficient matrix:

$$A\mathbf{x}^i = \mathbf{b}^i, \qquad i = 1, 2, \ldots, k.$$

We can write all of these linear systems as the single *matrix equation*

$$AX = B,$$

where the columns of B are the given right-side vectors, $\mathbf{b}^i$, and the columns of X are the corresponding solutions, $\mathbf{x}^i$. We can then solve this matrix equation in the same way that we solved $AX = I$ in demonstrating how to find the inverse of A.

As a simple example, suppose we are given the systems:

$$
\begin{array}{lll}
x + 2y = 5 & u + 2v = 1 & w + 2z = 8 \\
2x + 5y = 12 & 2u + 5v = 3 & 2w + 5z = 20
\end{array}
$$

The three systems can be written as the single matrix equation $AX = B$, with

$$A = \begin{bmatrix} 1 & 2 \\ 2 & 5 \end{bmatrix}, \qquad X = \begin{bmatrix} x & u & w \\ y & v & z \end{bmatrix}, \qquad \text{and} \qquad B = \begin{bmatrix} 5 & 1 & 8 \\ 12 & 3 & 20 \end{bmatrix}$$

To solve $AX = B$, we form the augmented matrix $[A \mid B]$ and transform it to row-reduced echelon form:

$$\begin{bmatrix} 1 & 2 & \vline & 5 & 1 & 8 \\ 2 & 5 & \vline & 12 & 3 & 20 \end{bmatrix} \rightarrow \begin{bmatrix} 1 & 0 & \vline & 1 & -1 & 0 \\ 0 & 1 & \vline & 2 & 1 & 4 \end{bmatrix}.$$

Thus the solutions of the three given systems are $(1, 2)$, $(-1, 1)$, and $(0, 4)$, respectively.

EXAMPLE 2.30   Find A^{-1} (if it exists) when

$$A = \begin{bmatrix} 1 & 0 & 1 & 1 \\ 0 & 0 & 1 & 0 \\ 1 & 1 & 1 & 0 \\ 1 & 0 & 0 & 2 \end{bmatrix}.$$

Solution

We adjoin the (fourth-order) identity matrix to get $[A \mid I]$,

$$\left[\begin{array}{cccc|cccc} 1 & 0 & 1 & 1 & 1 & 0 & 0 & 0 \\ 0 & 0 & 1 & 0 & 0 & 1 & 0 & 0 \\ 1 & 1 & 1 & 0 & 0 & 0 & 1 & 0 \\ 1 & 0 & 0 & 2 & 0 & 0 & 0 & 1 \end{array}\right],$$

and transform this matrix to row-reduced echelon form,

$$\left[\begin{array}{cccc|cccc} 1 & 0 & 0 & 0 & 2 & -2 & 0 & -1 \\ 0 & 1 & 0 & 0 & -2 & 1 & 1 & 1 \\ 0 & 0 & 1 & 0 & 0 & 1 & 0 & 0 \\ 0 & 0 & 0 & 1 & -1 & 1 & 0 & 1 \end{array}\right].$$

Consequently A^{-1} exists with

$$A^{-1} = \left[\begin{array}{cccc} 2 & -2 & 0 & -1 \\ -2 & 1 & 1 & 1 \\ 0 & 1 & 0 & 0 \\ -1 & 1 & 0 & 1 \end{array}\right].$$

EXAMPLE 2.31

Find the inverse of

$$A = \begin{bmatrix} 1 & -2 \\ -2 & 4 \end{bmatrix},$$

or show that no inverse exists.

Solution

We form the augmented matrix

$$[A \mid I] = \left[\begin{array}{cc|cc} 1 & -2 & 1 & 0 \\ -2 & 4 & 0 & 1 \end{array}\right],$$

and transform it to the form

$$\left[\begin{array}{cc|cc} 1 & 2 & 1 & 0 \\ 0 & 0 & 2 & 1 \end{array}\right].$$

Since A cannot be transformed into the identity matrix, A has no inverse.

METHOD OF INVERSES

We now return to the idea of solving the linear system $A\mathbf{x} = \mathbf{b}$ by using inverse matrices. Let A be an $n \times n$ nonsingular matrix. We have

$$A\mathbf{x} = \mathbf{b}$$

$$A^{-1}A\mathbf{x} = A^{-1}\mathbf{b} \quad \text{(multiplying both sides on the left by } A^{-1}\text{),}$$

$$I\mathbf{x} = A^{-1}\mathbf{b} \quad \text{(Theorem 1e of Section 2.3),}$$

$$\mathbf{x} = A^{-1}\mathbf{b} \quad \text{(Theorem 1f of Section 2.3).}$$

Hence we can solve a linear system whose coefficient matrix is square and nonsingular by finding the inverse and then multiplying it by the right-side vector. This type of solution procedure is called the **method of inverses.**

EXAMPLE 2.32

Solve the following system by the method of inverses:

$$
\begin{array}{rcr}
x_1 \quad\quad + x_3 &=& 2 \\
x_1 - x_2 \quad\quad &=& -1 \\
2x_2 + x_3 &=& 1.
\end{array}
$$

Solution

This linear system can be written in the form $A\mathbf{x} = \mathbf{b}$ by taking

$$
A = \begin{bmatrix} 1 & 0 & 1 \\ 1 & -1 & 0 \\ 0 & 2 & 1 \end{bmatrix}, \quad
\mathbf{x} = \begin{bmatrix} x_1 \\ x_2 \\ x_3 \end{bmatrix}, \quad \text{and} \quad
\mathbf{b} = \begin{bmatrix} 2 \\ -1 \\ 1 \end{bmatrix}.
$$

We must first find A^{-1}. We have

$$
[A \mid I] = \begin{bmatrix} 1 & 0 & 1 & 1 & 0 & 0 \\ 1 & -1 & 0 & 0 & 1 & 0 \\ 0 & 2 & 1 & 0 & 0 & 1 \end{bmatrix},
$$

which has the row-reduced echelon form

$$
\begin{bmatrix} 1 & 0 & 0 & -1 & 2 & 1 \\ 0 & 1 & 0 & -1 & 1 & 1 \\ 0 & 0 & 1 & 2 & -2 & -1 \end{bmatrix}.
$$

Thus

$$
A^{-1} = \begin{bmatrix} -1 & 2 & 1 \\ -1 & 1 & 1 \\ 2 & -2 & -1 \end{bmatrix}.
$$

Finally, we obtain $\mathbf{x} = A^{-1}\mathbf{b}$ by multiplying

$$
\begin{bmatrix} -1 & 2 & 1 \\ -1 & 1 & 1 \\ 2 & -2 & -1 \end{bmatrix}
\begin{bmatrix} 2 \\ -1 \\ 1 \end{bmatrix} =
\begin{bmatrix} -3 \\ -2 \\ 5 \end{bmatrix}.
$$

Hence $x_1 = -3$, $x_2 = -2$, and $x_3 = 5$.

NOTE The method of inverses cannot be used to solve the linear system $A\mathbf{x} = \mathbf{b}$ if A is not square or if A is not invertible.

EXAMPLE 2.33   We wish to solve a number of linear systems of the form $A\mathbf{x} = \mathbf{b}$, all of which have the same fourth-order nonsingular coefficient matrix. If we know that

$$A^{-1} = \begin{bmatrix} 1 & 0 & -1 & 0 \\ 2 & 0 & 0 & 1 \\ 0 & -1 & -2 & 0 \\ 0 & 0 & 1 & 1 \end{bmatrix},$$

give a formula for the solution vector $\mathbf{x}$ in terms of the components of the right-side vector $\mathbf{b}$.

Solution   We have $\mathbf{x} = A^{-1}\mathbf{b}$, so

$$\mathbf{x} = \begin{bmatrix} 1 & 0 & -1 & 0 \\ 2 & 0 & 0 & 1 \\ 0 & -1 & -2 & 0 \\ 0 & 0 & 1 & 1 \end{bmatrix} \begin{bmatrix} b_1 \\ b_2 \\ b_3 \\ b_4 \end{bmatrix} = \begin{bmatrix} b_1 - b_3 \\ 2b_1 + b_4 \\ -b_2 - 2b_3 \\ b_3 + b_4 \end{bmatrix}.$$

Thus $\mathbf{x} = (b_1 - b_3, 2b_1 + b_4, -b_2 - 2b_3, b_3 + b_4)$.

 COMPUTATIONAL NOTE   The method of inverses is not as efficient as Gauss–Jordan elimination for solving linear systems. It requires more operations—additions, subtractions, multiplications, and divisions—to complete the solution process. However, as you can see from these examples, if the inverse is somehow known beforehand, it is an extremely easy task to find the solution of the system by the method of inverses.

2.4 EXERCISES

In Exercises 1–4 write the given linear system in the matrix form $A\mathbf{x} = \mathbf{b}$.

1. $x - y + 3z = 1$
$\quad x \quad\;\; - z = 0$
$-2x + y \quad\quad = -1$

2. $2x_1 - 3x_2 \quad\quad + x_4 = 6$
$\quad\quad x_2 - x_3 + 3x_4 = 0$
$\quad -x_1 \quad\quad\quad + x_4 = 0$

3. $x = y - z$
$\quad y = x - z$

4. $x_1 = 2$
$\quad x_2 = 3 - x_1$
$\quad x_1 + x_2 + 1 = 4$

In Exercises 5–8 verify that matrices A and B are inverses of each other.

5. $A = \begin{bmatrix} 1 & 2 \\ 3 & 5 \end{bmatrix}, \quad B = \begin{bmatrix} -5 & 2 \\ 3 & -1 \end{bmatrix}$

6. $A = \begin{bmatrix} 1 & 0 \\ 4 & 2 \end{bmatrix}, \quad B = \begin{bmatrix} 1 & 0 \\ -2 & \frac{1}{2} \end{bmatrix}$

7. $A = \begin{bmatrix} 1 & 2 & -11 \\ 0 & -1 & 4 \\ 0 & 0 & 1 \end{bmatrix}$, $B = \begin{bmatrix} 1 & 2 & 4 \\ 0 & -1 & 4 \\ 0 & 0 & 1 \end{bmatrix}$

8. $A = \begin{bmatrix} 1 & 0 & -1 \\ 3 & -1 & 0 \\ 0 & -1 & 1 \end{bmatrix}$, $B = \frac{1}{2}\begin{bmatrix} -1 & 1 & -1 \\ -3 & 1 & -3 \\ -3 & 1 & -1 \end{bmatrix}$

In Exercises 9–16 find the inverse, if it exists, of the given matrix. If the matrix has no inverse, state this fact.

9. $\begin{bmatrix} 2 & 0 \\ -3 & 1 \end{bmatrix}$

10. $\begin{bmatrix} -1 & 2 \\ 3 & -4 \end{bmatrix}$

11. $\begin{bmatrix} 1 & 2 & -1 \\ 0 & 1 & 2 \\ 0 & 0 & 1 \end{bmatrix}$

12. $\begin{bmatrix} 1 & 0 & -2 \\ 2 & -1 & 0 \\ 1 & 0 & -2 \end{bmatrix}$

13. $\begin{bmatrix} 1 & -1 & 0 \\ -1 & 1 & 1 \\ 0 & -1 & 1 \end{bmatrix}$

14. $\begin{bmatrix} 1 & 0 & 0 \\ -1 & 2 & 0 \\ -2 & -1 & 1 \end{bmatrix}$

15. $\begin{bmatrix} 1 & 0 & 1 & 0 \\ 0 & 1 & 0 & 1 \\ 0 & 0 & 1 & 1 \\ 1 & 1 & 0 & 0 \end{bmatrix}$

16. $\begin{bmatrix} 1 & -1 & 0 & 0 \\ 0 & 1 & -1 & 0 \\ 0 & 0 & 1 & -1 \\ 1 & 0 & 0 & 1 \end{bmatrix}$

In Exercises 17 and 18 find the value of c for which A has no inverse.

17. $A = \begin{bmatrix} 1 & -2 \\ 3 & c \end{bmatrix}$

18. $A = \begin{bmatrix} 1 & -1 & c \\ 0 & 2 & 1 \\ -1 & 0 & 1 \end{bmatrix}$

In Exercises 19–22 solve the linear system $A\mathbf{x} = \mathbf{b}$ when the inverse, A^{-1}, and $\mathbf{b}$ are as given.

19. $A^{-1} = \begin{bmatrix} 1 & 2 \\ -2 & 3 \end{bmatrix}$, $\mathbf{b} = \begin{bmatrix} 1 \\ -2 \end{bmatrix}$

20. $A^{-1} = \begin{bmatrix} 0 & 1 \\ 1 & 0 \end{bmatrix}$, $\mathbf{b} = \begin{bmatrix} -1 \\ 1 \end{bmatrix}$

21. $A^{-1} = \begin{bmatrix} -1 & \frac{1}{2} & 0 \\ \frac{1}{2} & -1 & \frac{1}{2} \\ 0 & \frac{1}{2} & -1 \end{bmatrix}$, $\mathbf{b} = \begin{bmatrix} -3 \\ 1 \\ 2 \end{bmatrix}$

22. $A^{-1} = \begin{bmatrix} 1 & -1 & 3 & 1 \\ 0 & 1 & 0 & 2 \\ -2 & -1 & -3 & -1 \\ 1 & -1 & 0 & 0 \end{bmatrix}$, $\mathbf{b} = \begin{bmatrix} 1 \\ 2 \\ 3 \\ 4 \end{bmatrix}$.

In Exercises 23–28 use the method of inverses, if possible, to solve the following systems

23. $2x - 3y = 3$
$3x - 5y = 1$

24. $Ax = b$, where $A = \begin{bmatrix} 1 & -3 \\ 2 & 4 \end{bmatrix}$, $x = \begin{bmatrix} x_1 \\ x_2 \end{bmatrix}$, and $b = \begin{bmatrix} 3 \\ -1 \end{bmatrix}$

25. $x - 3y + 4z = 4$
$2x + 2y \quad\quad = 0$
$\quad\quad y - 2z = 2$

26. $-x_1 + x_3 = 1$
$x_1 + x_2 = -1$
$x_2 - x_3 = 0$

27. $Ax = b$, where $A = \begin{bmatrix} 1 & -1 & 2 \\ 2 & -1 & 1 \\ 0 & -1 & 3 \end{bmatrix}$, $x = \begin{bmatrix} x_1 \\ x_2 \\ x_3 \end{bmatrix}$, $b = \begin{bmatrix} 1 \\ 0 \\ 0 \end{bmatrix}$

28. $x - 3y + z = 2$
$2x + y - 2z = 7$

29. Show that the inverse of the 2×2 matrix

$$\begin{bmatrix} a & b \\ c & d \end{bmatrix} \text{ is the matrix } \frac{1}{ad - bc}\begin{bmatrix} d & -b \\ -c & a \end{bmatrix}.$$

provided that $ad - bc \neq 0$.

In Exercises 30 and 31 use the result of Exercise 29 to find the inverses quickly.

30. $\begin{bmatrix} 3 & -2 \\ 1 & 4 \end{bmatrix}$

31. $\begin{bmatrix} 1 & -1 \\ 0 & 4 \end{bmatrix}$

32. Show that if A and B have inverses and are of the same order, then AB has an inverse and in fact $(AB)^{-1} = B^{-1}A^{-1}$.

33. Show that the inverse of the nth-order diagonal matrix

$$D = \begin{bmatrix} a_1 & 0 & \cdots & 0 \\ 0 & a_2 & \cdots & 0 \\ \vdots & \vdots & \ddots & \vdots \\ 0 & 0 & \cdots & a_n \end{bmatrix} \text{ is } D^{-1} = \begin{bmatrix} 1/a_1 & 0 & \cdots & 0 \\ 0 & 1/a_2 & \cdots & 0 \\ \vdots & \vdots & \ddots & \vdots \\ 0 & 0 & \cdots & 1/a_n \end{bmatrix},$$

provided that the main diagonal entries a_1 are all nonzero.

In Exercises 34 and 35 use computer programs for finding inverses and multiplying matrices to solve the given system by the method of inverses.

34. $Ax = b$, where $A = \begin{bmatrix} 1 & 0 & 0 & 0 & 0 \\ 0 & 0 & 0 & 1 & 0 \\ 0 & 1 & 0 & 0 & 0 \\ 0 & 0 & 0 & 0 & 1 \\ 0 & 0 & 1 & 0 & 0 \end{bmatrix}$, $b = \begin{bmatrix} 1 \\ 0 \\ 0 \\ 0 \\ 0 \end{bmatrix}$

35. $A\mathbf{x} = \mathbf{b}$, where A is the square matrix of order 10 with 1's on and above the main diagonal and 0's below it, and $\mathbf{b}$ is the 10 vector, all of whose entries are 1.

NOTE   Exercises 34 and 35 can both be easily solved by hand, using Gauss–Jordan elimination rather than the method of inverses.

···

2.5   THEORY OF LINEAR SYSTEMS

In this section we investigate the nature of solutions of systems of linear equations rather than techniques for finding them. In doing so, we introduce an important special case: the *homogeneous* linear system.

We will continue to write a linear system in the matrix form, $A\mathbf{x} = \mathbf{b}$, developed in the last section. In such an equation, A will be an $m \times n$ matrix, $\mathbf{x}$ an n-vector, and $\mathbf{b}$ an m-vector. If A is square, of course $m = n$.

THEOREM 1   Every system of linear equations $A\mathbf{x} = \mathbf{b}$ has no solution, exactly one solution, or infinitely many solutions.

Proof   We have already seen examples of each of the three cases, so we need only show that there are no other possibilities. Assume that $\mathbf{y}$ and $\mathbf{z}$ are solutions with $\mathbf{y} \neq \mathbf{z}$. Since these are vectors in $\mathbf{R}^n$, the line they determine is given by the two-point form (Section 1.4),

$$\mathbf{x}(t) = (1 - t)\mathbf{y} + t\mathbf{z}$$
$$= \mathbf{y} + t(\mathbf{z} - \mathbf{y}).$$

Each point $\mathbf{x}(t)$ on this line is also a solution of $A\mathbf{x} = \mathbf{b}$, as seen by

$$A(\mathbf{x}(t)) = A(\mathbf{y} + t(\mathbf{z} - \mathbf{y}))$$
$$= A\mathbf{y} + A(t(\mathbf{z} - \mathbf{y}))$$
$$= \mathbf{b} + tA(\mathbf{z} - \mathbf{y})$$
$$= \mathbf{b} + t(A\mathbf{z} - A\mathbf{y})$$
$$= \mathbf{b} + t(\mathbf{b} - \mathbf{b}) = \mathbf{b}.$$

Consequently, if there is more than one solution to $A\mathbf{x} = \mathbf{b}$, then there are infinitely many.

DEFINITION

The linear system $A\mathbf{x} = \mathbf{b}$ is said to be **homogeneous** if $\mathbf{b} = \mathbf{0}$, the vector consisting solely of zeros. If $\mathbf{b} \neq \mathbf{0}$, then the system is said to be **nonhomogeneous**.

For example, the linear systems

$$\begin{bmatrix} 1 & -1 & 2 & 0 \\ 3 & 0 & 1 & 5 \end{bmatrix} \begin{bmatrix} x_1 \\ x_2 \\ x_3 \\ x_4 \end{bmatrix} = \begin{bmatrix} 0 \\ 0 \end{bmatrix}$$

and

$$\begin{array}{rcrcrcl} -x & - & 3y & + & z & = & 0 \\ 2x & & & - & 2z & = & 0 \\ & & 3y & + & z & = & 0 \end{array}$$

are *homogeneous*, but

$$\begin{array}{rcrcrcl} x_1 & - & 2x_2 & + & x_3 & = & 0 \\ 3x_1 & + & x_2 & - & x_3 & = & 1 \end{array}$$

is *nonhomogeneous*.

NOTE   The zero vector, **0**, is a solution of every homogeneous linear system $A\mathbf{x} = \mathbf{0}$, since $A\mathbf{0} = \mathbf{0}$. The solution $\mathbf{x} = \mathbf{0}$ is called the **trivial solution** of the homogeneous system. Any other solution is called a **nontrivial** one.

···

THEOREM 2   Every homogeneous linear system $A\mathbf{x} = \mathbf{0}$ has either exactly one solution or infinitely many solutions.

Proof   Since every homogeneous linear system has at least one solution (the trivial one), the result follows immediately from Theorem 1.

We can probe a little more deeply into this subject by considering the structure of the row-reduced echelon form for the augmented matrix of the given system. The key to this line of investigation is the notion of *rank*.

DEFINITION

The **rank** of a matrix A is the number of nonzero rows in the row-reduced echelon form of A.

NOTE   The rank of a matrix is *well defined* due to the uniqueness of the row-reduced echelon form (Theorem 3 of Section 2.2). No matter what sequence of elementary row operations is performed to put the given matrix in row-reduced echelon form, there will always be the same number of nonzero rows.

···

EXAMPLE 2.34   Find the rank of

$$A = \begin{bmatrix} 1 & -1 & 2 & 1 \\ 0 & 1 & 1 & -2 \\ 1 & -3 & 0 & 5 \end{bmatrix}.$$

Solution   The row-reduced echelon form of A is

$$\begin{bmatrix} 1 & 0 & 3 & -1 \\ 0 & 1 & 1 & -2 \\ 0 & 0 & 0 & 0 \end{bmatrix},$$

which has two nonzero rows. Hence the rank of A is 2.

NOTE   Given the linear system $A\mathbf{x} = \mathbf{b}$, the rank of the coefficient matrix A is always less than or equal to the rank of the augmented matrix for the system $[A \mid \mathbf{b}]$. This follows from the fact that, when the latter is transformed into row-reduced echelon form, its first n columns are the row-reduced form of the former. It is consequently impossible for the row-reduced echelon form of A to have more nonzero rows than that of $[A \mid \mathbf{b}]$. In other words, it is impossible for the rank of the coefficient matrix to be greater than that of the augmented matrix.

THEOREM 3   For the linear system $A\mathbf{x} = \mathbf{b}$, where A is an $m \times n$ matrix, let p be the rank of A and let q be the rank of $[A \mid \mathbf{b}]$. Then $A\mathbf{x} = \mathbf{b}$ has:

    a. no solution if $p < q$;
    b. a unique solution if $p = q = n$;
    c. infinitely many solutions if $p = q$ and $p < n$.

Proof

    3a. If p is less than q, then there must be one row in the row-reduced echelon form of $[A \mid \mathbf{b}]$ that reads

$$[0 \quad 0 \quad \cdots \quad 0 \mid 1].$$

    This corresponds to the equation $0 = 1$ and hence there is no solution of the system.

    3b. If p, q, and n are equal, then after discarding the zero rows of the row-reduced echelon form, we obtain the equivalent linear system

$$
\begin{aligned}
x_1 \qquad\qquad\qquad &= d_1 \\
x_2 \qquad\qquad &= d_2 \\
\ddots\ \qquad &\ \ \vdots \\
x_n &= d_n,
\end{aligned}
$$

which has the unique solution $(d_1, d_2, \ldots, d_n)$.

3c. If $p = q$ and $p < n$, then, after discarding the zero rows of the row-reduced echelon form, we obtain the equivalent linear system (with $r = p = q$):

$$
\begin{aligned}
x_1 \qquad\qquad &+ c_{1,r+1}x_{r+1} + \cdots + c_{1n}x_n = d_1 \\
x_2 \qquad &+ c_{2,r+1}x_{r+1} + \cdots + c_{2n}x_n = d_2 \\
\ddots\ \qquad &\qquad\ \vdots \qquad\qquad\quad \vdots \qquad \vdots \\
x_r &+ c_{r,r+1}x_{r+1} + \cdots + c_{rn}x_n = d_r.
\end{aligned}
$$

To see that this system has infinitely many solutions, take, for example, $x_{r+1} = t,\ x_{r+2} = \cdots = x_n = 0$ to get the solutions $(d_1 - c_{1,r+1}t, \ldots, d_r - c_{r,r+1}t, t, 0, 0, \ldots, 0)$.

For the homogeneous linear system $A\mathbf{x} = \mathbf{0}$, the augmented matrix is $[A \mid \mathbf{0}]$. Since the zeros in the last column of this matrix remain unchanged when it is transformed into row-reduced echelon form, the rank of A is always equal to the rank of $[A \mid \mathbf{0}]$. Thus the following corollaries are immediate consequences of Theorem 3. We leave their proofs as exercises.

COROLLARY I   The homogeneous linear system $A\mathbf{x} = \mathbf{0}$, where A is an $m \times n$ matrix, has:

a. a unique solution (the trivial one) if the rank of A is equal to n;

b. infinitely many solutions if the rank of A is less than n.

COROLLARY 2   A homogeneous linear system with more unknowns than equations has infinitely many solutions.

EXAMPLE 2.35   By inspection of the following systems, what can we say concerning the number of solutions they may have?

$$
\textbf{a.}\quad
\begin{aligned}
x_1 + 2x_2 - \ x_3 &= 0 \\
2x_1 - \ x_2 + 3x_3 &= 0
\end{aligned}
$$

b. $2x_1 + x_2 + x_3 = 1$
$\quad -x_1 - x_2 + 2x_3 = 0$

c. $x_1 \qquad\qquad + 2x_4 = \quad 1$
$\qquad x_2 \qquad - x_4 = -1$
$\qquad\qquad x_3 + x_4 = \quad 0.$

Solution

System (a) is a homogeneous one with more unknowns than equations. By Corollary 2 of Theorem 3 it has infinitely many solutions. System (b) is a nonhomogeneous one with more unknowns than equations. By Theorem 3 it has either no solution or infinitely many solutions. System (c) is in row-reduced echelon form. The corresponding augmented matrix is

$$\left[\begin{array}{cccc|c} 1 & 0 & 0 & 2 & 1 \\ 0 & 1 & 0 & -1 & -1 \\ 0 & 0 & 1 & 1 & 0 \end{array}\right].$$

We see that the ranks of the coefficient and augmented matrices are both 3, whereas the number of variables n is 4. Hence by Theorem 3 there are infinitely many solutions.

In Example 2.35 we can solve the system of part (c) to obtain the one-parameter family of solutions $(1 - 2t, -1 + t, -t, t)$. The next theorem tells us that we could have deduced the number of parameters in the solution from the ranks of the coefficient and augmented matrices.

THEOREM 4   Consider the system $A\mathbf{x} = \mathbf{b}$, where A is an $m \times n$ matrix with rank equal to $r < n$. If the augmented matrix $[A \mid \mathbf{b}]$ also has rank $= r$, then the system has an $(n - r)$-parameter family of solutions.

Proof   From Theorem 3 we know that $A\mathbf{x} = \mathbf{b}$ has infinitely many solutions. By discarding the zero rows of the row-reduced echelon form of $[A \mid \mathbf{b}]$ and assuming that the first r columns of A are the ones transformed into those that contain the leading ones, we obtain the equivalent system

$$x_1 \qquad\qquad + c_{1,r+1}x_{r+1} + \cdots + c_{1n}x_n = d_1$$
$$x_2 \qquad\quad + c_{2,r+1}x_{r+1} + \cdots + c_{2n}x_n = d_2$$
$$\ddots \qquad\qquad \vdots \qquad\qquad \vdots \quad \vdots$$
$$x_r + c_{r,r+1}x_{r+1} + \cdots + c_{rn}x_n = d_r.$$

Since the variables $x_{r+1}, \ldots, x_n$ can be taken as parameters, we obtain an $(n - r)$-parameter family of solutions.

COMPUTATIONAL NOTE   You should be wary of using a computer to determine the rank of a matrix—you may get an erroneous answer. Because a computer has to round off nonterminating decimals (such as the fraction $\frac{1}{3}$), small errors may be introduced in transforming a given matrix into row-reduced echelon form. These small errors could possibly result in nonzero entries occurring in the latter where in fact there should be zeros, or vice-versa. It is therefore possible for the computed rank to be either smaller or larger than it actually is.

··

We now restrict ourselves to linear systems with the same number of equations as unknowns. Since in such cases the coefficient matrix is square, we call these **square linear systems.**

THEOREM 5 If A is an $n \times n$ matrix, the following conditions are equivalent:

    a. A is invertible.

    b. $A\mathbf{x} = \mathbf{b}$ has a unique solution for any $\mathbf{b}$.

    c. $A\mathbf{x} = \mathbf{0}$ has only the trivial solution.

    d. The rank of A is n.

    e. The row-reduced echelon form of A is I, the identity matrix.

Proof We will show the equivalence of these statements by showing that (a) implies (b), (b) implies (c), (c) implies (d), (d) implies (e), and (e) implies (a). Thus any two statements will imply each other.

    (a) implies (b). We first note that the vector $A^{-1}\mathbf{b}$ is a solution, since $A(A^{-1}\mathbf{b}) = (AA^{-1})\mathbf{b} = I\mathbf{b} = \mathbf{b}$, as desired. Now suppose $\mathbf{y}$ is also a solution. Then $A\mathbf{y} = \mathbf{b}$, so $A^{-1}(A\mathbf{y}) = A^{-1}\mathbf{b}$, or $(A^{-1}A)\mathbf{y} = A^{-1}\mathbf{b}$; but this yields $\mathbf{y} = A^{-1}\mathbf{b}$ as well. Thus the solution is unique.

    (b) implies (c). Since $A\mathbf{x} = \mathbf{0}$ is a special case of $A\mathbf{x} = \mathbf{b}$, the former has a unique solution. Since $\mathbf{0}$ is a solution, it is the only solution.

    (c) implies (d). This is Corollary 1 of Theorem 3.

    (d) implies (e). Since A has n rows and its rank is n, its row-reduced echelon form has n nonzero rows, each with a leading one and no zero rows. However, since the matrix has n columns as well, the only nonzero element in column j is a 1 in the jth row. Thus the row-reduced echelon form is the identity matrix.

    (e) implies (a). This follows immediately from the procedure for matrix inverses presented in Section 2.4.

The next theorem gives a relationship between the solution of a homogeneous linear system, $A\mathbf{x} = \mathbf{0}$, and that of the corresponding nonhomogeneous system, $A\mathbf{x} = \mathbf{b}$. This result is of fundamental importance in the study of linear differential equations (see Section 8.8) and also appears in a variety of other applications.

THEOREM 6 Let A be an $m \times n$ matrix and let $\mathbf{b}$ be an n-vector. Suppose $\mathbf{x}_0$ is one *particular* solution of $A\mathbf{x} = \mathbf{b}$ and S is the set of *all* solutions to $A\mathbf{x} = \mathbf{0}$. Then all solutions of $A\mathbf{x} = \mathbf{b}$ have the form $\mathbf{x} = \mathbf{x}_0 + \mathbf{s}$, where $\mathbf{s}$ is in the set S.

Proof First we show that if a vector is of the form $\mathbf{x}_0 + \mathbf{s}$, then it must be a solution of $A\mathbf{x} = \mathbf{b}$. We have

$$A(\mathbf{x}_0 + \mathbf{s}) = A\mathbf{x}_0 + A\mathbf{s} = \mathbf{b} + \mathbf{0} = \mathbf{b},$$

so $\mathbf{x}_0 + \mathbf{s}$ satisfies the equation.

Conversely, if some vector (say, $\mathbf{y}$) is a solution of $A\mathbf{x} = \mathbf{b}$, then it must be equal to $\mathbf{x}_0 + \mathbf{s}$, for some $\mathbf{s}$ in S. To see this, for the given $\mathbf{y}$ define $\mathbf{s} = \mathbf{y} - \mathbf{x}_0$. Then $\mathbf{y} = \mathbf{x}_0 + \mathbf{s}$ and $\mathbf{s}$ is in S, as desired, because

$$A\mathbf{s} = A(\mathbf{y} - \mathbf{x}_0) = A\mathbf{y} - A\mathbf{x}_0 = \mathbf{b} - \mathbf{b} = \mathbf{0}.$$

EXAMPLE 2.36   Consider the linear system of Example 2.35, part (c):

$$
\begin{aligned}
x_1 \qquad\qquad\ + 2x_4 &= \ \ 1 \\
x_2 \qquad\ - \ \ x_4 &= -1 \\
x_3 + \ \ x_4 &= \ \ 0
\end{aligned}
$$

You can easily verify that $\mathbf{x}_0 = (1, -1, 0, 0)$ is a particular solution of this system. Moreover, the corresponding homogeneous system

$$
\begin{aligned}
x_1 \qquad\qquad\ + 2x_4 &= 0 \\
x_2 \qquad\ - \ \ x_4 &= 0 \\
x_3 + \ \ x_4 &= 0
\end{aligned}
$$

has $(-2t, t, -t, t)$ as a one-parameter family of solutions. Thus, by Theorem 6, the general solution to the nonhomogeneous system is the set of all vectors of the form:

$$(1, -1, 0, 0) + (-2t, t, -t, t) = (1 - 2t, -1 + t, -t, t).$$

2.5 EXERCISES

In Exercises 1 and 2 determine which systems are homogeneous.

1. $x_1 = 2x_2 + \ x_3$
$\quad\ x_2 = \ \ x_3 - \ x_4$
$\quad\ x_3 = \ \ x_1 - 4x_4$

2. $x + y + 3z = 0$
$\quad x - y + 2 \ \ = 0$
$\qquad\quad\ z - 3 \ = 0$

In Exercises 3–6 find the rank of the given matrix.

3. $\begin{bmatrix} 1 & 2 & -1 \\ 3 & -6 & 2 \end{bmatrix}$

4. $\begin{bmatrix} 1 & 0 & -1 \\ 3 & 1 & 1 \\ -1 & -1 & -3 \end{bmatrix}$

5. $\begin{bmatrix} 1 & 0 & 1 \\ -2 & 1 & 1 \\ 1 & 1 & 2 \end{bmatrix}$

6. $\begin{bmatrix} 2 & 4 & 0 & -2 \\ 0 & 1 & -1 & 0 \\ 1 & 2 & -1 & -1 \end{bmatrix}$

In Exercises 7 and 8 consider the linear system $A\mathbf{x} = \mathbf{b}$, where

$$A = \begin{bmatrix} a & b \\ c & d \end{bmatrix}, \qquad \mathbf{b} = \begin{bmatrix} e \\ f \end{bmatrix}, \qquad \text{and} \qquad A \neq 0.$$

7. Determine conditions on the constants a, b, c, d, e, and f so that
   (a) the rank of A is 2,
   (b) the rank of A is 1, but the rank of $[A \mid \mathbf{b}]$ is 2,
   (c) the rank of A and the rank of $[A \mid \mathbf{b}]$ are each 1.

8. Using the results of Exercise 7, give conditions on a, b, c, d, e, and f so that the linear system has
   (a) no solution,
   (b) one solution,
   (c) infinitely many solutions.

In Exercises 9–14 determine whether the given statement is true or false.

9. A linear system with more unknowns than equations always has infinitely many solutions.

10. A homogeneous linear system with the same number of equations as unknowns always has a unique solution.

11. If a linear system has at least two solutions, it must have infinitely many.

12. If a linear system has no solution, the rank of the coefficient matrix must be less than the number of equations.

13. A homogeneous linear system with two equations and four unknowns must have a two-parameter family of solutions.

14. If a linear system has the same number of equations as unknowns and the coefficient matrix is invertible, the system has exactly one solution.

In Exercises 15–18 use the results of Section 2.5 to say as much as possible concerning the number of solutions of the given linear system without actually solving it.

15. $\begin{aligned} 2x + 3y - z &= 0 \\ y + 2z - 3w &= 0 \\ 2x - 4z + w &= 0 \end{aligned}$

16. $\begin{aligned} 2x + 3y - z &= 1 \\ y + 2z - 3w &= 0 \\ 2x - 4z + w &= 0 \end{aligned}$

17. $\begin{aligned} 2x_1 + x_2 - x_3 &= 0 \\ 3x_1 - 2x_2 + x_3 &= 0 \\ x_1 - x_2 + x_3 &= 0 \end{aligned}$

18. $\begin{aligned} 2x - 3y &= 2 \\ -x + 4y &= 3 \\ 4x + 3y &= 7 \end{aligned}$

19. Let A, X, and B be $n \times n$ matrices with A nonsingular. If $AX = B$, show that $(AXA^{-1})A = B$.

20. Let A and B be invertible matrices with $AB = I$. Show that $(A^T)^{-1} = B^T$.

21. Suppose that the augmented matrix of a linear system is given by

$$\left[\begin{array}{ccc|c} 1 & 0 & 0 & 2 \\ 0 & 1 & 0 & 3 \\ 0 & 0 & x & y \end{array}\right].$$

For what values of x and y is there
(a) no solution?
(b) exactly one solution?
(c) infinitely many solutions?

22–25. Use a computer program to transform a matrix to row-reduced echelon form to find the rank of the matrices in Exercises 3–6. Do your results agree with the ones obtained by hand?

26. Find the rank of the matrix

$$A = \begin{bmatrix} 1 & \frac{1}{2} & \frac{1}{3} & \frac{1}{4} & \frac{1}{5} \\ \frac{1}{2} & \frac{1}{3} & \frac{1}{4} & \frac{1}{5} & \frac{1}{6} \\ \frac{1}{3} & \frac{1}{4} & \frac{1}{5} & \frac{1}{6} & \frac{1}{7} \\ \frac{1}{4} & \frac{1}{5} & \frac{1}{6} & \frac{1}{7} & \frac{1}{8} \\ \frac{1}{5} & \frac{1}{6} & \frac{1}{7} & \frac{1}{8} & \frac{1}{9} \end{bmatrix},$$

using a computer program to transform a matrix to row-reduced echelon form. (The rank of A is 5, but your computed answer may differ. See the Computational Note after Theorem 4.)

27. Prove Corollaries 1 and 2 of Theorem 3.

CHAPTER 2 REVIEW

In this chapter *matrices* were introduced to provide an efficient means of solving systems of linear equations. Then operations on matrices (such as *addition* and *multiplication*) were defined, and the concept of the *matrix inverse* was discussed. Finally, theorems were given to provide additional insight into the relationship between matrices and solutions of linear systems.
You should be able to:

1. identify systems of linear equations, as well as their augmented and coefficient matrices, and write them in matrix form;

2. solve linear systems using Gauss–Jordan and Gaussian elimination and the method of inverses;

3. determine the sum, difference, and product of matrices;

4. determine scalar multiples and powers of matrices;

5. determine the negative, transpose, and inverse of a matrix;

6. determine the rank of a matrix;

7. deduce, in some cases, information about solutions of linear systems without actually solving them.

KEY WORDS

linear equation

solution to a linear system

standard form

equivalent linear systems

entry, row, and column of a matrix

elementary row operations on a matrix

row-reduced echelon form and leading 1

echelon form of a matrix

back substitution

column and row vector

equal matrices

scalar multiple of a matrix

product of matrices

square matrices and order and main diagonal of a square matrix

diagonal matrix

transpose of a matrix

coefficient matrix of a linear system

noninvertible (or singular) matrices

method of inverses to solve a linear system

trivial and nontrivial solutions

square linear system

system of linear equations (or linear system)

inconsistent and consistent linear systems

elementary operations on a linear system

matrix (and matrices)

augmented matrix for a linear system

row equivalent matrices

Gauss–Jordan and Gaussian elimination

leading entry

$m \times n$ matrix (dimensions of a matrix)

corresponding entries of a matrix

sum and difference of matrices

negative of a matrix

zero matrix

powers of a matrix

identity matrix

coefficients and right-side constants

invertible (or nonsingular) matrices

inverse of a matrix

homogeneous and nonhomogeneous linear systems

rank of a matrix

KEY THEOREMS

Theorem   If two augmented matrices are row equivalent, the corresponding linear systems have the same solutions.

Theorem   Every matrix can be transformed by a finite sequence of elementary row operations into a unique row-reduced echelon form.

Theorem   Every system of linear equations has either no solution, exactly one solution, or infinitely many solutions.

Theorem   Let $A\mathbf{x} = \mathbf{b}$ be a linear system, where A is an $m \times n$ matrix, and let p and q be the ranks of the coefficient and augmented matrices, respectively. Then, the system has no solution if $q > p$, a unique solution if $q = p = n$, and infinitely many solutions if $q = p < n$.

Theorem   If A is a square matrix of order n, the following conditions are equivalent:

   a. A is invertible;

   b. $A\mathbf{x} = \mathbf{b}$ has a unique solution;

   c. $A\mathbf{x} = \mathbf{0}$ has only the trivial solution;

   d.   The rank of A is n;

   e.   The row-reduced echelon form of A is I.

Theorem   All solutions to a nonhomogeneous linear system, $A\mathbf{x} = \mathbf{b}$, are of the form $\mathbf{x} = \mathbf{x}_0 + \mathbf{s}$, where $\mathbf{x}_0$ is a particular solution of $A\mathbf{x} = \mathbf{b}$ and $\mathbf{s}$ is the general solution of the corresponding homogeneous system, $A\mathbf{x} = \mathbf{0}$.

SELF-TEST

Answers appear at the end of this chapter.

1. Determine whether $(t, 2t, 3t)$, where t is an arbitrary real number, is a solution of the linear system

$$x + y - z = 0$$
$$3x - 2y + z = 1. \qquad \text{[Section 2.1]}$$

In Problems 2 and 3, solve the given linear system by Gauss–Jordan elimination.

2. $x + 2y + z = 1$
   $x + 2y + 3z = 5$
   $\phantom{x + {}}2y + 2z = 2$

3. $x_1 - 2x_2 + x_3 - x_4 = 1$
   $2x_1 - 3x_2 + 2x_4 = 2$

   [Section 2.2]

4. Solve the linear system of Problem 2 by Gaussian elimination.   [Section 2.2]

In Problems 5–7, let

$$A = \begin{bmatrix} 1 & 0 \\ -1 & 2 \end{bmatrix} \quad \text{and} \quad B = \begin{bmatrix} 2 & 0 & 1 \\ 3 & -1 & 0 \end{bmatrix}.$$

Perform the indicated operations.

5. $3A - I$ 6. AB 7. B^T [Section 2.3]

8. Find the inverse of the matrix

$$C = \begin{bmatrix} 1 & 2 & 3 \\ 0 & 1 & 4 \\ 0 & 0 & 1 \end{bmatrix}.$$

                                                                      [Section 2.4]

9. Solve the following linear system by the method of inverses.

$$2x - y = -1$$
$$-x + y = 2$$ [Section 2.4]

10. Suppose that the row-reduced echelon form of a linear system $Ax = b$ has the following form, where a and b are arbitrary constants.

$$\begin{bmatrix} 1 & 0 & 0 & | & a \\ 0 & 1 & b & | & 0 \\ 0 & 0 & 0 & | & 0 \end{bmatrix}.$$

(a) What is the rank of A?

(b) How many solutions does $Ax = b$ have?

(c) Is A invertible?   [Section 2.5]

REVIEW EXERCISES

In Exercises 1–8, solve the given linear system by the methods of Gauss–Jordan elimination and Gaussian elimination.

1. $x - 2y = -1$
    $-2x + 4y = 2$

2. $-x + y - z = 1$
    $x \quad\quad + z = 3$

3. $x_1 - 2x_2 + x_3 = 0$
    $-x_1 + 2x_2 - x_3 = 0$

4. $x_1 + 2x_2 - 3x_3 = -4$
    $2x_1 - x_2 + x_3 = 5$
    $2x_2 + 3x_3 = 1$

5. $u + 2v - 3w = 4$
    $- v + w = -1$
    $2u \quad\quad - w = 3$

6. $u + v - w = 1$
    $v + 2w = 0$
    $u + 4v + 5w = 2$

7. $x + y + 2z - w = 1$
    $-x + 2y + z + w = 2$
    $-x \quad\quad - z + w = 3$

8. $x - y + z - w = 0$
    $y - z + 2w = 0$
    $2x - y + z \quad\quad = 0$

In Exercises 9–20 let

$$A = \begin{bmatrix} 1 & 0 & 1 \\ 0 & 1 & 3 \\ 0 & 0 & 3 \end{bmatrix} \quad \text{and} \quad B = \begin{bmatrix} -1 & 1 & 1 \\ 1 & 3 & 1 \\ 2 & 0 & -1 \end{bmatrix},$$

and determine each of the following:

9. $2A - 3B$

10. $2A + B - 3I$

11. AB

12. BA

13. A^{-1}

14. B^{-1}

15. $(A^T B)^{-1}$

16. $(A^T A)^{-1}$

17. rank of A

18. rank of B

19. $-A^2$

20. A^3

21–28. Write each of the linear systems in Exercises 1–8 in matrix form and solve it by the method of inverses, if possible. If the method of inverses cannot be used, explain why.

In Exercises 29–32 the given matrix is the augmented matrix of a linear system. Without solving the system, determine how many solutions it has. If there are infinitely many solutions, determine the number of parameters present.

29. $\begin{bmatrix} 1 & 0 & 0 & | & 0 \\ 0 & 1 & 0 & | & 0 \\ 0 & 0 & 1 & | & 0 \end{bmatrix}$

30. $\begin{bmatrix} 1 & 0 & 2 & 3 & | & 0 \\ 0 & 1 & -1 & 4 & | & 1 \end{bmatrix}$

31. $\begin{bmatrix} 1 & 2 & 3 & | & 0 \\ 4 & 5 & 6 & | & 0 \end{bmatrix}$

32. $\begin{bmatrix} 1 & 2 & 3 & | & 0 \\ 0 & 0 & 0 & | & 1 \\ 0 & 0 & 0 & | & 0 \end{bmatrix}$

ANSWERS TO SELF-TEST

1. No, $(t, 2t, 3t)$ does not satisfy the second equation.

2. $(1, -1, 2)$

3. $(1 + 3s - 7t, 2s - 4t, s, t)$

4. $(1, -1, 2)$

5. $\begin{bmatrix} 2 & 0 \\ -3 & 5 \end{bmatrix}$

6. $\begin{bmatrix} 2 & 0 & 1 \\ 4 & -2 & -1 \end{bmatrix}$

7. $\begin{bmatrix} 2 & 3 \\ 0 & -1 \\ 1 & 0 \end{bmatrix}$

8. $C^{-1} = \begin{bmatrix} 1 & -2 & 5 \\ 0 & 1 & -4 \\ 0 & 0 & 1 \end{bmatrix}$

9. $(1, 3)$

10. (a) 2

(b) Infinitely many solutions (Theorem 3c)

(c) No (Theorem 5)

3 DETERMINANTS

Associated with each square matrix is a number known as the *determinant* of the matrix. This number carries information about the matrix—most importantly, whether or not the matrix is invertible. When it is invertible, using determinants allows us to express the inverse in terms of the original matrix, which serves as an alternative to the inversion procedure described in Section 2.4. Similarly, the solution to a square system of linear equations can sometimes be expressed using determinants in a procedure known as *Cramer's rule*. Other uses of determinants will be encountered in later chapters.

3.1 DEFINITION OF DETERMINANT

In Section 1.2 we defined the determinant of a 2×2 matrix for the sole purpose of providing a simple means of remembering the definition of cross product. In this section we will extend the definition to square matrices of any order. We first give a convention that will make this description easier.

NOTATION For an $m \times n$ matrix A, the *submatrix* A_{ij} is obtained by deleting the ith row and jth column of A.

EXAMPLE 3.1 Find A_{23}, where

$$A = \begin{bmatrix} 1 & 1 & 0 & 2 \\ 2 & 1 & 1 & 1 \\ 3 & 0 & 0 & -1 \\ 1 & 1 & 2 & 1 \end{bmatrix}.$$

112

Solution   We mentally line out the second row and the third column,

$$
\begin{bmatrix}
1 & 1 & \vdots 0 & 2 \\
\text{--2----1----} & \vdots & \text{----1} \\
3 & 0 & \vdots 0 & -1 \\
1 & 1 & \vdots 2 & 1
\end{bmatrix},
$$

obtaining

$$
A_{23} = \begin{bmatrix}
1 & 1 & 2 \\
3 & 0 & -1 \\
1 & 1 & 1
\end{bmatrix}.
$$

DEFINITION   Let $A = [a_{ij}]$ be an $n \times n$ matrix. The **determinant** of A is given by

i. $\det A = a_{11}$ if $n = 1$,

ii. $\det A = a_{11}a_{22} - a_{12}a_{21}$ if $n = 2$,

iii. $\det A = a_{11} \det A_{11} - a_{12} \det A_{12} + \cdots + (-1)^{1+n}a_{1n} \det A_{1n}$

$$
= \sum_{j=1}^{n} (-1)^{1+j}a_{1j} \det A_{1j}
$$

if $n > 2$.

In words, case (iii) may be described as follows: Multiply each entry a_{1j} in the first row by $\det A_{1j}$, the determinant of the submatrix formed by deleting the row and the column that contain a_{1j}. Starting with a positive sign, alternate the signs of these terms—plus, minus, plus, minus, and so on. Finally, add all such terms.

NOTE   The definition of determinant is a *recursive definition*. An explicit rule is given for $n = 1$ and $n = 2$. For $n = 3$ each A_{1j} is of order 2, so $\det A_{1j}$ has been defined. For $n = 4$ each A_{1j} is of order 3, so $\det A_{1j}$ can be computed by the $n = 3$ case. This process may be continued for all positive integers n.

··

EXAMPLE 3.2   Compute $\det A$, where A is given by

$$
A = \begin{bmatrix}
2 & 1 & -3 \\
-3 & -2 & 0 \\
2 & 1 & 2
\end{bmatrix}.
$$

Solution

Since A is order 3, the definition of determinant yields

$$\det A = a_{11} \det A_{11} - a_{12} \det A_{12} + a_{13} \det A_{13}$$

$$= (2) \det \begin{bmatrix} -2 & 0 \\ 1 & 2 \end{bmatrix} - (1) \det \begin{bmatrix} -3 & 0 \\ 2 & 2 \end{bmatrix} + (-3) \det \begin{bmatrix} -3 & -2 \\ 2 & 1 \end{bmatrix}$$

$$= (2)[(-2)(2) - (1)(0)] - (1)[(-3)(2) - (2)(0)]$$
$$+ (-3)[(-3)(1) - (2)(-2)]$$

$$= 2(-4) - (1)(-6) + (-3)(1)$$

$$= -5.$$

EXAMPLE 3.3

Compute $\det A$, where

$$A = \begin{bmatrix} 2 & 1 & -3 & 1 \\ -3 & -2 & 0 & 2 \\ 2 & 1 & 0 & -1 \\ 1 & 0 & 1 & 2 \end{bmatrix}.$$

Solution

We have

$$\det A = (2) \det A_{11} - (1) \det A_{12} + (-3) \det A_{13} - (1) \det A_{14}.$$

This looks innocent enough, but note that each A_{1j} is of order 3, and computation of $\det A_{1j}$ requires repeating the process of Example 3.2. This must be done four times to finish the job. These values are as follows: $\det A_{11} = 0$, $\det A_{12} = 1$, $\det A_{13} = 2$, and $\det A_{14} = 1$. The final result is $\det A = -8$.

Fortunately, there are easier ways to compute $\det A$. In the remainder of this section and in Section 3.2, we will discuss some ways to simplify this process considerably.

DEFINITION

Let A be a square matrix of order n and let A_{ij} denote the submatrix (of order $n - 1$) obtained from A by deleting its ith row and jth column. Then

i. the **i, j-minor** of A is given by $\det(A_{ij})$, and
ii. the **i, j-cofactor** of A is given by $(-1)^{i+j} \det(A_{ij})$.

EXAMPLE 3.4

Find the 2,3-minor and the 2,3-cofactor of

$$A = \begin{bmatrix} 2 & -2 & 1 & 1 \\ 1 & 3 & 3 & 2 \\ 1 & 0 & 9 & 1 \\ 3 & 4 & 2 & 0 \end{bmatrix}.$$

Solution

The 2,3-minor is

$$\det A_{23} = \det \begin{bmatrix} 2 & -2 & 1 \\ 1 & 0 & 1 \\ 3 & 4 & 0 \end{bmatrix} = -10.$$

The 2,3-cofactor is

$$(-1)^{2+3} \det A_{23} = -\det \begin{bmatrix} 2 & -2 & 1 \\ 1 & 0 & 1 \\ 3 & 4 & 0 \end{bmatrix} = 10.$$

S P O T L I G H T

ORIGINS OF DETERMINANTS

You might be wondering why the determinant is defined as it is, as a certain combination of matrix entries. Historically, determinants arose from the study of systems of linear equations. To see how, consider the 2×2 linear system

$$ax + by = e$$
$$cx + dy = f.$$

Solving this system by Gauss–Jordan elimination, we see that if a unique solution exists, it is given by

$$x = \frac{ed - bf}{ad - bc}, \qquad y = \frac{af - ec}{ad - bc}.$$

These somewhat complicated formulas can be remembered more easily if we express them in terms of determinants:

$$x = \frac{\det \begin{bmatrix} e & b \\ f & d \end{bmatrix}}{\det \begin{bmatrix} a & b \\ c & d \end{bmatrix}}, \qquad y = \frac{\det \begin{bmatrix} a & e \\ c & f \end{bmatrix}}{\det \begin{bmatrix} a & b \\ c & d \end{bmatrix}}.$$

As you will see in Section 3.3, if an $n \times n$ linear system has a unique solution, that solution can be expressed in a similar way—as a ratio of determinants of matrices of order n. This technique has come to be known as *Cramer's rule* for solving systems of linear equations. The extension from the 2×2 case to general $n \times n$ linear systems helped motivate an intensive study of determinants in the mid-nineteenth century.

Note that Equation (iii) in the definition of the determinant may be written as

$$\det A = a_{11}[(-1)^{1+1} \det A_{11}] + a_{12}[(-1)^{1+2} \det A_{12}]$$
$$+ \cdots + a_{1n}[(-1)^{1+n} \det A_{1n}]$$

$$= \sum_{j=1}^{n} a_{1j}[(-1)^{1+j} \det A_{1j}].$$

In other words, $\det A$ is the sum of the products of each entry in the first row with the cofactor for that position. For this reason Equation (iii) of the definition is called **cofactor expansion** along the first row (or simply **expansion** along the first row).

The next theorem states that cofactor expansion along *any* row (or down any column) of a square matrix will also yield the determinant. The theorem will be assumed without proof.

THEOREM I Let A be a matrix of order n with $n \geq 2$. Let i be a fixed row number. Then

$$\det A = (-1)^{i+1} a_{i1} \det A_{i1} + (-1)^{i+2} a_{i2} \det A_{i2} + \cdots + (-1)^{i+n} a_{in} \det A_{in}$$

$$= \sum_{k=1}^{n} (-1)^{i+k} a_{ik} \det A_{ik}.$$

Let j be a fixed column number. Then

$$\det A = (-1)^{1+j} a_{1j} \det A_{1j} + (-1)^{2+j} a_{2j} \det A_{2j} + \cdots + (-1)^{n+j} a_{nj} \det A_{nj}$$

$$= \sum_{k=1}^{n} (-1)^{k+j} a_{kj} \det A_{kj}.$$

In computations, it is convenient to choose a row or column with as many zeros as possible, because it is not necessary to compute the cofactors of zero entries—the product will always be zero.

EXAMPLE 3.5 Compute $\det A$ by expansion along the row or column with the maximum number of zeros, where

$$A = \begin{bmatrix} 2 & 1 & -3 & 1 \\ -3 & -2 & 0 & 2 \\ 2 & 1 & 0 & -1 \\ 1 & 0 & 1 & 2 \end{bmatrix}.$$

Solution

This is just Example 3.3 again, but expanded down the third column:

$$\det A = (-1)^{1+3}(-3) \det A_{13} + (-1)^{2+3}(0) \det A_{23}$$
$$+ (-1)^{3+3}(0) \det A_{33} + (-1)^{4+3}(1) \det A_{43}$$
$$= -3 \det A_{13} - \det A_{43}.$$

The computation still requires evaluation of the determinants of two 3×3 matrices, but that is half as much work as was required before. Of course, the numerical value of $\det A$ is still -8.

NOTE It is easy to obtain the sign factors $(-1)^{i+j}$ in the terms of the expansion of a determinant, since the signs alternate when moving horizontally or vertically from one position to another, as indicated in the following arrays for $n = 4$ and $n = 5$:

$$\begin{bmatrix} + & - & + & - \\ - & + & - & + \\ + & - & + & - \\ - & + & - & + \end{bmatrix}, \qquad \begin{bmatrix} + & - & + & - & + \\ - & + & - & + & - \\ + & - & + & - & + \\ - & + & - & + & - \\ + & - & + & - & + \end{bmatrix}.$$

For example, if you wish to expand down the jth column, start in the upper left corner and proceed plus, minus, plus, minus, and so on, along the first row to the top of the jth column. That establishes the first position sign, and they alternate thereafter. If you are expanding along the ith row, alternate signs down the first column to the ith row and proceed from there.

EXAMPLE 3.6

Compute $\det A$, where

$$A = \begin{bmatrix} 2 & 9 & 3 & 2 \\ 4 & 0 & 0 & 6 \\ 3 & -1 & 1 & 2 \\ 5 & 0 & 0 & 1 \end{bmatrix}.$$

Solution

For the simplest computation we should expand along the second row, the fourth row, the second column, or the third column, since all contain two zeros. We will use the fourth row. Going down the first column—plus, minus, plus, minus—we see that the first position is negative, and therefore

$$\det A = -(5) \det A_{41} + (0) \det A_{42} - (0) \det A_{43} + (1) \det A_{44}$$

$$= -5 \det \begin{bmatrix} 9 & 3 & 2 \\ 0 & 0 & 6 \\ -1 & 1 & 2 \end{bmatrix} + \det \begin{bmatrix} 2 & 9 & 3 \\ 4 & 0 & 0 \\ 3 & -1 & 1 \end{bmatrix}.$$

The second row is the best for each of these third-order determinants,

$$\det \begin{bmatrix} 9 & 3 & 2 \\ 0 & 0 & 6 \\ -1 & 1 & 2 \end{bmatrix} = -(6) \det \begin{bmatrix} 9 & 3 \\ -1 & 1 \end{bmatrix} = -6(12) = -72,$$

$$\det \begin{bmatrix} 2 & 9 & 3 \\ 4 & 0 & 0 \\ 3 & -1 & 1 \end{bmatrix} = -(4) \det \begin{bmatrix} 9 & 3 \\ -1 & 1 \end{bmatrix} = -4(12) = -48.$$

Therefore
$$\det A = -5(-72) + (-48) = 312.$$

The freedom to compute determinants along any row or column makes the computation of determinants of certain special matrices extremely simple.

THEOREM 2   If a square matrix has a zero row or column, its determinant is zero.

Proof   Expand along the zero row or column.

THEOREM 3   If a square matrix has two equal rows (or two equal columns), its determinant is zero.

Proof   Let n be the order of A. If $n = 2$, A has the form

$$A = \begin{bmatrix} a_{11} & a_{12} \\ a_{11} & a_{12} \end{bmatrix},$$

so that $\det A = a_{11}a_{12} - a_{11}a_{12} = 0$.

For $n > 2$, let A be a matrix of order n with two equal rows. Since n is at least 3, we can choose a row (say, the ith) that is not one of the two equal rows. Expand along this row to obtain

$$\det A = (-1)^{i+1}a_{i1} \det A_{i1} + (-1)^{i+2}a_{i2} \det A_{i2}$$
$$+ \cdots + (-1)^{i+n}a_{in} \det A_{in}.$$

Now each of the matrices A_{ij} has two equal rows (deleting a column does not disturb the equality of the rows). If $n = 3$, each of the A_{ij} is of order $n - 1 = 2$, and by the first case, $\det A_{ij} = 0$ for each j. But then $\det A = 0$ as well. If $n > 3$, repeat the process just described as often as needed until you get down to the 2×2 case in each term. Thus $\det A = 0$.

A similar argument holds if two columns are equal.

SPOTLIGHT

VOLUME OF A PARALLELEPIPED

Determinants can make it easier to remember certain formulas. The complicated cross product formula of Section 1.2 is a case in point. You will see two other examples later in this chapter: In Section 3.3, formulas for the solution of certain linear systems and for the inverse of a matrix are expressed in terms of determinants. Geometrically, a 3×3 determinant provides us with a simple way to express the volume of a "box-shaped" region. More specifically, if three vectors, $\mathbf{u}$, $\mathbf{v}$, and $\mathbf{w}$, in $\mathbf{R}^3$ do not all lie in a plane, they determine a *parallelepiped* (see the figure below). The volume of this solid is given by the absolute value of $\mathbf{u} \cdot (\mathbf{v} \times \mathbf{w})$ (see Exercise 33 of Section 1.2), which in turn can be expressed as $|\det A|$, where

$$A = \begin{bmatrix} u_1 & u_2 & u_3 \\ v_1 & v_2 & v_3 \\ w_1 & w_2 & w_3 \end{bmatrix}.$$

(See Exercise 32.)

EXAMPLE 3.7

Compute det B for

$$B = \begin{bmatrix} 2 & 1 & 3 & 2 \\ 1 & 4 & 2 & 1 \\ 3 & -2 & 1 & 3 \\ 2 & 1 & 0 & 2 \end{bmatrix}.$$

Solution

The first and last columns are equal, so det $B = 0$.

Recall from Section 2.3 that the *main diagonal* of a square matrix A of order n consists of the entries a_{ii} for $i = 1$ to n. A square matrix is called **upper triangular** if all the entries below the main diagonal are zero and **lower triangular** if all the entries above the main diagonal are zero. Of course, a **diagonal matrix** (Section 2.3) is both upper and lower triangular.

As examples, let

$$A = \begin{bmatrix} 1 & 2 \\ 0 & 3 \end{bmatrix} \quad \text{and} \quad B = \begin{bmatrix} 2 & 0 & 0 \\ 1 & 2 & 0 \\ -1 & 1 & 3 \end{bmatrix}.$$

Here A is upper triangular and B is lower triangular.

THEOREM 4   If a matrix A of order n is upper triangular, lower triangular, or diagonal, then $\det A = a_{11}a_{22} \cdots a_{nn}$, the product of the entries on the main diagonal.

Proof   Suppose A is upper triangular. Then A has the form

$$\begin{bmatrix} a_{11} & a_{12} & \cdots & a_{1n} \\ 0 & a_{22} & \cdots & a_{2n} \\ \vdots & & \ddots & \vdots \\ 0 & 0 & \cdots & a_{nn} \end{bmatrix}.$$

Expansion down the first column gives at most one nonzero term

$$\det A = (-1)^{1+1}a_{11} \det A_{11} = a_{11} \det \begin{bmatrix} a_{22} & \cdots & a_{2n} \\ \vdots & \ddots & \vdots \\ 0 & \cdots & a_{nn} \end{bmatrix}.$$

The resulting matrix A_{11} is still upper triangular, so the same argument gives

$$\det A = a_{11}a_{22} \det \begin{bmatrix} a_{33} & \cdots & a_{3n} \\ \vdots & \ddots & \vdots \\ 0 & \cdots & a_{nn} \end{bmatrix}.$$

Continue this process until the submatrix is 2×2, so that

$$\det A = a_{11}a_{22} \cdots a_{n-2,n-2} \det \begin{bmatrix} a_{n-1,n-1} & a_{n-1,n} \\ 0 & a_{nn} \end{bmatrix}$$

$$= a_{11}a_{22} \cdots a_{nn}.$$

Since a diagonal matrix is upper triangular, the same result holds for a diagonal matrix. In case A is lower triangular, the same argument applies except that expansion is done along first rows.

EXAMPLE 3.8   Compute det A, where

$$A = \begin{bmatrix} 2 & 4 & 2 & 1 \\ 0 & 1 & 0 & 3 \\ 0 & 0 & 3 & 1 \\ 0 & 0 & 0 & -2 \end{bmatrix}.$$

Solution   By Theorem 4, det $A = (2)(1)(3)(-2) = -12$.

THEOREM 5   If I is an identity matrix of any order, then det $I = 1$.

Proof   Since I is diagonal, det I is the product of the main diagonal entries. Since each of these is 1, the product is also 1.

3.1 EXERCISES

1. Find the 3,1 and 3,2 minors and cofactors of

$$A = \begin{bmatrix} 2 & -1 & 3 \\ 1 & 2 & 1 \\ 0 & -3 & -4 \end{bmatrix}.$$

2. Find the 2,2 and 2,3 minors and cofactors of the matrix A in Exercise 1.

In Exercises 3–17 evaluate the determinant of each matrix.

3. $[3]$

4. $[-5]$

5. $\begin{bmatrix} 2 & 1 \\ 3 & -1 \end{bmatrix}$

6. $\begin{bmatrix} 4 & -6 \\ -2 & 3 \end{bmatrix}$

7. $\begin{bmatrix} 2 & 1 & 1 \\ 3 & 0 & -1 \\ 4 & 5 & 2 \end{bmatrix}$

8. $\begin{bmatrix} 1 & -2 & 3 \\ 2 & -2 & 4 \\ 0 & 3 & 2 \end{bmatrix}$

9. $\begin{bmatrix} 2 & 4 & 2 \\ 1 & 5 & 1 \\ 3 & -7 & 3 \end{bmatrix}$

10. $\begin{bmatrix} 1 & 3 & -1 & 2 \\ 2 & 1 & 0 & -4 \\ 3 & 1 & 0 & 4 \\ -1 & 2 & 0 & 3 \end{bmatrix}$

11. $\begin{bmatrix} 2 & 3 & 4 & 5 \\ 0 & 3 & 4 & 5 \\ 0 & 0 & 4 & 5 \\ 0 & 0 & 0 & 5 \end{bmatrix}$

12. $\begin{bmatrix} 2 & 0 & 2 & 0 \\ 1 & 4 & 0 & 0 \\ 2 & 3 & 1 & 0 \\ 2 & 5 & -1 & 0 \end{bmatrix}$

13. $\begin{bmatrix} a & ab \\ b & a^2 + b^2 \end{bmatrix}$

14. $\begin{bmatrix} ak & c \\ bk & d \end{bmatrix}$

15. $\begin{bmatrix} ak & ck \\ bk & dk \end{bmatrix}$ **16.** $\begin{bmatrix} \lambda - 2 & 3 \\ 4 & \lambda - 1 \end{bmatrix}$ **17.** $\begin{bmatrix} t - 3 & -1 \\ 3 & t \end{bmatrix}$

In Exercises 18–22 evaluate the determinant of the given matrix "by inspection."

18. $\begin{bmatrix} 1 & 2 & 3 \\ 0 & 4 & 5 \\ 0 & 0 & 6 \end{bmatrix}$ **19.** $\begin{bmatrix} 2 & 0 & 0 \\ 0 & 4 & 0 \\ 3 & 0 & -1 \end{bmatrix}$ **20.** $\begin{bmatrix} -1 & 0 & 0 \\ 0 & 3 & 0 \\ 0 & 0 & 5 \end{bmatrix}$

21. $\begin{bmatrix} 1 & 2 & 3 & 4 \\ 0 & 5 & 0 & 6 \\ 1 & 2 & 3 & 4 \\ 7 & 0 & 8 & 0 \end{bmatrix}$ **22.** $\begin{bmatrix} -1 & 1 & -3 & -1 \\ 3 & 2 & 2 & 3 \\ 0 & 1 & 1 & 0 \\ 2 & 1 & 0 & 2 \end{bmatrix}$

23. Verify one special case of Theorem 1 by proving that for matrices of order 3, computation of the determinant by expansion down the second column yields the same result as expansion along the first row.

24. Determine a formula analogous to Theorem 4 for the determinant of a matrix A of order n such that $a_{ij} = 0$ unless $i + j = n + 1$.

25. Determine a formula analogous to Theorem 4 for the determinant of a matrix A of order n such that $a_{ij} = 0$ if $i + j > n + 1$. Such a matrix is upper triangular in appearance but not upper triangular with respect to the main diagonal.

26. Let a_{ij} be constants for all integers i and j from 1 to n, let x be a variable, let

$$A(x) = \begin{bmatrix} x + a_{11} & a_{12} & \cdots & a_{1n} \\ a_{21} & x + a_{22} & \cdots & a_{2n} \\ \vdots & \vdots & & \vdots \\ a_{n1} & a_{n2} & \cdots & x + a_{nn} \end{bmatrix},$$

and let $f(x) = \det A(x)$ for each x. Prove that $f(x)$ is an nth degree polynomial.

27. Find all zeros (values of x such that $f(x) = 0$) of the polynomial

$$f(x) = \det \begin{bmatrix} x - 2 & 1 \\ 3 & x \end{bmatrix}.$$

28. Find all zeros of the polynomial

$$f(t) = \det \begin{bmatrix} t - 2 & 4 & -3 \\ 0 & t - 3 & 2 \\ 0 & 0 & t \end{bmatrix}.$$

29. For matrices of *order 3 only*, there is another convenient way to compute determinants that is sometimes called the "basket-weave method." First construct a 3×5 array by writing down the entries of the given matrix and adjoining (on the right) its first and second columns (see Figure 3.1). Then form the products of the three numbers on each of the six diagonals indicated, and set $\det A$ equal to the sum of the circled products 1, 2, and 3 minus the sum of circled products 4, 5, and 6. Show that this procedure does indeed yield the determinant of A.

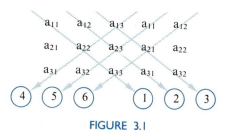

FIGURE 3.1

30. In Chapter 1 the *cross product* of two 3-vectors $\mathbf{v} = (a_1, a_2, a_3)$, $\mathbf{w} = (b_1, b_2, b_3)$ was defined to be

$$\mathbf{v} \times \mathbf{w} = (a_2 b_3 - a_3 b_2, a_3 b_1 - a_1 b_3, a_1 b_2 - a_2 b_1)$$

or, using $\mathbf{i}, \mathbf{j}, \mathbf{k}$ notation,

$$\mathbf{v} \times \mathbf{w} = (a_2 b_3 - a_3 b_2)\mathbf{i} + (a_3 b_1 - a_1 b_3)\mathbf{j} + (a_1 b_2 - a_2 b_1)\mathbf{k}.$$

Show that this vector can be conveniently remembered as the "determinant" of the matrix

$$\begin{bmatrix} \mathbf{i} & \mathbf{j} & \mathbf{k} \\ a_1 & a_2 & a_3 \\ b_1 & b_2 & b_3 \end{bmatrix}.$$

(The word *determinant* is in quotes because $\mathbf{i}, \mathbf{j}$, and $\mathbf{k}$ are vectors, so the result is a vector, not a scalar.)

31. Prove that $(A^T)_{ij} = (A_{ji})^T$, for any i, j, and square matrix A.

32. Let $\mathbf{u}, \mathbf{v}$, and $\mathbf{w}$ be vectors in $\mathbf{R}^3$, and let A be the matrix with these vectors as rows:

$$A = \begin{bmatrix} u_1 & u_2 & u_3 \\ v_1 & v_2 & v_3 \\ w_1 & w_2 & w_3 \end{bmatrix}.$$

Show that $|\det A|$ is the *volume of the parallelepiped* determined by the vectors $\mathbf{u}$, $\mathbf{v}$, and $\mathbf{w}$ (see the second Spotlight in this section).

33. Using the formula in Exercise 32, compute the volume of the parallelepiped determined by the vectors $(1, 2, 3)$, $(0, -4, 5)$, and $(0, 0, 6)$.

34. The set of points S given below is the set of vertices of a parallelepiped in $\mathbf{R}^3$. Verify this fact by translating a vertex to the origin, and then computing the volume of the parallelepiped.

$$S = \{(-1, 1, 3), (1, 1, 3), (0, 2, 3), (2, 2, 3), (0, 2, 6),$$
$$(2, 2, 6), (1, 3, 6), (3, 3, 6)\}$$

PROPERTIES OF DETERMINANTS

In this section we will examine some of the basic properties of determinants. Some of these properties will enable us to compute determinants more quickly. Others will provide important relationships between the value of a determinant and the invertibility of the corresponding matrix or the uniqueness of the solution to a related linear system.

THEOREM I   Let A and A' be square matrices that are the same except that in one row (or in one column) of A' each entry of A is multiplied by a scalar c. Then $\det A' = c \det A$. (That is, c can be "factored out" of the row or column.)

Proof   Suppose the difference occurs in the ith row. then

$$A = \begin{bmatrix} a_{11} & \cdots & a_{1n} \\ \vdots & & \vdots \\ a_{i1} & \cdots & a_{in} \\ \vdots & & \vdots \\ a_{n1} & \cdots & a_{nn} \end{bmatrix} \quad \text{and} \quad A' = \begin{bmatrix} a_{11} & \cdots & a_{1n} \\ \vdots & & \vdots \\ ca_{i1} & \cdots & ca_{in} \\ \vdots & & \vdots \\ a_{n1} & \cdots & a_{nn} \end{bmatrix}.$$

Compute $\det A'$ by expansion along the ith row. Then

$$\det A' = (-1)^{i+1}ca_{i1} \det A'_{i1} + (-1)^{i+2}ca_{i2} \det A'_{i2}$$
$$+ \cdots + (-1)^{i+n}ca_{in} \det A'_{in}$$
$$= \sum_{j=1}^{n} (-1)^{i+j}ca_{ij} \det A'_{ij} = c \sum_{j=1}^{n} (-1)^{i+j}a_{ij} \det A_{ij},$$

since c factors out of each term and since $A_{ij} = A'_{ij}$ for each j (the ith row has been deleted). The last expression is just $c \det A$. A similar argument holds for columns.

EXAMPLE 3.9   Compute $\det A$, given that

$$A = \begin{bmatrix} 2 & 0 & 4 \\ 1 & -6 & 2 \\ 3 & 9 & 12 \end{bmatrix}.$$

Solution

In the computation of det A, 2 can be factored out of the first row, 3 out of the third row, 3 out of the second column, and 2 out of the third column:

$$\det \begin{bmatrix} 2 & 0 & 4 \\ 1 & -6 & 2 \\ 3 & 9 & 12 \end{bmatrix} = 2 \det \begin{bmatrix} 1 & 0 & 2 \\ 1 & -6 & 2 \\ 3 & 9 & 12 \end{bmatrix} = 2(3) \det \begin{bmatrix} 1 & 0 & 2 \\ 1 & -6 & 2 \\ 1 & 3 & 4 \end{bmatrix}$$

$$= 2(3^2) \det \begin{bmatrix} 1 & 0 & 2 \\ 1 & -2 & 2 \\ 1 & 1 & 4 \end{bmatrix} = 2^2(3^2) \det \begin{bmatrix} 1 & 0 & 1 \\ 1 & -2 & 1 \\ 1 & 1 & 2 \end{bmatrix}.$$

Expanding (say, along the first row), we still need to compute two 2×2 determinants, as in the original problem. Now, however, the numbers are easier to work with than before. The answer is $36(-2) = -72$.

The same idea can be used in reverse to clear the fractions in a matrix before computing its determinant. Because integers are easier to work with, there is less chance of error. The rule here is to examine each row (or column) for the smallest integer that will clear the fractions (for the *least common denominator*), multiply the row by that number, and compensate by multiplying the determinant by the reciprocal of that number.

EXAMPLE 3.10

Compute det A, where

$$A = \begin{bmatrix} \frac{1}{3} & 0 & \frac{3}{4} \\ \frac{2}{5} & -1 & \frac{3}{2} \\ \frac{1}{8} & -\frac{3}{4} & \frac{5}{4} \end{bmatrix}.$$

Solution

Since 12 will clear the first row, 10 the second, and 8 the last, we multiply the determinant by $\frac{1}{12}$ and the first row by 12, and proceed similarly with the other rows.

$$\det A = \left(\frac{1}{12}\right)\left(\frac{1}{10}\right)\left(\frac{1}{8}\right) \det \begin{bmatrix} 4 & 0 & 9 \\ 4 & -10 & 15 \\ 1 & -6 & 10 \end{bmatrix}.$$

Now -2 can be factored out of the second column:

$$\det A = \left(\frac{1}{12}\right)\left(\frac{1}{10}\right)\left(\frac{1}{8}\right)(-2) \det \begin{bmatrix} 4 & 0 & 9 \\ 4 & 5 & 15 \\ 1 & 3 & 10 \end{bmatrix}.$$

The final value is det $A = (-1/480)(83) = -83/480$.

EXAMPLE 3.11   Compute det A for the matrix

$$A = \begin{bmatrix} 2 & 1 & 3 & -4 \\ -1 & 0 & 1 & 2 \\ -3 & 2 & -1 & 6 \\ 4 & 1 & 4 & -8 \end{bmatrix}.$$

Solution   Your first thought might be to expand along the second row or column because of the zero. This is a reasonable approach, but if you notice that the last column is -2 times the first and use Theorem 1 to factor out -2, you obtain

$$\det A = -2 \det \begin{bmatrix} 2 & 1 & 3 & 2 \\ -1 & 0 & 1 & -1 \\ -3 & 2 & -1 & -3 \\ 4 & 1 & 4 & 4 \end{bmatrix}.$$

Now the first and last columns are equal, and by Theorem 3 of Section 3.1, $\det A = 0$.

The method of Example 3.11 is important enough to be identified as a theorem.

THEOREM 2   If a square matrix A has a row that is a scalar multiple of another row (or a column that is a scalar multiple of another column), then $\det A = 0$.

Proof   Use Theorem 1 to factor out the scalar multiple, and apply Theorem 3 of the last section.

THEOREM 3   If two rows (or columns) of a square matrix are interchanged, then the determinant changes by a factor of -1; that is, its sign is reversed.

Proof   Let A and A' be matrices of order n such that A' is obtained from A by interchanging two of its rows.
      If $n = 2$, there are only two rows to interchange:

$$A = \begin{bmatrix} a_{11} & a_{12} \\ a_{21} & a_{22} \end{bmatrix} \quad \text{and} \quad A' = \begin{bmatrix} a_{21} & a_{22} \\ a_{11} & a_{12} \end{bmatrix}.$$

Then $\det A' = a_{21}a_{12} - a_{11}a_{22} = -(a_{11}a_{22} - a_{21}a_{12}) = -\det A$.
      For $n > 2$, expand $\det A'$ along any row not involved in the interchange.

Suppose the ith row is one such; then

$$\det A' = (-1)^{i+1}a_{i1} \det A'_{i1} + (-1)^{i+2}a_{i2} \det A'_{i2}$$
$$+ \cdots + (-1)^{i+n}a_{in} \det A'_{in}.$$

Each submatrix A'_{ij} is the same as the corresponding submatrix A_{ij} of A, except that two rows of A_{ij} are interchanged to obtain A'_{ij}. (Taking out a row not involved in the interchange leaves two rows still interchanged.)

If $n = 3$, each of these A'_{ij} is order 2, and by the $n = 2$ case, each term in the expansion of $\det A'$ is the opposite of the corresponding term in the expansion of $\det A$ along the same row. But then $\det A' = -\det A$. If $n > 3$, repeat the process just given as often as necessary to reduce to the order 2 case. A similar argument holds if two columns are interchanged.

So far we have seen how determinants of square matrices behave with respect to two of the three elementary row operations:

i. If a row is multiplied by a scalar, the determinant is multiplied by the same scalar.

ii. If two rows are interchanged, the sign of the determinant is reversed.

The third elementary row operation, addition of a scalar multiple of one row to another, has no effect at all on the determinant. It is also the most useful for simplifying computation of determinants.

THEOREM 4 If a scalar multiple of one row (or column) is added to another row (or column) of a square matrix, the determinant is unchanged.

Proof Let A and A' be matrices of order n, where A and A' agree except in the ith row, and $\mathbf{a}'_i = c\mathbf{a}_k + \mathbf{a}_i$; in other words, the ith row of A' is the sum of the ith row of A and c times the kth row of A, where $k \neq i$. We compute $\det A'$ by expansion along the ith row:

$$\det A' = (-1)^{i+1}a'_{i1} \det A'_{i1} + \cdots + (-1)^{i+n}a'_{in} \det A'_{in}$$
$$= \sum_{j=1}^{n} (-1)^{i+j}a'_{ij} \det A'_{ij}.$$

By hypothesis, $a'_{ij} = ca_{kj} + a_{ij}$ for each j, and since the ith row has been deleted, $A'_{ij} = A_{ij}$ for each j. Then

$$\det A' = \sum_{j=1}^{n} (-1)^{i+j}(ca_{kj} + a_{ij}) \det A_{ij}$$

$$= c\sum_{j=1}^{n} (-1)^{i+j}a_{kj} \det A_{ij} + \sum_{j=1}^{n} a_{ij}(-1)^{i+j} \det A_{ij}$$

$$= c \det A'' + \det A,$$

where A'' is obtained from A by replacing the ith row by its own kth row. But then A'' has two equal rows (the ith and the kth), and by Theorem 3 of Section 3.1, $\det A'' = 0$. This leaves just $\det A' = \det A$.

A similar proof holds for columns.

As we noted in the last section, $\det A$ is easy to compute if a row or column has lots of zeros. The usefulness of Theorem 4 is that if A does not have many zeros, we can "create them"—we can find a new matrix with the same determinant that does have many zeros.

EXAMPLE 3.12   One last time we compute the determinant of the matrix given in Example 3.3,

$$A = \begin{bmatrix} 2 & 1 & -3 & 1 \\ -3 & -2 & 0 & 2 \\ 2 & 1 & 0 & -1 \\ 1 & 0 & 1 & 2 \end{bmatrix}.$$

Solution   We have already noticed (Example 3.5) that it is most convenient to expand down the third column. Before doing so, however, the remaining work can be cut in half if we first add 3 times the last row to the first. Then we have

$$A' = \begin{bmatrix} 5 & 1 & 0 & 7 \\ -3 & -2 & 0 & 2 \\ 2 & 1 & 0 & -1 \\ 1 & 0 & 1 & 2 \end{bmatrix},$$

and, by Theorem 4,

$$\det A = \det A' = (-1)(1) \det A'_{43} = -\det \begin{bmatrix} 5 & 1 & 7 \\ -3 & -2 & 2 \\ 2 & 1 & -1 \end{bmatrix}.$$

To finish the computation as simply as possible, we use Theorem 4 again to obtain a new matrix with the same determinant as A'_{43} but with two zeros in some row or column. Any row or column would work, but a quick inspection shows that the second column or the third row requires minimal computation. We choose the third row. Adding -2 times the second column to the first, and just adding the second column to the third, we obtain

$$\det \begin{bmatrix} 5 & 1 & 7 \\ -3 & -2 & 2 \\ 2 & 1 & -1 \end{bmatrix} = \det \begin{bmatrix} 3 & 1 & 8 \\ 1 & -2 & 0 \\ 0 & 1 & 0 \end{bmatrix} = -(1) \det \begin{bmatrix} 3 & 8 \\ 1 & 0 \end{bmatrix} = 8.$$

Then $\det A = -8$.

COMPUTATIONAL NOTE It is somewhat surprising to realize that even with the aid of high-speed computers, simplifications similar to those in Example 3.12 must be made to compute det A when A is of even modest size. For example, the computation time of an order 25 matrix using only the definition of determinant would be astronomically long, even using the fastest computers available (or on the drawing board). For this reason, computer programs to calculate the determinant of a matrix are based on the following idea. Use Theorems 3 and 4 to bring the matrix into upper triangular form, keeping track of the number of row interchanges used. By Theorem 4 of Section 3.1 the determinant is then just ε times the product of the main diagonal entries, where $\varepsilon = 1$ if there are an even number of row interchanges, and $\varepsilon = -1$ otherwise.

EXAMPLE 3.13

Find det A by using the process described in the preceding Computational Note, where

$$A = \begin{bmatrix} 1 & 2 & 0 & -2 \\ 0 & 0 & 2 & -1 \\ 0 & -1 & 1 & 0 \\ 1 & 3 & 4 & 1 \end{bmatrix}.$$

Solution

By interchanging the second and third rows, and then adding multiples of each row to the bottom row, we obtain the sequence of equations

$$\det A = -\det \begin{bmatrix} 1 & 2 & 0 & -2 \\ 0 & -1 & 1 & 0 \\ 0 & 0 & 2 & -1 \\ 1 & 3 & 4 & 1 \end{bmatrix} = -\det \begin{bmatrix} 1 & 2 & 0 & -2 \\ 0 & -1 & 1 & 0 \\ 0 & 0 & 2 & -1 \\ 0 & 1 & 4 & 3 \end{bmatrix}$$

$$= -\det \begin{bmatrix} 1 & 2 & 0 & -2 \\ 0 & -1 & 1 & 0 \\ 0 & 0 & 2 & -1 \\ 0 & 0 & 5 & 3 \end{bmatrix} = -\det \begin{bmatrix} 1 & 2 & 0 & -2 \\ 0 & -1 & 1 & 0 \\ 0 & 0 & 2 & -1 \\ 0 & 0 & 0 & \frac{11}{2} \end{bmatrix}.$$

Finally, $\det A = -(1)(-1)(2)(\frac{11}{2}) = 11$.

Perhaps the most important single property of the determinant of a matrix is that it tells us whether or not the matrix in question is invertible.

THEOREM 5   A square matrix A is invertible if and only if $\det A \neq 0$.

Proof   Let B be the row-reduced echelon form for A. Since B is obtained from A by elementary row operations, repeated application of Theorems 1, 3, and 4 implies that $\det B = c \det A$ for some nonzero scalar c. This implies that $\det A \neq 0$ if and only if $\det B \neq 0$. But B is a row-reduced square matrix and as such is either the identity matrix or has at least one row of zeros. In other words, $\det B = 1$ if B is the identity matrix, and $\det B = 0$ otherwise. Thus we have $\det A \neq 0$ if and only if B is the identity matrix. By Theorem 5 of Section 2.5, this is equivalent to A being invertible.

EXAMPLE 3.14   Determine whether or not the matrix A is invertible, where A is given by

$$A = \begin{bmatrix} 2 & 1 & -3 & 1 \\ -3 & -2 & 0 & 2 \\ 2 & 1 & 0 & -1 \\ 1 & 0 & 1 & 2 \end{bmatrix}.$$

Solution   Since A is the matrix of Example 3.12, $\det A = -8 \neq 0$, and thus A is invertible.

NOTE   The fact that $\det A \neq 0$ implies that A is invertible does not tell us what the matrix A^{-1} looks like. Determinants can also be used for this purpose as we will see in Section 3.3.

The following theorem is an important direct consequence of Theorem 5. Its proof is left as an exercise.

THEOREM 6   Let A be a square matrix. Then the linear system $A\mathbf{x} = \mathbf{b}$ has a unique solution for every $\mathbf{b}$ if and only if $\det A \neq 0$.

The next theorem is basic in the study of linear algebra. Its proof is fairly complicated, however, and we will omit it.

THEOREM 7   If A and B are square matrices of the same order, then

$$\det AB = (\det A)(\det B).$$

Recall that the *transpose* of a matrix is the matrix obtained by interchanging its rows and columns. In other words, for each i, the *ith row* of a matrix A

is the *i*th *column* of the matrix A^T, and the *j*th *column* of *A* is the *j*th *row* of A^T. The next theorem tells us that the determinant of a matrix does not change when we take its transpose.

THEOREM 8   If *A* is a square matrix, then $\det A^T = \det A$.

Proof   Let *n* be the order of *A*. If $n = 2$,

$$A = \begin{bmatrix} a_{11} & a_{12} \\ a_{21} & a_{22} \end{bmatrix} \quad \text{and} \quad A^T = \begin{bmatrix} a_{11} & a_{21} \\ a_{12} & a_{22} \end{bmatrix}.$$

Hence $\det A = a_{11}a_{22} - a_{21}a_{12} = \det A^T$.

For larger values of *n*, the proof is based on the fact that the expansion of $\det A^T$ repeatedly along first *rows* yields exactly the same terms and in the same order as the expansion of $\det A$ repeatedly down first *columns*.

EXAMPLE 3.15   Suppose *A* and *B* are square matrices of the same order, with $\det A = 3$ and $\det B = 5$. Find $\det(AB^T)$.

Solution

$$\begin{aligned} \det(AB^T) &= \det(A)\det(B^T) \quad &\text{(Theorem 7)} \\ &= \det(A)\det(B) \quad &\text{(Theorem 8)} \\ &= (3)(5) = 15. \end{aligned}$$

Careful analysis of Theorem 5 (and its supporting Theorem 5 of Section 2.5) shows only that if $\det A \neq 0$, there exists a matrix *B* such that $AB = I$. This is because the left side of $[A \mid I]$ will row reduce to the identity matrix, and the resulting right side of $[I \mid B]$ must satisfy $AB = I$. Since matrix multiplication is in general *not* commutative, we cannot be sure that $BA = I$ as well, in order to conclude that *B* is the inverse of *A*. Theorems 7 and 8 can be used to prove this important fact, which will complete the proof of Theorem 2 of Section 2.4. We restate the result here and prove it.

THEOREM 9   Let *A* be a square matrix. If a square matrix *B* exists such that $AB = I$, then $BA = I$ as well, and thus $B = A^{-1}$.

Proof   We first show that there is some square matrix *C*, such that $CA = I$. Consider the system

$$A^T X = I.$$

Since $(\det A)(\det B) = \det(AB) = \det I = 1$, we see that $\det A \neq 0$. However, $\det A^T = \det A$, so $\det A^T \neq 0$ as well. Thus, the row-reduced form of A^T is also equal to I, so we can row-reduce the system $[A^T \mid I]$ to obtain $[I \mid D]$ with $A^T D = I$. But then $(A^T D)^T = I^T$, and, from properties of transpose,

$$D^T A = I.$$

Thus there is a matrix C, namely $C = D^T$, such that $CA = I$.

Finally, we show that $C = B$ (which implies that $B = A^{-1}$):

$$C = CI = C(AB) = (CA)B = IB = B.$$

3.2 EXERCISES

Evaluate the determinant of each matrix in Exercises 1–9. Use the theorems of this section to simplify the work.

1. $\begin{bmatrix} 1 & 2 & 3 \\ 1 & 3 & 7 \\ 1 & 4 & 13 \end{bmatrix}$

2. $\begin{bmatrix} 1 & 0 & 0 & 3 \\ 0 & 3 & 9 & 21 \\ 0 & 0 & 2 & 1 \\ 2 & 0 & 0 & 7 \end{bmatrix}$

3. $\begin{bmatrix} 1 & 0 & 0 & 0 \\ 0 & 0 & 3 & 0 \\ 0 & 2 & 0 & 0 \\ 0 & 0 & 0 & 4 \end{bmatrix}$

4. $\begin{bmatrix} 2 & -1 & 3 & -4 \\ 1 & 0 & 2 & 1 \\ -4 & 2 & -6 & 8 \\ 3 & 1 & 2 & 0 \end{bmatrix}$

5. $\begin{bmatrix} \frac{1}{3} & \frac{3}{5} & \frac{2}{5} \\ \frac{3}{8} & \frac{1}{2} & \frac{1}{4} \\ \frac{1}{3} & \frac{2}{3} & \frac{1}{2} \end{bmatrix}$

6. $\begin{bmatrix} 0.02 & -0.10 & 0.08 \\ 0.03 & 0.06 & -0.09 \\ 0 & 0.07 & 0.20 \end{bmatrix}$

7. $\begin{bmatrix} x & 2x & -3x \\ x & x-1 & -3 \\ 0 & 0 & 2x-1 \end{bmatrix}$:

8. $\begin{bmatrix} t & -1 & 3 \\ 1 & t-3 & 1 \\ 2 & 3 & t-1 \end{bmatrix}$

9. $\begin{bmatrix} \lambda-1 & 0 & 0 & 0 \\ 2 & 0 & \lambda+1 & 0 \\ 1 & \lambda-2 & 0 & 0 \\ 2 & 3 & 9 & \lambda+2 \end{bmatrix}$

10. Determine which of the matrices in Exercises 1–6 are invertible. Do *not* compute the inverse.

11. For which values of the variables in Exercises 7–9 are the matrices invertible?

12. Prove Theorem 6: If A is a square matrix, the linear system $A\mathbf{x} = \mathbf{b}$ has a unique solution if and only if $\det A \neq 0$.

In Exercises 13–16 use determinants to determine whether or not the given linear system has a unique solution.

13. $x - y = 4$
$x + y = 1$

14. $-x + y = 0$
$x - y = 0$

15. $x + y + z = 1$
$\quad x - y + 2z = 0$
$\quad\quad 2y - z = 1$

16. $x + y \quad\quad = 1$
$\quad\quad\quad y + z = 0$
$\quad x \quad\quad + z = 1$

17. Let A be a 3×3 matrix with $\det A = 4$. Find $\det(A^T)$ and $\det(2A)$.

18. Prove that, for a square matrix A of order n and scalar c, $\det(cA) = c^n \det A$.

19. Without expanding the determinant, show that for any values of s, t, and u,

$$\det \begin{bmatrix} s & s+1 & s+2 \\ t & t+1 & t+2 \\ u & u+1 & u+2 \end{bmatrix} = 0.$$

20. Let A' be obtained from the square matrix A by interchanging pairs of rows (columns) m times. Express $\det A'$ in terms of $\det A$ and m.

21. Use Exercise 20 to find a formula for $\det A'$ in terms of $\det A$, where A' is obtained from A by reversing the order in which the rows appear. That is, $\mathbf{a}'_i = \mathbf{a}_{n+1-i}$ for each i.

22. Compute and leave in simplest factored form:

$$\det \begin{bmatrix} 1 & s & s^2 \\ 1 & t & t^2 \\ 1 & u & u^2 \end{bmatrix}.$$

23. We know that in general $AB \neq BA$. However, prove that $\det AB = \det BA$ for all square nth-order matrices A and B.

24. Prove that $\det A^m = (\det A)^m$, where A^m denotes the mth power of A.

25. If $A^T A = I$, prove that $\det A = \pm 1$.

26. If S is an invertible matrix, and $B = S^{-1}AS$, prove that $\det A = \det B$.

Exercises 27–29 give an extremely simple method of finding a linear equation of the hyperplane that contains n independent points in $\mathbf{R}^n$ (see Section 1.4).

27. Let (a_1, a_2) and (b_1, b_2) be distinct points in $\mathbf{R}^2$. Prove that the equation

$$\det \begin{bmatrix} x & y & 1 \\ a_1 & a_2 & 1 \\ b_1 & b_2 & 1 \end{bmatrix} = 0$$

is a linear equation that describes the line that contains (a_1, a_2) and (b_1, b_2). (*Hint:* What happens when you make the substitution $x = a_1$ and $y = a_2$? Similarly for $x = b_1$ and $y = b_2$? Finally, expand along the top row.)

28. Let (a_1, a_2, a_3), (b_1, b_2, b_3), and (c_1, c_2, c_3) be noncollinear points in $\mathbf{R}^3$. Prove that the equation

$$\det \begin{bmatrix} x & y & z & 1 \\ a_1 & a_2 & a_3 & 1 \\ b_1 & b_2 & b_3 & 1 \\ c_1 & c_2 & c_3 & 1 \end{bmatrix} = 0$$

is a linear equation that describes the plane containing (a_1, a_2, a_3), (b_1, b_2, b_3), and (c_1, c_2, c_3).

29. Generalize the result of Exercises 27 and 28 to the case of n independent points in $\mathbf{R}^n$.

30. Compute the determinant of the following matrix by three different methods: (a) the definition, (b) triangularization, (c) any other. Which seems to be the best method?

$$A = \begin{bmatrix} 2 & 1 & -1 & 3 \\ 1 & 2 & 0 & 3 \\ -1 & 2 & -2 & 1 \\ 0 & 3 & 1 & 4 \end{bmatrix}.$$

31. Find the determinant of the matrix A using a computer:

$$A = \begin{bmatrix} 0.21 & -2.31 & 2.01 & 3.00 \\ 0.03 & -1.20 & 1.02 & 0.00 \\ 1.21 & 1.00 & -0.64 & -0.54 \\ 1.02 & 0.04 & 0.57 & 0.19 \end{bmatrix}.$$

3.3 CRAMER'S RULE

In this section we present new techniques for solving two old problems: solving a square linear system and inverting a matrix. In Sections 2.2 and 2.4 these problems were dealt with by using row reduction. Here we solve them with determinants.

CRAMER'S RULE

Consider a system of linear equations $A\mathbf{x} = \mathbf{b}$, where A is a square matrix of order n. We denote by $A(i)$ the matrix obtained from A by replacing its ith column $\mathbf{a}^i$ by $\mathbf{b}$. For example, if the given linear system is

$$\begin{aligned} x_1 - x_2 + x_3 &= 6 \\ 2x_1 \quad\quad + 3x_3 &= 1 \\ 2x_2 - x_3 &= 0, \end{aligned}$$

then

$$A = \begin{bmatrix} 1 & -1 & 1 \\ 2 & 0 & 3 \\ 0 & 2 & -1 \end{bmatrix}, \quad A(1) = \begin{bmatrix} 6 & -1 & 1 \\ 1 & 0 & 3 \\ 0 & 2 & -1 \end{bmatrix},$$

$$A(2) = \begin{bmatrix} 1 & 6 & 1 \\ 2 & 1 & 3 \\ 0 & 0 & -1 \end{bmatrix}, \quad A(3) = \begin{bmatrix} 1 & -1 & 6 \\ 2 & 0 & 1 \\ 0 & 2 & 0 \end{bmatrix}.$$

The following procedure for solving a square linear system is called **Cramer's rule** and will be justified later in the section.

THEOREM I Let $A\mathbf{x} = \mathbf{b}$ be a square linear system of n equations with $\det A \neq 0$. Then the solution to this system is given by

$$x_1 = \frac{\det A(1)}{\det A}, \quad x_2 = \frac{\det A(2)}{\det A}, \quad \ldots, \quad x_n = \frac{\det A(n)}{\det A}.$$

EXAMPLE 3.16

Solve the linear system of equations

$$2x_1 + 3x_2 - x_3 = 1$$
$$x_1 + 4x_2 + 2x_3 = 2$$
$$3x_1 - x_2 - x_3 = 3.$$

Solution

Letting A be the coefficient matrix,

$$A = \begin{bmatrix} 2 & 3 & -1 \\ 1 & 4 & 2 \\ 3 & -1 & -1 \end{bmatrix},$$

and, noting that $\det A = 30 \neq 0$, Cramer's rule asserts that

$$x_1 = \frac{\det A(1)}{\det A} = \frac{\det \begin{bmatrix} 1 & 3 & -1 \\ 2 & 4 & 2 \\ 3 & -1 & -1 \end{bmatrix}}{\det A} = \frac{36}{30} = \frac{6}{5},$$

$$x_2 = \frac{\det A(2)}{\det A} = \frac{\det \begin{bmatrix} 2 & 1 & -1 \\ 1 & 2 & 2 \\ 3 & 3 & -1 \end{bmatrix}}{\det A} = -\frac{6}{30} = -\frac{1}{5},$$

$$x_3 = \frac{\det A(3)}{\det A} = \frac{\det \begin{bmatrix} 2 & 3 & 1 \\ 1 & 4 & 2 \\ 3 & -1 & 3 \end{bmatrix}}{\det A} = \frac{24}{30} = \frac{4}{5}.$$

EXAMPLE 3.17

Solve the linear system of equations

$$x_1 + 2x_2 + x_3 = 3$$
$$3x_1 + x_2 - x_3 = -1$$
$$5x_1 \qquad - 3x_3 = -5.$$

Solution   Letting A be the coefficient matrix, we see that $\det A = 0$. Cramer's rule does not apply. The row-reduction methods of Section 2.2 should be used.

ADJOINT FORM OF INVERSE

From Theorem 5 of Section 3.2, we know that if a square matrix A is invertible, then $\det A \neq 0$. If $\det A \neq 0$, we not only can invert A; we can also obtain an explicit formula for the inverse with the aid of determinants. Toward this goal we introduce the following matrix.

DEFINITION

For a square matrix A, the **(classical) adjoint** of A is the matrix

$$\text{Adj } A = [c_{ij}], \qquad \text{where} \quad c_{ij} = (-1)^{i+j} \det A_{ji}.$$

In other words, the ijth entry of Adj A is the jith cofactor of A.

EXAMPLE 3.18   Find Adj A, where

$$A = \begin{bmatrix} 1 & -1 & 1 \\ 2 & 0 & 2 \\ 1 & 2 & 3 \end{bmatrix}.$$

Solution   We compute three of the entries of $C = \text{Adj } A$ explicitly:

$$c_{11} = (-1)^{1+1} \det A_{11} = (-1)^{1+1} \det \begin{bmatrix} 0 & 2 \\ 2 & 3 \end{bmatrix} = (-1)^2(-4) = -4,$$

$$c_{12} = (-1)^{1+2} \det A_{21} = (-1)^{1+2} \det \begin{bmatrix} -1 & 1 \\ 2 & 3 \end{bmatrix} = (-1)^3(-5) = 5,$$

$$c_{23} = (-1)^{2+3} \det A_{32} = (-1)^{2+3} \det \begin{bmatrix} 1 & 1 \\ 2 & 2 \end{bmatrix} = (-1)^5(0) = 0.$$

The other entries are computed in a similar manner to give

$$\text{Adj } A = \begin{bmatrix} -4 & 5 & -2 \\ -4 & 2 & 0 \\ 4 & -3 & 2 \end{bmatrix}.$$

The next theorem provides another method for computing the adjoint. You'll probably find this way easier to remember.

THEOREM 2 For a square matrix A, Adj A can be formed by replacing each entry of A^T with the corresponding cofactor of A^T.

Proof This fact is an immediate consequence of Exercise 31 of Section 3.1 and Theorem 8 of Section 3.2. That is, $\det (A^T)_{ij} = \det (A_{ji})^T = \det A_{ji}$.

As an example of the preceding theorem, we look again at Example 3.18. First form the transpose

$$A^T = \begin{bmatrix} 1 & 2 & 1 \\ -1 & 0 & 2 \\ 1 & 2 & 3 \end{bmatrix}.$$

Then replace each entry with its cofactor:

$$\text{Adj } A = \begin{bmatrix} -4 & 5 & -2 \\ -4 & 2 & 0 \\ 4 & -3 & 2 \end{bmatrix}.$$

The major importance of Adj A stems from the very simple product that results when it is multiplied by the original matrix A. As a numerical example, we again use the matrix of Example 3.18. We have

$$A(\text{Adj } A) = \begin{bmatrix} 1 & -1 & 1 \\ 2 & 0 & 2 \\ 1 & 2 & 3 \end{bmatrix} \begin{bmatrix} -4 & 5 & -2 \\ -4 & 2 & 0 \\ 4 & -3 & 2 \end{bmatrix} = \begin{bmatrix} 4 & 0 & 0 \\ 0 & 4 & 0 \\ 0 & 0 & 4 \end{bmatrix}$$

$$= 4 \begin{bmatrix} 1 & 0 & 0 \\ 0 & 1 & 0 \\ 0 & 0 & 1 \end{bmatrix} = 4I.$$

Also,

$$(\text{Adj } A)A = \begin{bmatrix} -4 & 5 & -2 \\ -4 & 2 & 0 \\ 4 & -3 & 2 \end{bmatrix} \begin{bmatrix} 1 & -1 & 1 \\ 2 & 0 & 2 \\ 1 & 2 & 3 \end{bmatrix} = \begin{bmatrix} 4 & 0 & 0 \\ 0 & 4 & 0 \\ 0 & 0 & 4 \end{bmatrix}$$

$$= 4 \begin{bmatrix} 1 & 0 & 0 \\ 0 & 1 & 0 \\ 0 & 0 & 1 \end{bmatrix} = 4I.$$

Since $\det A = 4$, the preceding equations can be written

$$(\text{Adj } A)A = A(\text{Adj } A) = (\det A)I.$$

Dividing through by $\det A$, we see that the inverse of A is given by $(1/\det A)\,\mathrm{Adj}\,A$. This fact suggests the following theorem, which yields the *adjoint form* of the inverse of a matrix. The proof will be delayed until later in the section.

THEOREM 3   Let A be a square matrix, with $\det A \neq 0$. Then A is invertible, and

$$A^{-1} = \frac{1}{\det A}\,\mathrm{Adj}\,A.$$

We have already seen one example of this theorem. As another, we show how to compute the inverse of a 2×2 matrix by inspection.

EXAMPLE 3.19

Find the inverse of A, where

$$A = \begin{bmatrix} a_{11} & a_{12} \\ a_{21} & a_{22} \end{bmatrix}.$$

Solution

If $\det A = a_{11}a_{22} - a_{21}a_{12} = 0$, A has no inverse. If $\det A \neq 0$, we proceed to compute $C = \mathrm{Adj}\,A$:

$$c_{11} = (-1)^{1+1} \det A_{11} = (1) \det [a_{22}] = a_{22}$$
$$c_{12} = (-1)^{1+2} \det A_{21} = (-1) \det [a_{12}] = -a_{12}$$
$$c_{21} = (-1)^{2+1} \det A_{12} = (-1) \det [a_{21}] = -a_{21}$$
$$c_{22} = (-1)^{2+2} \det A_{22} = (1) \det [a_{11}] = a_{11}.$$

Clearly, these entries can all be found with no computation. Now by Theorem 3,

$$A^{-1} = \frac{1}{\det A}\,\mathrm{Adj}\,A = \frac{1}{\det A} \begin{bmatrix} a_{22} & -a_{12} \\ -a_{21} & a_{11} \end{bmatrix}.$$

Thus, to find the inverse of a 2×2 matrix A, interchange its diagonal entries, take the negatives of its off-diagonal entries, and divide each entry in the resulting matrix by $\det A$.

 COMPUTATIONAL NOTE   Solving a square system of equations by Cramer's rule and likewise computing the inverse of a matrix by using the classical adjoint of the matrix are numerically inefficient methods. Although the situation is not nearly as bad as for computing determinants directly from the definition, the number of multiplications required to solve an $n \times n$ system by Cramer's rule is approximately n times the number needed to do this by row operations. For this reason, Cramer's rule and the adjoint form of the inverse are more useful as theoretical tools than for computational purposes.

THEORY OF CRAMER'S RULE

We turn now to the proofs of the theorems we have been discussing. In a square linear system $A\mathbf{x} = \mathbf{b}$, the notation $A(i)$ used in Cramer's rule is sometimes not explicit enough. When there is any doubt, we will write the matrix explicitly as a matrix of columns:

$$A(i) = [\mathbf{a}^1 \cdots \mathbf{a}^{i-1} \; \mathbf{b} \; \mathbf{a}^{i+1} \cdots \mathbf{a}^n].$$

Expressed as columns in this manner, Theorem 4 of Section 3.2 can be written as follows. Let A be an $n \times n$ matrix, and let d be a scalar. Then for any $k \neq i$,

$$\det [\mathbf{a}^1 \cdots \mathbf{a}^{i-1} \; \mathbf{a}^i + d\mathbf{a}^k \; \mathbf{a}^{i+1} \cdots \mathbf{a}^n] = \det A.$$

Combined with Theorem 1 of Section 3.2, we can extend this as follows: let A be an $n \times n$ matrix and c and d be scalars. Then for any $k \neq i$,

$$\det [\mathbf{a}^1 \cdots \mathbf{a}^{i-1} \; c\mathbf{a}^i + d\mathbf{a}^k \; \mathbf{a}^{i+1} \cdots \mathbf{a}^n] = c \det A.$$

Repeated application of this fact gives the proof of the following result.

LEMMA* Let A be a square matrix of order n, and let $\mathbf{b}$ be a column vector with $\mathbf{b} = c_1\mathbf{a}^1 + \cdots + c_n\mathbf{a}^n$, a linear combination of the columns of A. Then for each i,

$$\det A(i) = c_i \det A.$$

As an illustration of the Lemma, let

$$A = \begin{bmatrix} 2 & 3 & -1 \\ 1 & 2 & 4 \\ 3 & -1 & 0 \end{bmatrix},$$

and notice that

$$\begin{bmatrix} 1 \\ -5 \\ 11 \end{bmatrix} = 3\begin{bmatrix} 2 \\ 1 \\ 3 \end{bmatrix} - 2\begin{bmatrix} 3 \\ 2 \\ -1 \end{bmatrix} - \begin{bmatrix} -1 \\ 4 \\ 0 \end{bmatrix}.$$

---

* A *lemma* is a theorem that is used solely for the purpose of proving another theorem (which usually immediately follows it in the text).

Then by the Lemma,

$$\det \begin{bmatrix} 2 & 1 & -1 \\ 1 & -5 & 4 \\ 3 & 11 & 0 \end{bmatrix} = -2 \det \begin{bmatrix} 2 & 3 & -1 \\ 1 & 2 & 4 \\ 3 & -1 & 0 \end{bmatrix}.$$

Consider the square system of linear equations $A\mathbf{x} = \mathbf{b}$:

$$
\begin{aligned}
a_{11}x_1 + a_{12}x_2 + \cdots + a_{1n}x_n &= b_1 \\
a_{21}x_1 + a_{22}x_2 + \cdots + a_{2n}x_n &= b_2 \\
\vdots \qquad \vdots \qquad\qquad \vdots \quad\; \vdots \\
a_{n1}x_1 + a_{n2}x_2 + \cdots + a_{nn}x_n &= b_n.
\end{aligned}
\tag{1}
$$

Another way of viewing this system is to note that each element in the first column is multiplied by x_1, each in the second by x_2, and so on. The system can thus be written as

$$\mathbf{a}^1 x_1 + \mathbf{a}^2 x_2 + \cdots + \mathbf{a}^n x_n = \mathbf{b}. \tag{2}$$

In words, $\mathbf{x} = (x_1, x_2, \ldots, x_n)$ is a solution to the system if and only if it is a vector such that x_1 times the first column of A plus x_2 times the second column, and so on, yields the column $\mathbf{b}$. We now restate and prove Cramer's rule.

THEOREM I   Let $A\mathbf{x} = \mathbf{b}$ be a square system of linear equations. Each component x_i of $\mathbf{x}$ satisfies the equation $x_i \det A = \det A(i)$. If $\det A \neq 0$, the solution is unique, and for each i,

$$x_i = \frac{\det A(i)}{\det A}.$$

Proof   From Equation (2), any solution $\mathbf{x}$ gives the column vector $\mathbf{b}$ as a linear combination of the columns of A,

$$\mathbf{b} + x_1 \mathbf{a}^1 + \cdots + x_n \mathbf{a}^n.$$

From the Lemma, $\det A(i) = x_i \det A$. If $\det A \neq 0$, divide both sides by $\det A$ to obtain the final equation. That the solution is *unique* in case $\det A \neq 0$ follows from Theorem 6 of Section 3.2.

The proof of Theorem 3 requires no special preparation—only a careful look at the products $(\text{Adj } A)A$ and $A(\text{Adj } A)$. We restate Theorem 3 in a slightly more general manner and prove it.

THEOREM 3 For a square matrix A,

$$A(\text{Adj } A) = (\text{Adj } A)A = (\det A)I.$$

Proof We compute $(\text{Adj } A)A$ and leave $A(\text{Adj } A)$ as an exercise. Let $A = [a_{ij}]$ be of order n, and let $C = (\text{Adj } A)A$. Then by the definition of matrix multiplication,

$$
\begin{aligned}
c_{ij} &= (-1)^{i+1}(\det A_{1i})a_{1j} + \cdots + (-1)^{i+n}(\det A_{ni})a_{nj} \\
&= (-1)^{i+1}a_{1j} \det A_{1i} + \cdots + (-1)^{i+n}a_{nj} \det A_{ni}.
\end{aligned}
$$

If $i = j$ (for an element on the main diagonal),

$$
\begin{aligned}
c_{ii} &= (-1)^{i+1}a_{1i} \det A_{1i} + \cdots + (-1)^{i+n}a_{ni} \det A_{ni} \\
&= \det A,
\end{aligned}
$$

since this is just the expansion of $\det A$ down the ith column. If $i \neq j$ (for an off-diagonal element),

$$
\begin{aligned}
c_{ij} &= (-1)^{i+1}a_{1j} \det A_{1i} + \cdots + (-1)^{i+n}a_{nj} \det A_{ni} \\
&= \det A',
\end{aligned}
$$

where A' is obtained from A by replacing the ith column by the jth column. (The expansion of $\det A'$ is down the ith column.) However, A' has two equal columns (the ith and jth), and therefore $\det A' = 0$. That is, $c_{ij} = 0$ for any off-diagonal element.

3.3 EXERCISES

In Exercises 1–8 determine whether or not Cramer's rule can be used to solve the given system. If it can be used, solve the system by Cramer's rule. If it cannot be used, explain why not.

1. $\quad x - 2y = 3$
$\quad\quad 2x + y = 1$

2. $\quad -x + y = 1$
$\quad\quad\quad x - y = 2$

3. $\quad x + y - z = 1$
$\quad\quad 2x \quad\quad - 3z = 0$
$\quad\quad\quad 2y + z = 1$

4. $\quad x - y + z = 1$
$\quad\quad 2x + 3y + z = 2$
$\quad\quad\quad x + 2y \quad = 3$

5. $3x - y + z = 2$
$\quad 2x + y - z = 1$
$\quad\; x + 5y - 3z = 3$

6. $\quad x + 2y - z = 3$
$\quad\quad 2x \quad\quad - z = 1$
$\quad\quad\quad x + 6y - z = 5$

7. $3w + x + y + z = 2$
$\quad\quad\; x + 2y + 3z = 1$
$\quad 3w \quad\quad\quad - 2z = 2$
$\quad\quad 2x + 4y + 7z = 1$

8. $2w + x \quad\quad - z = 1$
$\quad\quad x + y \quad = 2$
$\quad\quad\quad\; y + z = 2$

In Exercises 9–14 compute the adjoint of each matrix, and find the inverse of the matrix if it exists.

9. $\begin{bmatrix} 1 & 2 \\ -3 & 4 \end{bmatrix}$

10. $\begin{bmatrix} 1 & -1 \\ 1 & -2 \end{bmatrix}$

11. $\begin{bmatrix} 1 & -1 & 0 \\ 0 & 2 & -2 \\ 0 & 0 & 1 \end{bmatrix}$

12. $\begin{bmatrix} 2 & 1 & 0 \\ 1 & 2 & 2 \\ 3 & -1 & 4 \end{bmatrix}$

13. $\begin{bmatrix} 1 & 2 & -1 \\ 2 & 1 & 4 \\ 1 & 5 & -7 \end{bmatrix}$

14. $\begin{bmatrix} 1 & -2 & 0 \\ -2 & 1 & -2 \\ 0 & -2 & 1 \end{bmatrix}$

15. Use the adjoint form of the inverse to derive an explicit formula for computing the inverse of a 3×3 matrix.

16. Prove directly that $A(\text{Adj } A) = (\det A)I$ for any square matrix A.

17. A matrix A is *symmetric* if $A = A^T$. Prove that if A is symmetric and invertible, then A^{-1} is also symmetric and invertible.

18. Prove that the inverse of an invertible upper triangular matrix of order 3 is invertible and upper triangular. (The statement is in fact true for invertible upper triangular matrices in general.)

19. Prove that $\det A \neq 0$ if and only if $\det (\text{Adj } A) \neq 0$. (*Hint:* Use Theorem 3.)

20. Prove that $\det (\text{Adj } A) = (\det A)^{n-1}$, $n > 1$. (*Note:* You need Exercise 19 to prove the $\det A = 0$ case.)

 21. Solve the system of equations by Cramer's rule using a computer:

$$0.02w - 0.23x + 0.40y - \quad z = 3.02$$
$$2.10w + \quad x - 1.30y \quad = 4.11$$
$$1.16w - 0.15x \quad - 2.35z = 0$$
$$3.91x + 0.63y - \quad z = 2.34.$$

 22. Find the adjoint of the matrix using a computer:

$$\begin{bmatrix} 2.11 & 2.11 & -3.04 & 1.11 \\ -0.02 & 1.23 & 2.22 & 1.02 \\ 0.14 & -0.06 & 1.21 & -1.08 \\ 1.32 & 0.20 & 0 & 3.90 \end{bmatrix}.$$

(*Hint:* If A is invertible, $\text{Adj } A = (\det A)A^{-1}$. Otherwise use the definition of Adj A and a program for determinants.)

CHAPTER 3 REVIEW

In this chapter the *determinant* of an $n \times n$ matrix was defined and properties of determinants were discussed. Determinants were then used to solve certain square linear systems (by *Cramer's rule*) and to determine the inverse of a square matrix.

You should be able to:

1. calculate determinants of square matrices of all orders;
2. use elementary row (and column) operations to reduce the work involved in finding a determinant;
3. solve certain linear systems by Cramer's rule;
4. find the adjoint of a square matrix and its inverse by using the adjoint.

KEY WORDS

determinant of a matrix

cofactor expansion of a determinant

Cramer's rule for solving a linear system

minor and cofactor of a matrix

upper triangular, lower triangular, and diagonal matrices

adjoint of a square matrix

KEY THEOREMS

Theorem If a square matrix A is diagonal, lower triangular, or upper triangular, then $\det A = a_{11}a_{22}\cdots a_{nn}$.

Theorem If a square matrix B is obtained from a square matrix A by:

i. multiplying a row (or column) of A by the scalar c, then $\det B = c \det A$;

ii. interchanging two rows (or columns) of A, then $\det B = -\det A$;

iii. adding a scalar multiple of one row (or column) of A to another, then $\det B = \det A$.

Theorem If a matrix has a row (or column) that is either zero or a scalar multiple of another row (or column), then its determinant is zero.

Theorem Let A be a square matrix. Then $\det A \neq 0$ is equivalent to either of the following:

i. A is invertible; *or*

ii. $A\mathbf{x} = \mathbf{b}$ has a unique solution for every $\mathbf{b}$.

Theorem If A and B are square matrices of the same order, then $\det AB = (\det A)(\det B)$.

Theorem If A is a square matrix, then $\det A = \det (A^T)$.

Theorem Let $A\mathbf{x} = \mathbf{b}$ be a linear system where A is an $n \times n$ matrix with $\det A \neq 0$ and let $A(i)$ denote the $n \times n$ matrix obtained from A by replacing its ith column by $\mathbf{b}$. Then

$$x_i = \frac{\det A(i)}{\det A} \quad \text{for } i = 1, 2, \ldots, n.$$

Theorem   Let A be a square matrix with det $A \neq 0$ and let Adj A denote the adjoint of A. Then

$$A^{-1} = \left(\frac{1}{\det A}\right) \text{Adj } A.$$

SELF-TEST

Answers appear at the end of this chapter.

Find the determinant of each matrix in Exercises 1–3. Use any facts about determinants that simplify the work.

1. $\begin{bmatrix} 3 & 2 & 1 \\ 1 & 4 & 2 \\ 2 & 5 & 3 \end{bmatrix}$

2. $\begin{bmatrix} 0.02 & 0.03 & 0.01 \\ -0.01 & 0.06 & 0.02 \\ 0.02 & -0.09 & -0.03 \end{bmatrix}$

3. $\begin{bmatrix} 0 & 2 & 0 & 0 \\ 3 & 0 & 0 & 0 \\ 0 & 0 & 0 & 2 \\ 0 & 0 & 3 & 0 \end{bmatrix}$

4. Solve the equation $\det \begin{bmatrix} 1-t & 2 & 0 \\ 0 & 2-t & 2 \\ 0 & 0 & 3-t \end{bmatrix} = 0.$ [Sections 3.1, 3.2]

Solve the linear systems in Exercises 5 and 6 by Cramer's rule.

5. $\begin{aligned} x - y &= 2 \\ 3x + 4y &= 3 \end{aligned}$

6. $\begin{aligned} 2x + y + z &= 3 \\ 3x \quad\quad - z &= 2 \\ 4x + 5y + 2z &= 1 \end{aligned}$ [Section 3.3]

7. Find all values of t for which the matrix A is invertible:

$$A = \begin{bmatrix} 2 & 1 & 1 \\ 3 & 1 & t \\ 1 & t & -2 \end{bmatrix}.$$ [Section 3.2]

8. If det $(A^2) = 7$, show that A is invertible.   [Section 3.2]

9. Determine Adj A where A is the matrix of Exercise 1.   [Section 3.3]

10. If A and B are *similar* (that is, if there exists an invertible matrix P such that $B = P^{-1}AP$), show that det $A =$ det B.   [Section 3.2]

REVIEW EXERCISES

In Exercises 1–8 evaluate the determinant of the given matrix in the easiest way you can.

1. $\begin{bmatrix} 1 & 1 & 0 \\ 1 & 0 & 1 \\ 0 & 1 & 1 \end{bmatrix}$

2. $\begin{bmatrix} 1 & -1 & 1 \\ -1 & 0 & -1 \\ 0 & 1 & -1 \end{bmatrix}$

3. $\begin{bmatrix} 1 & 0 & 0 \\ 1 & 2 & 0 \\ 1 & 2 & 3 \end{bmatrix}$

4. $\begin{bmatrix} 1 & 2 & 3 \\ 3 & 2 & 1 \\ 2 & 4 & 6 \end{bmatrix}$ **5.** $\begin{bmatrix} 1 & x & y \\ 1 & x^2 & y^2 \\ 1 & x^3 & y^3 \end{bmatrix}$ **6.** $\begin{bmatrix} 1 & x & y \\ 1 & x^2 & y^2 \\ 0 & 2x & 2y \end{bmatrix}$

7. $\begin{bmatrix} 1 & 0 & 2 & 1 \\ 0 & 1 & -1 & 0 \\ 2 & 0 & 5 & 2 \\ -1 & 0 & -2 & 0 \end{bmatrix}$ **8.** $\begin{bmatrix} 1 & 1 & 1 & 1 \\ 0 & 1 & 2 & 0 \\ 1 & 0 & 2 & 1 \\ -1 & -1 & -1 & 0 \end{bmatrix}$

9. Use the results of Exercises 1–4 to determine in which cases the inverse of the given matrix exists.

10. Let A be a square matrix such that $A^3 = I$. Find the value of det A.

11. Let A be an invertible matrix with det $A = 2$. Find det (A^{-1}) and det $(A^T)^{-1}$.

12. Can you give an example of a linear system that can be solved by Cramer's rule but not by the method of inverses? If not, why not?

13–16. For each matrix A in Exercises 1–4, determine Adj A and, if it exists, A^{-1}.

In Exercises 17–22 solve the given linear system by Cramer's rule, if possible.

17. $\begin{aligned} x - 2y &= 1 \\ 2x - y &= 1 \end{aligned}$ **18.** $\begin{aligned} x + y &= 2 \\ 2x - 3y &= -1 \end{aligned}$

19. $\begin{aligned} x_1 + x_2 &= 1 \\ x_1 + x_3 &= 0 \\ x_2 + x_3 &= 0 \end{aligned}$ **20.** $\begin{aligned} x_1 - x_2 \quad\;\; &= 1 \\ x_1 \quad\;\; - x_3 &= 2 \\ x_2 - x_3 &= 3 \end{aligned}$

21. $\begin{aligned} x \quad\;\; + z \quad\;\; &= 0 \\ y \quad\;\; + w &= 0 \\ x \quad\;\; - z \quad\;\; &= 0 \\ y + 2z + w &= 0 \end{aligned}$ **22.** $\begin{aligned} x + y \quad\quad\;\; &= 0 \\ y + z \quad\;\; &= 0 \\ z + w &= 0 \\ x \quad\quad\;\; + w &= 0 \end{aligned}$

ANSWERS TO SELF-TEST

1. 5 **2.** 0 **3.** 36 **4.** $\{1, 2, 3\}$

5. $(\frac{11}{7}, -\frac{3}{7})$ **6.** $(\frac{4}{3}, -\frac{5}{3}, 2)$

7. $\{(2 + \sqrt{6})/2, (2 - \sqrt{6})/2\}$ **8.** det $A = \pm\sqrt{7} \neq 0$

9. $\begin{bmatrix} 2 & -1 & 0 \\ 1 & 7 & -5 \\ -3 & -11 & 10 \end{bmatrix}$

10. $\begin{aligned} \det B &= \det (P^{-1}AP) \\ &= \det (P^{-1}) \det A \det P \\ &= \det A \det (P^{-1}) \det P \\ &= \det A \det (P^{-1}P) \\ &= \det A \det I \\ &= \det A \end{aligned}$

INDEPENDENCE AND BASIS IN R^m

We have sometimes referred to $\mathbf{R}^m$, the set of all m-vectors, as "m-dimensional space." In this chapter we will make the notion of *dimension* more precise. To do so, we first introduce the concepts of *linear independence* and *basis* in $\mathbf{R}^m$. The ideas and tools developed here will be used throughout the rest of this text.

LINEAR DEPENDENCE AND INDEPENDENCE

As we have seen in Section 1.4, two 3-vectors, $\mathbf{v}_1$ and $\mathbf{v}_2$, determine a plane in $\mathbf{R}^3$, as long as one of them is not a scalar multiple of the other. Moreover, a third 3-vector, $\mathbf{w}$, will lie in the plane of $\mathbf{v}_1$ and $\mathbf{v}_2$ if in some sense it is "dependent" on these vectors. More precisely, this will occur if (and only if) we can find scalars c_1 and c_2 such that $\mathbf{w} = c_1\mathbf{v}_1 + c_2\mathbf{v}_2$. Figure 4.1 illustrates this statement. In this section we will investigate such dependence relations in the more general setting of $\mathbf{R}^m$.

DEFINITION

A **linear combination** of the vectors $\mathbf{v}_1, \mathbf{v}_2, \ldots, \mathbf{v}_n$ in $\mathbf{R}^m$ is an expression of the form

$$c_1\mathbf{v}_1 + c_2\mathbf{v}_2 + \cdots + c_n\mathbf{v}_n,$$

where the c_i are scalars.

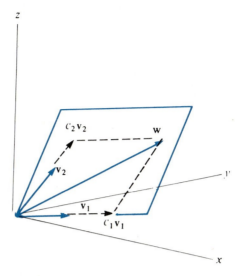

FIGURE 4.1

EXAMPLE 4.1

Let $\mathbf{v} = (1, 2, 3)$, $\mathbf{v}_1 = (1, 0, 0)$, $\mathbf{v}_2 = (1, 1, 0)$, and $\mathbf{v}_3 = (1, 1, 1)$. Since

$$(1, 2, 3) = (-1)(1, 0, 0) + (-1)(1, 1, 0) + 3(1, 1, 1),$$

we can express $\mathbf{v}$ as a linear combination of the given vectors, $\mathbf{v} = c_1\mathbf{v}_1 + c_2\mathbf{v}_2 + c_3\mathbf{v}_3$, by taking $c_1 = -1$, $c_2 = -1$, and $c_3 = 3$.

EXAMPLE 4.2

Let $\mathbf{v}_1 = (-2, 0, 1)$, $\mathbf{v}_2 = (1, -1, 2)$, and $\mathbf{v}_3 = (4, -2, 3)$. Express $\mathbf{v}_3$ as a linear combination of the other two vectors.

Solution

We need to find constants c_1 and c_2 such that

$$(4, -2, 3) = c_1(-2, 0, 1) + c_2(1, -1, 2).$$

To determine c_1 and c_2, we first perform the scalar multiplication and addition on the right, obtaining

$$(4, -2, 3) = (-2c_1, 0, c_1) + (c_2, -c_2, 2c_2)$$
$$= (-2c_1 + c_2, -c_2, c_1 + 2c_2).$$

Now, equating corresponding components yields the linear system

$$\begin{aligned}
4 &= -2c_1 + c_2 \\
-2 &= - c_2 \\
3 &= c_1 + 2c_2.
\end{aligned}$$

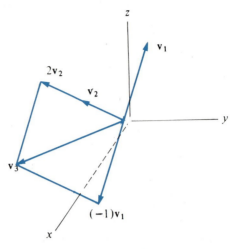

FIGURE 4.2

This system has solution: $c_1 = -1$, $c_2 = 2$. Thus

$$(4, -2, 3) = (-1)(-2, 0, 1) + (2)(1, -1, 2).$$

(Figure 4.2 gives a picture of the situation.)

DEFINITION

A set of two or more vectors is **linearly dependent** in $\mathbf{R}^m$ if one of them can be expressed as a linear combination of the others. A set consisting of a single vector is **linearly dependent** if that vector is equal to **0**. If a set of vectors is *not* linearly dependent, it is said to be **linearly independent**.

EXAMPLE 4.3

Show that the vectors

$$\mathbf{v}_1 = (-2, 0, 1), \qquad \mathbf{v}_2 = (1, -1, 2), \qquad \text{and} \qquad \mathbf{v}_3 = (4, -2, 3)$$

are linearly dependent.

Solution

We have already shown in Example 4.2 that $\mathbf{v}_3 = (-1)\mathbf{v}_1 + 2\mathbf{v}_2$. Thus, by definition, the vectors are linearly dependent. (Geometrically, we have shown that $\mathbf{v}_3$ lies in the plane determined by $\mathbf{v}_1$ and $\mathbf{v}_2$, as shown in Figure 4.2.)

NOTE   In Example 4.3 we can also express $\mathbf{v}_1$ as a linear combination of $\mathbf{v}_2$ and $\mathbf{v}_3$, and $\mathbf{v}_2$ as a linear combination of $\mathbf{v}_1$ and $\mathbf{v}_3$. You might try to find the appropriate scalars involved.

The following theorem gives another characterization of linear dependence/independence. This characterization leads to a systematic way of testing a set of m-vectors for linear dependence.

The vector equation $c_1\mathbf{v}_1 + c_2\mathbf{v}_2 + \cdots + c_n\mathbf{v}_n = \mathbf{0}$, where the $\mathbf{v}_i$ are given, always has the solution $c_1 = c_2 = \cdots = c_n = 0$. Theorem 1 states that this is the *only* solution if the vectors are linearly independent.

THEOREM 1   Let $\mathcal{S} = \{\mathbf{v}_1, \mathbf{v}_2, \ldots, \mathbf{v}_n\}$. Then $\mathcal{S}$ is a linearly dependent set if and only if the vector equation $c_1\mathbf{v}_1 + c_2\mathbf{v}_2 + \cdots + c_n\mathbf{v}_n = \mathbf{0}$ has a solution with at least one $c_i \neq 0$.

Proof   When $n = 1$, the vector equation is $c_1\mathbf{v}_1 = \mathbf{0}$. This has a solution $c_1 \neq 0$ if and only if $\mathbf{v}_1 = \mathbf{0}$, as desired. Now consider the case when $n > 1$.

First, assume the set $\mathcal{S}$ is linearly dependent. Then we can express one of the vectors (say, $\mathbf{v}_1$) as a linear combination of the others. In other words, there exist $k_2, \ldots, k_n$ such that $\mathbf{v}_1 = k_2\mathbf{v}_2 + \cdots + k_n\mathbf{v}_n$. This implies that $1\mathbf{v}_1 - k_2\mathbf{v}_2 - \cdots - k_n\mathbf{v}_n = \mathbf{0}$. By letting $c_1 = 1$ and $c_i = -k_i$ for $i > 1$, we have a solution of the vector equation $c_1\mathbf{v}_1 + c_2\mathbf{v}_2 + \cdots + c_n\mathbf{v}_n = \mathbf{0}$ with at least one $c_i \neq 0$, namely, $c_1 = 1$.

Now assume that the vector equation $c_1\mathbf{v}_1 + c_2\mathbf{v}_2 + \cdots + c_n\mathbf{v}_n = \mathbf{0}$ has a solution with at least one $c_i \neq 0$. Say it is c_1 that is not zero. Then dividing both sides by c_1 yields

$$\mathbf{v}_1 + (c_2/c_1)\mathbf{v}_2 + \cdots + (c_n/c_1)\mathbf{v}_n = \mathbf{0},$$

and transposing all terms except $\mathbf{v}_1$ to the right-hand side, we obtain

$$\mathbf{v}_1 = -(c_2/c_1)\mathbf{v}_2 - \cdots - (c_n/c_1)\mathbf{v}_n.$$

Thus we have expressed one of the vectors as a linear combination of the others. (The same argument holds if some other c_i is not zero.) Hence the set is linearly dependent.

NOTE   The solution $c_1 = c_2 = \cdots = c_n = 0$ of the vector equation $c_1\mathbf{v}_1 + c_2\mathbf{v}_2 + \cdots + c_n\mathbf{v}_n = \mathbf{0}$ is called the **trivial solution.** Using this terminology, we can state the following corollary to Theorem 1.

··

COROLLARY   The set $\{\mathbf{v}_1, \mathbf{v}_2, \ldots, \mathbf{v}_n\}$ is linearly independent if and only if the vector equation $c_1\mathbf{v}_1 + c_2\mathbf{v}_2 + \cdots + c_n\mathbf{v}_n = \mathbf{0}$ has only the trivial solution.

EXAMPLE 4.4   Determine whether or not the set of vectors $\{\mathbf{v}_1, \mathbf{v}_2, \mathbf{v}_3\}$ is linearly dependent, where

$$\mathbf{v}_1 = (1, 0, -1), \qquad \mathbf{v}_2 = (0, 1, 2), \qquad \mathbf{v}_3 = (1, 1, 3).$$

Solution   We wish to determine whether or not the vector equation $c_1\mathbf{v}_1 + c_2\mathbf{v}_2 + c_3\mathbf{v}_3 = \mathbf{0}$ has a nontrivial solution. This equation is equivalent to

$$c_1(1, 0, -1) + c_2(0, 1, 2) + c_3(1, 1, 3) = (0, 0, 0),$$

$$(c_1, 0, -c_1) + (0, c_2, 2c_2) + (c_3, c_3, 3c_3) = (0, 0, 0),$$

or

$$(c_1 + c_3, c_2 + c_3, -c_1 + 2c_2 + 3c_3) = (0, 0, 0).$$

Now, equating corresponding components, we get the system of equations:

$$\begin{aligned} c_1 \quad\quad + \ c_3 &= 0 \\ c_2 + \ c_3 &= 0 \\ -c_1 + 2c_2 + 3c_3 &= 0. \end{aligned}$$

In matrix form this system is

$$\begin{bmatrix} 1 & 0 & 1 \\ 0 & 1 & 1 \\ -1 & 2 & 3 \end{bmatrix} \begin{bmatrix} c_1 \\ c_2 \\ c_3 \end{bmatrix} = \begin{bmatrix} 0 \\ 0 \\ 0 \end{bmatrix}.$$

So we may restate the original problem as, "Does the linear system, $A\mathbf{c} = \mathbf{0}$, where

$$A = \begin{bmatrix} 1 & 0 & 1 \\ 0 & 1 & 1 \\ -1 & 2 & 3 \end{bmatrix}, \qquad \mathbf{c} = \begin{bmatrix} c_1 \\ c_2 \\ c_3 \end{bmatrix}, \qquad \text{and} \qquad \mathbf{0} = \begin{bmatrix} 0 \\ 0 \\ 0 \end{bmatrix},$$

have a nontrivial solution?" If so, the set $\{\mathbf{v}_1, \mathbf{v}_2, \mathbf{v}_3\}$ is linearly dependent.

Transforming the augmented matrix for this system to row-reduced echelon form, we obtain

$$\left[\begin{array}{ccc|c} 1 & 0 & 0 & 0 \\ 0 & 1 & 0 & 0 \\ 0 & 0 & 1 & 0 \end{array}\right].$$

Hence the system has the unique solution $c_1 = 0$, $c_2 = 0$, $c_3 = 0$, and the set $\{\mathbf{v}_1, \mathbf{v}_2, \mathbf{v}_3\}$ is linearly independent by the corollary to Theorem 1.

Notice that in Example 4.4 the first, second, and third columns of the matrix A are, respectively, the given vectors $\mathbf{v}_1$, $\mathbf{v}_2$, and $\mathbf{v}_3$. This is no coincidence and suggests the next procedure, which will be justified by Theorem 4 (given toward the end of this section). We first introduce the following terminology.

DEFINITION

An **elementary vector** in $\mathbf{R}^m$ is one that has one component equal to 1 and all other components equal to 0. If the 1 occurs as the ith component, the elementary vector is denoted $\mathbf{e}_i$. A column of a matrix whose entries form an elementary vector is called an **elementary column.**

For example, if

$$A = \begin{bmatrix} 1 & 2 & 0 & 0 & 1 \\ 0 & 0 & 1 & 0 & 2 \\ 0 & 0 & 0 & 1 & 3 \end{bmatrix},$$

then the first, third, and fourth columns of A are *elementary columns*. Specifically $\mathbf{a}^1 = \mathbf{e}_1$, $\mathbf{a}^3 = \mathbf{e}_2$, and $\mathbf{a}^4 = \mathbf{e}_3$.

TEST FOR LINEAR DEPENDENCE/INDEPENDENCE

Let $S = \{\mathbf{v}_1, \mathbf{v}_2, \ldots, \mathbf{v}_n\}$ be a set of m-vectors. Construct the $m \times n$ matrix A whose ith column $\mathbf{a}^i$ consists of the components of $\mathbf{v}_i$. The set S is linearly independent if and only if the row-reduced echelon form of A contains only distinct elementary columns.

EXAMPLE 4.5

Determine whether or not the set

$$\{(-1, 0, 2, 1), \quad (3, 1, -2, 0), \quad (0, 1, 4, 3)\}$$

is linearly dependent.

Solution

The 4×3 matrix A with these vectors as columns is

$$A = \begin{bmatrix} -1 & 3 & 0 \\ 0 & 1 & 1 \\ 2 & -2 & 4 \\ 1 & 0 & 3 \end{bmatrix}.$$

The row-reduced echelon form of A is

$$\begin{bmatrix} 1 & 0 & 3 \\ 0 & 1 & 1 \\ 0 & 0 & 0 \\ 0 & 0 & 0 \end{bmatrix}.$$

Since the third column is not an elementary column, this set is linearly dependent.

COMPUTATIONAL NOTE   It is possible to use a computer program to help you determine whether a set of vectors is or is not linearly independent. To do so, form the matrix A described in the test for linear dependence/independence and use a program to transform A into row-reduced echelon form. Then inspect the resulting matrix. If it consists entirely of distinct elementary vectors, the original set is linearly independent; otherwise the set is linearly dependent. Unfortunately, rounding errors inherent in computer calculations (see the Computational Note of Section 2.5) could result in nonzero entries appearing in the row-reduced echelon form where in fact zeros should occur, or vice versa. Even if the program is written to minimize errors of this sort (for example, by working with fractions rather than converting them to decimals), it is possible that an incorrect answer will be obtained.

Since the set of vectors of Example 4.5 is linearly dependent, one of them can be expressed as a linear combination of the others. In fact, we have

$$(0, 1, 4, 3) = (3)(-1, 0, 2, 1) + (1)(3, 1, -2, 0).$$

Notice that the coefficients (3 and 1) of the linear combination appear in the third column of the row-reduced echelon matrix. When a set of m-vectors is linearly dependent, these *coefficients of dependence* can always be found in this way. To be more explicit, we state the following procedure (justified by Theorem 5 at the end of this section).

DETERMINING COEFFICIENTS OF DEPENDENCE

Let $\mathcal{S} = \{\mathbf{v}_1, \mathbf{v}_2, \ldots, \mathbf{v}_n\}$ be a set of vectors in $\mathbf{R}^m$. Construct the $m \times n$ matrix A with $\mathbf{a}^i = \mathbf{v}_i$. Transform A to its row-reduced echelon form, B. Suppose a column of B, say $\mathbf{b}^j$, is either a nonelementary column or is a duplication of a preceding elementary column. Then we have

$$\mathbf{v}_j = b_{1j}\mathbf{w}_1 + b_{2j}\mathbf{w}_2 + \cdots + b_{kj}\mathbf{w}_k,$$

where $\mathbf{w}_1, \mathbf{w}_2, \ldots, \mathbf{w}_k$ are the vectors in $\mathcal{S}$ that have been transformed into the distinct elementary columns of B that precede the jth column.

EXAMPLE 4.6   Let $\mathbf{v}_1 = (1, -1, 2)$, $\mathbf{v}_2 = (-2, 3, 0)$, $\mathbf{v}_3 = (0, 1, 4)$, $\mathbf{v}_4 = (1, 0, 1)$, and $\mathbf{v}_5 = (2, 1, 1)$. Show that the set $\{\mathbf{v}_1, \mathbf{v}_2, \mathbf{v}_3, \mathbf{v}_4, \mathbf{v}_5\}$ is linearly dependent, and express at least one of the $\mathbf{v}_i$ as a linear combination of the others.

Solution   The matrix A with $\mathbf{a}^i = \mathbf{v}_i$ is

$$A = \begin{bmatrix} 1 & -2 & 0 & 1 & 2 \\ -1 & 3 & 1 & 0 & 1 \\ 2 & 0 & 4 & 1 & 1 \end{bmatrix}.$$

The row-reduced echelon form of A is the matrix

$$B = \begin{bmatrix} 1 & 0 & 2 & 0 & -1 \\ 0 & 1 & 1 & 0 & 0 \\ 0 & 0 & 0 & 1 & 3 \end{bmatrix}.$$

We see that v_1, v_2, and v_4 have been transformed into the elementary vectors e_1, e_2, and e_3, respectively. (Thus v_1, v_2, and v_4 become w_1, w_2, and w_3 in the terminology of the procedure.) We conclude that

$$v_3 = 2v_1 + 1v_2 \qquad \text{(from the third column of } B)$$

and

$$v_5 = (-1)v_1 + (0)v_2 + 3v_4 \qquad \text{(from the fifth column of } B).$$

This can be easily checked by computing

$$2(1, -1, 2) + (-2, 3, 0) = (0, 1, 4)$$

and

$$(-1)(1, -1, 2) + 3(1, 0, 1) = (2, 1, 1).$$

Had Example 4.6 been phrased simply, "Is the set $\{v_1, v_2, v_3, v_4, v_5\}$ linearly independent?" the following theorem would have enabled us to answer immediately, "No." We leave the proof as Exercise 27.

THEOREM 2 Let $\mathcal{S}$ be a set of n vectors in $\mathbf{R}^m$ with $n > m$. Then $\mathcal{S}$ is linearly dependent.

EXAMPLE 4.7

Show that the vectors $v_1 = (1, -1)$, $v_2 = (0, 1)$, and $v_3 = (2, 1)$ form a linearly dependent set.

Solution

Here the given set consists of three 2-vectors, so by Theorem 2 (with $m = 2$, $n = 3$), it is linearly dependent.

If we are given a subset of $\mathbf{R}^m$ containing exactly m vectors, the following theorem supplies us with a relatively simple alternative to our previous test for linear independence. It is a consequence of Theorem 4 (following Example 4.8), since for a square matrix A, $\det A \neq 0$ if and only if A is invertible, which is true if and only if its row-reduced form is the identity matrix.

THEOREM 3   Let $\mathcal{S} = \{\mathbf{v}_1, \mathbf{v}_2, \ldots, \mathbf{v}_m\}$ be a set of vectors in $\mathbf{R}^m$, and let A be the $m \times m$ matrix with $\mathbf{a}^i = \mathbf{v}_i$. Then $\mathcal{S}$ is linearly independent if and only if $\det A \neq 0$.

EXAMPLE 4.8   Determine whether or not the set $\mathcal{S} = \{\mathbf{v}_1, \mathbf{v}_2, \mathbf{v}_3\}$ is linearly dependent, where $\mathbf{v}_1 = (1, 2, 3)$, $\mathbf{v}_2 = (-1, 0, 1)$, and $\mathbf{v}_3 = (0, 0, 1)$.

Solution   We form the matrix A with $\mathbf{a}^i = \mathbf{v}_i$:

$$A = \begin{bmatrix} 1 & -1 & 0 \\ 2 & 0 & 0 \\ 3 & 1 & 1 \end{bmatrix}.$$

Now, $\det A = 2 \neq 0$, so by Theorem 3, $\mathcal{S}$ is linearly independent.

To summarize our discussion so far, suppose we are given a set of n vectors in $\mathbf{R}^m$.

i. If $n > m$, then the set is linearly dependent.

ii. If $n = m$, we can test for linear independence by using the row-reduction procedure or by using determinants.

iii. If $n < m$, we can test for linear independence by using the row-reduction procedure.

We close this section with two theorems, the results of which have already been used in our computational work.

THEOREM 4   Let $\mathcal{S} = \{\mathbf{v}_1, \mathbf{v}_2, \ldots, \mathbf{v}_n\}$ be a set of m-vectors. Let A be the $m \times n$ matrix with $\mathbf{a}^i = \mathbf{v}_i$. Let B be the row-reduced echelon form of A. Then the set $\mathcal{S}$ is linearly independent if and only if the columns of B are the elementary vectors $\mathbf{e}_1, \mathbf{e}_2, \ldots, \mathbf{e}_n$.

Proof   The set $\mathcal{S}$ will be linearly independent if and only if the vector equation $c_1\mathbf{v}_1 + c_2\mathbf{v}_2 + \cdots + c_n\mathbf{v}_n = \mathbf{0}$ has a unique solution, the trivial one. By the definition of matrix multiplication and equality of vectors, this will be true if and only if $A\mathbf{c} = \mathbf{0}$ (where $\mathbf{c} = (c_1, c_2, \ldots, c_n)$) has a unique solution. But by Theorem 3 of Section 2.5, the latter is true if and only if the rank A is equal to n—that is, if and only if the number of nonzero rows of B is equal to n. Since in row-reduced echelon form each nonzero row has a leading 1, and each column with a leading 1 has all other entries equal to 0, the rank of A is n if and only if the columns of B are the m-vectors $\mathbf{e}_1, \mathbf{e}_2, \ldots, \mathbf{e}_n$, as desired.

THEOREM 5   Let $\mathcal{S} = \{\mathbf{v}_1, \mathbf{v}_2, \ldots, \mathbf{v}_n\}$ be a set of m-vectors. Let A be the $m \times n$ matrix with columns $\mathbf{a}^i = \mathbf{v}_i$. Let B be the row-reduced echelon form of A, and suppose that one of its columns, $\mathbf{b}^j$, is not an elementary vector or is a duplication of a preceding elementary vector. Then $\mathcal{S}$ is linearly dependent, and we have $\mathbf{v}_j = b_{1j}\mathbf{w}_1 + b_{2j}\mathbf{w}_2 + \cdots + b_{kj}\mathbf{w}_k$, where $\mathbf{w}_1, \mathbf{w}_2, \ldots, \mathbf{w}_k$ are vectors in $\mathcal{S}$ that have been transformed into $\mathbf{e}_1, \mathbf{e}_2, \ldots, \mathbf{e}_k$, the distinct elementary columns of B that precede the jth column.

Proof   By Theorem 4, the set $\mathcal{S}$ is linearly dependent. Now, the vector equation $c_1\mathbf{v}_1 + c_2\mathbf{v}_2 + \cdots + c_n\mathbf{v}_n = \mathbf{0}$ is equivalent to $A\mathbf{c} = \mathbf{0}$, which in turn is equivalent to $B\mathbf{c} = \mathbf{0}$. For the sake of notational convenience, suppose $\mathbf{w}_i = \mathbf{v}_i(i = 1, 2, \ldots, k)$. Then the first k equations of $B\mathbf{c} = \mathbf{0}$ become

$$
\begin{aligned}
c_1 \qquad\quad + \cdots + b_{1j}c_j + \cdots &= 0 \\
c_2 \qquad + \cdots + b_{2j}c_j + \cdots &= 0 \\
\ddots \qquad\qquad \vdots \qquad\quad \vdots \\
c_k + \cdots + b_{kj}c_j + \cdots &= 0.
\end{aligned}
$$

Letting $c_j = 1$ and $c_i = 0$ for all variables other than $c_1, \ldots, c_k$, we have a solution $c_1 = -b_{1j}, c_2 = -b_{2j}, \ldots, c_k = -b_{kj}, c_j = 1$ and $c_i = 0$ for $i > k$ and $i \neq j$. But the c_i were the coefficients needed to express

$$c_1\mathbf{v}_1 + c_2\mathbf{v}_2 + \cdots + c_n\mathbf{v}_n = \mathbf{0}.$$

On replacing the c_i with the solution values, we have

$$
\begin{aligned}
-b_{1j}\mathbf{v}_1 - b_{2j}\mathbf{v}_2 - \cdots - b_{kj}\mathbf{v}_k + 0\mathbf{v}_{k+1} + \cdots + 0\mathbf{v}_{j-1} + 1\mathbf{v}_j + 0\mathbf{v}_{j+1} \\
+ \cdots + 0\mathbf{v}_n = \mathbf{0},
\end{aligned}
$$

or, solving for $\mathbf{v}_j$,

$$\mathbf{v}_j = b_{1j}\mathbf{v}_1 + b_{2j}\mathbf{v}_2 + \cdots + b_{kj}\mathbf{v}_k,$$

as desired.

4.1 EXERCISES

In Exercises 1–4 let $\mathbf{u} = (1, 0, -1)$ and $\mathbf{v} = (-2, 1, 1)$.

1. Write $\mathbf{w}_1 = (-1, 2, -1)$ as a linear combination of $\mathbf{u}$ and $\mathbf{v}$.

2. Show that $\mathbf{w}_2 = (-1, 1, 1)$ *cannot* be written as a linear combination of $\mathbf{u}$ and $\mathbf{v}$.

3. For what value of c is the vector $(1, 1, c)$ a linear combination of $\mathbf{u}$ and $\mathbf{v}$?

4. If the vector (x_1, x_2, x_3) is a linear combination of $\mathbf{u}$ and $\mathbf{v}$, find an equation relating x_1, x_2, and x_3.

In Exercises 5–14 determine whether the given set of vectors is linearly dependent or independent. If it is linearly dependent, express one of the vectors in the set as a linear combination of the others.

5. $\{(1, 2, 3), (-1, 0, 1), (0, 1, 2)\}$

6. $\{(-1, 1, 2), (3, 3, 1), (1, 2, 2)\}$

7. $\{(0, 1, 2, 3), (0, -1, 2, -1), (0, 1, 0, 1)\}$

8. $\{(-1, -2, 2, 1), (0, 0, 0, 0), (1, 2, 3, 4)\}$

9. $\{(1, 2), (-1, 2)\}$

10. $\{(1, -1, 0), (2, 4, 0)\}$

11. $\{(1, 0), (0, 1), (1, 1)\}$

12. $\{(2, 0, 1), (1, -2, 0), (4, -4, 1), (1, 1, 1)\}$

13. $\{(1, -1, 1, -1), (0, 1, 0, 1), (1, 1, 0, 0), (2, 1, 1, 1)\}$

14. $\{(1, 1, 1, 1), (0, 1, 2, 3), (0, -1, -2, -3), (1, 0, -1, -2)\}$

In Exercises 15–18 let $\mathcal{S} = \{\mathbf{v}_1, \mathbf{v}_2, \mathbf{v}_3, \mathbf{v}_4, \mathbf{v}_5\}$ be vectors in $\mathbf{R}^3$ and let A be the 3×5 matrix with ith column $\mathbf{a}^i = \mathbf{v}_i$. Suppose that the row-reduced echelon form of A is

$$\begin{bmatrix} 1 & 2 & 0 & -1 & 0 \\ 0 & 0 & 1 & 3 & 0 \\ 0 & 0 & 0 & 0 & 1 \end{bmatrix}.$$

Are the indicated sets linearly dependent or independent? If linearly dependent, express one vector as a linear combination of the others.

15. $\{\mathbf{v}_1, \mathbf{v}_2, \mathbf{v}_3\}$

16. $\{\mathbf{v}_1, \mathbf{v}_3, \mathbf{v}_4\}$

17. $\{\mathbf{v}_1, \mathbf{v}_4, \mathbf{v}_5\}$

18. $\{\mathbf{v}_3, \mathbf{v}_4, \mathbf{v}_5\}$

In Exercises 19–26 use the definition or Theorems 2 or 3 to determine whether or not the given set is linearly dependent.

19. $\{(1, 2), (0, 2), (1, 0), (-1, 1)\}$

20. $\{(1, -1, 0), (1, 1, 1), (0, -2, 1), (1, 4, -2)\}$

21. $\{(1, 2), (3, -1)\}$

22. $\{(1, 0, 1), (-1, 1, 0), (0, 1, 1)\}$

23. $\{(1, -1, 3, 0), (1, 0, 1, 0), (0, 0, 0, 0), (4, 2, 3, -6)\}$

24. $\{(2, 0, 0, 0), (2, 1, 0, 0), (-1, 3, -2, 0), (1, -2, 4, -3)\}$

25. $\{(1, 1, 1), (2, 2, 2)\}$

26. $\{(1, 0, -1, 0), (3, 6, 1, 2), (-2, 0, 2, 0)\}$

27. Prove Theorem 2: If $\mathcal{S}$ is a set of n vectors in $\mathbf{R}^m$ with $n > m$, then $\mathcal{S}$ is linearly dependent. (*Hint:* Apply Theorem 3, Corollary 2 of Section 2.5.)

28. Let $\mathbf{v}_1, \mathbf{v}_2, \ldots, \mathbf{v}_n$ be arbitrary vectors in $\mathbf{R}^m$. Show that $\{\mathbf{0}, \mathbf{v}_1, \mathbf{v}_2, \ldots, \mathbf{v}_n\}$ is linearly dependent.

29. Show that any set of distinct elementary vectors in $\mathbf{R}^m$ is linearly independent.

30. Let $\{\mathbf{v}_1, \mathbf{v}_2, \mathbf{v}_3, \mathbf{v}_4\}$ be a linearly independent set in $\mathbf{R}^m$. Show that $\{\mathbf{v}_1, \mathbf{v}_2, \mathbf{v}_3\}$ is also linearly independent.

31. Let $\mathcal{S}$ be a linearly independent set in $\mathbf{R}^m$. Show that any subset of $\mathcal{S}$ is also linearly independent.

32. Let $\{\mathbf{v}_1, \mathbf{v}_2, \ldots, \mathbf{v}_n\}$ be a linearly dependent set in $\mathbf{R}^m$. Let $\mathbf{x}$ be any vector in $\mathbf{R}^m$. Show that $\{\mathbf{v}_1, \ldots, \mathbf{v}_n, \mathbf{x}\}$ is linearly dependent.

33. Let $\mathbf{u}$ and $\mathbf{v}$ be nonzero orthogonal vectors in $\mathbf{R}^m$. Show that they are linearly independent. (*Hint:* Use the fact that $\mathbf{u} \cdot (c\mathbf{u} + d\mathbf{v}) = c\mathbf{u} \cdot \mathbf{u}$ and $\mathbf{v} \cdot (c\mathbf{u} + d\mathbf{v}) = d\mathbf{v} \cdot \mathbf{v}$.)

34. Show that a set $\{\mathbf{u}_1, \mathbf{u}_2, \ldots, \mathbf{u}_n\}$ of mutually orthogonal nonzero vectors in $\mathbf{R}^m$ is linearly independent. (See Exercise 33.)

In Exercises 35–37, using a computer program to transform a matrix to row-reduced echelon form, determine whether or not the following sets of vectors are linearly dependent. If a set is linearly dependent, express one of its vectors as a linear combination of the others.

35. $\{(1, 2, 3, 4, 5), (3, 4, 5, 2, 1), (2, 3, 1, 5, 4), (4, 5, 2, 1, 3), (5, 1, 4, 3, 2)\}$

36. $\{(1, 1, 1, 1, 1, 1), (2, 2, 2, 2, 2, 3), (3, 3, 3, 3, 3, 5), (4, 4, 4, 4, 4, 8)\}$

37. $\{\mathbf{v}_1, \mathbf{v}_2, \mathbf{v}_3, \mathbf{v}_4\}$, where $\mathbf{v}_1 = (0.216, 0.531, 0.870, 1.213)$, $\mathbf{v}_2 = (0.760, 1.432, 1.614, 0.666)$, $\mathbf{v}_3 = (0.228, 0.370, -1.026, -1.760)$, $\mathbf{v}_4 = (-0.050, 0, -0.450, 0)$.

4.2 SUBSPACES OF $\mathbf{R}^m$

As we saw in Chapter 1, lines, planes, and certain other regions in $\mathbf{R}^m$ can be described as linear combinations of vectors. In this section we see that these regions are *subspaces* of $\mathbf{R}^m$ formed by the *span* of a set of vectors.

SUBSPACES OF $\mathbf{R}^m$

DEFINITION

A **subspace** of $\mathbf{R}^m$ is a nonempty set, $\mathbf{S}$, of vectors satisfying the following two conditions:

    i. if $\mathbf{v}_1$ and $\mathbf{v}_2$ are both in $\mathbf{S}$, then $\mathbf{v}_1 + \mathbf{v}_2$ is in $\mathbf{S}$,

    ii. If $\mathbf{v}$ is in $\mathbf{S}$, then $c\mathbf{v}$ is in $\mathbf{S}$ for any scalar c.

NOTE If $\mathbf{v}_1$ and $\mathbf{v}_2$ are vectors in a subspace $\mathbf{S}$, and c_1 and c_2 are scalars, then $c_1\mathbf{v}_1 + c_2\mathbf{v}_2$ is also in $\mathbf{S}$. This results from the fact that both $c_1\mathbf{v}_1$ and $c_2\mathbf{v}_2$ are in $\mathbf{S}$ due to condition (ii) of the definition, and therefore their sum is in $\mathbf{S}$ from condition (i). More generally, if $\mathbf{v}_1, \mathbf{v}_2, \ldots, \mathbf{v}_n$ are vectors in a subspace $\mathbf{S}$, then every linear combination of the $\mathbf{v}_i$ is in $\mathbf{S}$ (Exercise 27 at the end of this section).

Taking $c = 0$ in condition (ii), we see that every subspace must contain the vector $\mathbf{0}$. In fact it is easily checked (Exercise 4) that $\mathbf{S} = \{\mathbf{0}\}$, the set containing only the zero vector, is itself a subspace of $\mathbf{R}^m$. Moreover, since $\mathbf{R}^m$ is a subset of itself, $\mathbf{R}^m$ is a subspace as well. Thus for each m, $\mathbf{R}^m$ has at least two subspaces, $\{\mathbf{0}\}$ and $\mathbf{R}^m$. Other examples of subspaces are given in Examples 4.9 through 4.11.

EXAMPLE 4.9

Show that the set $\mathbf{S}$ consisting of all 4-vectors whose first and last components are equal to zero is a subspace of $\mathbf{R}^4$.

Solution

All vectors in $\mathbf{S}$ are of the form $(0, a, b, 0)$, where a and b are scalars. To verify that $\mathbf{S}$ is a subspace, we check conditions (i) and (ii) of the definition.

i. We have $(0, a_1, b_1, 0) + (0, a_2, b_2, 0) = (0, a_1 + a_2, b_1 + b_2, 0)$. Since this is a 4-vector whose first and last components are zero, the sum of two vectors in $\mathbf{S}$ is in $\mathbf{S}$.

ii. We have $c(0, a, b, 0) = (0, ca, cb, 0)$. Thus any scalar multiple of a vector in $\mathbf{S}$ is in $\mathbf{S}$.

Consequently $\mathbf{S}$ is a subspace of $\mathbf{R}^4$.

EXAMPLE 4.10

Let A be an $m \times n$ matrix. Show that the set $\mathbf{S}$ of all solutions of the homogeneous linear system $A\mathbf{x} = \mathbf{0}$ is a subspace of $\mathbf{R}^n$.

Solution

To verify that $\mathbf{S}$ is a subspace, we check conditions (i) and (ii) of the definition.

i. Let $\mathbf{x}_1$ and $\mathbf{x}_2$ be vectors in $\mathbf{R}^n$ that are solutions of $A\mathbf{x} = \mathbf{0}$. That is, $A\mathbf{x}_1 = \mathbf{0}$ and $A\mathbf{x}_2 = \mathbf{0}$. Now, $A(\mathbf{x}_1 + \mathbf{x}_2) = A\mathbf{x}_1 + A\mathbf{x}_2 = \mathbf{0} + \mathbf{0} = \mathbf{0}$ (by Theorem 1g of Section 2.3). So $\mathbf{x}_1 + \mathbf{x}_2$ is a solution of $A\mathbf{x} = \mathbf{0}$ and hence is in $\mathbf{S}$.

ii. Let $\mathbf{x}$ be a solution of $A\mathbf{x} = \mathbf{0}$, and let c be a scalar. Then $A(c\mathbf{x}) = c(A\mathbf{x}) = c\mathbf{0} = \mathbf{0}$ (by Theorem 1n of Section 2.3). Hence $c\mathbf{x}$ is a solution of $A\mathbf{x} = \mathbf{0}$ and hence is in $\mathbf{S}$.

Thus $\mathbf{S}$ is a subspace of $\mathbf{R}^n$.

NOTE   The subspace of Example 4.10 is called the **solution space** of the given homogeneous linear system. It is *not* true (Exercise 12) that the set of all solutions of a *nonhomogeneous* linear system forms a subspace.

EXAMPLE 4.11

Show that the set $\mathbf{S}$ of all linear combinations of the vectors $\mathbf{u}_1 = (1, 2, 3)$ and $\mathbf{u}_2 = (1, 0, -1)$ is a subspace of $\mathbf{R}^3$.

Solution

A vector is in **S** if and only if it is of the form $a\mathbf{u}_1 + b\mathbf{u}_2$ for some scalars a and b. We check conditions (i) and (ii).

i. Let $\mathbf{v}_1$ and $\mathbf{v}_2$ be vectors in **S**. Then there are scalars a_1, b_1 and a_2, b_2 such that $\mathbf{v}_1 = a_1\mathbf{u}_1 + b_1\mathbf{u}_2$ and $\mathbf{v}_2 = a_2\mathbf{u}_1 + b_2\mathbf{u}_2$. Now $\mathbf{v}_1 + \mathbf{v}_2 = (a_1 + a_2)\mathbf{u}_1 + (b_1 + b_2)\mathbf{u}_2$, a linear combination of $\mathbf{u}_1$ and $\mathbf{u}_2$, so $\mathbf{v}_1 + \mathbf{v}_2$ is in **S**.

ii. Let $\mathbf{v}$ be a vector in **S**, $\mathbf{v} = a\mathbf{u}_1 + b\mathbf{u}_2$, and let c be a scalar. Then $c\mathbf{v} = (ca)\mathbf{u}_1 + (cb)\mathbf{u}_2$, a linear combination of $\mathbf{u}_1$ and $\mathbf{u}_2$, so $c\mathbf{v}$ is in **S**.

Thus **S** is a subspace of $\mathbf{R}^3$.

NOTE   As we saw in Section 1.4, the set of all linear combinations of two noncollinear vectors in $\mathbf{R}^m$ describes a plane. The plane generated in Example 4.11 is pictured in Figure 4.3. Note that it contains the origin, as must all subspaces of $\mathbf{R}^m$.

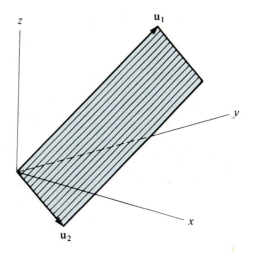

FIGURE 4.3

In the solution to Example 4.11 we made no use of the fact that $\mathbf{u}_1$ and $\mathbf{u}_2$ were specific vectors in $\mathbf{R}^3$. Consequently a similar argument can be used to prove the following theorem. We leave its proof as Exercise 28.

THEOREM I   Let $\mathcal{T} = \{\mathbf{v}_1, \mathbf{v}_2, \ldots, \mathbf{v}_n\}$ be a set of vectors in $\mathbf{R}^m$. The set of all linear combinations, $c_1\mathbf{v}_1 + c_2\mathbf{v}_2 + \cdots + c_n\mathbf{v}_n$, of the vectors in $\mathcal{T}$ is a subspace of $\mathbf{R}^m$.

NOTE   The subspace described in Theorem 1 is called the **subspace generated by** (or **spanned by**) $\mathcal{T}$. For example, the subspace of $\mathbf{R}^3$ described in Example 4.11 is the *subspace generated by* $(1, 2, 3)$ *and* $(1, 0, -1)$.

....

We now give an example of a subset of $\mathbf{R}^m$ that is *not* a subspace.

EXAMPLE 4.12

Show that the set **S** of all vectors of the form $(a, 1)$ is not a subspace of $\mathbf{R}^2$.

Solution

We will show that condition (i) of the definition does not hold for **S**. To do so, we need only find two vectors in **S** whose sum is not in **S**. Let $\mathbf{v}_1 = (2, 1)$ and $\mathbf{v}_2 = (3, 1)$. Then $\mathbf{v}_1$ and $\mathbf{v}_2$ are each in **S** (since their second components are equal to 1), but $\mathbf{v}_1 + \mathbf{v}_2 = (5, 2)$ is not in **S**. (You can also show that condition (ii) fails to hold.)

SPAN OF A SET OF VECTORS

DEFINITION

Let **S** be a subspace of $\mathbf{R}^m$. A subset $\mathcal{T}$ of **S** **spans S** if every vector in **S** can be written as a linear combination of elements of $\mathcal{T}$.

EXAMPLE 4.13

Show that the set of m distinct elementary vectors in $\mathbf{R}^m$, $\mathcal{T} = \{\mathbf{e}_1, \mathbf{e}_2, \ldots, \mathbf{e}_m\}$, spans $\mathbf{R}^m$.

Solution

Let $\mathbf{w} = (w_1, w_2, \ldots, w_m)$ be an arbitrary vector in $\mathbf{R}^m$. Then

$$\mathbf{w} = w_1\mathbf{e}_1 + w_2\mathbf{e}_2 + \cdots + w_m\mathbf{e}_m$$

is a linear combination of the $\mathbf{e}_i$, as desired.

EXAMPLE 4.14

Let **S** be the subspace of $\mathbf{R}^4$ consisting of all vectors whose first and last components are zero (see Example 4.9). Show that the vectors $\mathbf{v}_1 = (0, 1, -1, 0)$ and $\mathbf{v}_2 = (0, 2, -1, 0)$ span **S**.

Solution

Let $\mathbf{w} = (0, a, b, 0)$ be an arbitrary vector in **S**. We must show that there are constants c_1 and c_2 such that $\mathbf{w} = c_1\mathbf{v}_1 + c_2\mathbf{v}_2$, that is,

$$(0, a, b, 0) = c_1(0, 1, -1, 0) + c_2(0, 2, -1, 0).$$

Performing the scalar multiplications and additions on the right side and equating corresponding components yields the two equations

$$c_1 + 2c_2 = a,$$
$$-c_1 - c_2 = b.$$

(Equating the first, as well as the last, components just gives $0 = 0$.) Since this system has the solution $c_1 = -(a + 2b)$, $c_2 = a + b$, $\{\mathbf{v}_1, \mathbf{v}_2\}$ spans $\mathbf{S}$. For example, to express the vector $(0, 3, 7, 0)$ as a linear combination of $\mathbf{v}_1$ and $\mathbf{v}_2$, take $c_1 = -(3 + 2(7)) = -17$ and $c_2 = 3 + 7 = 10$:

$$(0, 3, 7, 0) = -17(0, 1, -1, 0) + 10(0, 2, -1, 0).$$

If $\mathbf{S}$ is a subspace of $\mathbf{R}^m$ *generated by a set* $\mathcal{T}$ of vectors in $\mathbf{S}$, then by definition $\mathcal{T}$ spans $\mathbf{S}$. The following theorem simplifies the task of determining proper subsets of $\mathcal{T}$ that also span $\mathbf{S}$.

THEOREM 2 Let $\mathbf{S}$ be the subspace of $\mathbf{R}^m$ spanned by the set $\mathcal{T} = \{\mathbf{v}_1, \mathbf{v}_2, \ldots, \mathbf{v}_n\}$. If one of the vectors in $\mathcal{T}$ is a linear combination of the others, then the subset of $\mathcal{T}$ obtained by deleting that vector still spans $\mathbf{S}$.

Proof For the sake of notational convenience, suppose that $\mathbf{v}_n$ is a linear combination of the other vectors in $\mathcal{T}$; that is, $\mathbf{v}_n = c_1\mathbf{v}_1 + c_2\mathbf{v}_2 + \cdots + c_{n-1}\mathbf{v}_{n-1}$. Let $\mathbf{w}$ be an arbitrary vector in $\mathbf{S}$. Since $\mathcal{T}$ spans $\mathbf{S}$, we have $\mathbf{w} = a_1\mathbf{v}_1 + a_2\mathbf{v}_2 + \cdots + a_n\mathbf{v}_n$ for some scalars $a_1, a_2, \ldots, a_n$. But then

$$\mathbf{w} = a_1\mathbf{v}_1 + a_2\mathbf{v}_2 + \cdots + a_{n-1}\mathbf{v}_{n-1} + a_n(c_1\mathbf{v}_1 + c_2\mathbf{v}_2 + \cdots + c_{n-1}\mathbf{v}_{n-1})$$

$$= (a_1 + a_n c_1)\mathbf{v}_1 + (a_2 + a_n c_2)\mathbf{v}_2 + \cdots + (a_{n-1} + a_n c_{n-1})\mathbf{v}_{n-1}.$$

Thus $\mathbf{w}$ is a linear combination of $\{\mathbf{v}_1, \mathbf{v}_2, \ldots, \mathbf{v}_{n-1}\}$, the subset of $\mathcal{T}$ obtained by deleting $\mathbf{v}_n$, and this subset spans $\mathbf{S}$, as desired.

EXAMPLE 4.15

Let $\mathbf{S}$ be the subspace of $\mathbf{R}^3$ generated by the set $\mathcal{T} = \{(1, -1, 0), (-2, 1, 2), (-1, -1, 4)\}$. Show that the subset $\mathcal{T}_0 = \{(1, -1, 0), (-2, 1, 2)\}$ spans $\mathbf{S}$.

Solution

To apply Theorem 2, we show that $(-1, -1, 4)$ is a linear combination of $(1, -1, 0)$ and $(-2, 1, 2)$ by using Theorem 5 of Section 4.1. We construct the matrix

$$\begin{bmatrix} 1 & -2 & -1 \\ -1 & 1 & -1 \\ 0 & 2 & 4 \end{bmatrix}$$

and transform it to the row-reduced echelon form

$$\begin{bmatrix} 1 & 0 & 3 \\ 0 & 1 & 2 \\ 0 & 0 & 0 \end{bmatrix}.$$

Consequently $(-1, -1, 4) = 3(1, -1, 0) + 2(-2, 1, 2)$ (from the third column of the latter matrix). Thus $\mathcal{T}_0$ spans **S**.

NOTE   Geometrically, we have shown that the three vectors of Example 4.15 lie in a plane—the plane generated by $(1, -1, 0)$ and $(-2, 1, 2)$, illustrated in Figure 4.4.

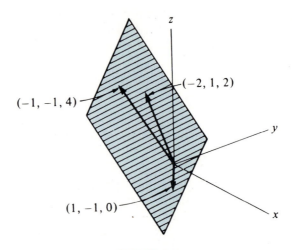

FIGURE 4.4

4.2 EXERCISES

In Exercises 1–6 show that the given sets of vectors are subspaces of **R**^{m}.

1. The set of all scalar multiples of the vector $(1, 2, 3)$ (of **R**^{3})

2. The set of all m-vectors whose first component is 0 (of **R**^{m})

3. The set of all linear combinations of the vectors $(1, 0, 1, 0)$ and $(0, 1, 0, 1)$ (of **R**^{4})

4. The set consisting solely of the zero vector, $\{\mathbf{0}\}$ (of **R**^{m})

5. The set of all vectors of the form $(a, b, a - b, a + b)$ (of **R**^{4})

6. The set of all vectors (x, y, z) such that $x + y + z = 0$ (of **R**^{3})

In Exercises 7–12 show that the given sets of vectors do *not* form subspaces of **R**^{m}.

7. The set consisting of the single vector $(1, 1, 1, 1)$ (of **R**^{4})

8. The set of all m-vectors whose first component is 2 (of **R**^{m})

9. The set of all m-vectors *except* the m-vector $\mathbf{0}$ (of **R**^{m})

10. The set of all 4-vectors *except* the vector $(1, 1, 1, 1)$ (of **R**^{4})

11. The set of all 3-vectors the sum of whose components is 1 (of **R**^{3})

12. The set of all solutions of $Ax = \mathbf{b}$, $\mathbf{b} \neq \mathbf{0}$ (of **R**^{n}, where A is $m \times n$)

In Exercises 13–16 determine which sets span $\mathbf{R}^3$.

13. $\{(1, 2, 3), (-1, 0, 1), (0, 1, 2)\}$

14. $\{(-1, 1, 2), (3, 3, 1), (1, 2, 2)\}$

15. $\{(1, 2, -1), (1, 0, 1)\}$

16. $\{(1, 0, 0), (0, 1, 0), (0, 0, 1), (1, 1, 1)\}$

17. Show that $\{\mathbf{e}_2, \mathbf{e}_3\}$ spans the subspace of $\mathbf{R}^4$ consisting of all vectors whose first and last components are zero.

18. Show that $\{(1, 1, 0), (1, 0, 1)\}$ spans the subspace of $\mathbf{R}^3$ consisting of all vectors (x, y, z) such that $x = y + z$.

19–24. Find a set that spans each of the subspaces in Exercises 1 through 6.

25. Show that $\{(0, 1, 1)\}$ spans the solution space of $A\mathbf{x} = \mathbf{0}$, where

$$A = \begin{bmatrix} 1 & -1 & 1 \\ 2 & 1 & -1 \\ 0 & -1 & 1 \end{bmatrix}.$$

26. Show that $\{(1, 1, 0), (-2, 0, 1)\}$ spans the solution space of $A\mathbf{x} = \mathbf{0}$, where

$$A = \begin{bmatrix} 1 & -1 & 2 \\ -2 & 2 & -4 \end{bmatrix}.$$

27. Show that if $\mathbf{v}_1, \mathbf{v}_2, \ldots, \mathbf{v}_n$ are vectors in a subspace $\mathbf{S}$ of $\mathbf{R}^m$, then $\mathbf{w} = c_1\mathbf{v}_1 + c_2\mathbf{v}_2 + \cdots + c_n\mathbf{v}_n$ is in $\mathbf{S}$ for all scalars $c_1, c_2, \ldots, c_n$.

28. Prove Theorem 1: Let $\mathscr{T} = \{\mathbf{v}_1, \mathbf{v}_2, \ldots, \mathbf{v}_n\}$ be a set of vectors in $\mathbf{R}^m$, and show that the set of all linear combinations of the vectors in $\mathscr{T}$ is a subspace of $\mathbf{R}^m$.

4.3 BASIS AND DIMENSION

In Section 4.1 we considered sets that are linearly independent and in Section 4.2 sets that span a given subspace. Subsets of a subspace that are linearly independent *and* span the subspace are of particular importance in linear algebra. In this section we investigate the properties of such sets.

BASIS FOR A SUBSPACE

DEFINITION

Let $\mathbf{S}$ be a subspace of $\mathbf{R}^m$. A set $\mathscr{T}$ of vectors in $\mathbf{S}$ is a **basis** for $\mathbf{S}$ if

i. $\mathscr{T}$ is linearly independent and

ii. $\mathscr{T}$ spans $\mathbf{S}$.

EXAMPLE 4.16   Show that the set of elementary vectors in $\mathbf{R}^m$, $\mathscr{T} = \{e_1, e_2, \ldots, e_m\}$, is a basis for $\mathbf{R}^m$.

Solution   We must show that $\mathscr{T}$ is linearly independent and that $\mathscr{T}$ spans $\mathbf{R}^m$. To demonstrate the former, we show that the vector equation

$$c_1 e_1 + c_2 e_2 + \cdots + c_m e_m = \mathbf{0}$$

has only the trivial solution. Now, each e_i has a 1 in its ith position and zeros everywhere else. Hence, if we perform the scalar multiplications and additions on the left side of this equation, we obtain

$$(c_1, c_2, \ldots, c_m) = (0, 0, \ldots, 0).$$

Equating corresponding components gives $c_i = 0$ for all i, so $\mathscr{T}$ is linearly independent. Since we have already shown in Example 4.13 that $\mathscr{T}$ spans $\mathbf{R}^m$, this set is indeed a basis for $\mathbf{R}^m$.

NOTE The basis, $\{e_1, e_2, \ldots, e_m\}$ is called the **standard basis** for $\mathbf{R}^m$. For example, the standard basis for $\mathbf{R}^2$ is $\{e_1, e_2\} = \{(1, 0), (0, 1)\}$, and the standard basis for $\mathbf{R}^3$ is $\{(1, 0, 0), (0, 1, 0), (0, 0, 1)\}$.

··

EXAMPLE 4.17   Show that the set $\mathscr{T} = \{(1, 0, 0), (1, 2, 0), (1, 2, 3)\}$ is a basis for $\mathbf{R}^3$.

Solution   We can show that $\mathscr{T}$ is linearly independent and that $\mathscr{T}$ spans $\mathbf{R}^3$ at the same time. Let $\mathbf{w} = (w_1, w_2, w_3)$ be an arbitrary vector in $\mathbf{R}^3$. Then $\mathscr{T}$ spans $\mathbf{R}^3$ if there exist scalars c_1, c_2, c_3 such that

$$c_1(1, 0, 0) + c_2(1, 2, 0) + c_3(1, 2, 3) = (w_1, w_2, w_3).$$

Performing the scalar multiplications and equating corresponding components yields the linear system

$$\begin{aligned}
c_1 + c_2 + c_3 &= w_1 \\
2c_2 + 2c_3 &= w_2 \\
3c_3 &= w_3,
\end{aligned}$$

which has the coefficient matrix

$$A = \begin{bmatrix} 1 & 1 & 1 \\ 0 & 2 & 2 \\ 0 & 0 & 3 \end{bmatrix}.$$

Now, $\det A = 6$. Since $\det A$ is nonzero, Theorem 3 of Section 4.1 tells us that $\mathscr{T}$ is linearly independent, and also (by Theorem 6 of Section 3.2) that our linear system has a (unique) solution. Hence $\mathscr{T}$ spans $\mathbf{R}^3$ as well.

Now suppose that we wish to find a basis for the subspace **S** generated by a set $\mathcal{T}$ of vectors in **S**. By definition, $\mathcal{T}$ spans **S**, but $\mathcal{T}$ is not necessarily linearly independent. The following theorem and accompanying procedure show us how to *reduce a spanning set to a basis*.

THEOREM 1 Let **S** be the subspace of $\mathbf{R}^m$ generated by the set $\mathcal{T} = \{\mathbf{v}_1, \mathbf{v}_2, \ldots, \mathbf{v}_n\}$. Then there is a subset of $\mathcal{T}$ that is a basis for **S**.

PROCEDURE FOR REDUCING A SPANNING SET TO A BASIS

Let A be the $m \times n$ matrix with columns $\mathbf{a}^i = \mathbf{v}_i$, and let B be its row-reduced echelon form. If $\mathbf{w}_1, \mathbf{w}_2, \ldots, \mathbf{w}_s$ are the vectors in $\mathcal{T}$ that are transformed into the distinct elementary columns of B, then $\mathcal{T}_0 = \{\mathbf{w}_1, \mathbf{w}_2, \ldots, \mathbf{w}_s\}$ is a basis for **S**.

Proof By Theorem 4 of Section 4.1, $\mathcal{T}_0$ is linearly independent, and by Theorem 5 of that section, the other vectors in $\mathcal{T}$ are linear combinations of those in $\mathcal{T}_0$. But then, by Theorem 2 of Section 4.2, $\mathcal{T}_0$ spans **S** as well. Hence $\mathcal{T}_0$ is a basis for **S**.

EXAMPLE 4.18

Find a basis for the subspace, **S**, of $\mathbf{R}^4$ generated by $\mathcal{T} = \{\mathbf{v}_1, \mathbf{v}_2, \mathbf{v}_3, \mathbf{v}_4, \mathbf{v}_5\}$, where $\mathbf{v}_1 = (1, 0, -1, 1)$, $\mathbf{v}_2 = (-2, 0, 2, -2)$, $\mathbf{v}_3 = (1, 1, 1, 1)$, $\mathbf{v}_4 = (1, -1, -3, 1)$, and $\mathbf{v}_5 = (1, -1, 1, -1)$.

Solution

We apply the procedure, constructing the matrix A with $\mathbf{a}^i = \mathbf{v}_i$,

$$A = \begin{bmatrix} 1 & -2 & 1 & 1 & 1 \\ 0 & 0 & 1 & -1 & -1 \\ -1 & 2 & 1 & -3 & 1 \\ 1 & -2 & 1 & 1 & -1 \end{bmatrix},$$

which has the row-reduced echelon form

$$B = \begin{bmatrix} 1 & -2 & 0 & 2 & 0 \\ 0 & 0 & 1 & -1 & 0 \\ 0 & 0 & 0 & 0 & 1 \\ 0 & 0 & 0 & 0 & 0 \end{bmatrix}.$$

We see that $\mathbf{v}_1$, $\mathbf{v}_3$, and $\mathbf{v}_5$ have been transformed into the distinct elementary vectors $\mathbf{e}_1$, $\mathbf{e}_2$, and $\mathbf{e}_3$ ($\mathbf{v}_1$, $\mathbf{v}_3$, and $\mathbf{v}_5$ are the $\mathbf{w}_1$, $\mathbf{w}_2$, and $\mathbf{w}_3$ of the procedure), so $\{\mathbf{v}_1, \mathbf{v}_3, \mathbf{v}_5\}$ is a basis for **S**.

In Theorem 1 we learned how to *reduce a spanning set to a basis*. This raises the question, "Is it possible to *extend a linearly independent set to a basis*?" More precisely, if $\mathcal{T}$ is a linearly independent subset of a subspace **S**, "Are there vectors in **S** that can be added to the set $\mathcal{T}$ to obtain a basis for **S**?" The answer is "Yes," and if we have a spanning set for **S**, there is a straightforward computational technique to construct this basis. The next theorem and accompanying procedure describe the process.

THEOREM 2   Let **S** be the subspace of $\mathbf{R}^m$ generated by the set $\mathcal{T} = \{\mathbf{v}_1, \mathbf{v}_2, \ldots, \mathbf{v}_n\}$, and let $\mathcal{T}_1 = \{\mathbf{u}_1, \mathbf{u}_2, \ldots, \mathbf{u}_s\}$ be a linearly independent subset of **S**. Then there is a subset $\mathcal{T}_2$ of $\mathcal{T}$ such that the union of $\mathcal{T}_1$ and $\mathcal{T}_2$ is a basis for **S**.

PROCEDURE FOR EXTENDING AN INDEPENDENT SET TO A BASIS

Let A be the $m \times (s + n)$ matrix, with columns $\mathbf{a}^i = \mathbf{u}_i$ for $i = 1, \ldots, s$ and $\mathbf{a}^{s+j} = \mathbf{v}_j$ for $j = 1, \ldots, n$. In other words, the first s columns of A are the vectors in $\mathcal{T}_1$ and the last n columns are the vectors in $\mathcal{T}$. Let B be the row-reduced echelon matrix for A, and let $\mathbf{w}_1, \mathbf{w}_2, \ldots, \mathbf{w}_t$ be those vectors in $\mathcal{T}$ that are transformed into distinct elementary columns of B other than $\mathbf{e}_1, \mathbf{e}_2, \ldots, \mathbf{e}_s$. Then $\mathcal{T}_2 = \{\mathbf{w}_1, \mathbf{w}_2, \ldots, \mathbf{w}_t\}$.

Proof   By the procedure of Theorem 1, the columns of A that correspond to the distinct elementary columns of B form a basis for **S**. Since the first s columns of A are linearly independent, they are transformed into $\mathbf{e}_1, \ldots, \mathbf{e}_s$, the first s columns of B. Thus each element of $\mathcal{T}_1$ is in the basis. Let $\mathcal{T}_2$ be the set of other basis vectors. Then $\mathcal{T}_2$ is a subset of $\mathcal{T}$, since each element comes from the last n columns of A. So the union of $\mathcal{T}_1$ and $\mathcal{T}_2$ is just the set of all basis elements.

EXAMPLE 4.19   Let $\mathcal{T}_1 = \{\mathbf{v}_1, \mathbf{v}_2\}$ be the set of linearly independent vectors in $\mathbf{R}^4$ with $\mathbf{v}_1 = (1, 0, -1, 1)$ and $\mathbf{v}_2 = (1, 1, 1, -1)$. Extend this set to form a basis for $\mathbf{R}^4$.

Solution   Since $\mathbf{R}^4$ is spanned by the set $\mathcal{T} = \{\mathbf{e}_1, \mathbf{e}_2, \mathbf{e}_3, \mathbf{e}_4\}$, we may find a subset $\mathcal{T}_2$ of this set $\mathcal{T}$

such that the union of $\mathcal{T}_1$ and $\mathcal{T}_2$ is a basis for $\mathbf{R}^4$. Following the procedure of Theorem 2, we construct the matrix

$$\left[\begin{array}{rr|rrrr} 1 & 1 & 1 & 0 & 0 & 0 \\ 0 & 1 & 0 & 1 & 0 & 0 \\ -1 & 1 & 0 & 0 & 1 & 0 \\ 1 & -1 & 0 & 0 & 0 & 1 \end{array}\right]$$

and transform it to row-reduced echelon form:

$$\left[\begin{array}{rr|rrrr} 1 & 0 & 0 & 1 & 0 & 1 \\ 0 & 1 & 0 & 1 & 0 & 0 \\ 0 & 0 & 1 & -2 & 0 & -1 \\ 0 & 0 & 0 & 0 & 1 & 1 \end{array}\right].$$

We see that $\mathbf{e}_1$ and $\mathbf{e}_3$ of the set $\mathcal{T}$ have been transformed into distinct elementary columns of the latter matrix, so the set $\{\mathbf{v}_1, \mathbf{v}_2, \mathbf{e}_1, \mathbf{e}_3\}$ is a basis for $\mathbf{R}^4$.

We will now state an important theorem concerning bases, to be proved in Chapter 5.

THEOREM 3 Let $\mathbf{S}$ be a subspace of $\mathbf{R}^m$. Then any basis for $\mathbf{S}$ contains the same number of vectors as any other basis.

For example, we have shown (Example 4.16) that there is a basis for $\mathbf{R}^m$ containing m vectors. As a result, Theorem 3 tells us that *every* basis for $\mathbf{R}^m$ must contain m vectors.

DIMENSION OF A SUBSPACE

DEFINITION

Let $\mathbf{S}$ be a subspace of $\mathbf{R}^m$. The **dimension** of $\mathbf{S}$ is equal to the number of vectors in any basis for $\mathbf{S}$.

NOTE If we find one basis for a subspace that contains n vectors, and then determine another basis for the same subspace, Theorem 3 guarantees that it too will contain n vectors. Consequently we say that the concept of dimension is "well defined."

EXAMPLE 4.20   Find the dimension of $\mathbf{R}^2$, $\mathbf{R}^3$, and $\mathbf{R}^m$.

Solution   Bases for these subspaces are given by $\{\mathbf{e}_1, \mathbf{e}_2\}$, $\{\mathbf{e}_1, \mathbf{e}_2, \mathbf{e}_3\}$, and $\{\mathbf{e}_1, \mathbf{e}_2, \ldots, \mathbf{e}_m\}$, respectively. Thus $\mathbf{R}^2$ has dimension 2, $\mathbf{R}^3$ has dimension 3, and in general $\mathbf{R}^m$ has dimension m.

EXAMPLE 4.21   Find the dimension of the subspaces $\mathbf{S}$ of $\mathbf{R}^4$ consisting of all vectors whose first and last components are equal to 0 (see Example 4.9).

Solution   In Example 4.14 we showed that the vectors $\mathbf{v}_1 = (0, 1, -1, 0)$ and $\mathbf{v}_2 = (0, 2, -1, 0)$ span $\mathbf{S}$. By transforming the matrix

$$\begin{bmatrix} 0 & 0 \\ 1 & 2 \\ -1 & -1 \\ 0 & 0 \end{bmatrix}$$

to row-reduced echelon form, we can show that they are also linearly independent. Hence $\{\mathbf{v}_1, \mathbf{v}_2\}$ is a basis for $\mathbf{S}$. Thus $\mathbf{S}$ has dimension equal to 2.

One of the more important subspaces of $\mathbf{R}^n$ is the *solution space* of a homogeneous linear system—the set of all solutions of $A\mathbf{x} = \mathbf{0}$ for a given matrix A (see Example 4.10). The next theorem and its accompanying procedure describe how we can find a basis for, and hence the dimension of, these subspaces. This is a process we will use several times in the chapters that follow.

THEOREM 4   Let A be an $m \times n$ matrix, and let B be its row-reduced echelon form. Let the rank of A (the number of nonzero rows of B) be equal to r. Then the dimension of the solution space of $A\mathbf{x} = \mathbf{0}$ is $n - r$.

Before we prove Theorem 4, we will present a procedure for determining a *basis* of $n - r$ vectors for the solution space and an example illustrating this procedure. The proof of the theorem will simply be a verification that the set given in the procedure is in fact a basis.

PROCEDURE FOR FINDING A BASIS FOR THE SOLUTION SPACE OF A$\mathbf{x} = \mathbf{0}$

Let A be an $m \times n$ matrix with rank equal to r.

     i. Solve $A\mathbf{x} = \mathbf{0}$ by Gauss–Jordan elimination (Section 2.2). There will be $n - r$ parameters in the solution.

ii. Obtain $n - r$ vectors by setting, in turn, each parameter in the solution equal to 1 as the rest are set equal to 0.

iii. The $n - r$ vectors of step (ii) are a basis for the solution space.

···

EXAMPLE 4.22   Find a basis for and the dimension of the solution space of the homogeneous linear system $A\mathbf{x} = \mathbf{0}$, where

$$A = \begin{bmatrix} 1 & 1 & 5 & 0 & 1 \\ 1 & -1 & 1 & -2 & -1 \\ 1 & 1 & 5 & 1 & 1 \\ 0 & 2 & 4 & 3 & 2 \\ 0 & 1 & 2 & 2 & 1 \end{bmatrix}.$$

Solution   Following the procedure just given, we first transform A to its row-reduced echelon form:

$$B = \begin{bmatrix} 1 & 0 & 3 & 0 & 0 \\ 0 & 1 & 2 & 0 & 1 \\ 0 & 0 & 0 & 1 & 0 \\ 0 & 0 & 0 & 0 & 0 \\ 0 & 0 & 0 & 0 & 0 \end{bmatrix}.$$

The rank of A is 3, so the dimension of the solution space is $5 - 3 = 2$. Translating B into a system of equations and solving for $x_1, x_2,$ and x_4 (the variables that correspond to the leading 1's of B), we obtain

$$x_1 = -3x_3,$$
$$x_2 = -2x_3 - x_5,$$
$$x_4 = \quad 0.$$

Thus the solution of $A\mathbf{x} = \mathbf{0}$ is

$$(-3s, -2s - t, s, 0, t).$$

Setting $s = 1$ and $t = 0$, we get $\mathbf{v}_1 = (-3, -2, 1, 0, 0)$; setting $t = 1$ and $s = 0$, we get $\mathbf{v}_2 = (0, -1, 0, 0, 1)$. Then $\{\mathbf{v}_1, \mathbf{v}_2\}$ is a basis for the solution space of $A\mathbf{x} = \mathbf{0}$.

We now prove Theorem 4.

Proof of Theorem 4   By Theorem 4 of Section 2.5, the solution set of $A\mathbf{x} = \mathbf{0}$ contains $(n - r)$ parameters. Recalling the setting with $d_1 = d_2 = \cdots = d_r = 0$, we assume that the first r columns of A are the ones transformed into a set of distinct elementary columns of B. This gives the system of equations

$$
\begin{aligned}
x_1 \qquad\quad\ + b_{1,r+1}x_{r+1} + \cdots + b_{1,n}x_n &= 0 \\
x_2 \quad\ + b_{2,r+1}x_{r+1} + \cdots + b_{2,n}x_n &= 0 \\
\ddots \qquad \vdots \qquad\qquad \vdots \qquad \vdots \\
x_r + b_{r,r+1}x_{r+1} + \cdots + b_{r,n}x_n &= 0.
\end{aligned}
$$

Solving for $x_1, \ldots, x_r$, and setting $x_{r+1} = t_1$, $x_{r+2} = t_2, \ldots, x_n = t_{n-r}$, we obtain the solution

$$
\begin{aligned}
x_1 &= -b_{1,r+1}t_1 - b_{1,r+2}t_2 - \cdots - b_{1,n}t_{n-r} \\
x_2 &= -b_{2,r+1}t_1 - b_{2,r+2}t_2 - \cdots - b_{2,n}t_{n-r} \\
\vdots &\quad\ \vdots \qquad\ \vdots \qquad\qquad \vdots \\
x_r &= -b_{r,r+1}t_1 - b_{r,r+2}t_2 - \cdots - b_{r,n}t_{n-r} \\
x_{r+1} &= \quad 1t_1 + \qquad 0t_2 + \cdots + \qquad 0t_{n-r} \\
x_{r+2} &= \quad 0t_1 + \qquad 1t_2 + \cdots + \qquad 0t_{n-r} \\
\vdots &\quad\ \vdots \qquad\ \vdots \qquad \ddots \qquad \vdots \\
x_n &= \quad 0t_1 + \qquad 0t_2 + \cdots + \qquad 1t_{n-r}.
\end{aligned} \tag{1}
$$

Letting $\mathbf{v}_i$ be the vector obtained by setting $t_i = 1$ and the rest of the parameters equal to 0, we may write the solution given by System (1) in vector form as

$$
\mathbf{x} = t_1\mathbf{v}_1 + t_2\mathbf{v}_2 + \cdots + t_{n-r}\mathbf{v}_{n-r}. \tag{2}
$$

Each vector $\mathbf{v}_i$ is in the solution space, and since any solution $\mathbf{x}$ can be written as a linear combination of $\mathcal{T} = \{\mathbf{v}_1, \ldots, \mathbf{v}_{n-r}\}$, $\mathcal{T}$ spans the solution space. Moreover, $\mathcal{T}$ is linearly independent. To see this, note that each vector $\mathbf{v}_i$ has 1 as its $(r + i)$th component and 0 for each of the other last $(n - r)$ components. Set $\mathbf{0}$ equal to a linear combination of the vectors in $\mathcal{T}$, say,

$$
\mathbf{0} = c_1\mathbf{v}_1 + c_2\mathbf{v}_2 + \cdots + c_{n-r}\mathbf{v}_{n-r}.
$$

Looking only at the $(r + i)$th component, we have

$$
0 = c_1(0) + \cdots + c_{i-1}(0) + c_i(1) + c_{i+1}(0) + \cdots + c_{n-r}(0),
$$

from which $0 = c_i$. Since this holds for each i, $\mathcal{T}$ is linearly independent. Thus $\mathcal{T}$ spans the solution space and is linearly independent, so it is a basis.

___

NOTE   We can generalize the procedure of Theorem 4 to provide us with a method for finding the dimension of, and a basis for, any subspace in which we can give the general form for its vectors. This technique is illustrated in the next example.

EXAMPLE 4.23   Find a basis for, and the dimension of, the subspace **S** of $\mathbf{R}^4$ of all vectors (a, b, c, d) for which $d = a + b + c$.

Solution   The general form of a vector in **S** is $(a, b, c, a + b + c)$. Notice that three parameters, a, b, and c, appear in this general vector. Comparing this situation to that of Theorem 4 leads us to believe that the dimension of **S** is 3. Moreover, imitating the procedure associated with this theorem, we can obtain a basis $\{\mathbf{v}_1, \mathbf{v}_2, \mathbf{v}_3\}$ for **S** by setting

$$a = 1, b = 0, c = 0: \quad \mathbf{v}_1 = (1, 0, 0, 1);$$
$$a = 0, b = 1, c = 0: \quad \mathbf{v}_2 = (0, 1, 0, 1);$$
$$a = 0, b = 0, c = 1: \quad \mathbf{v}_3 = (0, 0, 1, 1).$$

It is easy to check that $\{\mathbf{v}_1, \mathbf{v}_2, \mathbf{v}_3\}$ *is* a basis for **S**, so the dimension of **S** is indeed 3.

S P O T L I G H T

DEGREES OF FREEDOM

In $\mathbf{R}^m$, a vector is uniquely determined by its m components. In a proper subspace **S** of $\mathbf{R}^m$, certain constraints are placed upon these components. The number of components of a vector in **S** that can vary independently of the others (that is, the number of *degrees of freedom* in **S**) is intuitively the dimension of this subspace. For example, in the subspace of $\mathbf{R}^4$ consisting of all vectors with first and last components equal to zero, only two components (the second and third) are free to vary. Intuitively, the dimension of this subspace is 2.

The idea of degrees of freedom is used in physics to analyze certain mechanical systems. If a system consists of n particles, then (since each particle has three coordinates) $3n$ coordinates will describe the position of all particles in the system. If n is large, analyzing the motion of the system, for example, can be very cumbersome. However, due to interactions among the particles, there may be far fewer independent coordinates—far fewer degrees of freedom. These independent, or *generalized*, coordinates can be used to reduce the complexity of the system, enabling one to analyze it more easily. (For more information about this, see the *Feynman Lectures on Physics* by Richard Feynman, Volume 1, Chapter 39 (Addison-Wesley, 1963).)

The following theorem gives some useful relationships among the notions of *linear dependence, span, basis,* and *dimension.* We will only prove part (i), but the proofs of the other parts are just as easy.

THEOREM 5   Let **S** be a subspace of $\mathbf{R}^m$ with dimension equal to n and let $\mathscr{T}$ be a subset of **S**.

  i. If the number of vectors in $\mathscr{T}$ is greater than n, then $\mathscr{T}$ is not linearly independent.

  ii. If the number of vectors in $\mathscr{T}$ is less than n, then $\mathscr{T}$ does not span **S**.

  iii. If the number of vectors in $\mathscr{T}$ is equal to n, then $\mathscr{T}$ is linearly independent if and only if it spans **S**.

Proof of (i)   Suppose that the number of vectors in $\mathscr{T}$ is greater than n and $\mathscr{T}$ is linearly independent. By Theorem 2, $\mathscr{T}$ can be extended to a basis for **S**. But this gives a basis for **S** with more than n vectors, which contradicts Theorem 3. Hence $\mathscr{T}$ must be linearly dependent.

This theorem can sometimes be used to simplify the task of answering questions concerning linear dependence, span, or basis.

EXAMPLE 4.24

  a. Do the vectors $(0, 1, 2, 3)$, $(1, 3, 2, 6)$, $(-1, 4, 0, 1)$ span $\mathbf{R}^4$?

  b. Show that the set $\{(0, 1, 2, 0), (0, 1, 0, 0), (0, 1, 1, 0)\}$ is linearly dependent.

  c. Show that the set $\{(1, 0, 0), (1, 2, 0), (1, 2, 3)\}$ is a basis for $\mathbf{R}^3$.

Solution

  a. Apply part (ii) of Theorem 5. We are given three vectors in $\mathbf{R}^4$, which has dimension 4 (Example 4.20). Consequently the given vectors do not span $\mathbf{R}^4$.

  b. The given vectors all lie in the subspace **S** of $\mathbf{R}^4$, consisting of all vectors with first and last components equal to zero. By Example 4.21, the dimension of **S** is 2. Consequently, by part (i) of Theorem 5, the given set is linearly dependent.

  c. By part (iii) of Theorem 5, we need only show that the given set of three vectors is linearly independent (since the dimension of $\mathbf{R}^3$ is three). Let A be the 3×3 matrix with columns formed by the given vectors. Since $\det A = 6 \neq 0$, the set is linearly independent by Theorem 3 of Section 4.1.

The solution technique of part (c) of Example 4.24 is important enough to state as a theorem.

THEOREM 6   Let $\mathscr{T} = \{\mathbf{v}_1, \mathbf{v}_2, \ldots, \mathbf{v}_m\}$ be a set of vectors in $\mathbf{R}^m$. Then $\mathscr{T}$ is a basis for $\mathbf{R}^m$ if and only if $\det A \neq 0$, where A is the $m \times m$ matrix with $\mathbf{a}^i = \mathbf{v}_i$.

Proof   By Theorem 3 of Section 4.1, $\mathcal{T}$ is linearly independent if and only if det $A \neq 0$. But since $\mathbf{R}^m$ has dimension m, by part (iii) of Theorem 5, $\mathcal{T}$ is a basis for $\mathbf{R}^m$ if and only if it is linearly independent. Consequently $\mathcal{T}$ is a basis for $\mathbf{R}^m$ if and only if det $A \neq 0$.

EXAMPLE 4.25

Determine whether or not the set $\mathcal{T} = \{\mathbf{v}_1, \mathbf{v}_2, \mathbf{v}_3, \mathbf{v}_4\}$ is a basis for $\mathbf{R}^4$, where $\mathbf{v}_1 = (1, 0, 0, 0)$, $\mathbf{v}_2 = (-1, 1, 3, 2)$, $\mathbf{v}_3 = (1, 2, 3, 1)$, and $\mathbf{v}_4 = (1, 1, 2, 1)$.

Solution

Here we have four vectors in $\mathbf{R}^4$, so we can apply Theorem 6. To do so, we construct the matrix A with $\mathbf{a}^i = \mathbf{v}_i$:

$$A = \begin{bmatrix} 1 & -1 & 1 & 1 \\ 0 & 1 & 2 & 1 \\ 0 & 3 & 3 & 2 \\ 0 & 2 & 1 & 1 \end{bmatrix}.$$

Since det $A = 0$, $\mathcal{T}$ is not a basis for $\mathbf{R}^4$.

4.3 EXERCISES

In Exercises 1–8 determine whether or not the given set forms a basis for the indicated subspace.

1. $\{(1, 2, 3), (-1, 0, 1), (0, 1, 2)\}$ for $\mathbf{R}^3$
2. $\{(-1, 1, 2,), (3, 3, 1), (1, 2, 2)\}$ for $\mathbf{R}^3$
3. $\{(1, -1, 0), (0, 1, -1)\}$ for the subspace of $\mathbf{R}^3$ of all (x, y, z) such that $x + y + z = 0$
4. $\{(1, 1, 0), (1, 1, 1)\}$ for the subspace of $\mathbf{R}^3$ of all (x, y, z) such that $y = x + z$
5. $\{(1, 1, 0, 0), (0, 0, 1, 1)\}$ for the subspace of $\mathbf{R}^4$ of all vectors of the form $(a, a + b, b, b)$.
6. $\{(1, 2, -1, 3), (0, 0, 0, 0)\}$ for the subspace of $\mathbf{R}^4$ of all vectors of the form $(a, b, a - b, a + b)$
7. $\{(1, -1), (-2, 3)\}$ for the subspace of $\mathbf{R}^2$ generated by the set $(1, 2)$ and $(2, 1)$
8. $\{(2, 0, 2), (1, 1, 1)\}$ for the subspace of $\mathbf{R}^3$ generated by $(1, 0, 1)$ and $(0, 1, 0)$

In Exercises 9–12 find a basis for the subspace generated by the given set of vectors.

9. $\{(1, -1, 1), (2, 0, 1), (1, 1, 0)\}$
10. $\{(1, 0, 1, 0), (0, -1, 1, 2), (2, -1, 3, 2), (1, 1, 1, 1)\}$
11. $\{(-1, 1, 2, -1), (1, 0, -1, 1), (-1, 2, 3, -1), (1, 1, 0, 1)\}$
12. $\{(1, -1, 0), (0, 1, 1), (1, 0, 2)\}$

In Exercises 13–16 extend the given set to form a basis for $\mathbf{R}^m$.

13. $\{(1, 1, 0), (1, -1, 1)\}$ for $\mathbf{R}^3$

14. $\{(-1, 1, 0, 0), (1, -1, 1, 0), (0, 0, 1, 2)\}$ for $\mathbf{R}^4$

15. $\{(1, 0, 0, 1), (0, 1, 1, 0,), (1, 1, 1, 2)\}$ for $\mathbf{R}^4$

16. $\{(-1, -1, 2, 1), (2, 1, -1, -2), (0, -1, 4, 0)\}$ for $\mathbf{R}^4$

In Exercises 17–20 explain why the given statement is true "by inspection."

17. The set $\{(1, 0, 3), (-1, 1, 0), (1, 2, 4), (0, -1, -2)\}$ is linearly dependent.

18. The set $\{(1, -1, 2), (0, 1, 1)\}$ does not span $\mathbf{R}^3$.

19. If the set $\{\mathbf{v}_1, \mathbf{v}_2, \mathbf{v}_3, \mathbf{v}_4\}$ of vectors in $\mathbf{R}^4$ is linearly independent, then it spans $\mathbf{R}^4$.

20. The set $\{(0, 1, -1, 0), (0, -1, 2, 0)\}$ is linearly independent, and so it spans the subspace of $\mathbf{R}^4$ of all vectors of the form $(0, a, b, 0)$.

In Exercises 21–28 find a basis for and the dimension of the indicated subspace.

21. The solution space of the homogeneous linear system

$$
\begin{aligned}
x_1 - 2x_2 + x_3 &= 0 \\
x_2 - x_3 + x_4 &= 0 \\
x_1 - x_2 + x_4 &= 0
\end{aligned}
$$

22. The solution space of

$$
\begin{aligned}
x_1 - 3x_2 + x_3 &- x_5 = 0 \\
x_1 - 2x_2 + x_3 - x_4 &= 0 \\
x_1 - x_2 + x_3 - 2x_4 + x_5 &= 0
\end{aligned}
$$

23. The subspace of $\mathbf{R}^3$ of all vectors of the form $(x, 2x, z)$

24. The subspace of $\mathbf{R}^3$ of all vectors of the form $(x + y, y, x)$

25. The subspace of $\mathbf{R}^4$ of all vectors of the form $(x, -y, x - 2y, 3y)$

26. The subspace of $\mathbf{R}^4$ of all vectors of the form $(0, x, y, x - y)$

27. The subspace of $\mathbf{R}^4$ generated by the set

$$\{(1, -1, 3, 1), (-2, 0, -2, 0), (0, -1, 2, 1)\}$$

28. The subspace of $\mathbf{R}^3$ generated by the set

$$\{(1, -1, 0), (0, 1, 1), (2, -1, 1), (1, 0, 1)\}$$

29. Let $\mathbf{v}$ be a nonzero vector in $\mathbf{R}^m$. Find the dimension of the subspace $\mathbf{S}$ of $\mathbf{R}^m$ consisting of all scalar multiples of $\mathbf{v}$. (Geometrically, $\mathbf{S}$ is the line through the origin containing $\mathbf{v}$.)

30. Let $\mathbf{v}_1$ and $\mathbf{v}_2$ be two noncollinear vectors in $\mathbf{R}^m$. Find the dimension of the subspace $\mathbf{S}$ of $\mathbf{R}^m$ generated by $\mathbf{v}_1$ and $\mathbf{v}_2$. (Geometrically, $\mathbf{S}$ is the plane containing the given vectors.)

31. Let $\{\mathbf{v}_1, \mathbf{v}_2, \ldots, \mathbf{v}_n\}$ be a basis for a subspace $\mathbf{S}$ of $\mathbf{R}^m$. Show that every vector $\mathbf{x}$ in $\mathbf{S}$ may be written as a *unique* linear combination of the $\mathbf{v}_i$.

4.4 RANK OF A MATRIX

In Section 2.5 the notion of the *rank* of a matrix was defined as the number of nonzero rows in its row-reduced echelon form. In this section we introduce the concepts of *column rank* and *row rank* of a matrix and show that these three types of rank are in fact one and the same.

DEFINITION

Let A be an $m \times n$ matrix. The subspace of $\mathbf{R}^m$ generated by the columns of the matrix, $\mathbf{a}^1, \mathbf{a}^2, \ldots, \mathbf{a}^n$, is called the **column space** of A. Its dimension is called the **column rank** of A.

EXAMPLE 4.26

Let A be given by

$$A = \begin{bmatrix} 1 & -1 & 3 & -1 & 2 \\ 2 & 2 & 1 & -1 & 1 \\ 1 & 0 & 1 & -15 & -1 \\ 1 & 0 & 2 & 4 & 2 \end{bmatrix}.$$

Describe the column space of A, and find a basis for it. Determine the column rank of A.

Solution

The column space of A is the subspace of $\mathbf{R}^4$ generated by the vectors $\mathbf{v}_1 = (1, 2, 1, 1)$, $\mathbf{v}_2 = (-1, 2, 0, 0)$, $\mathbf{v}_3 = (3, 1, 1, 2)$, $\mathbf{v}_4 = (-1, -1, -15, 4)$, and $\mathbf{v}_5 = (2, 1, -1, 2)$. To find a basis for this subspace, we follow the procedure of Theorem 1 of Section 4.3 and transform A into the row-reduced echelon form

$$B = \begin{bmatrix} 1 & 0 & 0 & -34 & -4 \\ 0 & 1 & 0 & 24 & 3 \\ 0 & 0 & 1 & 19 & 3 \\ 0 & 0 & 0 & 0 & 0 \end{bmatrix}.$$

Since the vectors $\mathbf{v}_1$, $\mathbf{v}_2$, and $\mathbf{v}_3$ have been transformed into distinct elementary vectors, the set $\{\mathbf{v}_1, \mathbf{v}_2, \mathbf{v}_3\}$ is a basis for the column space of A. Moreover, the dimension (the number of vectors in a basis) of the latter is 3, and so the column rank of A is 3.

Note that for the matrix of Example 4.26, the column rank and the rank (the number of nonzero rows in the row-reduced echelon form) are equal. The following theorem states that this is no coincidence.

THEOREM 1 The dimension of the column space of a matrix A is equal to the rank of A.

Proof Let B be the row-reduced echelon form of A. According to the procedure of Theorem 1 of Section 4.3, we need only show that the number of distinct elementary columns of B is equal to the number of nonzero rows of B. The last statement follows immediately from the definition of row-reduced echelon form. The number of nonzero rows is the same as the number of *leading ones,* and the number of leading ones is in turn the same as the number of distinct elementary columns.

By considering the rows of a matrix instead of its columns, we can develop a theorem similar to the one just given.

DEFINITION

Let A be an $m \times n$ matrix. The subspace of $\mathbf{R}^n$ generated by the rows of the matrix $\mathbf{a}_1, \mathbf{a}_2, \ldots, \mathbf{a}_m$ is called the **row space** of A. Its dimension is called the **row rank** of A.

EXAMPLE 4.27

As in Example 4.26, let

$$A = \begin{bmatrix} 1 & -1 & 3 & -1 & 2 \\ 2 & 2 & 1 & -1 & 1 \\ 1 & 0 & 1 & -15 & -1 \\ 1 & 0 & 2 & 4 & 2 \end{bmatrix}.$$

Describe the row space of A, and find a basis for it. Determine the row rank of A.

Solution

The row space of A is the subspace of $\mathbf{R}^5$ generated by the vectors $\mathbf{w}_1 = (1, -1, 3, -1, 2)$, $\mathbf{w}_2 = (2, 2, 1, -1, 1)$, $\mathbf{w}_3 = (1, 0, 1, -15, -1)$, and $\mathbf{w}_4 = (1, 0, 2, 4, 2)$. To find a basis for it, we again apply Theorem 1 of Section 4.3. We must first form the matrix whose ith column is $\mathbf{w}_i$. This matrix is simply the transpose of A,

$$A^T = \begin{bmatrix} 1 & 2 & 1 & 1 \\ -1 & 2 & 0 & 0 \\ 3 & 1 & 1 & 2 \\ -1 & -1 & -15 & 4 \\ 2 & 1 & -1 & 2 \end{bmatrix},$$

which has the row-reduced echelon form

$$B = \begin{bmatrix} 1 & 0 & 0 & \frac{2}{3} \\ 0 & 1 & 0 & \frac{1}{3} \\ 0 & 0 & 1 & -\frac{1}{3} \\ 0 & 0 & 0 & 0 \\ 0 & 0 & 0 & 0 \end{bmatrix}.$$

Since $\mathbf{w}_1$, $\mathbf{w}_2$, and $\mathbf{w}_3$ are transformed to $\mathbf{e}_1$, $\mathbf{e}_2$, and $\mathbf{e}_3$, the set $\{\mathbf{w}_1, \mathbf{w}_2, \mathbf{w}_3\}$ is a basis for the row space of A, and so the dimension of the latter is 3. Thus the row rank of A is 3.

In Example 4.26 we found that the *column rank* of this matrix A is also 3. It is in fact always true that the row rank of a given matrix is the same as its column rank. As an intermediate step toward obtaining this result, we prove the following theorem.

THEOREM 2 Let A be an $m \times n$ matrix, and suppose that C is the matrix obtained from A by performing an elementary row operation on the latter. Then the row space of A is the same as the row space of C.

Proof Let the rows of A be $\mathbf{a}_1, \mathbf{a}_2, \ldots, \mathbf{a}_m$, and let those of C be $\mathbf{c}_1, \mathbf{c}_2, \ldots, \mathbf{c}_m$. Then the row space of A is the subspace $\mathbf{S}_1$ of $\mathbf{R}^n$ generated by $\mathscr{T}_1 = \{\mathbf{a}_1, \mathbf{a}_2, \ldots, \mathbf{a}_m\}$, and the row space of C is the subspace $\mathbf{S}_2$ generated by $\mathscr{T}_2 = \{\mathbf{c}_1, \mathbf{c}_2, \ldots, \mathbf{c}_m\}$. We must show that under each of the three elementary row operations, $\mathbf{S}_1 = \mathbf{S}_2$.

i. Suppose C is obtained from A by interchanging two rows. Then $\mathscr{T}_1 = \mathscr{T}_2$, and so $\mathbf{S}_1 = \mathbf{S}_2$.

ii. Suppose C is obtained from A by multiplication of row i by the nonzero scalar k. Then

$$\mathscr{T}_2 = \{\mathbf{a}_1, \ldots, \mathbf{a}_{i-1}, k\mathbf{a}_i, \mathbf{a}_{i+1}, \ldots, \mathbf{a}_m\}.$$

If $\mathbf{v}$ is in the row space of C, $\mathbf{S}_2$, then there exist scalars $c_1, c_2, \ldots, c_m$ such that

$$\mathbf{v} = c_1\mathbf{a}_1 + \cdots + c_{i-1}\mathbf{a}_{i-1} + c_i(k\mathbf{a}_i) + c_{i+1}\mathbf{a}_{i+1} + \cdots + c_m\mathbf{a}_m$$
$$= c_1\mathbf{a}_1 + \cdots + c_{i-1}\mathbf{a}_{i-1} + (c_ik)\mathbf{a}_i + c_{i+1}\mathbf{a}_{i+1} + \cdots + c_m\mathbf{a}_m,$$

a linear combination of $\mathscr{T}_1$. Thus $\mathbf{v}$ lies in the row space of A, $\mathbf{S}_1$. Similarly, we can show that if $\mathbf{w}$ is in $\mathbf{S}_1$, it must also be in $\mathbf{S}_2$. Hence $\mathbf{S}_1 = \mathbf{S}_2$.

iii. Suppose C is obtained from A by the addition of k times row j to row i. Then

$$\mathscr{T}_2 = \{\mathbf{a}_1, \ldots, \mathbf{a}_{i-1}, k\mathbf{a}_j + \mathbf{a}_i, \mathbf{a}_{i+1}, \ldots, \mathbf{a}_m\}.$$

By an argument similar to the one in part (ii), we can show that in this case as well, $\mathbf{S}_1 = \mathbf{S}_2$.

NOTE By applying Theorem 2 successively a finite number of times, we see that any finite sequence of row operations on a matrix A results in a matrix C that has the same row space as A. In other words, any two matrices that are *row equivalent* have the same row space.

EXAMPLE 4.28   Show that the matrices

$$A = \begin{bmatrix} 5 & 3 & 4 \\ 2 & 2 & -4 \\ 3 & 2 & 1 \\ -1 & 2 & 1 \end{bmatrix} \quad \text{and} \quad C = \begin{bmatrix} 3 & 1 & 8 \\ 1 & 4 & -3 \\ -2 & -1 & -3 \\ 1 & 1 & -2 \end{bmatrix}$$

have the same row space and find a basis for this subspace of $\mathbf{R}^3$.

Solution   By performing the computations, we see that both A and C have the row-reduced echelon form

$$B = \begin{bmatrix} 1 & 0 & 0 \\ 0 & 1 & 0 \\ 0 & 0 & 1 \\ 0 & 0 & 0 \end{bmatrix}.$$

Thus A and C are row equivalent to the same matrix, so they are row equivalent to one another and have the same row space.

This row space is also that of B—the one generated by $\{(1,0,0),\ (0,1,0),\ (0,0,1),\ (0,0,0)\}$. A basis for this subspace is clearly $\{(1,0,0),\ (0,1,0),\ (0,0,1)\}$.

NOTE   Examples 4.27 and 4.28 demonstrate two essentially different methods of producing bases for the row space of a matrix. The first provides a basis that consists of a *subset* of the given rows. The second provides a basis that ordinarily is *not* a subset of the original rows. However, the elements of the latter are in some sense simpler than those of the former.

....

We will now give the main result of this section.

THEOREM 3   Let A be a matrix. The row rank, the column rank, and the rank of A are all equal.

Proof   By Theorem 1, the column rank of A is equal to the rank of A. Thus it suffices to show that the row rank of A is equal to the rank of A.

Let B be the row-reduced echelon form of A. According to Theorem 2 and the definition of rank, we need only show that the nonzero rows of B form a basis for the row space of B. Since these nonzero rows span the row space, it is sufficient to prove that the nonzero rows of B form a linearly independent set. We leave the verification of this last statement as Exercise 19.

4.4 EXERCISES

In Exercises 1–6 verify in each case that the row rank is equal to the column rank by explicitly finding the dimensions of the row space and the column space of the given matrix.

1. $\begin{bmatrix} 1 & 2 & 0 \\ 0 & 0 & 1 \\ 0 & 0 & 0 \end{bmatrix}$

2. $\begin{bmatrix} 1 & 2 & 1 \\ 2 & 1 & -1 \end{bmatrix}$

3. $\begin{bmatrix} 1 & 0 & -1 \\ -1 & 0 & 1 \end{bmatrix}$

4. $\begin{bmatrix} 0 & 0 \\ -1 & 2 \\ 2 & -4 \end{bmatrix}$

5. $\begin{bmatrix} 1 & -1 & 3 \\ 0 & 1 & 1 \\ 1 & 1 & 0 \\ 2 & -1 & 1 \end{bmatrix}$

6. $\begin{bmatrix} 1 & -1 & 1 & 0 \\ 1 & 1 & 0 & 0 \\ 1 & 0 & 0 & 1 \end{bmatrix}$

7–12. Find a basis for the column space of each matrix of Exercises 1–6.

13–18. Find a basis for the row space of each matrix of Exercises 1–6.

19. Prove that the set of nonzero rows of the row-reduced echelon form of a matrix is linearly independent.

The method of finding a basis for the row space of a matrix A given by Example 4.27 has the advantage of introducing no new mathematical tools, simply finding a basis for the column space of A^T. There is an alternate approach that parallels the development based on the idea of *column operations*. Exercises 20 through 30 explore this idea.

20. Define the three **elementary column operations** on a matrix (see Section 2.2).

21. Define the notion of a matrix being in **column-reduced echelon form.**

22. For a matrix A, prove that a column operation on A has exactly the same effect as the corresponding row operation on A^T followed by transposing the resulting matrix.

23. If A is a matrix and B is its column-reduced echelon form, prove that the rows of A that correspond to a distinct set of elementary rows of B form a basis for the row space of A.

24–29. Find a basis for the row space of each matrix in Exercises 1–6 using Exercise 23.

30. Recall that to solve the equation $A\mathbf{x} = \mathbf{b}$, we augment A by $\mathbf{b}$, obtaining $[A \mid \mathbf{b}]$, and row-reduce. We can use this procedure to solve the equation $\mathbf{x}^T A = \mathbf{b}^T$ by solving $A^T \mathbf{x} = \mathbf{b}$ since $(\mathbf{x}^T A)^T = A^T \mathbf{x}$. Devise a technique (using column operations) to solve the problem $\mathbf{x}^T A = \mathbf{b}^T$ without using the transpose of A.

CHAPTER 4 REVIEW

The concepts of *linear independence*, *spanning set*, *basis*, and *dimension* were introduced in this chapter (in the context of $\mathbf{R}^m$) and the relationships among them were explored. As an application of these ideas, the notion of *rank* was considered here in more depth than was the case in Chapter 2.

You should be able to:

1. determine whether a set of vectors in $\mathbf{R}^m$ is linearly independent or dependent, and if dependent, express one of its vectors as a linear combination of the others;

2. determine whether or not a subset of $\mathbf{R}^m$ is a subspace;

3. determine whether or not a set of vectors spans a given subspace of $\mathbf{R}^m$;

4. determine whether or not a set of vectors is a basis for a given subspace of $\mathbf{R}^m$;

5. find a basis for, and the dimension of, a given subspace of $\mathbf{R}^m$ (including the solution space of a homogeneous linear system);

6. extend a linearly independent set to a basis for a given subspace of $\mathbf{R}^m$;

7. reduce a set that generates a subspace of $\mathbf{R}^m$ to a basis for that subspace;

8. find a basis for the column space and row space of a matrix.

KEY WORDS

linear combination of vectors

elementary vectors and columns

solution space of a homogeneous linear system

basis for a subspace

dimension of a subspace

linearly dependent and linearly independent vectors

subspace of $\mathbf{R}^m$

subspace spanned (or generated) by a set of vectors

standard basis for $\mathbf{R}^m$

column space, column rank, row space, and row rank of a matrix

KEY THEOREMS

Theorem $\mathcal{S} = \{\mathbf{v}_1, \mathbf{v}_2, \ldots, \mathbf{v}_n\}$ is a linearly dependent set if and only if the equation $c_1\mathbf{v}_1 + c_2\mathbf{v}_2 + \cdots + c_n\mathbf{v}_n = \mathbf{0}$ has a solution with at least one $c_i \neq 0$.

Theorem If a set $\mathcal{T}$ generates a subspace **S**, then there is a subset of $\mathcal{T}$ that is a basis for **S**.

Theorem If a subset $\mathcal{T}$ of a subspace **S** is linearly independent, then there is a set containing $\mathcal{T}$ which is a basis for **S**.

Theorem Every basis for a subspace **S** contains the same number of vectors.

Theorem Let **S** be a subspace of $\mathbf{R}^m$ with dimension n and let $\mathcal{T}$ be a subset of **S** containing s vectors.

    i. If $s > n$, then $\mathcal{T}$ is linearly dependent.

    ii. If $s < n$, then $\mathcal{T}$ does not span **S**.

    iii. If $s = n$ and $\mathcal{T}$ is linearly independent, then it spans **S**.

    iv. If $s = n$ and $\mathcal{T}$ spans **S**, then it is linearly independent.

Theorem The row rank, the column rank, and the rank of a matrix are all equal.

SELF-TEST

Answers appear at the end of this chapter.

1. Determine whether the set of vectors

$$\{(1, 0, 1, 0), \ (0, 1, -1, 1), \ (2, 1, 1, 1)\}$$

is linearly independent or linearly dependent. If it is linearly dependent, express one of the vectors as a linear combination of the others. [Section 4.1]

2. Determine whether or not the set of vectors

$$\{(1, 1, 1), \ (0, 1, 2), \ (1, 2, 3)\}$$

spans $\mathbf{R}^3$. [Section 4.2]

3. Show that the set of all vectors of the form $(a + b, \ a - b, \ c, \ 0)$ is a subspace of $\mathbf{R}^4$. [Section 4.2]

4. Determine whether or not the set of all vectors with first component equal to 1 is a subspace of $\mathbf{R}^m$. [Section 4.2]

5. Show that the set

$$\{(1, 1, 0, 0), \ (1, -1, 0, 0), \ (0, 0, 1, 0)\}$$

is a basis for the subspace of Problem 3. [Section 4.3]

6. Use the result of Problem 5 to determine the dimension of the subspace of Problem 3. [Section 4.3]

7. Determine the dimension of, and a basis for, the solution space of the homogeneous linear system $A\mathbf{x} = \mathbf{0}$, where

$$A = \begin{bmatrix} 1 & 0 & 1 & 2 \\ 0 & 1 & 1 & 1 \end{bmatrix}.$$ [Section 4.3]

8. Let **S** be a subspace of **R**^{5} of dimension 3, and let $\mathcal{T}$ be a subset of **S**. Are the following statements true or false?

(a) If $\mathcal{T}$ contains two vectors, then it cannot span **S**.

(b) If $\mathcal{T}$ contains three vectors, then it must be a basis for **S**.

(c) If $\mathcal{T}$ contains four vectors, then it must be linearly dependent.   [Section 4.3]

9. Let

$$B = \begin{bmatrix} 1 & 2 & 3 & 4 \\ 0 & 1 & 2 & 3 \\ 1 & 1 & 1 & 1 \end{bmatrix}.$$

Determine a basis for, and the dimension of, the following subspaces:

(a) The column space of B.   (b) The row space of B.   [Section 4.4]

10. If $\mathcal{S} = \{\mathbf{u}, \mathbf{v}, \mathbf{w}\}$ is a basis for **R**^{3}, show that $\mathcal{T} = \{\mathbf{u}, \mathbf{v}\}$ is a basis for the subspace **S** generated by **u** and **v**.   [Section 4.3]

REVIEW EXERCISES

In Exercises 1–4 determine whether the given set of vectors is linearly independent or dependent. If it is dependent, express one of the vectors as a linear combination of the others.

1. $\{(1, 0, 0), (1, 2, 0), (1, 2, 3)\}$

2. $\{(1, 2, -1), (0, 1, -1), (2, 1, 1)\}$

3. $\{(1, 0, 1, 0), (-1, 1, 2, -1), (3, -1, 0, 1)\}$

4. $\{(1, 2, 3, 4, 5), (5, 4, 3, 2, 1)\}$

In Exercises 5–8 determine whether or not the given set of vectors is a subspace of **R**^{m}.

5. The set of all vectors in **R**^{4} of the form $(0, x, -2x, x + 2y)$.

6. The set of all vectors in **R**^{4} with first two components zero and third component equal to the square of the fourth.

7. The set of all vectors **x** in **R**^{3} for which $\mathbf{x} + (0, 0, 1)$ is not **0**.

8. The set of all vectors **x** in **R**^{3} such that $\mathbf{x} \cdot (0, 0, 1) = 0$.

In Exercises 9 and 10, determine if the given vector **x** spans the indicated subspace **S**.

9. $\mathbf{x} = (-1, 1, 1)$, **S** is the solution space of $A\mathbf{x} = \mathbf{0}$ where

$$A = \begin{bmatrix} 1 & 0 & 1 \\ 0 & 1 & -1 \end{bmatrix}.$$

10. $\mathbf{x} = (0, 1, 0)$, **S** is the row space of the matrix

$$\begin{bmatrix} 1 & 0 & -1 \\ -1 & 0 & 1 \end{bmatrix}.$$

In Exercises 11–14, determine if the given set is a basis for the indicated subspace.

11. $\{(1, 0, 0), (1, 2, 0), (1, 2, 3)\}$ for $\mathbf{R}^3$.

12. $\{(1, 0, 0, 0), (1, 1, 0, 0), (1, 1, 1, 0), (1, 1, 1, 1)\}$ for $\mathbf{R}^4$.

13. $\{(1, 2, 0, 0), (1, 0, 2, 0)\}$ for the subspace of $\mathbf{R}^4$ of all vectors of the form $(x, x + y, x - y, 0)$.

14. $\{(1, 0, 1, 0), (0, 1, 2, 1)\}$ for the subspace of $\mathbf{R}^4$ of all vectors of the form $\{(x, y, x + y + z, y - z)\}$.

In Exercises 15–21, find a basis for, as well as the dimension of, each of the given subspaces.

15. The subspace of $\mathbf{R}^4$ of all vectors with second component equal to the sum of the other components.

16. The subspace of $\mathbf{R}^5$ of all vectors with first and second component zero and last component equal to the sum of the first four.

17. The subspace of $\mathbf{R}^3$ consisting of all vectors $\mathbf{x}$ such that $\mathbf{x} \cdot (1, 2, 3) = 0$.

18. The solution space of $A\mathbf{x} = \mathbf{0}$, where

$$A = \begin{bmatrix} 1 & 2 & 3 & 4 \\ 0 & 1 & 2 & 3 \end{bmatrix}.$$

19. The row space of the matrix

$$B = \begin{bmatrix} 1 & 1 & 1 & 1 & 1 \\ 0 & 1 & 1 & 1 & 1 \\ 0 & 0 & 1 & 1 & 1 \end{bmatrix}.$$

20. The column space of the matrix B of Exercise 19.

21. The subspace of $\mathbf{R}^4$ generated by the vectors $(1, -1, 2, 0)$, $(-1, -1, 4, -2)$, $(2, -1, 1, 1)$, $(2, -3, 7, -1)$.

22. Find a basis for $\mathbf{R}^3$ that contains the vectors $(1, 2, 3)$ and $(3, 2, 1)$.

ANSWERS TO SELF-TEST

1. Linearly dependent: $(2, 1, 1, 1) = 2(1, 0, 1, 0) + (0, 1, -1, 1)$.

2. It does not span.

3. The given set of vectors (call it $\mathcal{S}$) is a subset of $\mathbf{R}^4$. We must show that it is closed under addition and multiplication by scalars. First take two typical vectors in $\mathcal{S}$ and add them:

$$(a_1 + b_1, a_1 - b_1, c_1, 0) + (a_2 + b_2, a_2 - b_2, c_2, 0)$$
$$= (a_1 + b_1 + a_2 + b_2, a_1 - b_1 + a_2 - b_2, c_1 + c_2, 0)$$
$$= ((a_1 + a_2) + (b_1 + b_2), (a_1 + a_2) - (b_1 + b_2), c_1 + c_2, 0).$$

Since this is a vector in $\mathcal{S}$, that set is closed under addition.

The given set S is closed under multiplication by scalars as well, because

$$k(a + b, a - b, c, 0) = (k(a + b), k(a - b), kc, 0)$$
$$= (ka + kb, ka - kb, kc, 0),$$

which is a vector in S.

4. It is not a subspace.

5. We must show that the given set (call it T) is linearly independent and spans the set S of Problem 3. To show linear independence, apply the test for linear independence given in Section 4.1. To show that T spans, notice that we can write an arbitrary vector in S, $(x + y, x - y, z, 0)$, as the following linear combination of those in T:

$$(x + y, x - y, z, 0) = x(1, 1, 0, 0) + y(1, -1, 0, 0) + z(0, 0, 1, 0).$$

6. Its dimension is 3.

7. Its dimension is 2; a basis is $\{(-1, -1, 1, 0), (-2, -1, 0, 1)\}$.

8. (a) True.   (b) False.   (c) True.

9. (a) Its dimension is 2; a basis is $\{(1, 0, 1), (2, 1, 1)\}$.

   (b) Its dimension is 2; a basis is $\{(1, 2, 3, 4), (0, 1, 2, 3)\}$.

10. By definition, T spans $\mathbf{S}$, so we need only show that T is linearly independent. Suppose it is not. Then there are scalars a and b, not both zero, such that $a\mathbf{u} + b\mathbf{v} = \mathbf{0}$. But then $a\mathbf{u} + b\mathbf{v} + 0\mathbf{w} = \mathbf{0}$, which implies (since either a or b is nonzero) that $\{\mathbf{u}, \mathbf{v}, \mathbf{w}\}$ is linearly dependent, a contradiction. Thus T is linearly independent and therefore is a basis for $\mathbf{S}$.

5

VECTOR SPACES

In the preceding chapters we developed rules of arithmetic for, and some of the properties of, special vectors: vectors in R^m. Here we study more general collections of vectors, that is, other sets with similar arithmetic properties. A *vector space* is a mathematical structure that collects these entities under one name and allows us to study their common properties all at once.

5.1 VECTOR SPACES AND SUBSPACES

A *vector space* is intuitively a set that, together with operations of addition and multiplication by scalars, behaves in essentially the same manner as $\mathbf{R}^m$. The following definition makes this precise.

DEFINITION

A set $\mathbf{V}$ together, with operations called *addition* and *multiplication by scalars* (or *scalar multiplication*), is a **vector space** if the following properties are satisfied for every choice of elements $\mathbf{u}$, $\mathbf{v}$, and $\mathbf{w}$ in $\mathbf{V}$ and scalars c and d in $\mathbf{R}$. The elements of $\mathbf{V}$ are called **vectors**.

a. $\mathbf{u} + \mathbf{v}$ is in $\mathbf{V}$ ($\mathbf{V}$ is closed under addition).

b. $c\mathbf{u}$ is in $\mathbf{V}$ ($\mathbf{V}$ is closed under scalar multiples).

c. $\mathbf{u} + \mathbf{v} = \mathbf{v} + \mathbf{u}$ (addition is commutative).

d. $(\mathbf{u} + \mathbf{v}) + \mathbf{w} = \mathbf{u} + (\mathbf{v} + \mathbf{w})$ (addition is associative).

e. There is an element, denoted by $\mathbf{0}$, in $\mathbf{V}$ such that $\mathbf{u} + \mathbf{0} = \mathbf{u}$ for all $\mathbf{u}$ in $\mathbf{V}$ ($\mathbf{0}$ is the **zero** of $\mathbf{V}$).

f. For each $\mathbf{u}$ in $\mathbf{V}$, there is an element, denoted by $-\mathbf{u}$ in $\mathbf{V}$, such that $\mathbf{u} + (-\mathbf{u}) = \mathbf{0}$ ($-\mathbf{u}$ is the **negative** of $\mathbf{u}$).

g. $(cd)\mathbf{u} = c(d\mathbf{u})$.

h. $(c + d)\mathbf{u} = c\mathbf{u} + d\mathbf{u}$ ⎫

i. $c(\mathbf{u} + \mathbf{v}) = c\mathbf{u} + c\mathbf{v}$ ⎬   (distributive laws).

j. $1\mathbf{u} = \mathbf{u}$.

NOTE   Since we have restricted the scalars to be the set of real numbers, $\mathbf{V}$ is called a vector space *over the real numbers*. If the scalars were allowed to be complex numbers, the same definition would apply, and the resulting structure would be called a vector space *over the complex numbers*. In Section 7.3 we will study such vector spaces.

··

EXAMPLE 5.1   The set $\mathbf{R}^m$ of all m-vectors, with the usual operations of addition and multiplication by scalars, is a vector space.

This statement is a consequence of Theorem 2 of Section 1.4.

EXAMPLE 5.2   The set $\mathbf{M}^{m,n}$ of all $m \times n$ matrices, with the usual operations of addition and scalar multiplication, is a vector space.

This statement is a consequence of Theorem 1 of Section 2.3.

EXAMPLE 5.3   The set $\mathbf{R}^+$ of positive real numbers is *not* a vector space under the usual operations of addition and multiplication, since it is not closed under scalar multiplication. For example, 2 is in $\mathbf{R}^+$ and -1 is a real number, but $(-1)(2)$ is not in $\mathbf{R}^+$. Thus property (b) is not satisfied. (Neither is property (f), but all we need show is that *one* property fails to hold.)

EXAMPLE 5.4   The set $\mathbf{C}$ of all complex numbers, with the usual operations of addition of complex numbers and multiplication by real numbers, is a vector space (over the real numbers).

Solution   Recall that $\mathbf{C}$ is the set of all elements of the form $z = a + bi$, where a and b are real numbers. Two complex numbers $a + bi$ and $c + di$ are equal if and only if $a = c$ and $b = d$. Addition is defined by

$$(a + bi) + (c + di) = (a + c) + (b + d)i$$

and scalar multiplication by $k(a + bi) = (ka) + (kb)i$. Since each of these is an element of $\mathbf{C}$, properties (a) and (b) of the definition of vector space are satisfied.

c. For $z_1 = a_1 + b_1 i$ and $z_2 = a_2 + b_2 i$,

$$\begin{aligned}
z_1 + z_2 &= (a_1 + b_1 i) + (a_2 + b_2 i) = (a_1 + a_2) + (b_1 + b_2)i \\
&= (a_2 + a_1) + (b_2 + b_1)i = (a_2 + b_2 i) + (a_1 + b_1 i) \\
&= z_2 + z_1.
\end{aligned}$$

d. For $z_1 = a_1 + b_1 i$, $z_2 = a_2 + b_2 i$, and $z_3 = a_3 + b_3 i$,

$$\begin{aligned}
(z_1 + z_2) + z_3 &= [(a_1 + b_1 i) + (a_2 + b_2 i)] + (a_3 + b_3 i) \\
&= [(a_1 + a_2) + (b_1 + b_2)i] + (a_3 + b_3 i)
\end{aligned}$$

$$= [(a_1 + a_2) + a_3] + [(b_1 + b_2) + b_3]i$$
$$= [a_1 + (a_2 + a_3)] + [b_1 + (b_2 + b_3)]i$$
$$= (a_1 + b_1 i) + [(a_2 + a_3) + (b_2 + b_3)i]$$
$$= (a_1 + b_1 i) + [(a_2 + b_2 i) + (a_3 + b_3 i)]$$
$$= z_1 + (z_2 + z_3).$$

e. $0 + 0i$ serves as the zero element, since

$$(a + bi) + (0 + 0i) = (a + 0) + (b + 0)i = a + bi.$$

f. For $z = a + bi$,

$$-z = -a - bi,$$

since $(a + bi) + (-a - bi) = 0 + 0i$.

g. For $z = a + bi$ and real numbers c and d,

$$(cd)(a + bi) = (cd)a + (cd)bi$$
$$= c(da) + c(db)i$$
$$= c(da + dbi) = c(d(a + bi)).$$

h. For $z = a + bi$ and real numbers c and d,

$$(c + d)(a + bi) = (c + d)a + (c + d)bi$$
$$= (ca + da) + (cb + db)i$$
$$= (ca + cbi) + (da + dbi) = c(a + bi) + d(a + bi).$$

i. For $z_1 = a_1 + b_1 i$, $z_2 = a_2 + b_2 i$, and real number c,

$$c(z_1 + z_2) = c[(a_1 + b_1 i) + (a_2 + b_2 i)]$$
$$= c[(a_1 + a_2) + (b_1 + b_2)i]$$
$$= c(a_1 + a_2) + c(b_1 + b_2)i$$
$$= (ca_1 + ca_2) + (cb_1 + cb_2)i$$
$$= (ca_1 + cb_1 i) + (ca_2 + cb_2 i)$$
$$= c(a_1 + b_1 i) + c(a_2 + b_2 i)$$
$$= cz_1 + cz_2.$$

j. For $z = a + bi$,

$$1z = 1(a + bi) = (1a) + (1b)i = a + bi = z.$$

Thus **C** is a vector space.

NOTE Example 5.4 did not mention that two complex numbers can be *multiplied*. This added structure is ignored when **C** is being viewed as a vector space over the real numbers. You may notice that special characteristics of *m*-vectors and matrices have been ignored as well. In the study of vector spaces, only properties (a) through (j) and their consequences are examined.

The next example deals with *polynomials*. Recall that a **polynomial p** is an expression of the form

$$\mathbf{p}(x) = a_n x^n + a_{n-1} x^{n-1} + \cdots + a_1 x + a_0,$$

where each a_i is a real number called the *coefficient* of x^i. The *degree* of **p** is n, provided $a_n \neq 0$. Two polynomials are *equal* if all their corresponding coefficients are equal. The *sum* of two polynomials is the polynomial obtained by adding their corresponding coefficients, that is, adding "like" terms. The *product* of a polynomial by a scalar is the polynomial obtained by multiplying each of its coefficients by the scalar.

EXAMPLE 5.5

Let **P** be the set of all polynomials in x with the usual addition and scalar multiplication. Then **P** is a vector space.

Solution

Clearly, properties (a) and (b) of the definition of vector space are satisfied by these operations. The **0** for **P** is the zero polynomial, the one whose graph is the x-axis. The negative of **p**, $-\mathbf{p}$, is the polynomial with each coefficient replaced by its negative. (The graph of $-\mathbf{p}$ is the reflection of the graph of **p** in the x-axis.) The other vector space properties follow easily; the arguments are very similar to those of Example 5.4.

EXAMPLE 5.6

(*From calculus*) Let $\mathbf{C}(-\infty, \infty)$ denote the set of all continuous real-valued functions of a real variable x. Addition of **f** and **g** in $\mathbf{C}(-\infty, \infty)$ and multiplication by a scalar c are defined in the usual way for functions, by

    i. $(\mathbf{f} + \mathbf{g})(x) = \mathbf{f}(x) + \mathbf{g}(x)$, for all real numbers x,

    ii. $(c\mathbf{f})(x) = c(\mathbf{f}(x))$, for all real numbers x.

Then $\mathbf{C}(-\infty, \infty)$ is a vector space over **R**.

Solution

Vector space properties (a) and (b) follow from the fact that the sum of two continuous functions is continuous, and a scalar multiple of a continuous function is continuous. Further properties rest on the fact that two functions are equal if they agree at every point. That is, $\mathbf{f} = \mathbf{g}$ if and only if $\mathbf{f}(x) = \mathbf{g}(x)$ for each real number x. For example, to check property (c), let **f** and **g** be in $\mathbf{C}(-\infty, \infty)$. Then for each real number x,

$$(\mathbf{f} + \mathbf{g})(x) = \mathbf{f}(x) + \mathbf{g}(x) = \mathbf{g}(x) + \mathbf{f}(x) = (\mathbf{g} + \mathbf{f})(x).$$

Thus $\mathbf{f} + \mathbf{g} = \mathbf{g} + \mathbf{f}$. The other properties are also easily checked. It should be noted that **0** is the zero function of Example 5.5, and for **f** in $\mathbf{C}(-\infty, \infty)$, the negative of **f** is the function $-\mathbf{f}$, defined by $(-\mathbf{f})(x) = -(\mathbf{f}(x))$ for each x. As in Example 5.5, the graph of $-\mathbf{f}$ is the reflection of the graph of **f** in the x-axis.

NOTE In Example 5.6 we showed that $\mathbf{C}(-\infty, \infty)$, the set of all continuous real-valued functions defined on the real line, is a vector space. In a similar manner it can be shown that the following sets of real-valued functions are also vector spaces:

C(a, b), the set of all continuous functions on the open interval (a, b);

C$[a, b]$, the set of all continuous functions on the closed interval $[a, b]$;

C$^\infty(a, b)$, the set of all functions that have derivatives of all orders on (a, b);

C$^\infty(-\infty, \infty)$, the set of all functions that have derivatives of all orders on the entire real line.

SPOTLIGHT

WHY VECTOR SPACES?

At this point a logical question to ask is, "Why are abstract vector spaces so important?" As you read more about them, work the exercises, and see their applications, you may be able to supply some specific reasons yourself. Here, in general terms, is an answer to the question posed above.

- Abstract vector spaces can be applied to help solve concrete, real-world problems. Two of these applications are given in Chapter 8: approximation of continuous functions (Section 8.7) and linear differential equations (Section 8.8).

- Vector spaces allow us to see the common underlying structure of certain seemingly different mathematical entities. For example, you have already seen that matrices and polynomials, which on the surface appear to have little in common, share many of the same properties.

- Some vector spaces, such as **R**^{n}, are easier to study than others. By standard techniques (for example, using *coordinates*, as in Section 5.3), it may be possible to rephrase a problem that involves an abstract vector space in terms of a simpler one.

- In many areas of mathematics and its applications, there are one or more underlying vector spaces that give the structure of the entity being studied. With this in mind, the mathematician (or other expert) can make use of the extensive information available about vector spaces to pursue difficult questions without needing to "reinvent the wheel."

To put it simply, vector spaces supply mathematicians and those that use mathematics with powerful tools to attack many problems.

In the previous examples, the operations of addition and scalar multiplication have been defined in a "natural" way—in the way you'd expect. However, in describing a vector space we can define these operations in any way we like. As long as the given set under these operations satisfies properties (a)–(j) of the definition, it is a vector space; if these properties are not satisfied, then it is not a vector space. The next example illustrates this point.

EXAMPLE 5.7   Let **V** be the set $\mathbf{R}^2$ of all ordered pairs of real numbers. Let addition be defined in the usual way, but define scalar multiplication, *, by

$$c * \mathbf{v} = |c|\, \mathbf{v}$$

for any scalar c and ordered pair **v**. That is, $c * \mathbf{v}$ is the usual scalar product of the *absolute value* of c times **v**. Show that **V** is *not* a vector space.

Solution   All we need show is that *one* of properties (a) through (j) fails to hold for *some* choice of elements of **V** and *some* choice of scalars. We show that property (h) fails when $c = 1, d = -1$, and $\mathbf{u} = (2, 3)$. Since

$$(c + d) * \mathbf{u} = (1 + -1) * (2, 3) = 0 * (2, 3) = 0(2, 3) = (0, 0),$$

but $c * \mathbf{u} + d * \mathbf{u} = 1 * \mathbf{u} + (-1) * \mathbf{u} = 1\mathbf{u} + 1\mathbf{u} = \mathbf{u} + \mathbf{u} = (4, 6),$

we have that $(c + d) * \mathbf{u} \neq c * \mathbf{u} + d * \mathbf{u}$. Thus **V** is not a vector space under these operations. (You can check that the rest of the properties *do* hold.)

The next theorem gives three additonal properties common to all vector spaces. Although all of them may look "obvious," these properties must be proved from our existing knowledge of vector spaces.

THEOREM 1   Let **V** be a vector space, let **v** be an element of **V**, and let c be a scalar. Then

a. $0\mathbf{v} = \mathbf{0}$,

b. $(-1)\mathbf{v} = -\mathbf{v}$,

c. $c\mathbf{0} = \mathbf{0}$.

Proof (**a**) Since $0\mathbf{v} = (0 + 0)\mathbf{v} = 0\mathbf{v} + 0\mathbf{v}$ by property (h), and $0\mathbf{v} = 0\mathbf{v} + \mathbf{0}$ by property (e), we have

$$\mathbf{0} + 0\mathbf{v} = 0\mathbf{v} + 0\mathbf{v},$$

because addition is commutative. By property (f), every element has a negative, so there is an element $-(0\mathbf{v})$ that we add to both sides:

$$(\mathbf{0} + 0\mathbf{v}) + -(0\mathbf{v}) = (0\mathbf{v} + 0\mathbf{v}) + -(0\mathbf{v}).$$

Applying property (d), and then (f) and (e), we have

$$\mathbf{0} + (0\mathbf{v} + -(0\mathbf{v})) = 0\mathbf{v} + (0\mathbf{v} + -(0\mathbf{v}))$$
$$\mathbf{0} + \mathbf{0} = 0\mathbf{v} + \mathbf{0}$$
$$\mathbf{0} = 0\mathbf{v}.$$

(b) We first show that $(-1)\mathbf{v}$ "acts like" $-\mathbf{v}$; that is, $(-1)\mathbf{v} + \mathbf{v} = 0$. Using properties (h) and (j), we compute

$$(-1)\mathbf{v} + \mathbf{v} = (-1)\mathbf{v} + 1\mathbf{v} = (-1 + 1)\mathbf{v} = 0\mathbf{v} = \mathbf{0},$$

the last equality being part (a) of this theorem. But also, $-\mathbf{v} + \mathbf{v} = \mathbf{0}$, and thus we have

$$(-1)\mathbf{v} + \mathbf{v} = -\mathbf{v} + \mathbf{v}.$$

Adding $-\mathbf{v}$ to both sides of the equation, we obtain

$$(-1)\mathbf{v} = -\mathbf{v}.$$

(c) Since $c(\mathbf{0}) = c(\mathbf{0} + \mathbf{0}) = c(\mathbf{0}) + c(\mathbf{0})$, we have

$$\mathbf{0} + c(\mathbf{0}) = c(\mathbf{0}) + c(\mathbf{0}).$$

Now adding $-c(\mathbf{0})$ to each side, we have the result.

Since polynomials are examples of continuous functions, the set **P** of Example 5.5 can be viewed as a subset of $\mathbf{C}(-\infty, \infty)$ of Example 5.6. Furthermore, the operations are "compatible," which means that addition of polynomials is the same as addition of functions, and multiplication by scalars is also the same in both vector spaces. We say that **P** is a *subspace* of $\mathbf{C}(-\infty, \infty)$.

DEFINITION

Let **V** be a vector space, and let **W** be a subset of **V**. Then **W** is a **subspace** of **V** if **W** is a vector space using the operations of **V**.

NOTE By virtue of the next theorem, this definition of subspace agrees with that of Section 4.2 in the case of subspaces of $\mathbf{R}^m$. Thus all of the subspaces of $\mathbf{R}^m$ considered in Chapter 4 are subspaces under this new definition.

THEOREM 2 Let **V** be a vector space, and let **W** be a nonempty subset of **V**. Then **W** is a subspace of **V** if and only if vector space properties (a) and (b) are satisfied; that is, if and only if **W** is closed under addition and scalar multiplication.

Proof   If $\mathbf{W}$ is a subspace, *all* of properties (a) through (j) must be satisfied, so in particular (a) and (b) must hold. Conversely, suppose that (a) and (b) are satisfied. Properties (c), (d), and (g) through (j) are "inherited" from $\mathbf{V}$. In other words, since they hold for all vectors in $\mathbf{V}$, and $\mathbf{W}$ is a subset of $\mathbf{V}$, they hold for all vectors in $\mathbf{W}$ as well.

Since $\mathbf{V}$ is a vector space, there is an element $\mathbf{0}$ in $\mathbf{V}$, by property (e), and for each $\mathbf{w}$ in $\mathbf{W}$, $-\mathbf{w}$ is in $\mathbf{V}$, by property (f). We need to know that each of these is in $\mathbf{W}$ to complete the argument that $\mathbf{W}$ is a subspace. By property (b), for any $\mathbf{w}$ in $\mathbf{W}$, it follows that $0\mathbf{w}$ and $(-1)\mathbf{w}$ are in $\mathbf{W}$. But from Theorem 1, these are seen to be simply $\mathbf{0}$ and $-\mathbf{w}$. Hence $\mathbf{W}$ is a subspace.

We close this section with three additional examples—the first giving a subspace of the vector space $\mathbf{M}^{m,n}$, the second a subspace of $\mathbf{P}$, and the third a *subset* of $\mathbf{P}$ that is *not* a *subspace* of $\mathbf{P}$.

EXAMPLE 5.8

Let $\mathbf{D}^{2,2}$ be the set of all 2×2 *diagonal* matrices. Then $\mathbf{D}^{2,2}$ is a subspace of $\mathbf{M}^{2,2}$, the set of *all* 2×2 matrices.

Solution

By Theorem 2, all we need show is that $\mathbf{D}^{2,2}$ satisfies properties (a) and (b). (In other words, all we need show is that $\mathbf{D}^{2,2}$ is *closed* under the addition and scalar multiplication of $\mathbf{M}^{2,2}$.) To do this, let A and B be diagonal matrices, and let c be a scalar. Then A and B have the form

$$A = \begin{bmatrix} a_1 & 0 \\ 0 & a_2 \end{bmatrix}, \qquad B = \begin{bmatrix} b_1 & 0 \\ 0 & b_2 \end{bmatrix},$$

and we have

$$A + B = \begin{bmatrix} a_1 + b_1 & 0 \\ 0 & a_2 + b_2 \end{bmatrix}$$

and

$$cA = \begin{bmatrix} ca_1 & 0 \\ 0 & ca_2 \end{bmatrix}.$$

Since $A + B$ and cA are diagonal matrices—that is, elements of $\mathbf{D}^{2,2}$—the set $\mathbf{D}^{2,2}$ is a subspace of $\mathbf{M}^{2,2}$.

NOTE   By Theorem 2, often the easiest way to show that a set with two operations is a vector space is to show that it is a subspace of a known vector space. The following is an example of this technique.

EXAMPLE 5.9

Let $\mathbf{P}_n$ be the set of all polynomials of degree less than or equal to n, together with the usual polynomial arithmetic. Then $\mathbf{P}_n$ is a vector space for each positive integer n.

Solution

We show that $\mathbf{P}_n$ is a subspace of $\mathbf{P}$ for each positive integer n. Since adding two polynomials does not introduce new terms of degree higher than those of the polynomials, property (a) is satisfied. Likewise, multiplication by a scalar leaves the degree unchanged unless the scalar is zero, which *lowers* the degree. Thus property (b) is satisfied as well, and $\mathbf{P}_n$ is a subspace of $\mathbf{P}$. Since $\mathbf{P}_n$ is a subspace of $\mathbf{P}$, it is a vector space in its own right.

EXAMPLE 5.10

Let $\mathbf{P}_3'$ be the set of all polynomials of degree *exactly equal* to three together with $\mathbf{0}$. Then $\mathbf{P}_3'$ is *not* a subspace of $\mathbf{P}_3$, the set of all polynomials of degree *less than or equal* to three.

Solution

To show this we need only show that either property (a) or (b) *fails* to hold for a specific choice of polynomials and/or scalars. For example, let $\mathbf{p}(x) = x^3 + 2x - 1$, and let $\mathbf{q}(x) = -x^3 + x^2 - 3$. Then

$$(\mathbf{p} + \mathbf{q})(x) = x^2 + 2x - 4,$$

which is a polynomial of degree two. Thus $\mathbf{p}$ and $\mathbf{q}$ are two elements in $\mathbf{P}_3'$ whose sum is not in $\mathbf{P}_3'$, so $\mathbf{P}_3'$ is not a subspace of $\mathbf{P}_3$.

5.1 EXERCISES

In Exercises 1–12 determine whether or not the given set is a vector space under the given operations. If it is not a vector space, list all properties that fail to hold.

1. The set of all 2×3 matrices whose second column consists of zeros: the usual matrix operations.

2. The set of all 2×3 matrices the sum of whose entries is 1: the usual matrix operations.

3. The set of all polynomials of degree greater than or equal to three together with the polynomial $\mathbf{0}$: the usual polynomial operations.

4. The set of all polynomials of degree less than or equal to two, the sum of whose coefficients is 0: the usual operations.

5. The set of all polynomials with positive coefficients.

6. The set of all vectors in $\mathbf{R}^3$: the usual scalar multiplication but addition of $\mathbf{u}$ and $\mathbf{v}$ defined by taking the cross product of $\mathbf{u}$ and $\mathbf{v}$.

In Exercises 7 and 8 the *only* elements of the set are the formal symbols $\mathbf{x}$ and $\mathbf{y}$, with no special properties other than those listed in the exercises.

7. The set $\{\mathbf{x}, \mathbf{y}\}$, where $\mathbf{x} \neq \mathbf{y}$, with addition: $\mathbf{x} + \mathbf{x} = \mathbf{x}$, $\mathbf{x} + \mathbf{y} = \mathbf{y} + \mathbf{x} = \mathbf{y}$, $\mathbf{y} + \mathbf{y} = \mathbf{x}$, and scalar multiplication: $c\mathbf{x} = \mathbf{x}$, $c\mathbf{y} = \mathbf{x}$ for all scalars c.

8. Same as Exercise 7 except scalar multiplication: $c\mathbf{x} = \mathbf{x}$, $c\mathbf{y} = \mathbf{y}$ for all scalars c.

9. The set of all continuous functions with the property that the function evaluated at 0 is 0.

10. The set of all continuous functions with the property that the function evaluated at 0 is 1.

11. The set of all continuous functions with the property that the function is 0 at every integer; for example, $\mathbf{f}(x) = \sin(\pi x)$.

12. The set consisting solely of a single element $\mathbf{0}$, with addition: $\mathbf{0} + \mathbf{0} = \mathbf{0}$, and scalar multiplication: $c\mathbf{0} = \mathbf{0}$ for all scalars c.

In Exercises 13–16 determine whether or not the given set is a subspace of $\mathbf{M}^{2,2}$ (see Example 5.2).

13. The set of all 2×2 matrices the sum of whose entries is zero.

14. The set of all 2×2 matrices whose determinant is zero.

15. The set of all 2×2 matrices of the form

$$\begin{bmatrix} a & b \\ c & 1 \end{bmatrix},$$

where a, b, and c are real numbers.

16. The set of all 2×2 matrices of the form

$$\begin{bmatrix} a & b \\ b & a + b \end{bmatrix},$$

where a and b are real numbers.

In Exercises 17 and 18 determine whether or not the given set is a subspace of $\mathbf{M}^{n,n}$, the space of all square matrices of order n.

17. The diagonal matrices of order n.

18. The matrices of order n with trace equal to 0. (The **trace** of a square matrix is the sum of the diagonal elements.)

In Exercises 19 and 20 determine whether or not the given set is a subspace of $\mathbf{C}(-\infty, \infty)$ (see Example 5.6).

19. The set of all functions of the form $a \sin x + b \cos x$, where a and b are real numbers.

20. The set $\mathbf{D}$ of all differentiable functions.

21. If $\mathcal{S} = \{\mathbf{v}_1, \mathbf{v}_2, \ldots, \mathbf{v}_n\}$ is a set of vectors in a vector space $\mathbf{V}$, show that

$$c_1\mathbf{v}_1 + c_2\mathbf{v}_2 + \cdots + c_n\mathbf{v}_n$$

is also in $\mathbf{V}$, where the c_i are arbitrary scalars.

5.2   LINEAR INDEPENDENCE, BASIS, AND DIMENSION

In Chapter 4 we studied the concepts of *linear independence, span, basis,* and *dimension* in the context of Euclidean *m*-space $\mathbf{R}^m$. In this section we generalize these ideas to arbitrary vector spaces.

LINEAR INDEPENDENCE

We will follow Theorem 1 of Section 4.1 for the extended definition of linear independence of a set of vectors. Consequently the definition given here is compatible with the one for the special case of vectors in $\mathbf{R}^m$.

DEFINITION

Let $\mathcal{S} = \{v_1, v_2, \ldots, v_n\}$ be a set of vectors in a vector space $\mathbf{V}$. The set $\mathcal{S}$ is **linearly independent** if the equation

$$c_1 v_1 + c_2 v_2 + \cdots + c_n v_n = 0$$

has *only* the trivial solution, $c_i = 0$, for all i. If $\mathcal{S}$ is not linearly independent, we say that it is **linearly dependent**. An *infinite* subset $\mathcal{S}$ of $\mathbf{V}$ is linearly independent if every nonempty finite subset of $\mathcal{S}$ is linearly independent.

NOTE If $\mathcal{S} = \{v_1, v_2, \ldots, v_n\}$ is a set of vectors in a vector space $\mathbf{V}$, a **linear combination** of the v_i is any expression v of the form

$$v = c_1 v_1 + c_2 v_2 + \cdots + c_n v_n,$$

where the c_i are scalars. By Exercise 21 of Section 5.1, all such expressions are in $\mathbf{V}$. Thus $\mathcal{S}$ is linearly independent if and only if the only linear combination of the v_i equal to zero is the one in which every scalar c_i is zero.

EXAMPLE 5.11 Determine whether or not the set

$$\mathcal{S} = \left\{ \begin{bmatrix} 1 & 0 \\ 0 & -1 \end{bmatrix}, \begin{bmatrix} 0 & 1 \\ 0 & -1 \end{bmatrix}, \begin{bmatrix} 0 & 0 \\ 1 & -1 \end{bmatrix} \right\}$$

is linearly independent in $\mathbf{M}^{2,2}$ (see Example 5.2).

Solution We consider the matrix equation

$$c_1 \begin{bmatrix} 1 & 0 \\ 0 & -1 \end{bmatrix} + c_2 \begin{bmatrix} 0 & 1 \\ 0 & -1 \end{bmatrix} + c_3 \begin{bmatrix} 0 & 0 \\ 1 & -1 \end{bmatrix} = \begin{bmatrix} 0 & 0 \\ 0 & 0 \end{bmatrix}.$$

If the only solution is the trivial one, $c_1 = c_2 = c_3 = 0$, then $\mathcal{S}$ is linearly independent. To determine whether or not this is the case, we perform the scalar multiplications and additions on the left side to obtain

$$\begin{bmatrix} c_1 & c_2 \\ c_3 & -c_1 - c_2 - c_3 \end{bmatrix} = \begin{bmatrix} 0 & 0 \\ 0 & 0 \end{bmatrix}$$

and then equate corresponding entries. This yields the homogeneous system of four linear equations

$$
\begin{aligned}
c_1 \qquad\qquad &= 0 \\
c_2 \qquad &= 0 \\
c_3 &= 0 \\
-c_1 - c_2 - c_3 &= 0.
\end{aligned}
$$

From the first three equations, it is clear that the only solution of this system is the trivial one. Thus $\mathcal{S}$ is linearly independent.

EXAMPLE 5.12

Determine whether or not the set

$$\mathcal{S} = \{x + 1, x - 1, 2x + 1\}$$

is linearly independent in $\mathbf{P}_1$, the space of polynomials of degree less than or equal to one.

Solution

We consider the equation

$$c_1(x + 1) + c_2(x - 1) + c_3(2x + 1) = 0.$$

Combining terms of the same degree, we have

$$(c_1 + c_2 + 2c_3)x + (c_1 - c_2 + c_3) = 0.$$

This yields the set of linear equations

$$
\begin{aligned}
c_1 + c_2 + 2c_3 &= 0 \\
c_1 - c_2 + \ c_3 &= 0.
\end{aligned}
$$

Since this system is a homogeneous system of linear equations with more unknowns than equations, it has (infinitely many) nonzero solutions. Hence the set $\mathcal{S}$ is linearly dependent.

EXAMPLE 5.13

(*From calculus*) Determine whether or not the set $\mathcal{S} = \{1, x, \sin x\}$ is linearly independent in the vector space, $\mathbf{C}(-\infty, \infty)$, of continuous functions (see Example 5.6).

Solution

Here we consider the equation

$$c_1(1) + c_2 x + c_3 \sin x = 0,$$

where 0 is the zero function. Since this equation must hold for *all* x, it holds for $x = 0$, $\pi/2$, and π. Successively substituting these values for x into the previous equation yields

$$c_1 \qquad\qquad\qquad = 0$$
$$c_1 + (\pi/2)c_2 + c_3 = 0$$
$$c_1 + (\pi)c_2 \qquad\quad = 0.$$

You can easily check that this homogeneous linear system has only the trivial solution. Hence $\mathcal{S}$ is linearly independent.

BASIS FOR A VECTOR SPACE

We will now investigate the notions of *span*, *basis*, and *dimension* of an arbitrary vector space. The definitions are virtually identical to the ones given in the context of $\mathbf{R}^m$ in Sections 4.2 and 4.3.

DEFINITION

Let $\mathbf{V}$ be a vector space. A set $\mathcal{S}$ of vectors in $\mathbf{V}$ **spans** (or **generates**) $\mathbf{V}$ if every vector in $\mathbf{V}$ can be expressed as a linear combination of a finite subset of $\mathcal{S}$.

EXAMPLE 5.14

Show that the set $\mathcal{S} = \{1, x - 1, x^2 - 1\}$ spans $\mathbf{P}_2$, the subspace of $\mathbf{P}$ consisting of all polynomials of degree less than or equal to two (see Example 5.9).

Solution

Let $\mathbf{p}(x) = a_0 + a_1 x + a_2 x^2$ be an arbitrary vector in $\mathbf{P}_2$. We must show that there exist scalars c_1, c_2, and c_3 such that

$$c_1(1) + c_2(x - 1) + c_3(x^2 - 1) = a_0 + a_1 x + a_2 x^2.$$

Collecting like terms on the left side, we obtain

$$(c_1 - c_2 - c_3) + c_2 x + c_3 x^2 = a_0 + a_1 x + a_2 x^2.$$

Since two polynomials are equal if and only if their corresponding coefficients are equal, this last equation implies that

$$c_1 - c_2 - c_3 = a_0$$
$$c_2 \qquad\quad = a_1$$
$$c_3 = a_2.$$

Now this linear system has a solution for c_1, c_2, and c_3 no matter what a_0, a_1, and a_2 are. Hence $\mathcal{S}$ spans $\mathbf{P}_2$.

DEFINITION

> Let $\mathbf{V}$ be a vector space. A set $\mathcal{S}$ of vectors in $\mathbf{V}$ is a **basis** for $\mathbf{V}$ if
>
> **i.** $\mathcal{S}$ is linearly independent and
>
> **ii.** $\mathcal{S}$ spans $\mathbf{V}$.

Just as the set $\{\mathbf{e}_1, \mathbf{e}_2, \ldots, \mathbf{e}_m\}$ provides a *standard basis* for $\mathbf{R}^m$, there are standard bases for $\mathbf{P}_n$, $\mathbf{P}$, and $\mathbf{M}^{m,n}$. The next theorem describes these especially simple bases.

THEOREM I   Each of the following is a basis for the respective vector space:

a. The set $\mathcal{S} = \{1, x, x^2, \ldots, x^n\}$ for $\mathbf{P}_n$, the vector space of all polynomials of degree less than or equal to n.

b. The set $\mathcal{S} = \{1, x, x^2, \ldots, x^n, \ldots\}$ for $\mathbf{P}$, the vector space of all polynomials.

c. The set $\{E^{ij} \mid i = 1, \ldots, m; j = 1, \ldots, n\}$, where E^{ij} is the $m \times n$ matrix with one in the i, j position and zeros elsewhere, for $\mathbf{M}^{m,n}$, the vector space of all $m \times n$ matrices.

Proof   We will prove only parts (a) and (b), leaving the proof of (c) as an exercise.

To prove part (a), we must show that $\mathcal{S}$ is linearly independent and spans $\mathbf{P}_n$. First consider the equation

$$c_1(1) + c_2(x) + \cdots + c_{n+1}(x^n) = \mathbf{0} = 0(1) + 0(x) + \cdots + 0(x^n).$$

Since two polynomials are equal if and only if their corresponding coefficients are equal, we must have $c_i = 0$ for $i = 1, \ldots, n + 1$. Hence $\mathcal{S}$ is linearly independent.

Now suppose $\mathbf{p}(x) = a_0 + a_1 x + \cdots + a_n x^n$ is an arbitrary member of $\mathbf{P}_n$. Then $\mathbf{p}$ is *already* expressed as a linear combination of $\mathcal{S}$ with the a_i as the scalars (since $a_0 = a_0 \cdot 1$). Thus $\mathcal{S}$ spans $\mathbf{P}_n$ as well.

The proof that $\mathcal{S} = \{1, x, x^2, \ldots\}$ is a basis for $\mathbf{P}$ is almost exactly the same as that for part (a). The important fact is that each polynomial has *some* degree, so that we can choose n to be the degree of the polynomial and we can express the polynomial as a linear combination of the *finite* set $\{1, x, \ldots, x^n\}$.

NOTE   Usually we study situations in which the spanning set $\mathcal{S}$ is itself finite and *all* the elements of $\mathcal{S}$ can be used to represent a given vector as a linear combination of elements of $\mathcal{S}$. If $\mathcal{S}$ is infinite, however, we must find for each vector a finite subset of $\mathcal{S}$ to use to represent that vector, since sums of infinitely many terms have no meaning in general vector spaces.

EXAMPLE 5.15   Find a basis for (a) $\mathbf{P}_3$ and (b) $\mathbf{M}^{2,3}$.

Solution   We apply Theorem 1 to these special cases:

a. $\mathcal{S} = \{1, x, x^2, x^3\}$ is a basis for $\mathbf{P}_3$.

b. $\mathcal{S} = \left\{ \begin{bmatrix} 1 & 0 & 0 \\ 0 & 0 & 0 \end{bmatrix}, \begin{bmatrix} 0 & 1 & 0 \\ 0 & 0 & 0 \end{bmatrix}, \begin{bmatrix} 0 & 0 & 1 \\ 0 & 0 & 0 \end{bmatrix}, \begin{bmatrix} 0 & 0 & 0 \\ 1 & 0 & 0 \end{bmatrix}, \begin{bmatrix} 0 & 0 & 0 \\ 0 & 1 & 0 \end{bmatrix}, \begin{bmatrix} 0 & 0 & 0 \\ 0 & 0 & 1 \end{bmatrix} \right\}$

is a basis for $\mathbf{M}^{2,3}$.

EXAMPLE 5.16   Show that the set

$$\mathcal{S} = \left\{ \begin{bmatrix} 1 & 0 \\ 0 & -1 \end{bmatrix}, \begin{bmatrix} 0 & 1 \\ 0 & -1 \end{bmatrix}, \begin{bmatrix} 0 & 0 \\ 1 & -1 \end{bmatrix} \right\}$$

is a basis for the subspace $\mathbf{S}$ of $\mathbf{M}^{2,2}$ consisting of all 2×2 matrices, the sum of whose entries is zero.

Solution   We know that $\mathbf{S}$ is a subspace of $\mathbf{M}^{2,2}$ by Exercise 13 of Section 5.1 and that $\mathcal{S}$ is certainly a subset of $\mathbf{S}$. Therefore all we need do is verify that $\mathcal{S}$ is linearly independent and spans $\mathbf{S}$. We have already shown the first part in Example 5.11. To show that $\mathcal{S}$ spans $\mathbf{S}$, let

$$A = \begin{bmatrix} a_1 & a_2 \\ a_3 & a_4 \end{bmatrix}, \qquad a_1 + a_2 + a_3 + a_4 = 0,$$

be an arbitrary vector in $\mathbf{S}$. We must find scalars c_1, c_2, and c_3 such that

$$c_1 \begin{bmatrix} 1 & 0 \\ 0 & -1 \end{bmatrix} + c_2 \begin{bmatrix} 0 & 1 \\ 0 & -1 \end{bmatrix} + c_3 \begin{bmatrix} 0 & 0 \\ 1 & -1 \end{bmatrix} = \begin{bmatrix} a_1 & a_2 \\ a_3 & a_4 \end{bmatrix}.$$

Performing the scalar multiplications and additions on the left side and equating corresponding entries yields the linear system

$$
\begin{aligned}
c_1 \quad\quad\quad\quad &= a_1 \\
c_2 \quad\quad &= a_2 \\
c_3 &= a_3 \\
-c_1 - c_2 - c_3 &= a_4.
\end{aligned}
$$

From the first three equations, we see that $c_1 = a_1$, $c_2 = a_2$, and $c_3 = a_3$. Substituting these values for the c_i into the fourth equation yields $-a_1 - a_2 - a_3 = a_4$, or equivalently, $a_1 + a_2 + a_3 + a_4 = 0$. But the latter is true by hypothesis, so A can be expressed as a linear combination of the vectors in $\mathcal{S}$ for all choices of the a_i; that is, the set $\mathcal{S}$ spans $\mathbf{S}$. Thus $\mathcal{S}$ is a basis for $\mathbf{S}$.

The following theorems are the counterparts of theorems in Chapter 4. For example, the next theorem is a generalization of Theorem 2 of Section 4.2. That theorem was stated in terms of a *subspace* of $\mathbf{R}^m$, whereas the next theorem refers to a *vector space*. Since any subspace of a vector space is also a vector space, the following result does encompass the former one. The proof is exactly the same as was given before and hence is omitted.

THEOREM 2   Let $\mathbf{V}$ be a vector space spanned by the set $\mathcal{T}$. If one of the vectors in $\mathcal{T}$ is a linear combination of the others, then the set obtained by deleting that vector still spans $\mathbf{V}$.

If $\mathcal{T}$ in Theorem 2 is a finite set, repeated application of Theorem 2 results in a set that both spans and is linearly independent—a basis. The result is true even if $\mathcal{T}$ is infinite, and we state this as the next theorem.

THEOREM 3   Let $\mathbf{V}$ be a vector space spanned by a set $\mathcal{T}$. Then there is a subset of $\mathcal{T}$ that is a basis for $\mathbf{V}$. In other words, a generating set may be reduced to a basis.

EXAMPLE 5.17

The set $\mathcal{T} = \{1, x + 1, x - 1\}$ generates $\mathbf{P}_1$, the vector space of polynomials of degree less than or equal to one (Exercise 15). Theorem 3 tells us that there is a subset of $\mathcal{T}$ that is a basis for $\mathbf{P}_1$. In fact, there are three of them. Each of $\mathcal{T}_1 = \{1, x + 1\}$, $\mathcal{T}_2 = \{1, x - 1\}$, and $\mathcal{T}_3 = \{x + 1, x - 1\}$ is a basis for $\mathbf{P}_1$ (Exercise 43).

THEOREM 4   Let $\mathbf{V}$ be a vector space spanned by a set $\mathcal{S}$, and let $\mathcal{T}_1$ be a linearly independent subset of $\mathbf{V}$. Then there is a subset $\mathcal{T}_2$ of $\mathcal{S}$ such that $\mathcal{T} = \mathcal{T}_1 \cup \mathcal{T}_2$, the union of $\mathcal{T}_1$ and $\mathcal{T}_2$, is a basis for $\mathbf{V}$. In other words, a linearly independent set may be extended to a basis.

Proof   We prove only the case in which $\mathcal{T}_1$ and $\mathcal{S}$ are finite sets. Consider the set $\mathcal{T}_1 \cup \mathcal{S}$. Fixing the order of the elements, we take first the elements of $\mathcal{T}_1$ and then those of $\mathcal{S}$, and we apply Theorem 2. Since $\mathcal{S}$ alone spans $\mathbf{V}$, the set $\mathcal{T}_1 \cup \mathcal{S}$ does as well. Since no element of $\mathcal{T}_1$ is a linear combination of the others ($\mathcal{T}_1$ is linearly independent), if $\mathcal{T}_1 \cup \mathcal{S}$ is not linearly independent, some element of $\mathcal{S}$ is a linearly dependent combination of the others and can be deleted. Repeated application of this process yields a set that spans $\mathbf{V}$, is linearly independent, and contains $\mathcal{T}_1$. The remaining elements constitute $\mathcal{T}_2$.

EXAMPLE 5.18   From Example 5.11 we know that

$$\mathcal{T}_1 = \left\{ \begin{bmatrix} 1 & 0 \\ 0 & -1 \end{bmatrix}, \begin{bmatrix} 0 & 1 \\ 0 & -1 \end{bmatrix}, \begin{bmatrix} 0 & 0 \\ 1 & -1 \end{bmatrix} \right\}$$

is a linearly independent subset of $\mathbf{M}^{2,2}$, and from Theorem 1c that the set

$$\mathcal{S} = \left\{ \begin{bmatrix} 1 & 0 \\ 0 & 0 \end{bmatrix}, \begin{bmatrix} 0 & 1 \\ 0 & 0 \end{bmatrix}, \begin{bmatrix} 0 & 0 \\ 1 & 0 \end{bmatrix}, \begin{bmatrix} 0 & 0 \\ 0 & 1 \end{bmatrix} \right\}$$

is a generating set (in fact a basis) for $\mathbf{M}^{2,2}$. Taking

$$\mathcal{T}_2 = \left\{ \begin{bmatrix} 0 & 0 \\ 0 & 1 \end{bmatrix} \right\}$$

to be the subset of $\mathcal{S}$, it can be shown that the union of $\mathcal{T}_1$ and $\mathcal{T}_2$ is a basis for $\mathbf{M}^{2,2}$ (Exercises 5 and 17).

DIMENSION OF A VECTOR SPACE

THEOREM 5   Let $\mathbf{V}$ be a vector space with bases $\mathcal{B}$ and $\mathcal{B}'$. Then $\mathcal{B}$ and $\mathcal{B}'$ have the same number of elements.

Proof   We consider only the case in which $\mathcal{B}$ and $\mathcal{B}'$ are finite sets. Let

$$\mathcal{B} = \{\mathbf{v}_1, \ldots, \mathbf{v}_n\} \quad \text{and} \quad \mathcal{B}' = \{\mathbf{w}_1, \ldots, \mathbf{w}_m\}.$$

We must show that $m = n$. It suffices to assume that $n \le m$ and show that $n = m$, since the two sets could be interchanged before starting the proof.

Since $\mathcal{B}$ spans $\mathbf{V}$, and $\mathbf{w}_1$ is in $\mathbf{V}$, there are scalars $c_1, \ldots, c_n$ such that

$$\mathbf{w}_1 = c_1 \mathbf{v}_1 + \cdots + c_n \mathbf{v}_n.$$

If every $c_i = 0$, $\mathbf{w}_1$ would be the zero vector, and thus $\mathbf{0}$ would be in $\mathcal{B}'$. It is easy to show (Exercise 45) that if $\mathbf{0}$ is in a set, the set is not linearly independent. But $\mathcal{B}'$ is linearly independent, so $\mathbf{w}_1 \ne \mathbf{0}$, and thus at least one $c_i \ne 0$. By renumbering the elements of $\mathcal{B}$, if necessary, we may assume that $c_1 \ne 0$. Now we solve for $\mathbf{v}_1$,

$$\mathbf{v}_1 = (1/c_1)\mathbf{w}_1 - (c_2/c_1)\mathbf{v}_2 - \cdots - (c_n/c_1)\mathbf{v}_n.$$

Thus $\mathbf{v}_1$ is a linear combination of the elements of the set

$$\mathcal{B}_1 = \{\mathbf{w}_1, \mathbf{v}_2, \ldots, \mathbf{v}_n\}.$$

Letting $\mathcal{B}^*$ be the set $\mathcal{B}$ together with $\mathbf{w}_1$, we have a spanning set (since $\mathcal{B}$ alone is a spanning set), with the element $\mathbf{v}_1$ as a linear combination of the others. By Theorem 2, the set $\mathcal{B}_1$, obtained by deleting $\mathbf{v}_1$, still spans $\mathbf{V}$.

Since $\mathscr{B}_1$ spans $\mathbf{V}$, and $\mathbf{w}_2$ is in $\mathbf{V}$, there are scalars $d_1, \ldots, d_n$ such that

$$\mathbf{w}_2 = d_1\mathbf{w}_1 + d_2\mathbf{v}_2 + \cdots + d_n\mathbf{v}_n.$$

If $d_2 = d_3 = \cdots = d_n = 0$, we would have $\mathbf{w}_2 = d_1\mathbf{w}_1$, from which it is easy to show that the set $\mathscr{B}'$ is not linearly independent. Thus one of the $d_i \neq 0$ for $i = 2, \ldots, n$. By renumbering the remaining elements of $\mathscr{B}$, if necessary, we may assume that $d_2 \neq 0$. Now we solve for $\mathbf{v}_2$:

$$\mathbf{v}_2 = -(d_1/d_2)\mathbf{w}_1 + (1/d_2)\mathbf{w}_2 - (d_3/d_2)\mathbf{v}_3 - \cdots - (d_n/d_2)\mathbf{v}_n.$$

Thus $\mathbf{v}_2$ is a linear combination of the elements of the set

$$\mathscr{B}_2 = (\mathbf{w}_1, \mathbf{w}_2, \mathbf{v}_3, \ldots, \mathbf{v}_n\}.$$

Letting $\mathscr{B}^*$ be the set $\mathscr{B}_1$ together with $\mathbf{w}_2$, we have a spanning set, with the element $\mathbf{v}_2$ as a linear combination of the others. By Theorem 2, the set $\mathscr{B}_2$ still spans $\mathbf{V}$.

Continuing this same idea, we can use each $\mathbf{w}_i$ to replace some $\mathbf{v}_i$ and obtain

$$\mathscr{B}_k = \{\mathbf{w}_1, \ldots, \mathbf{w}_k, \mathbf{v}_{k+1}, \ldots, \mathbf{v}_n\}$$

as a spanning set for $\mathbf{V}$ until $k = n$. In other words, we continue until we run out of elements of $\mathscr{B}$ and have

$$\mathscr{B}_n = \{\mathbf{w}_1, \ldots, \mathbf{w}_n\}$$

as a spanning set for $\mathbf{V}$. If $n < m$, we would have $\mathbf{w}_{n+1}$ in the span of $\mathscr{B}_n$. It follows easily that $\mathscr{B}'$ could not be linearly independent, a contradiction. Therefore $n = m$.

DEFINITION

> The **dimension** of a vector space $\mathbf{V}$ is equal to the number of vectors in a basis for $\mathbf{V}$.

NOTE   For the zero vector space $\{\mathbf{0}\}$ it is customary to take the empty set (no elements) as a basis. Thus the dimension of the zero vector space is 0. If a vector space $\mathbf{V}$ has a finite basis, or more generally, a finite generating set, it can be shown by an argument similar to the proof of Theorem 5 that all bases for $\mathbf{V}$ are finite and hence have the same number of elements. Such a vector space is said to be **finite dimensional**. If a basis for a vector space $\mathbf{V}$ is not finite, we say that $\mathbf{V}$ is **infinite dimensional**.

....

As a direct consequence of Theorem 1 and the definition of dimension, we have the following theorem.

THEOREM 6

a. $\mathbf{P}_n$ has dimension $n + 1$.

b. $\mathbf{P}$ is infinite dimensional.

c. $\mathbf{M}^{n,m}$ has dimension mn.

EXAMPLE 5.19 Find the dimension of the subspace $\mathbf{S}$ of $\mathbf{M}^{3,3}$ consisting of all 3×3 diagonal matrices.

Solution It can be easily shown that a basis for $\mathbf{S}$ is given by

$$\left\{ \begin{bmatrix} 1 & 0 & 0 \\ 0 & 0 & 0 \\ 0 & 0 & 0 \end{bmatrix}, \begin{bmatrix} 0 & 0 & 0 \\ 0 & 1 & 0 \\ 0 & 0 & 0 \end{bmatrix}, \begin{bmatrix} 0 & 0 & 0 \\ 0 & 0 & 0 \\ 0 & 0 & 1 \end{bmatrix} \right\}.$$

Therefore $\mathbf{S}$ has dimension equal to three.

We will close this section with a theorem that is sometimes useful in quickly answering questions concerning a given subset of a vector space: "Is it linearly independent?" "Does it span?" "Is it a basis?" We leave the proof as an exercise.

THEOREM 7 Let $\mathcal{S}$ be a subset of a vector space $\mathbf{V}$ of dimension n.

a. If $\mathcal{S}$ contains fewer than n vectors, it does not span $\mathbf{V}$.

b. If $\mathcal{S}$ contains more than n vectors, it is not linearly independent (it is linearly dependent).

c. If $\mathcal{S}$ contains exactly n vectors and is either linearly independent or spans $\mathbf{V}$, then it is both, so it is a basis for $\mathbf{V}$.

EXAMPLE 5.20 Demonstrate each of the following:

a. The set $\{x, 1 + x, 1 - x\}$ is linearly dependent in $\mathbf{P}_1$.

b. The set

$$\left\{ \begin{bmatrix} 1 & 1 & 1 \\ 0 & 0 & 0 \end{bmatrix}, \begin{bmatrix} 0 & 0 & 0 \\ 1 & 1 & 1 \end{bmatrix} \right\}$$

does not span $\mathbf{M}^{2,3}$.

Solution   These follow immediately from conditions (b) and (a) of Theorem 7, respectively. In part (a) the set consists of three elements and is a subset of the two-dimensional vector space $\mathbf{P}_1$. Therefore it is linearly dependent. In part (b) the given set contains two vectors in the six-dimensional $\mathbf{M}^{2,3}$, so it cannot span.

5.2 EXERCISES

In Exercises 1–12 determine whether or not the given set is linearly independent. If the set is linearly dependent, write one of its vectors as a linear combination of the others.

1. $\{1, 1 + x, 1 + x + x^2\}$ in $\mathbf{P}_2$

2. $\{1 + x^2, 1 + x + 2x^2, x + x^2\}$ in $\mathbf{P}_2$

3. $\{1, x - 1, x + 1\}$ in $\mathbf{P}_1$

4. $\{x - 1, x + 1\}$ in $\mathbf{P}_1$

5. $\left\{ \begin{bmatrix} 1 & 0 \\ 0 & -1 \end{bmatrix}, \begin{bmatrix} 0 & 1 \\ 0 & -1 \end{bmatrix}, \begin{bmatrix} 0 & 0 \\ 1 & -1 \end{bmatrix}, \begin{bmatrix} 0 & 0 \\ 0 & 1 \end{bmatrix} \right\}$ in $\mathbf{M}^{2,2}$

6. $\left\{ \begin{bmatrix} 1 & 2 \\ -1 & 0 \end{bmatrix}, \begin{bmatrix} 0 & -1 \\ 1 & 1 \end{bmatrix}, \begin{bmatrix} 1 & 0 \\ 1 & 2 \end{bmatrix} \right\}$ in $\mathbf{M}^{2,2}$

7. $\{1, \sin^2 x, \cos^2 x\}$ in $\mathbf{C}(-\infty, \infty)$

8. $\{1, e^x, e^{-x}\}$ in $\mathbf{C}(-\infty, \infty)$

9. $\left\{ \begin{bmatrix} 2 & 0 & 0 \\ 0 & -1 & 0 \\ 0 & 0 & 1 \end{bmatrix}, \begin{bmatrix} -2 & 0 & 0 \\ 0 & -1 & 0 \\ 0 & 0 & -1 \end{bmatrix} \right\}$ in the vector space of 3×3 diagonal matrices.

10. $\left\{ \begin{bmatrix} 2 & 0 & 0 \\ 1 & -1 & 0 \\ 2 & 1 & -1 \end{bmatrix}, \begin{bmatrix} 2 & 0 & 0 \\ 0 & -1 & 0 \\ 2 & 1 & -1 \end{bmatrix} \right\}$ in the vector space of 3×3 matrices with trace 0.

11. $\{2 + x - 3\sin x + \cos x, \; x + \sin x - 3\cos x, \; 1 - 2x + 3\sin x + \cos x, \; 2 - x - \sin x - \cos x, \; 2 + x - 3\cos x\}$ in the space of all linear combinations of the set $\{1, x, \sin x, \cos x\}$.

12. All 2×2 matrices with exactly two 1's and two 0's for entries in the space of all 2×2 matrices.

13–24. Determine whether or not each set in Exercises 1–12 generates (that is, spans) the indicated vector space. (Theorem 7 may come in handy here.)

25–36. Determine whether or not each set in Exercises 1–12 is a basis for the indicated vector space.

In Exercises 37–42 find the dimension of the given vector space.

37. $\mathbf{M}^{3,4}$ **38.** $\mathbf{P}_5$

39. The subspace of $\mathbf{M}^{2,2}$ consisting of all diagonal 2×2 matrices.

40. The subspace of $\mathbf{M}^{2,2}$ consisting of all 2×2 matrices whose diagonal entries are zero.

41. The subspace of $\mathbf{P}_1$ consisting of all polynomials $\mathbf{p}(x) = a_0 + a_1 x$ such that $a_0 + a_1 = 0$.

42. The subspace of $\mathbf{P}_3$ consisting of all polynomials

$$\mathbf{p}(x) = a_0 + a_1 x + a_2 x^2 + a_3 x^3,$$

with $a_2 = 0$.

43. Show that $\mathcal{T}_1 = \{1, x + 1\}$, $\mathcal{T}_2 = \{1, x - 1\}$, and $\mathcal{T}_3 = \{x + 1, x - 1\}$ are all bases for $\mathbf{P}_1$.

44. Prove part (c) of Theorem 1.

45. If $\mathcal{T}$ is a subset of a vector space $\mathbf{V}$, and $\mathbf{0}$ is in $\mathcal{T}$, prove that $\mathcal{T}$ is linearly dependent.

46. Let $\mathcal{S} = \{\mathbf{v}_1, \mathbf{v}_2, \ldots, \mathbf{v}_n\}$ be a linearly independent set in a vector space $\mathbf{V}$.
(a) Show that the set $\{\mathbf{v}_1, \mathbf{v}_2, \ldots, \mathbf{v}_m\}$ is linearly independent if $m < n$.
(b) More generally, show that any subset of $\mathcal{S}$ is linearly independent.

47. Let $\mathcal{S} = \{\mathbf{v}_1, \mathbf{v}_2, \ldots, \mathbf{v}_n\}$ be a linearly dependent set in a vector space $\mathbf{V}$.
(a) If $\mathbf{w}$ is a vector in $\mathbf{V}$, show that the set $\{\mathbf{w}, \mathbf{v}_1, \mathbf{v}_2, \ldots, \mathbf{v}_n\}$ is linearly dependent.
(b) More generally, if $\mathcal{T}$ is any set in $\mathbf{V}$ that contains $\mathcal{S}$, show that $\mathcal{T}$ is linearly dependent.

48. If $\mathcal{B}$ is a basis for a vector space $\mathbf{V}$, and $\mathbf{v}$ is in $\mathbf{V}$, show that the representation of $\mathbf{v}$ as a linear combination of elements of $\mathcal{B}$ is unique.

5.3 COORDINATE VECTORS

In this section we show how the theory of $\mathbf{R}^m$ developed in Chapters 1 through 4 can sometimes be usefully applied to more general finite dimensional vector spaces.

If $\mathcal{B}$ is a basis for a vector space $\mathbf{V}$, it follows from Exercise 48 of Section 5.2 that each vector $\mathbf{v}$ in $\mathbf{V}$ can be expressed as a linear combination of the elements of $\mathcal{B}$ in one and only one way.

DEFINITION

Let $\mathcal{B} = \{\mathbf{b}_1, \ldots, \mathbf{b}_n\}$ be a basis for a vector space $\mathbf{V}$, and let $\mathbf{v}$ be a vector in $\mathbf{V}$. The unique vector $\mathbf{x} = (x_1, \ldots, x_n)$ in $\mathbf{R}^n$ such that $\mathbf{v} = x_1 \mathbf{b}_1 + \cdots + x_n \mathbf{b}_n$ is called the **coordinate vector** of $\mathbf{v}$ with respect to the basis $\mathcal{B}$.

EXAMPLE 5.21

Find the coordinate vector of the polynomial $\mathbf{p}(x) = 1 + 2x - 3x^2$ with respect to the basis $\{1, x, x^2, x^3\}$ for $\mathbf{P}_3$, the vector space of polynomials of degree less than or equal to 3.

Solution

Since $\mathbf{p}(x) = 1(1) + 2(x) + (-3)(x^2) + 0(x^3)$, the coordinate vector is $(1, 2, -3, 0)$.

NOTE   If **V** is a vector space with basis $\mathcal{B} = \{\mathbf{b}_1, \dots, \mathbf{b}_n\}$, then the coordinate vector of each $\mathbf{b}_i$ is $\mathbf{e}_i$, *the ith standard basis vector for* $\mathbf{R}^n$.

....

THEOREM 1   Let **V** be a vector space with basis $\mathcal{B} = \{\mathbf{b}_1, \dots, \mathbf{b}_n\}$, let $\mathbf{v}, \mathbf{v}_1, \dots, \mathbf{v}_m$ be elements of **V**, and let $\mathbf{x}, \mathbf{x}_1, \dots, \mathbf{x}_m$ be their respective coordinate vectors in $\mathbf{R}^n$. Then **v** is a linear combination of $\mathbf{v}_1, \dots, \mathbf{v}_m$ if and only if **x** is a linear combination of $\mathbf{x}_1, \dots, \mathbf{x}_m$. That is, for scalars $d_1, \dots, d_m$,

$$\mathbf{v} = d_1\mathbf{v}_1 + \cdots + d_m\mathbf{v}_m$$

if and only if

$$\mathbf{x} = d_1\mathbf{x}_1 + \cdots + d_m\mathbf{x}_m.$$

Proof   For each $i = 1, \dots, m$, let $\mathbf{x}_i = (x_{i1}, \dots, x_{in})$ be the coordinate vector of $\mathbf{v}_i$, and let $\mathbf{x} = (x_1, \dots, x_n)$ be the coordinate vector of **v**, all with respect to the basis $\mathcal{B}$. That is,

$$\mathbf{v}_i = x_{i1}\mathbf{b}_1 + \cdots + x_{in}\mathbf{b}_n \qquad (i = 1, \dots, m)$$

and

$$\mathbf{v} = x_1\mathbf{b}_1 + \cdots + x_n\mathbf{b}_n.$$

Assuming that **v** is a linear combination of the $\mathbf{v}_i$, there are scalars d_i such that $\mathbf{v} = d_1\mathbf{v}_1 + \cdots + d_m\mathbf{v}_m$. Replacing each **v** in this equation by its representation as a linear combination of the basis vectors, we obtain

$$\mathbf{v} = d_1(x_{11}\mathbf{b}_1 + \cdots + x_{1n}\mathbf{b}_n) + \cdots + d_m(x_{m1}\mathbf{b}_1 + \cdots + x_{mn}\mathbf{b}_n),$$

or by regrouping terms

$$\mathbf{v} = (d_1x_{11} + \cdots + d_mx_{m1})\mathbf{b}_1 + \cdots + (d_1x_{1n} + \cdots + d_mx_{mn})\mathbf{b}_n.$$

This last equation expresses **v** as a linear combination of basis vectors, and such a representation is unique. But we also have $\mathbf{v} = x_1\mathbf{b}_1 + \cdots + x_n\mathbf{b}_n$. Thus $x_i = d_1x_{1i} + \cdots + d_mx_{mi}$ for each $i = 1, \dots, n$.

Taken collectively,

$$\mathbf{x} = (x_1, \dots, x_n) = (d_1x_{11} + \cdots + d_mx_{m1}, \dots, d_1x_{1n} + \cdots + d_mx_{mn})$$

$$= d_1(x_{11}, \dots, x_{1n}) + \cdots + d_m(x_{m1}, \dots, x_{mn})$$

$$= d_1\mathbf{x}_1 + \cdots + d_m\mathbf{x}_m.$$

The statements in this proof are reversible, and thus the result is proved.

By letting $\mathbf{x} = \mathbf{0}$ in Theorem 1, we obtain the following corollary.

COROLLARY Using the notation of Theorem 1, $\{v_1, \ldots, v_m\}$ is a linearly independent set in **V** if and only if $\{x_1, \ldots, x_m\}$ is a linearly independent set in $\mathbf{R}^m$.

EXAMPLE 5.22

Determine whether or not the set

$$\mathcal{S} = \left\{ \begin{bmatrix} 2 & 1 \\ 1 & 1 \end{bmatrix}, \begin{bmatrix} 1 & 0 \\ 1 & 0 \end{bmatrix}, \begin{bmatrix} 0 & 1 \\ -1 & 1 \end{bmatrix}, \begin{bmatrix} 5 & 1 \\ 4 & 3 \end{bmatrix} \right\}$$

is linearly independent in $\mathbf{M}^{2,2}$, the set of all 2×2 matrices.

Solution

The coordinate vectors of the elements of $\mathcal{S}$ with respect to the basis $\{E^{11}, E^{12}, E^{21}, E^{22}\}$ (Theorem 1c of Section 5.2) are

$$\begin{bmatrix} 2 & 1 \\ 1 & 1 \end{bmatrix} \leftrightarrow (2, 1, 1, 1),$$

$$\begin{bmatrix} 1 & 0 \\ 1 & 0 \end{bmatrix} \leftrightarrow (1, 0, 1, 0),$$

$$\begin{bmatrix} 0 & 1 \\ -1 & 1 \end{bmatrix} \leftrightarrow (0, 1, -1, 1),$$

$$\begin{bmatrix} 5 & 1 \\ 4 & 3 \end{bmatrix} \leftrightarrow (5, 1, 4, 3).$$

As a consequence of the corollary to Theorem 1, $\mathcal{S}$ is linearly independent if and only if $\mathcal{S}' = \{(2, 1, 1, 1), (1, 0, 1, 0), (0, 1, -1, 1), (5, 1, 4, 3)\}$ is linearly independent. But these are vectors in $\mathbf{R}^4$, so we may use the methods of Chapter 4. Since

$$\det \begin{bmatrix} 2 & 1 & 0 & 5 \\ 1 & 0 & 1 & 1 \\ 1 & 1 & -1 & 4 \\ 1 & 0 & 1 & 3 \end{bmatrix} = 0,$$

$\mathcal{S}'$ is linearly dependent. Hence $\mathcal{S}$ is as well.

NOTE Since the *order* in which basis elements are listed affects the coordinate vectors with respect to that basis, an order must be chosen and fixed throughout the computation. The order for $\{E^{ij}\}$ of Theorem 1 of Section 5.2 will be taken *along rows*. That is, let $i = 1$, and take $j = 1, \ldots, n$; then let $i = 2$, and take $j = 1, \ldots, n$; and so on.

EXAMPLE 5.23   Show that $\mathcal{B} = \{1, x - 1, (x - 1)^2\}$ is a basis for $\mathbf{P}_2$, the set of all polynomials of degree less than or equal to two, and express

$$\mathbf{p}(x) = 2x^2 + x - 1$$

as a linear combination of the elements of $\mathcal{B}$.

Solution   We know from Theorem 1a of Section 5.2 that $\mathcal{B}' = \{1, x, x^2\}$ is a basis for $\mathbf{P}_2$. The coordinate vectors of each element of $\mathcal{B}$ with respect to the basis $\mathcal{B}'$ are

$$1 \leftrightarrow (1, 0, 0),$$
$$x - 1 = -1 + x \leftrightarrow (-1, 1, 0),$$
$$(x - 1)^2 = 1 - 2x + x^2 \leftrightarrow (1, -2, 1).$$

Applying Theorem 1 to the coordinate vectors, we have that $\mathcal{B}$ is linearly independent, since

$$\det \begin{bmatrix} 1 & -1 & 1 \\ 0 & 1 & -2 \\ 0 & 0 & 1 \end{bmatrix} \neq 0.$$

Since $\mathcal{B}$ is linearly independent and $\mathbf{P}_2$ is a three dimensional vector space, $\mathcal{B}$ is a basis. For the second part we need the coordinate vector of $\mathbf{p}$,

$$2x^2 + x - 1 = -1 + x + 2x^2 \leftrightarrow (-1, 1, 2).$$

According to Theorem 1, the coefficients needed to express $(-1, 1, 2)$ as a linear combination of $\{(1, 0, 0), (-1, 1, 0), (1, -2, 1)\}$ are exactly the coefficients needed to express $\mathbf{p}$ as a linear combination of elements of $\mathcal{B}$. But this is also a problem from Chapter 4. We perform the row reduction

$$\begin{bmatrix} 1 & -1 & 1 & | & -1 \\ 0 & 1 & -2 & | & 1 \\ 0 & 0 & 1 & | & 2 \end{bmatrix} \rightarrow \begin{bmatrix} 1 & 0 & 0 & | & 2 \\ 0 & 1 & 0 & | & 5 \\ 0 & 0 & 1 & | & 2 \end{bmatrix},$$

and as a consequence of Theorem 5 of Section 4.1, we obtain from the last column

$$\mathbf{p}(x) = 2(1) + 5(x - 1) + 2(x - 1)^2.$$

EXAMPLE 5.24   Show that $\mathcal{S} = \{x^2 - 1, x^2 + 1\}$ is a linearly independent subset of $\mathbf{P}_3$, and extend $\mathcal{S}$ to a basis.

Solution   Letting $\mathcal{B} = \{1, x, x^2, x^3\}$, we have the coordinate vectors

$$x^2 - 1 = -1 + 0x + x^2 + 0x^3 \leftrightarrow (-1, 0, 1, 0),$$
$$x^2 + 1 = 1 + 0x + x^2 + 0x^3 \leftrightarrow (1, 0, 1, 0).$$

Forming the matrix of columns and row reducing, we have

$$\begin{bmatrix} -1 & 1 \\ 0 & 0 \\ 1 & 1 \\ 0 & 0 \end{bmatrix} \rightarrow \begin{bmatrix} 1 & 0 \\ 0 & 1 \\ 0 & 0 \\ 0 & 0 \end{bmatrix}.$$

Since the columns of the row-reduced matrix are distinct elementary columns, by Theorem 4 of Section 4.1, the coordinate vectors are linearly independent. Hence the original polynomials are as well.

To extend $\mathcal{S}$ to a basis, we augment the matrix of coordinate vector columns by a basis for $\mathbf{R}^4$ (specifically, the standard basis) and row reduce:

$$\left[\begin{array}{cc|cccc} -1 & 1 & 1 & 0 & 0 & 0 \\ 0 & 0 & 0 & 1 & 0 & 0 \\ 1 & 1 & 0 & 0 & 1 & 0 \\ 0 & 0 & 0 & 0 & 0 & 1 \end{array}\right] \rightarrow \left[\begin{array}{cc|cccc} 1 & 0 & -\frac{1}{2} & 0 & \frac{1}{2} & 0 \\ 0 & 1 & \frac{1}{2} & 0 & \frac{1}{2} & 0 \\ 0 & 0 & 0 & 1 & 0 & 0 \\ 0 & 0 & 0 & 0 & 0 & 1 \end{array}\right].$$

From the row-reduced matrix we see that $\mathbf{e}_2$ and $\mathbf{e}_4$ are needed to form a basis for $\mathbf{R}^4$. The corresponding polynomials are x and x^3. Thus $\{x^2 - 1, x^2 + 1, x, x^3\}$ is a basis for $\mathbf{P}_3$.

SPOTLIGHT

COMPUTERS AND ABSTRACT VECTOR SPACES

More abstract vector spaces such as the ones in Examples 5.3, 5.5, and 5.6 may be so complicated that computer assistance is needed to study them. Is it necessary to develop new software to handle these vector spaces? The answer is, "Of course not!"

Coordinate vectors provide the mechanism for converting the original situation into the familiar $\mathbf{R}^n$ case, where row-reduction, inverse, and other programs exist to resolve the problems at hand. The results must then be translated back to the original setting to complete the solution.

Examples 5.22, 5.23, and 5.24 have small integer coefficients to make the arithmetic easy to do by hand, but that is not necessary for computer applications. See Exercises 22 and 23, which demonstrate the idea with concrete examples.

5.3 EXERCISES

In Exercises 1–5 find the coordinate vector of **v** with respect to the given basis $\mathcal{B}$ for the vector space **V**.

1. $\mathbf{v} = 2 - x + 3x^3$, $\mathcal{B} = \{1, x, x^2, x^3\}$, $\mathbf{V} = \mathbf{P}_3$.

2. $\mathbf{v} = \begin{bmatrix} 1 & 2 & 1 \\ -1 & 1 & 2 \end{bmatrix}$, $\mathcal{B} = \{E^{ij} \mid i = 1, 2; j = 1, 2, 3\}$, $\mathbf{V} = \mathbf{M}^{2,3}$.

3. $\mathbf{v} = 2 - 5x$, $\mathcal{B} = \{x + 1, x - 1\}$, $\mathbf{V} = \mathbf{P}_1$.

4. $\mathbf{v} = \begin{bmatrix} -2 & 0 \\ 0 & 3 \end{bmatrix}$, $\mathcal{B} = \left\{ \begin{bmatrix} 1 & 0 \\ 0 & 0 \end{bmatrix}, \begin{bmatrix} 0 & 0 \\ 0 & 1 \end{bmatrix} \right\}$,

   V is the vector space of all 2×2 diagonal matrices.

5. $\mathbf{v} = \begin{bmatrix} -2 & 0 \\ 0 & 3 \end{bmatrix}$, $\mathcal{B} = \left\{ \begin{bmatrix} 3 & 1 \\ 0 & 2 \end{bmatrix}, \begin{bmatrix} 1 & -1 \\ 0 & 1 \end{bmatrix}, \begin{bmatrix} 2 & 2 \\ 0 & 0 \end{bmatrix} \right\}$,

   V is the vector space of all 2×2 matrices A such that $a_{21} = 0$.

In Exercises 6–12 use coordinate vectors to decide whether or not the given set is linearly independent. If it is linearly dependent, express one of the vectors as a linear combination of the others.

6. $\mathcal{S}_1 = \{x^2 + x - 1, x^2 - 2x + 3, x^2 + 4x - 5\}$ in $\mathbf{P}_2$.

7. $\mathcal{S}_2 = \{x^2 + x - 1, x^2 - 2x + 3, x^2 + 4x - 3\}$ in $\mathbf{P}_3$.

8. $\mathcal{S}_3 = \left\{ \begin{bmatrix} 2 & -1 \\ 1 & 0 \end{bmatrix}, \begin{bmatrix} 1 & 0 \\ 2 & 1 \end{bmatrix}, \begin{bmatrix} 3 & 1 \\ -1 & 1 \end{bmatrix}, \begin{bmatrix} -1 & 4 \\ 2 & 0 \end{bmatrix} \right\}$ in $\mathbf{M}^{2,2}$.

9. $\mathcal{S}_4 = \left\{ \begin{bmatrix} 1 & 2 \\ -1 & 0 \end{bmatrix}, \begin{bmatrix} 0 & -1 \\ 1 & 1 \end{bmatrix}, \begin{bmatrix} 1 & 0 \\ 1 & 2 \end{bmatrix} \right\}$ in $\mathbf{M}^{2,2}$.

10. $\mathcal{S}_5 = \left\{ \begin{bmatrix} 2 & 1 \\ 1 & 3 \end{bmatrix}, \begin{bmatrix} 3 & -1 \\ -1 & 2 \end{bmatrix}, \begin{bmatrix} 1 & 0 \\ 0 & 1 \end{bmatrix}, \begin{bmatrix} 2 & 2 \\ 2 & 4 \end{bmatrix} \right\}$,

   in the subspace of $\mathbf{M}^{2,2}$ of all matrices of the form $\begin{bmatrix} a & b \\ b & a + b \end{bmatrix}$.

11. $\{2 + x - 3\sin x + \cos x, \ x + \sin x - 3\cos x, \ 1 - 2x + 3\sin x + \cos x, \ 2 - x - \sin x - \cos x, \ 2 + x - 3\cos x\}$ in the space of all linear combinations of the set $\{1, x, \sin x, \cos x\}$.

12. All 2×2 matrices with exactly two 1's and two 0's for entries in the space of all 2×2 matrices.

In Exercises 13–17 the given set is linearly independent. Extend it to a basis; that is, include other elements (from the standard basis for the vector space) so that the resulting set is a basis.

13. $\{x^2 - x + 1, x - 1, x + 1\}$ in $\mathbf{P}_2$.

14. $\{x + 1, x^2 + x, x^3 + x^2 + x\}$ in $\mathbf{P}_3$.

15. $\left\{ \begin{bmatrix} 1 & -1 \\ 1 & 0 \end{bmatrix}, \begin{bmatrix} 1 & 1 \\ 0 & 0 \end{bmatrix}, \begin{bmatrix} 0 & 0 \\ 1 & 1 \end{bmatrix} \right\}$ in $\mathbf{M}^{2,2}$.

16. $\left\{ \begin{bmatrix} 1 & 2 \\ 3 & 1 \end{bmatrix}, \begin{bmatrix} 1 & 3 \\ 2 & 1 \end{bmatrix} \right\}$ in $\mathbf{M}^{2,2}$.

17. $\left\{\begin{bmatrix} 1 & 0 & -1 \\ 1 & -1 & 1 \end{bmatrix}, \begin{bmatrix} 0 & 1 & 1 \\ 2 & 1 & -1 \end{bmatrix}, \begin{bmatrix} 1 & -1 & 1 \\ 1 & -1 & 1 \end{bmatrix}, \begin{bmatrix} 0 & 1 & 0 \\ 0 & 1 & 0 \end{bmatrix}\right\}$ in $\mathbf{M}^{2,3}$.

In Exercises 18–21 the given set spans the space. Reduce it to an independent set which has the same span.

18. $\{x^2 - x + 1, x^2 + 2x + 1, x - 2, x + 2, x^2 - 1, x^2 + 1\}$ in $\mathbf{P}_2$.

19. $\{x + 1, x - 1, x^2 + 1, x^2 - 1, x^2 + x, x^2 - x\}$ in $\mathbf{P}_2$.

20. $\left\{\begin{bmatrix} 0 & 0 & 0 \\ 0 & -1 & 0 \\ 0 & 0 & 1 \end{bmatrix}, \begin{bmatrix} 2 & 0 & 0 \\ 0 & 1 & 0 \\ 0 & 0 & 1 \end{bmatrix}, \begin{bmatrix} 1 & 0 & 0 \\ 0 & 1 & 0 \\ 0 & 0 & 1 \end{bmatrix}, \begin{bmatrix} 2 & 0 & 0 \\ 0 & 2 & 0 \\ 0 & 0 & -1 \end{bmatrix}\right\}$ in the vector space of 3×3 diagonal matrices.

21. $\left\{\begin{bmatrix} 2 & 1 \\ 0 & 1 \end{bmatrix}, \begin{bmatrix} -1 & 1 \\ 0 & 0 \end{bmatrix}, \begin{bmatrix} 1 & 2 \\ 0 & 2 \end{bmatrix}, \begin{bmatrix} 1 & -1 \\ 0 & -1 \end{bmatrix}, \begin{bmatrix} 2 & -1 \\ 0 & -1 \end{bmatrix}\right\}$ in the set all 2×2 matrices A with $a_{21} = 0$.

In Exercises 22 and 23 decide if the functions in the given set are linearly independent or dependent. If they are dependent, express one of them as a linear combination of the others.

$$f_1(x) = 1.1x + 6.3x^2 - 3.1 \sin x$$
$$f_2(x) = 2.1x - 3.3x^2 + 5.2 \sin x$$
$$f_3(x) = 1.3x + 1.2x^2 - 1.1 \sin x$$
$$f_4(x) = 8.9x - 7.5x^2 + 13.4 \sin x$$

22. $S = \{f_1, f_2, f_3\}$

23. $S = \{f_1, f_2, f_3, f_4\}$

5.4 INNER PRODUCT SPACES

In Section 1.4 we introduced the notion of the *dot product* (also called *inner product* or *scalar product*) of two vectors in $\mathbf{R}^m$. In this section we generalize this concept to vectors in an arbitrary vector space. This in turn leads to a natural definition of *orthogonality* of such vectors and the study of bases that consist of orthogonal vectors.

In Theorem 3 of Section 1.4 we gave several important properties of the dot product in $\mathbf{R}^m$. We now use these properties to define our extended notion of dot product. Following standard convention, we use the alternate terminology *inner product* and also change the previous notation in the general definition.

DEFINITION

Let $\mathbf{V}$ be a vector space over the real numbers. An **inner product** for $\mathbf{V}$ is a function that associates with every pair of vectors, $\mathbf{u}$ and $\mathbf{v}$, in $\mathbf{V}$, a real number, $(\mathbf{u}, \mathbf{v})$, satisfying the following properties. For all $\mathbf{u}$, $\mathbf{v}$, and $\mathbf{w}$ in $\mathbf{V}$

and scalars c:

    a. $(\mathbf{u}, \mathbf{v}) = (\mathbf{v}, \mathbf{u})$,

    b. $(\mathbf{u}, \mathbf{v} + \mathbf{w}) = (\mathbf{u}, \mathbf{v}) + (\mathbf{u}, \mathbf{w})$,

    c. $(c\mathbf{u}, \mathbf{v}) = c(\mathbf{u}, \mathbf{v})$,

    d. $(\mathbf{u}, \mathbf{u}) \geq 0$ and $(\mathbf{u}, \mathbf{u}) = 0$, if and only if $\mathbf{u} = \mathbf{0}$.

A vector space **V** together with an inner product is called an **inner product space.**

EXAMPLE 5.25

Let $\mathbf{u} = (u_1, u_2, \ldots, u_m)$ and $\mathbf{v} = (v_1, v_2, \ldots, v_m)$ be arbitrary vectors in the vector space $\mathbf{R}^m$. Define

$$(\mathbf{u}, \mathbf{v}) = u_1 v_1 + u_2 v_2 + \cdots + u_m v_m,$$

the dot product of $\mathbf{u}$ and $\mathbf{v}$. Then $(\mathbf{u}, \mathbf{v})$ defines an inner product for $\mathbf{R}^m$.

Solution

By Theorem 3 of Section 1.4, this function satisfies the four properties of the definition and is therefore an inner product.

Example 5.25 verifies that the dot product for $\mathbf{R}^m$ defined in Section 1.4 is an inner product for this vector space. The next example demonstrates that one can define other inner products for $\mathbf{R}^m$—there may be more than one inner product for a given vector space.

EXAMPLE 5.26

Let $\mathbf{u} = (u_1, u_2, \ldots, u_m)$ and $\mathbf{v} = (v_1, v_2, \ldots, v_m)$ be arbitrary vectors in $\mathbf{R}^m$. Define

$$(\mathbf{u}, \mathbf{v}) = u_1 v_1 + 2u_2 v_2 + 3u_3 v_3 + \cdots + m u_m v_m.$$

Then $(\mathbf{u}, \mathbf{v})$ is an inner product for $\mathbf{R}^m$.

Solution

We must verify properties (a) through (d) of the definition:

a.
$$
\begin{aligned}
(\mathbf{u}, \mathbf{v}) &= u_1 v_1 + 2u_2 v_2 + \cdots + m u_m v_m \\
&= v_1 u_1 + 2v_2 u_2 + \cdots + m v_m u_m \\
&= (\mathbf{v}, \mathbf{u}).
\end{aligned}
$$

b.
$$
\begin{aligned}
(\mathbf{u}, \mathbf{v} + \mathbf{w}) &= u_1(v_1 + w_1) + 2u_2(v_2 + w_2) + \cdots + m u_m(v_m + w_m) \\
&= (u_1 v_1 + u_1 w_1) + (2u_2 v_2 + 2u_2 w_2) \\
&\quad + \cdots + (m u_m v_m + m u_m w_m) \\
&= (u_1 v_1 + 2u_2 v_2 + \cdots + m u_m v_m) + (u_1 w_1 + 2u_2 w_2 \\
&\quad + \cdots + m u_m v_m) \\
&= (\mathbf{u}, \mathbf{v}) + (\mathbf{u}, \mathbf{w}).
\end{aligned}
$$

c.
$$(c\mathbf{u}, \mathbf{v}) = cu_1v_1 + 2cu_2v_2 + \cdots + mcu_mv_m$$
$$= c(u_1v_1 + 2u_2v_2 + \cdots + mu_mv_m)$$
$$= c(\mathbf{u}, \mathbf{v}).$$

d.
$$(\mathbf{u}, \mathbf{u}) = u_1^2 + 2u_2^2 + \cdots + mu_m^2,$$

which is nonnegative, and if

$$u_1^2 + 2u_2^2 + \cdots + mu_m^2 = 0,$$

then

$$u_1 = u_2 = \cdots = u_m = 0.$$

Thus $\mathbf{u} = \mathbf{0}$, as desired.

NOTE Example 5.26 shows that inner products other than the dot product may be defined on $\mathbf{R}^m$. If no inner product on $\mathbf{R}^m$ is explicitly mentioned, we assume that the usual dot product is the inner product intended.

EXAMPLE 5.27 Let $\mathbf{p}$ and $\mathbf{q}$ be arbitrary vectors in $\mathbf{P}_1$, the vector space of all polynomials in x of degree less than or equal to one. Define

$$(\mathbf{p}, \mathbf{q}) = \mathbf{p}(0)\mathbf{q}(0) + \mathbf{p}(1)\mathbf{q}(1).$$

Then $(\mathbf{p}, \mathbf{q})$ is an inner product for $\mathbf{P}_1$.

Solution Properties (a), (b), and (c) are easily verified. For property (d) we have $(\mathbf{p}, \mathbf{p}) = [\mathbf{p}(0)]^2 + [\mathbf{p}(1)]^2$, which is nonnegative. Moreover, $(\mathbf{p}, \mathbf{p}) = 0$ if and only if $\mathbf{p}(0) = 0$ and $\mathbf{p}(1) = 0$. But the latter implies that $\mathbf{p}(x)$ is the zero polynomial, since the graph of $\mathbf{p}(x)$, a line, goes through the points $(0, 0)$ and $(1, 0)$ and is therefore the graph of the zero polynomial.

EXAMPLE 5.28 (*From calculus*) Let $\mathbf{f}$ and $\mathbf{g}$ be vectors in $\mathbf{C}[-1, 1]$, the vector space of all real-valued functions continuous on the interval $[-1, 1]$. (See the note following Example 5.6.) Define

$$(\mathbf{f}, \mathbf{g}) = \int_{-1}^{1} \mathbf{f}(x)\mathbf{g}(x)\, dx.$$

Then $(\mathbf{f}, \mathbf{g})$ is an inner product for $\mathbf{C}[-1, 1]$.

Solution   The verification that $(\mathbf{f}, \mathbf{g})$ is an inner product rests on some properties of the definite integral learned in calculus. Specifically, properties (b), (c), and (d) hold because, respectively,

$$\int_{-1}^{1} \mathbf{f}(x)(\mathbf{g}(x) + \mathbf{h}(x))\, dx = \int_{-1}^{1} \mathbf{f}(x)\mathbf{g}(x)\, dx + \int_{-1}^{1} \mathbf{f}(x)\mathbf{h}(x)\, dx,$$

$$\int_{-1}^{1} c\mathbf{f}(x)\mathbf{g}(x)\, dx = c \int_{-1}^{1} \mathbf{f}(x)\mathbf{g}(x)\, dx,$$

and

$$\int_{-1}^{1} (\mathbf{f}(x))^2\, dx \geq 0 \quad \text{with} \quad \int_{-1}^{1} (\mathbf{f}(x))^2\, dx = 0$$

if and only if $\mathbf{f}(x)$ is the zero function on $[-1, 1]$. Property (a) holds simply because $\mathbf{f}(x)\mathbf{g}(x) = \mathbf{g}(x)\mathbf{f}(x)$ for all x in $[-1, 1]$.

From the defining properties of the inner product, we can derive several other properties satisfied by all inner products. Some of these are presented in the next two theorems.

THEOREM I Let $\mathbf{V}$ be an inner product space. Then for all vectors $\mathbf{u}_1, \mathbf{u}_2,$ and $\mathbf{v}$ in $\mathbf{V}$ and scalars c_1 and c_2:

a. $(c_1\mathbf{u}_1 + c_2\mathbf{u}_2, \mathbf{v}) = (\mathbf{v}, c_1\mathbf{u}_1 + c_2\mathbf{u}_2) = c_1(\mathbf{u}_1, \mathbf{v}) + c_2(\mathbf{u}_2, \mathbf{v}),$
b. $(\mathbf{0}, \mathbf{v}) = (\mathbf{v}, \mathbf{0}) = 0.$

Proof

a. $(c_1\mathbf{u}_1 + c_2\mathbf{u}_2, \mathbf{v}) = (\mathbf{v}, c_1\mathbf{u}_1 + c_2\mathbf{u}_2)$ (property (a)),

$\qquad\qquad\qquad\quad = (\mathbf{v}, c_1\mathbf{u}_1) + (\mathbf{v}, c_2\mathbf{u}_2)$ (property (b)),

$\qquad\qquad\qquad\quad = (c_1\mathbf{u}_1, \mathbf{v}) + (c_2\mathbf{u}_2, \mathbf{v})$ (property (a)),

$\qquad\qquad\qquad\quad = c_1(\mathbf{u}_1, \mathbf{v}) + c_2(\mathbf{u}_2, \mathbf{v})$ (property (c)).

b. Exercise 32.

In Section 1.4 we saw that for any $\mathbf{u}$ in $\mathbf{R}^m$, the *length* or *norm* of $\mathbf{u}$, $\|\mathbf{u}\|$, is equal to $\sqrt{\mathbf{u} \cdot \mathbf{u}}$. In a general inner product space we use this for a definition. If $\mathbf{u}$ is a vector in an inner product space, then the **norm** of $\mathbf{u}$, denoted $\|\mathbf{u}\|$, is given by $\|\mathbf{u}\| = \sqrt{(\mathbf{u}, \mathbf{u})}$. The next theorem, known as the **Cauchy–Schwarz inequality,** relates the inner product of vectors $\mathbf{u}$ and $\mathbf{v}$ to the norms of $\mathbf{u}$ and $\mathbf{v}$.

THEOREM 2   Let **u** and **v** be vectors in an inner product space. Then

$$|(\mathbf{u}, \mathbf{v})| \leq \|\mathbf{u}\| \, \|\mathbf{v}\|,$$

where $|(\mathbf{u}, \mathbf{v})|$ denotes the *absolute value* of $(\mathbf{u}, \mathbf{v})$.

Proof   First note that if $\mathbf{u} = \mathbf{0}$, then $(\mathbf{u}, \mathbf{v}) = 0$, by Theorem 1b, and

$$\|\mathbf{u}\| = \sqrt{(\mathbf{u}, \mathbf{u})} = 0$$

as well. Thus the theorem holds if $\mathbf{u} = \mathbf{0}$.

Now assume that $\mathbf{u} \neq \mathbf{0}$. If x is an arbitrary real number, we have

$$
\begin{aligned}
0 &\leq (x\mathbf{u} + \mathbf{v}, x\mathbf{u} + \mathbf{v}) && \text{(property (d))} \\
&= x^2(\mathbf{u}, \mathbf{u}) + x(\mathbf{u}, \mathbf{v}) + x(\mathbf{v}, \mathbf{u}) + (\mathbf{v}, \mathbf{v}) && \text{(Theorem 1a),} \\
&= x^2(\mathbf{u}, \mathbf{u}) + 2x(\mathbf{u}, \mathbf{v}) + (\mathbf{v}, \mathbf{v}) && \text{(property (a)).}
\end{aligned}
$$

Now, setting $(\mathbf{u}, \mathbf{u}) = a$, $2(\mathbf{u}, \mathbf{v}) = b$, and $(\mathbf{v}, \mathbf{v}) = c$, this inequality becomes $ax^2 + bx + c \geq 0$, which is quadratic in x. In geometric terms, this implies that the parabola $y = ax^2 + bx + c$ must be contained in the region on and above the x-axis (Figures 5.1 and 5.2). But this in turn implies that the polynomial $ax^2 + bx + c$ cannot have two distinct real zeros, since, if it did, the curve would dip below the x-axis (Figure 5.3).

Now the equation $ax^2 + bx + c = 0$ fails to have two real roots (it has either no roots or one root) if and only if $b^2 - 4ac \leq 0$. But this implies that $b^2 \leq 4ac$, or

$$4(\mathbf{u}, \mathbf{v})^2 \leq 4(\mathbf{u}, \mathbf{u})(\mathbf{v}, \mathbf{v}),$$

or

$$(\mathbf{u}, \mathbf{v})^2 \leq (\mathbf{u}, \mathbf{u})(\mathbf{v}, \mathbf{v}).$$

Finally, taking square roots of both sides, we obtain

$$\sqrt{(\mathbf{u}, \mathbf{v})^2} \leq \sqrt{(\mathbf{u}, \mathbf{u})} \, \sqrt{(\mathbf{v}, \mathbf{v})}$$

or

$$|(\mathbf{u}, \mathbf{v})| \leq \|\mathbf{u}\| \, \|\mathbf{v}\|.$$

  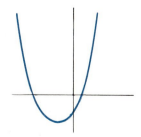

FIGURE 5.1 No real roots   FIGURE 5.2 One real root   FIGURE 5.3 Two real roots

EXAMPLE 5.29   With the inner products of Examples 5.25 and 5.28, the Cauchy–Schwarz inequality becomes

$$|u_1v_1 + u_2v_2 + \cdots + u_mv_m| \le \sqrt{u_1^2 + u_2^2 + \cdots + u_m^2}\,\sqrt{v_1^2 + v_2^2 + \cdots + v_m^2}$$

and

$$\left|\int_{-1}^{1} \mathbf{f}(x)\mathbf{g}(x)\,dx\right| \le \sqrt{\int_{-1}^{1} (\mathbf{f}(x))^2\,dx}\,\sqrt{\int_{-1}^{1} (\mathbf{g}(x))^2\,dx},$$

respectively.

ORTHOGONAL SETS AND ORTHOGONAL BASES

The standard basis for $\mathbf{R}^m$, $\{\mathbf{e}_1, \mathbf{e}_2, \ldots, \mathbf{e}_m\}$, has several computational advantages over an arbitrary one. The standard basis vectors are pairwise orthogonal, each has unit length, and it is a simple matter to write an arbitrary vector in $\mathbf{R}^m$ as a linear combination of these basis vectors. Unfortunately, subspaces of $\mathbf{R}^m$ do not ordinarily have bases that are subsets of $\{\mathbf{e}_1, \mathbf{e}_2, \ldots, \mathbf{e}_m\}$. Nevertheless we can construct bases for them—*orthonormal* ones—that do have the nice properties of the standard basis. The same situation applies to other inner product spaces as well. Here we will investigate the properties of orthonormal bases and present a method for replacing an arbitrary basis by an orthonormal one.

Recall that two vectors in $\mathbf{R}^m$ are *orthogonal* if their dot product is equal to zero. The next definition generalizes the concept of orthogonality to inner product spaces.

DEFINITION   Two vectors, $\mathbf{u}$ and $\mathbf{v}$, in an inner product space $\mathbf{V}$ are **orthogonal** if $(\mathbf{u}, \mathbf{v}) = 0$. A set of vectors, $\mathcal{S}$, is an **orthogonal set** if every pair of distinct vectors in $\mathcal{S}$ is orthogonal. A set $\mathcal{T}$ is an **orthonormal set** if it is orthogonal and every vector in $\mathcal{T}$ has norm equal to one.

EXAMPLE 5.30   Define an inner product on the vector space $\mathbf{M}^{2,2}$, the set of all 2×2 matrices, by

$$(A, B) = a_{11}b_{11} + a_{12}b_{12} + a_{21}b_{21} + a_{22}b_{22}.$$

In this inner product space show that the matrices

$$C_1 = \begin{bmatrix} 1 & -1 \\ 1 & 0 \end{bmatrix}, \qquad C_2 = \begin{bmatrix} -1 & -1 \\ 0 & 2 \end{bmatrix}, \qquad C_3 = \begin{bmatrix} 0 & 2 \\ 2 & 1 \end{bmatrix}$$

form an orthogonal set. (Exercise 4 will ask you to show that (A, B) is an inner product.)

Solution   We show that $(C_1, C_2) = (C_1, C_3) = (C_2, C_3) = 0$ by computing each of these inner products. We have

$$(C_1, C_3) = (1)(0) + (-1)(2) + (1)(2) + (0)(1) = 0,$$

as desired. Similarly, $(C_1, C_2) = 0$ and $(C_2, C_3) = 0$. Thus $\{C_1, C_2, C_3\}$ is orthogonal.

EXAMPLE 5.31   Show that $\mathscr{T} = \{\mathbf{v}_1, \mathbf{v}_2, \mathbf{v}_3\}$ is an orthonormal set in $\mathbf{R}^3$, where

$$\mathbf{v}_1 = (1/3, 2/3, -2/3),$$
$$\mathbf{v}_2 = (0, 1/\sqrt{2}, 1/\sqrt{2}),$$

and

$$\mathbf{v}_3 = (-4/(3\sqrt{2}), 1/(3\sqrt{2}), -1/(3\sqrt{2})).$$

Solution   We first check that $\mathscr{T}$ is orthogonal by verifying that the dot products $\mathbf{v}_1 \cdot \mathbf{v}_2$, $\mathbf{v}_1 \cdot \mathbf{v}_3$, and $\mathbf{v}_2 \cdot \mathbf{v}_3$ are all equal to 0. We then check the "length = 1" condition by verifying that $\|\mathbf{v}_1\| = \|\mathbf{v}_2\| = \|\mathbf{v}_3\| = 1$. Thus $\mathscr{T}$ is an orthonormal set.

The following theorem (which generalizes Exercise 34 of Section 4.1) gives an important property of orthogonal sets that do not contain the zero vector.

THEOREM 3   Let $\mathscr{T} = \{\mathbf{v}_1, \mathbf{v}_2, \ldots, \mathbf{v}_n\}$ be an orthogonal set of nonzero vectors in an inner product space $\mathbf{V}$. Then $\mathscr{T}$ is linearly independent.

Proof   Consider the equation $c_1\mathbf{v}_1 + c_2\mathbf{v}_2 + \cdots + c_n\mathbf{v}_n = \mathbf{0}$. We will show that all c_i must be zero. To accomplish this, form the inner product of both sides of this equation with each vector $\mathbf{v}_j$. For each j this yields the equation

$$c_1(\mathbf{v}_1, \mathbf{v}_j) + c_2(\mathbf{v}_2, \mathbf{v}_j) + \cdots + c_j(\mathbf{v}_j, \mathbf{v}_j) + \cdots + c_n(\mathbf{v}_n, \mathbf{v}_j) = (\mathbf{0}, \mathbf{v}_j).$$

However, because the set $\mathscr{T}$ is orthogonal, all inner products in this equation *except* $(\mathbf{v}_j, \mathbf{v}_j)$ are equal to zero. Thus the equation reduces to

$$c_j(\mathbf{v}_j, \mathbf{v}_j) = 0.$$

By hypothesis, $\mathbf{v}_j \neq \mathbf{0}$ for $j = 1, 2, \ldots, n$, so $(\mathbf{v}_j, \mathbf{v}_j) \neq \mathbf{0}$, and the last equation implies that $c_j = 0$ $(j = 1, 2, \ldots, n)$, as desired.

Since $\mathbf{0}$ cannot be a member of an ortho*normal* set ($\|\mathbf{0}\| = 0 \neq 1$), Theorem 3 has an immediate corollary.

COROLLARY   An orthonormal set of vectors is linearly independent.

As a consequence of this theorem and corollary, if an orthonormal set, or an orthogonal set of nonzero vectors, spans an inner product space, it is a basis for this space.

DEFINITION

> Let **V** be an inner product space. If a set $\mathscr{T}$ of vectors is a basis for **V**, and
>
> **i.** if $\mathscr{T}$ is *orthogonal*, it is called an **orthogonal basis** for **V**;
>
> **ii.** if $\mathscr{T}$ is *orthonormal*, it is called an **orthonormal basis** for **V**.

For example, it is easy to check that the standard basis for $\mathbf{R}^m$, $\{\mathbf{e}_1, \mathbf{e}_2, \ldots, \mathbf{e}_m\}$, is an orthonormal set (Exercise 31), and consequently it is an orthonormal basis for $\mathbf{R}^m$. However, the standard basis is certainly not the *only* orthonormal one for $\mathbf{R}^m$. For example, the set $\mathscr{T}$ of Example 5.31 is an orthonormal set containing three vectors. By Theorem 3 and part (iii) of Theorem 5 of Section 4.3, it is an orthonormal basis for $\mathbf{R}^3$.

The next theorem gives one of the major reasons why orthonormal bases are so desirable—they provide a simple way to write an arbitrary vector in a given vector space as a linear combination of the vectors in an orthonormal basis for that space.

THEOREM 4   Let $\mathscr{T} = \{\mathbf{v}_1, \mathbf{v}_2, \ldots, \mathbf{v}_n\}$ be an orthonormal basis for an inner product space **V**. Let **x** be an arbitrary vector in **V**. Then $\mathbf{x} = c_1\mathbf{v}_1 + c_2\mathbf{v}_2 + \cdots + c_n\mathbf{v}_n$, where $c_j = (\mathbf{x}, \mathbf{v}_j)$ $(j = 1, 2, \ldots, n)$.

Proof   Since $\mathscr{T}$ is a basis for **V**, we know that there exist scalars $c_1, c_2, \ldots, c_n$ such that $\mathbf{x} = c_1\mathbf{v}_1 + c_2\mathbf{v}_2 + \cdots + c_n\mathbf{v}_n$. Moreover,

$$(\mathbf{x}, \mathbf{v}_j) = ((c_1\mathbf{v}_1 + c_2\mathbf{v}_2 + \cdots + c_j\mathbf{v}_j + \cdots + c_n\mathbf{v}_n), \mathbf{v}_j)$$
$$= c_1(\mathbf{v}_1, \mathbf{v}_j) + c_2(\mathbf{v}_2, \mathbf{v}_j) + \cdots + c_j(\mathbf{v}_j, \mathbf{v}_j) + \cdots + c_n(\mathbf{v}_n, \mathbf{v}_j)$$
$$= c_j(\mathbf{v}_j, \mathbf{v}_j),$$

due to the orthogonality of $\mathscr{T}$. But $(\mathbf{v}_j, \mathbf{v}_j) = 1$, since $\mathscr{T}$ is orthonormal, so $(\mathbf{x}, \mathbf{v}_j) = c_j$, as desired.

EXAMPLE 5.32

Express $\mathbf{x} = (1, 2, 3)$ as a linear combination of the vectors

$$\mathbf{v}_1 = (1/3, 2/3, -2/3),$$
$$\mathbf{v}_2 = (0, 1/\sqrt{2}, 1/\sqrt{2}),$$

and

$$\mathbf{v}_3 = (-4/(3\sqrt{2}), 1/(3\sqrt{2}), -1/(3\sqrt{2})).$$

Solution

In Example 5.31 we showed that $\{\mathbf{v}_1, \mathbf{v}_2, \mathbf{v}_3\}$ is an orthonormal set. By Theorem 3, this set is linearly independent; by part (iii) of Theorem 5 of Section 4.3, it is a basis for $\mathbf{R}^3$. We may therefore apply Theorem 4 (with $\mathbf{V} = \mathbf{R}^3$) to obtain $\mathbf{x} = c_1\mathbf{v}_1 + c_2\mathbf{v}_2 + c_3\mathbf{v}_3,$

with

$$c_1 = \mathbf{x} \cdot \mathbf{v}_1 = -1/3,$$
$$c_2 = \mathbf{x} \cdot \mathbf{v}_2 = 5/\sqrt{2},$$

and

$$c_3 = \mathbf{x} \cdot \mathbf{v}_3 = -5/(3\sqrt{2}).$$

In the preceding discussion we have defined *orthonormal basis* and given some of its properties. Now we describe how to construct one from a known, but nonorthonormal, basis.

If the known basis is *orthogonal,* all we need do to produce an orthonormal one is to *normalize* each vector in the given basis—in other words, divide each vector by its length. This preserves the orthogonality of the given set and at the same time creates vectors of norm 1.

EXAMPLE 5.33

Find an orthonormal basis for the subspace **S** of $\mathbf{R}^4$ generated by the orthogonal set $\mathcal{T} = \{\mathbf{v}_1, \mathbf{v}_2, \mathbf{v}_3\}$, where

$$\mathbf{v}_1 = (0, 2, -1, 1), \qquad \mathbf{v}_2 = (0, 0, 1, 1), \qquad \mathbf{v}_3 = (-2, 1, 1, -1).$$

Solution

Since $\mathcal{T}$ is orthogonal, it is linearly independent, and since $\mathcal{T}$ generates **S**, it spans **S**. Hence $\mathcal{T}$ is an *orthogonal* basis for **S**. To obtain an ortho*normal* basis for **S**, all we need do is normalize each vector in $\mathcal{T}$. Form the set

$$\mathcal{T}' = \{(1/\|\mathbf{v}_1\|)\mathbf{v}_1, (1/\|\mathbf{v}_2\|)\mathbf{v}_2, (1/\|\mathbf{v}_3\|)\mathbf{v}_3\}$$
$$= \{(0, 2/\sqrt{6}, -1/\sqrt{6}, 1/\sqrt{6}), (0, 0, 1/\sqrt{2}, 1/\sqrt{2}),$$
$$(-2/\sqrt{7}, 1/\sqrt{7}, 1/\sqrt{7}, -1/\sqrt{7})\}.$$

Then $\mathcal{T}'$ is an orthonormal basis for **S**.

When the known basis is not orthogonal, the process of finding an orthonormal basis is more complicated but still mechanical. To motivate the procedure, we consider the special case of two vectors in $\mathbf{R}^m$.

Let **u** and **v** be nonzero vectors in $\mathbf{R}^m$, and let **x** be the vector along **v** such that the points $0, X,$ and U form a right triangle. The vector **x** is called the **orthogonal projection** of **u** onto **v** (Figure 5.4).

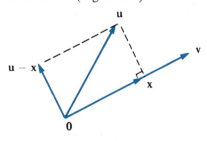

FIGURE 5.4

THEOREM 5   Let $\mathbf{u}$ and $\mathbf{v}$ be nonzero vectors in $\mathbf{R}^m$. The orthogonal projection of $\mathbf{u}$ onto $\mathbf{v}$ is given by

$$\mathbf{x} = \left(\frac{\mathbf{u} \cdot \mathbf{v}}{\mathbf{v} \cdot \mathbf{v}}\right)\mathbf{v}.$$

Proof   Since $\mathbf{u} - \mathbf{x}$ is equivalent to $\overrightarrow{XU}$, we need only check that $\mathbf{u} - \mathbf{x}$ is orthogonal to $\mathbf{v}$:

$$(\mathbf{u} - \mathbf{x}) \cdot \mathbf{v} = \left(\mathbf{u} - \left(\frac{\mathbf{u} \cdot \mathbf{v}}{\mathbf{v} \cdot \mathbf{v}}\right)\mathbf{v}\right) \cdot \mathbf{v} = \mathbf{u} \cdot \mathbf{v} - \left(\frac{\mathbf{u} \cdot \mathbf{v}}{\mathbf{v} \cdot \mathbf{v}}\right)\mathbf{v} \cdot \mathbf{v} = 0.$$

Note that the sets $\{\mathbf{v}, \mathbf{u}\}$ and $\{\mathbf{v}, \mathbf{u} - \mathbf{x}\}$ span the same set, because $\mathbf{u}$ is a linear combination of $\mathbf{v}$ and $\mathbf{u} - \mathbf{x}$, whereas $\mathbf{u} - \mathbf{x}$ is a linear combination of $\mathbf{v}$ and $\mathbf{u}$. Specifically, we have

$$\mathbf{u} = \left(\frac{\mathbf{u} \cdot \mathbf{v}}{\mathbf{v} \cdot \mathbf{v}}\right)\mathbf{v} + (1)(\mathbf{u} - \mathbf{x})$$

and

$$\mathbf{u} - \mathbf{x} = (1)\mathbf{u} + (-1)\left(\frac{\mathbf{u} \cdot \mathbf{v}}{\mathbf{v} \cdot \mathbf{v}}\right)\mathbf{v}.$$

It follows that if $\{\mathbf{v}, \mathbf{u}\}$ is a basis for a subspace of $\mathbf{R}^m$, then $\{\mathbf{v}, \mathbf{u} - \mathbf{x}\}$ is an *orthogonal* basis for the same subspace.

This process of replacing a basis by an orthogonal one for the same subspace can be generalized to any finite number of vectors in any inner product space. We will present the proof of this procedure after we demonstrate its use.

PROCEDURE FOR CONSTRUCTING AN ORTHOGONAL BASIS (GRAM–SCHMIDT PROCESS)

Compute the vectors in the basis as described in the following theorem.

THEOREM 6   Let $\mathcal{T} = \{\mathbf{u}_1, \mathbf{u}_2, \ldots, \mathbf{u}_n\}$ be a basis for an inner product space $\mathbf{V}$. Let $\mathcal{T}' = \{\mathbf{v}_1, \mathbf{v}_2, \ldots, \mathbf{v}_n\}$ be defined as follows:

$$\mathbf{v}_1 = \mathbf{u}_1,$$

$$\mathbf{v}_2 = \mathbf{u}_2 - \frac{(\mathbf{u}_2, \mathbf{v}_1)}{(\mathbf{v}_1, \mathbf{v}_1)}\mathbf{v}_1,$$

$$\mathbf{v}_3 = \mathbf{u}_3 - \frac{(\mathbf{u}_3, \mathbf{v}_1)}{(\mathbf{v}_1, \mathbf{v}_1)} \mathbf{v}_1 - \frac{(\mathbf{u}_3, \mathbf{v}_2)}{(\mathbf{v}_2, \mathbf{v}_2)} \mathbf{v}_2,$$

$$\vdots$$

$$\mathbf{v}_n = \mathbf{u}_n - \frac{(\mathbf{u}_n, \mathbf{v}_1)}{(\mathbf{v}_1, \mathbf{v}_1)} \mathbf{v}_1 - \frac{(\mathbf{u}_n, \mathbf{v}_2)}{(\mathbf{v}_2, \mathbf{v}_2)} \mathbf{v}_2 - \cdots - \frac{(\mathbf{u}_n, \mathbf{v}_{n-1})}{(\mathbf{v}_{n-1}, \mathbf{v}_{n-1})} \mathbf{v}_{n-1}.$$

Then the set $\mathcal{T}'$ is an orthogonal basis for **V**. An orthonormal basis for **V** is given by $\mathcal{T}'' = \{\mathbf{w}_1, \mathbf{w}_2, \ldots, \mathbf{w}_n\}$, where $\mathbf{w}_i = (1/\|\mathbf{v}_i\|)\mathbf{v}_i$ for $i = 1, 2, \ldots, n$.

Before proving this theorem we will present a couple of examples.

EXAMPLE 5.34

Let **S** be the subspace of $\mathbf{R}^5$ generated by the set $\mathcal{T} = \{\mathbf{u}_1, \mathbf{u}_2, \mathbf{u}_3\}$, where $\mathbf{u}_1 = (-1, -1, 1, 0, 0)$, $\mathbf{u}_2 = (0, -1, 0, 0, 1)$, and $\mathbf{u}_3 = (1, -1, 0, 1, 0)$. Find an orthonormal basis for **S**.

Solution

We apply the Gram–Schmidt process to $\mathcal{T}$.

$$\mathbf{v}_1 = \mathbf{u}_1 = (-1, -1, 1, 0, 0),$$

$$\mathbf{v}_2 = \mathbf{u}_2 - \left(\frac{\mathbf{u}_2 \cdot \mathbf{v}_1}{\mathbf{v}_1 \cdot \mathbf{v}_1}\right)\mathbf{v}_1$$

$$= (0, -1, 0, 0, 1) - (\tfrac{1}{3})(-1, -1, 1, 0, 0)$$

$$= (\tfrac{1}{3}, -\tfrac{2}{3}, -\tfrac{1}{3}, 0, 1),$$

$$\mathbf{v}_3 = \mathbf{u}_3 - \left(\frac{\mathbf{u}_3 \cdot \mathbf{v}_1}{\mathbf{v}_1 \cdot \mathbf{v}_1}\right)\mathbf{v}_1 - \left(\frac{\mathbf{u}_3 \cdot \mathbf{v}_2}{\mathbf{v}_2 \cdot \mathbf{v}_2}\right)\mathbf{v}_2$$

$$= (1, -1, 0, 1, 0) - (\tfrac{0}{3})(-1, -1, 1, 0, 0) - \left(\frac{1}{\frac{5}{3}}\right)(\tfrac{1}{3}, -\tfrac{2}{3}, -\tfrac{1}{3}, 0, 1)$$

$$= (\tfrac{4}{5}, -\tfrac{3}{5}, \tfrac{1}{5}, 1, -\tfrac{3}{5}).$$

The set $\mathcal{T}' = \{\mathbf{v}_1, \mathbf{v}_2, \mathbf{v}_3\}$ is an orthogonal basis for **S**. Normalizing each vector, we obtain an orthonormal basis $\mathcal{T}'' = \{\mathbf{w}_1, \mathbf{w}_2, \mathbf{w}_3\}$, where

$$\mathbf{w}_1 = (-\sqrt{3}/3, -\sqrt{3}/3, \sqrt{3}/3, 0, 0),$$
$$\mathbf{w}_2 = (\sqrt{15}/15, -2\sqrt{15}/15, -\sqrt{15}/15, 0, \sqrt{15}/5),$$
$$\mathbf{w}_3 = (2\sqrt{15}/15, -\sqrt{15}/10, \sqrt{15}/30, \sqrt{15}/6, -\sqrt{15}/10).$$

EXAMPLE 5.35

(*From calculus*) Find an orthonormal basis for $\mathbf{P}_2$, the vector space of all polynomials of degree less than or equal to 2, under the inner product $(\mathbf{p}, \mathbf{q}) = \int_{-1}^{1} \mathbf{p}(x)\mathbf{q}(x)\, dx$.

Solution

We know that a basis for $\mathbf{P}_2$ (the *standard basis* for $\mathbf{P}_2$) is given by $\mathcal{T} = \{\mathbf{u}_1, \mathbf{u}_2, \mathbf{u}_3\}$, where $\mathbf{u}_1 = 1$, $\mathbf{u}_2 = x$, and $\mathbf{u}_3 = x^2$ (Theorem 1 of Section 5.2). Applying Theorem 6,

$\mathcal{T}' = \{\mathbf{v}_1, \mathbf{v}_2, \mathbf{v}_3\}$ will be an *orthogonal* basis for $\mathbf{P}_2$ if

$$\mathbf{v}_1 = 1,$$

$$\mathbf{v}_2 = x - \left(\frac{\int_{-1}^{1} (x)(1)\, dx}{\int_{-1}^{1} (1)(1)\, dx} \right)(1) = x - \left(\frac{0}{2} \right)1 = x,$$

$$\mathbf{v}_3 = x^2 - \left(\frac{\int_{-1}^{1} (x^2)(1)\, dx}{\int_{-1}^{1} (1)(1)\, dx} \right)(1) - \left(\frac{\int_{-1}^{1} (x^2)(x)\, dx}{\int_{-1}^{1} (x)(x)\, dx} \right)(x)$$

$$= x^2 - \left(\frac{\frac{2}{3}}{2} \right)(1) - \left(\frac{0}{\frac{2}{3}} \right)(x) = x^2 - \tfrac{1}{3}.$$

Finally, $\mathcal{T}'' = \{\mathbf{w}_1, \mathbf{w}_2, \mathbf{w}_3\}$ is an ortho*normal* basis for $\mathbf{P}_2$ if we let

$$\mathbf{w}_1 = \left(\frac{1}{\sqrt{\int_{-1}^{1} (1)(1)\, dx}} \right)(1) = \frac{1}{\sqrt{2}},$$

$$\mathbf{w}_2 = \left(\frac{1}{\sqrt{\int_{-1}^{1} (x)(x)\, dx}} \right)(x) = \frac{1}{\sqrt{\frac{2}{3}}} x = \sqrt{\tfrac{3}{2}}\, x,$$

$$\mathbf{w}_3 = \left(\frac{1}{\sqrt{\int_{-1}^{1} (x^2 - \frac{1}{3})(x^2 - \frac{1}{3})\, dx}} \right)(x^2 - \tfrac{1}{3})$$

$$= \frac{1}{\frac{2}{3}\sqrt{\frac{2}{5}}} (x^2 - \tfrac{1}{3}) = \tfrac{1}{2}\sqrt{\tfrac{5}{2}}(3x^2 - 1).$$

The following two lemmas will be used in the proof of Theorem 6.

LEMMA I   Let $\mathcal{T}$ and $\mathcal{T}'$ be as in Theorem 6 and let

$$a_{ij} = \frac{(\mathbf{u}_i, \mathbf{v}_j)}{(\mathbf{v}_j, \mathbf{v}_j)}.$$

Then $(a_{ij}\mathbf{v}_j, \mathbf{v}_j) = (\mathbf{u}_i, \mathbf{v}_j)$, and, if $\mathbf{v}_i$ and $\mathbf{v}_j$ are orthogonal, $(a_{ij}\mathbf{v}_j, \mathbf{v}_i) = 0$.

Proof   We have

$$(a_{ij}\mathbf{v}_j, \mathbf{v}_j) = \frac{(\mathbf{u}_i, \mathbf{v}_j)}{(\mathbf{v}_j, \mathbf{v}_j)}(\mathbf{v}_j, \mathbf{v}_j) = (\mathbf{u}_i, \mathbf{v}_j).$$

If $\mathbf{v}_i$ and $\mathbf{v}_j$ are orthogonal, we have $(a_{ij}\mathbf{v}_j, \mathbf{v}_i) = a_{ij}(\mathbf{v}_j, \mathbf{v}_i) = 0$, as desired.

LEMMA 2 Let $\mathcal{T}$ and $\mathcal{T}'$ be as in Theorem 6. Then every vector in $\mathcal{T}'$ is nonzero.

Proof Suppose a vector in $\mathcal{T}'$ (say, $\mathbf{v}_k$) is zero. Then, using the notation of Lemma 1.

$$\mathbf{u}_k = a_{k1}\mathbf{v}_1 + a_{k2}\mathbf{v}_2 + \cdots + a_{k,k-1}\mathbf{v}_{k-1}.$$

Now each $\mathbf{v}_i$ is a linear combination of $\mathbf{u}_i$ and the $\mathbf{v}_j$ that precede it. By successive substitutions into this equation, we can write $\mathbf{u}_k$ as a linear combination of $\mathbf{u}_1, \mathbf{u}_2, \ldots, \mathbf{u}_{k-1}$. Hence $\{\mathbf{u}_1, \mathbf{u}_2, \ldots, \mathbf{u}_k\}$ is linearly dependent; by Exercise 47 of Section 5.2, so is $\mathcal{T}$, because $\mathcal{T}$ contains this set. But this is a contradiction, since $\mathcal{T}$ is a basis for $\mathbf{V}$, so all members of $\mathcal{T}'$ must be nonzero.

Proof of Theorem 6 We will first show that $\mathcal{T}'$ is an orthogonal set and then that it is a basis for $\mathbf{V}$. The fact that $\mathcal{T}''$ is an orthonormal basis for $\mathbf{V}$ then follows easily, and we will leave it as Exercise 33.

Using the notation and results of Lemma 1, we have

$$(\mathbf{v}_2, \mathbf{v}_1) = (\mathbf{u}_2 - a_{21}\mathbf{v}_1, \mathbf{v}_1) = (\mathbf{u}_2, \mathbf{v}_1) - (a_{21}\mathbf{v}_1, \mathbf{v}_1)$$
$$= (\mathbf{u}_2, \mathbf{v}_1) - (\mathbf{u}_2, \mathbf{v}_1) = 0.$$

Therefore $\mathbf{v}_2$ and $\mathbf{v}_1$ are orthogonal. Furthermore,

$$(\mathbf{v}_3, \mathbf{v}_1) = (\mathbf{u}_3 - a_{31}\mathbf{v}_1 - a_{32}\mathbf{v}_2, \mathbf{v}_1)$$
$$= (\mathbf{u}_3, \mathbf{v}_1) - (a_{31}\mathbf{v}_1, \mathbf{v}_1) - (a_{32}\mathbf{v}_2, \mathbf{v}_1)$$
$$= (\mathbf{u}_3, \mathbf{v}_1) - (\mathbf{u}_3, \mathbf{v}_1) - 0 = 0.$$

Therefore $\mathbf{v}_3$ and $\mathbf{v}_1$ are orthogonal. In a similar manner we can show that $\mathbf{v}_3$ and $\mathbf{v}_2$ are orthogonal and that $\mathbf{v}_4$ is orthogonal, successively, to $\mathbf{v}_1$, $\mathbf{v}_2$, and $\mathbf{v}_3$. For example,

$$(\mathbf{v}_4, \mathbf{v}_2) = (\mathbf{u}_4 - a_{41}\mathbf{v}_1 - a_{42}\mathbf{v}_2 - a_{43}\mathbf{v}_3, \mathbf{v}_2)$$
$$= (\mathbf{u}_4, \mathbf{v}_2) - (a_{41}\mathbf{v}_1, \mathbf{v}_2) - (a_{42}\mathbf{v}_2, \mathbf{v}_2) - (a_{43}\mathbf{v}_3, \mathbf{v}_2)$$
$$= (\mathbf{u}_4, \mathbf{v}_2) - 0 - (\mathbf{u}_4, \mathbf{v}_2) - 0 = 0$$

(by once again applying Lemma 1). We continue this process until we finally show that $\mathbf{v}_n$ is orthogonal, successively, to $\mathbf{v}_1, \mathbf{v}_2, \ldots, \mathbf{v}_{n-1}$. Thus $\mathcal{T}'$ is an orthogonal set.

To show that $\mathcal{T}'$ is a basis for $\mathbf{V}$, note that it contains n vectors, as does the known basis, $\mathcal{T}$. Thus by part (iii) of Theorem 7 of Section 5.2, all we need show is that $\mathcal{T}'$ is linearly independent. But since $\mathcal{T}'$ is an orthogonal set of nonzero vectors (Lemma 2), $\mathcal{T}'$ is linearly independent and therefore a basis—an orthogonal basis—for $\mathbf{V}$, by Theorem 3.

5.4 EXERCISES

In Exercises 1–6 determine whether or not each of the following is an inner product for the indicated vector space. If it is not, list all properties violated.

1. $(\mathbf{p}, \mathbf{q}) = a_0b_0 + a_1b_1$; $\mathbf{p}(x) = a_0 + a_1x$, $\mathbf{q}(x) = b_0 + b_1x$ in $\mathbf{P}_1$

2. $(\mathbf{p}, \mathbf{q}) = \mathbf{p}(0)\mathbf{q}(0)$; $\mathbf{p}$, $\mathbf{q}$ in $\mathbf{P}_1$

3. $(A, B) = \det(A)\det(B)$; A, B in $\mathbf{M}^{2,2}$

4. $(A, B) = a_{11}b_{11} + a_{12}b_{12} + a_{21}b_{21} + a_{22}b_{22}$; A, B in $\mathbf{M}^{2,2}$

5. $(\mathbf{f}, \mathbf{g}) = \mathbf{f}'(0)\mathbf{g}'(0)$; $\mathbf{f}$, $\mathbf{g}$ in $\mathbf{C}^1(-1, 1)$, the vector space of all differentiable functions on the interval $(-1, 1)$

6. $(\mathbf{f}, \mathbf{g}) = \left(\max_{[-1,1]} |\mathbf{f}(x)|\right)\left(\max_{[-1,1]} |\mathbf{g}(x)|\right)$; $\mathbf{f}$, $\mathbf{g}$ in $\mathbf{C}[-1, 1]$

In Exercises 7–16 determine whether or not the given set of vectors is (a) orthogonal or (b) orthonormal for the indicated vector space and inner product.

7. $\{(1/\sqrt{5}, 2/\sqrt{5}, 0), (-2/\sqrt{5}, 1/\sqrt{5}, 0), (0, 0, 1)\}$ for $\mathbf{R}^3$

8. $\{(0, 1, 0), (1, 0, 1), (0, 0, 1)\}$ for $\mathbf{R}^3$

9. $\{(1, 2, 2), (2, 1, -2), (1, -2, 2)\}$ for $\mathbf{R}^3$

10. $\{(0, \sin\theta, \cos\theta, 0), (0, \cos\theta, -\sin\theta, 0)\}$, where θ is any real number for the subspace of $\mathbf{R}^4$ of all vectors whose first and last components are zero.

11–12. $\{(\sqrt{2}/2, 0, \sqrt{2}/2)\}$ for the solution space of $A\mathbf{x} = \mathbf{0}$, where

11. $A = \begin{bmatrix} 1 & 0 & -1 \\ 0 & 1 & 0 \\ 1 & 1 & -1 \end{bmatrix}$ **12.** $A = \begin{bmatrix} 1 & 0 & -1 \\ 1 & 1 & -1 \\ 2 & 1 & -2 \end{bmatrix}$

13. $\{1, x, x^2\}$ in $\mathbf{P}_2$; $(\mathbf{p}, \mathbf{q}) = \int_0^1 \mathbf{p}(x)\mathbf{q}(x)\, dx$

14. $\{x/2, x^2 - 1\}$ in $\mathbf{P}_2$; $(\mathbf{p}, \mathbf{q}) = \mathbf{p}(0)\mathbf{q}(0) + \mathbf{p}(1)\mathbf{q}(1) + \mathbf{p}(-1)\mathbf{q}(-1)$

15. $\left\{ \begin{bmatrix} \frac{1}{3} & -\frac{2}{3} \\ \frac{2}{3} & 0 \end{bmatrix}, \begin{bmatrix} \frac{2}{3} & -\frac{1}{3} \\ -\frac{2}{3} & 0 \end{bmatrix}, \begin{bmatrix} \frac{2}{3} & \frac{2}{3} \\ \frac{1}{3} & 0 \end{bmatrix} \right\}$

in $\mathbf{M}^{2,2}$; $(A, B) = a_{11}b_{11} + a_{12}b_{12} + a_{21}b_{21} + a_{22}b_{22}$

16. $\{2/\pi, (\pi/2)\cos x\}$ in $\mathbf{C}[0, \pi/2]$, $(\mathbf{f}, \mathbf{g}) = \int_0^{\pi/2} \mathbf{f}(x)\mathbf{g}(x)\, dx$

In Exercises 17 and 18 express the given vector $\mathbf{x}$ as a linear combination of the vectors in the orthonormal basis $\{(\frac{1}{3}, -\frac{2}{3}, \frac{2}{3}), (-\frac{2}{3}, \frac{1}{3}, \frac{2}{3}), (\frac{2}{3}, \frac{2}{3}, \frac{1}{3})\}$.

17. $\mathbf{x} = (1, 2, 3)$ **18.** $\mathbf{x} = (-1, 0, 1)$

In Exercises 19 and 20 the given set is an *orthogonal* basis for a certain subspace of $\mathbf{R}^4$. Find an ortho*normal* basis for this subspace.

19. $\{(4, 5, 0, -2), (-2, 2, 0, 1), (1, 0, 0, 2)\}$

20. $\{(1, -1, 1, -1), (1, 1, 1, 1), (1, 0, -1, 0)\}$

In Exercises 21–26 the given set is a basis for a certain subspace of $\mathbf{R}^m$. Use the Gram–Schmidt process to find an orthonormal basis for this subspace.

21. $\{(1, -1, 0), (0, 1, -1)\}$

22. $\{(1, 0, -1), (1, 0, 1)\}$

23. $\{(-1, 1, 0, 0), (1, -1, 1, 0), (0, 0, 1, 2)\}$

24. $\{(-1, -1, 2, 1), (2, 1, -1, -2), (1, 1, 0, 1)\}$

25. $\{(1, 0, 0, 0), (1, 1, 0, 0), (1, 1, 1, 0), (1, 1, 1, 1)\}$

26. $\{(1, 1, 1, 0), (1, -1, 1, -1), (1, 0, 1, 1), (0, 1, 1, 1)\}$

In Exercises 27–30 apply the Gram–Schmidt process to find an orthonormal basis for the subspace of $\mathbf{C}[a, b]$ (under the inner product $(\mathbf{f}, \mathbf{g}) = \int_a^b \mathbf{f}(x)\mathbf{g}(x)\, dx$) spanned by the given set $\mathcal{T}$.

27. $\mathcal{S} = \{1, x, x^2, x^3\}$, $[a, b] = [-1, 1]$

28. $\mathcal{S} = \{1, x, x^2\}$, $[a, b] = [0, 1]$

29. $\mathcal{S} = \{1, e^x\}$, $[a, b] = [0, 1]$

30. $\mathcal{S} = \{1, e^{-x}\}$, $[a, b] = [0, 1]$

31. Show that the standard basis for $\mathbf{R}^m$, $\{\mathbf{e}_1, \mathbf{e}_2, \ldots, \mathbf{e}_n\}$, is an orthonormal set.

32. Prove (Theorem 1b) that for all vectors $\mathbf{v}$ in an inner product space $\mathbf{V}$, we have $(\mathbf{0}, \mathbf{v}) = (\mathbf{v}, \mathbf{0}) = 0$.

33. Using the fact that $\mathcal{T}'$ of Theorem 6 is an *orthogonal* basis for $\mathbf{V}$, show that the set $\mathcal{T}''$ of this theorem is an ortho*normal* basis for $\mathbf{V}$.

CHAPTER 5 REVIEW

In this chapter the notion of *vector space* was introduced and shown to be a generalization of $\mathbf{R}^m$. Then concepts such as linear *independence, subspace, dimension,* and *inner product* were defined and discussed in this broader context.

As well as knowing the definition of vector space and being able to verify that a set, together with some operations, is or is not a vector space, you should be able to determine in a given vector space:

1. if a set is linearly independent and, if it is not, to express one vector as a linear combination of others;

2. if a set spans the space;

3. if a set is a basis for the space;

4. the dimension of the space;

5. the coordinate vector of a vector with respect to a given basis for the space.

In an inner product space you should be able to:

6. determine whether or not a set of vectors is an orthogonal (or orthonormal) set;

7. use the Gram–Schmidt process to construct an orthonormal basis for the space.

KEY WORDS

vector	vector space
negative of a vector	zero vector
subspace of a vector space	linearly independent/dependent vectors
linear combination	span of a set of vectors
generate a vector space	basis for a vector space
dimension of a vector space	finite dimensional vector space
coordinate vector	inner product
inner product space	norm (length) of a vector
orthogonal set or basis	orthonormal set or basis
orthogonal projection	

KEY THEOREMS

Theorem A subset of a vector space is a subspace if and only if it is closed under addition and scalar multiplication.

Theorem A spanning set may be reduced to a basis.

Theorem A linearly independent set may be extended to a basis.

Theorem Any two bases for the same space have the same number of elements.

Theorem If a space has dimension n, then

   a. fewer than n vectors cannot span the space;

   b. more than n vectors cannot be linearly independent;

   c. exactly n vectors span the space if and only if they are linearly independent.

Theorem Coordinate vectors with respect to a basis may be used to replace the original vectors in many calculations.

Theorem Cauchy–Schwarz inequality: $|(\mathbf{u}, \mathbf{v})| \leq \|\mathbf{u}\| \, \|\mathbf{v}\|$.

Theorem Nonzero orthogonal vectors are linearly independent.

Theorem Gram–Schmidt process: A spanning set may be replaced by an orthonormal basis.

SELF-TEST

Answers appear at the end of this chapter.

In Exercises 1–4 determine whether or not the given set with the usual operations is a vector space. If it is not a vector space, give one property that fails to hold and verify that it fails with a *specific example*. [Section 5.1]

1. The set of all ordered pairs (a, b) with both a and b real numbers greater than or equal to zero.

2. The set of all 2×2 matrices A with real entries and $a_{12} = 0$.

3. The set of all polynomials with no x^2 term.

4. The set of all polynomials with degree exactly 4, together with the zero polynomial.

5. Verify explicitly that the set $\mathcal{T} = \{1 + x + x^2, x - x^2, 1 - x^2, 1 + x - x^2\}$ is linearly dependent in $\mathbf{P}_2$ by expressing $1 + x - x^2$ as a linear combination of the other three polynomials. [Section 5.2]

6. Verify that the set $\mathcal{T}$ of Exercise 5 is linearly dependent, using no calculations other than counting. [Section 5.2]

7. Use coordinate vectors to calculate the coefficients needed to express $\begin{bmatrix} 1 & 1 \\ 1 & 1 \end{bmatrix}$ as a linear combination of the basis for $\mathbf{M}^{2,2}$ given by

$$\left\{ \begin{bmatrix} 0 & 1 \\ 1 & 1 \end{bmatrix}, \begin{bmatrix} 1 & 0 \\ 1 & 1 \end{bmatrix}, \begin{bmatrix} 1 & 1 \\ 0 & 1 \end{bmatrix}, \begin{bmatrix} 1 & 1 \\ 1 & 0 \end{bmatrix} \right\}.$$ [Section 5.3]

8. Prove that in any inner product space **V**, if **v** and **w** are orthogonal, then any scalar multiple of **v** is orthogonal to any scalar multiple of **w**. [Section 5.4]

9. Verify that $f(x) = x^2 - 2x$ and $g(x) = x - 1$ are orthogonal in the inner product space of $\mathbf{P}_2$, with

$$(f, g) = f(0)g(0) + f(1)g(1) + f(2)g(2).$$ [Section 5.4]

10. The vectors $\mathbf{v}_1 = (1, 1, 1, 0)$ and $\mathbf{v}_2 = (1, 0, -1, 1)$ are orthogonal in $\mathbf{R}^4$, and $\mathbf{v}_3 = (1, 1, 1, 1)$ is linearly independent from them. Use the Gram–Schmidt process to replace $\mathbf{v}_3$ with a third vector **w** that is orthogonal to both $\mathbf{v}_1$ and $\mathbf{v}_2$ and that together with them generates the same subspace as the original three vectors. [Section 5.4]

REVIEW EXERCISES

In Exercises 1–6 determine whether or not the given set is a vector space under the indicated operations. If it is not a vector space, list each property that fails to hold.

1. The set of all polynomials of degree less than three whose coefficients have a sum of 1: the usual operations.

2. The set of all polynomials of degree less than three whose coefficients have a sum of 0: the usual operations.

3. Those functions in $\mathbf{C}[0, 3]$ that have value 0 at every point in the subinterval $[1, 2]$: the operations of Example 5.6.

4. Those functions in $\mathbf{C}[0, 3]$ that have value *not* equal to 0 at every point in the subinterval $[1, 2]$: the operations of Example 5.6.

5. The set of all 2×3 matrices of the form

$$\begin{bmatrix} a & b & a - b \\ a + 2b & ab & b - a \end{bmatrix}.$$

6. The set of all 2×3 matrices of the form

$$\begin{bmatrix} a & b & a - b \\ a + 2b & a + b & b - a \end{bmatrix}.$$

In Exercises 7–10 determine whether or not the given set is linearly independent. If the set is linearly dependent, express one of its vectors as a linear combination of the others. (*Note*: It is easier to use coordinate vectors with respect to the appropriate standard basis.)

7. $\{x + 2x^2 + 3x^3, -x + 2x^2 - x^3, x + x^3\}$ in $\mathbf{P}_3$.

8. $\left\{ \begin{bmatrix} 2 & 0 \\ 0 & 1 \end{bmatrix}, \begin{bmatrix} 1 & -2 \\ 0 & 0 \end{bmatrix}, \begin{bmatrix} 4 & -4 \\ 0 & 1 \end{bmatrix}, \begin{bmatrix} 1 & 1 \\ 0 & 1 \end{bmatrix} \right\}$ in $\mathbf{M}^{2,2}$.

9. $\{x - \cos x, 3x + x^2 + \cos x + 2 \sin x, -2x + 2 \cos x\}$ in $\mathbf{C}^\infty$.

10. $\left\{ \begin{bmatrix} -1 & 1 \\ 2 & -1 \end{bmatrix}, \begin{bmatrix} 1 & 0 \\ -1 & 1 \end{bmatrix}, \begin{bmatrix} -1 & 2 \\ 3 & -1 \end{bmatrix}, \begin{bmatrix} 1 & 1 \\ 0 & 1 \end{bmatrix} \right\}$ in $\mathbf{M}^{2,2}$.

In Exercises 11–16 determine whether or not the given set is orthogonal in the indicated inner product space.

11. $\{(2, -1, 1, 0), (0, 1, 1, 0), (1, 1, -1, -2)\}$ in $\mathbf{R}^4$, usual dot product.

12. The same set as in Exercise 11 but with the inner product of Example 5.26, with $m = 4$.

13. $\{1, 1 - 2x\}$ in $\mathbf{P}_1$: $(\mathbf{p}, \mathbf{q}) = \int_0^1 \mathbf{p}(x)\mathbf{q}(x)\, dx$.

14. $\{1, 1 - 2x, 3x^2 - 1\}$ in $\mathbf{P}_2$: $(\mathbf{p}, \mathbf{q}) = \int_0^1 \mathbf{p}(x)\mathbf{q}(x)\, dx$.

15. $\{x - 2, x^2 - x\}$ in $\mathbf{P}_2$: $(\mathbf{p}, \mathbf{q}) = \mathbf{p}(0)\mathbf{q}(0) + \mathbf{p}(1)\mathbf{q}(1) + \mathbf{p}(2)\mathbf{q}(2)$.

16. $\{x, \sin x\}$ in $\mathbf{C}^*[-\pi/2, \pi/2]$: $(\mathbf{f}, \mathbf{g}) = \int_{-\pi/2}^{\pi/2} \mathbf{f}(x)\mathbf{g}(x)\, dx$.

In Exercises 17–20 determine whether or not each of the following is an inner product for the indicated vector space.

17. $(\mathbf{p}, \mathbf{q}) = \mathbf{p}(1)\mathbf{q}(1) + \mathbf{p}(2)\mathbf{q}(2)$ on $\mathbf{P}_2$.

18. $(\mathbf{p}, \mathbf{q}) = \mathbf{p}(1)\mathbf{q}(1) + \mathbf{p}(2)\mathbf{q}(2) + \mathbf{p}(3)\mathbf{q}(3)$ on $\mathbf{P}_2$.

19. $(A, B) = \sum_{i=1}^{3} a_{ii}b_{ii}$ on $\mathbf{M}^{3,3}$ (the sum of the products of the diagonal entries of A and B).

20. $(A, B) = \sum_{i=1}^{2} \sum_{j=1}^{2} ija_{ij}b_{ij}$ on $\mathbf{M}^{2,2}$.

In Exercises 21 and 22 use the Gram–Schmidt process to find an orthonormal basis for the space generated by the set in the given inner product space.

21. $\{(1, 1, 1, 0), (1, 1, 0, 1), (1, 0, 1, 1)\}$ in $\mathbf{R}^4$.

22. $\{e^x, e^{-x}\}$ in $\mathbf{C}[0, 1]$ with $(\mathbf{f}, \mathbf{g}) = \int_0^1 \mathbf{f}(x)\mathbf{g}(x)\, dx$.

23. Prove that $(A, B) = \sum\limits_{i=1}^{m} \sum\limits_{j=1}^{n} ija_{ij}b_{ij}$ is an inner product on $\mathbf{M}^{m,n}$.

24. Show by example that vectors in a vector space may be orthogonal with respect to one inner product and not orthogonal with respect to another.

ANSWERS TO SELF-TEST

1. Not a vector space; not closed under multiplication by scalars: $-1(1, 0) = (-1, 0)$

2. Vector space

3. Vector space

4. Not a vector space; not closed under addition:

$$(x + x^4) + (x - x^4) = 2x$$

5. $(1/3)(1 + x + x^2) + (2/3)(x - x^2) + (2/3)(1 - x^2) = 1 + x - x^2$

6. The set contains 4 vectors, but the vector space is of dimension 3.

7.

$$\left[\begin{array}{cccc|c} 0 & 1 & 1 & 1 & 1 \\ 1 & 0 & 1 & 1 & 1 \\ 1 & 1 & 0 & 1 & 1 \\ 1 & 1 & 1 & 0 & 1 \end{array}\right] \rightarrow \left[\begin{array}{cccc|c} 1 & 0 & 0 & 0 & 1/3 \\ 0 & 1 & 0 & 0 & 1/3 \\ 0 & 0 & 1 & 0 & 1/3 \\ 0 & 0 & 0 & 1 & 1/3 \end{array}\right],$$

so that

$$\begin{bmatrix} 1 & 1 \\ 1 & 1 \end{bmatrix} = \frac{1}{3}\begin{bmatrix} 0 & 1 \\ 1 & 1 \end{bmatrix} + \frac{1}{3}\begin{bmatrix} 1 & 0 \\ 1 & 1 \end{bmatrix} + \frac{1}{3}\begin{bmatrix} 1 & 1 \\ 0 & 1 \end{bmatrix} + \frac{1}{3}\begin{bmatrix} 1 & 1 \\ 1 & 0 \end{bmatrix}.$$

8. $(a\mathbf{v}, b\mathbf{w}) = ab(\mathbf{v}, \mathbf{w}) = ab(0) = 0$

9. Simply evaluate each function as required by the formula

$$\begin{array}{lll} f(0) = & 0 & f(1) = -1 & f(2) = 0, \\ g(0) = -1 & g(1) = & 0 & g(2) = 1. \end{array}$$

Thus $(f, g) = f(0)g(0) + f(1)g(1) + f(2)g(2) = 0 + 0 + 0 = 0.$

10.

$$\begin{aligned} \mathbf{w} &= (1, 1, 1, 1) - \frac{(1, 1, 1, 1) \cdot (1, 1, 1, 0)}{(1, 1, 1, 0) \cdot (1, 1, 1, 0)}(1, 1, 1, 0) \\ &\quad - \frac{(1, 1, 1, 1) \cdot (1, 0, -1, 1)}{(1, 0, -1, 1) \cdot (1, 0, -1, 1)}(1, 0, -1, 1) \\ &= (1, 1, 1, 1) - 1(1, 1, 1, 0) - (1/3)(1, 0, -1, 1) \\ &= (-1/3, 0, 1/3, 2/3) \end{aligned}$$

6

LINEAR TRANSFORMATIONS

In this chapter we will discuss *linear transformations*—special types of functions between two vector spaces. You will see that matrices play an important role in this subject, and that many questions concerning a linear transformation can be answered by considering a corresponding matrix. Nevertheless, linear transformations are very important in their own right. In fact, in more advanced studies of linear algebra, they often provide the underlying theme for the entire course.

6.1 DEFINITION OF A LINEAR TRANSFORMATION

Let us first recall some terminology about *functions*. A **function** or **map** from one set to another is an association of each element of the first set with a unique element of the second. If $\mathcal{A}$ and $\mathcal{B}$ are sets, the notation

$$f: \mathcal{A} \to \mathcal{B}$$

denotes a function from $\mathcal{A}$ to $\mathcal{B}$. For a in $\mathcal{A}$, the unique element of $\mathcal{B}$ associated with a is called the **image** of a under f and is denoted by $f(a)$.

For our purposes, the sets $\mathcal{A}$ and $\mathcal{B}$ will be vector spaces $\mathbf{V}$ and $\mathbf{W}$. We will not be interested in all functions from $\mathbf{V}$ to $\mathbf{W}$, but only those that are *linear* in the following sense.

DEFINITION

Let $\mathbf{V}$ and $\mathbf{W}$ be vector spaces. A **linear transformation** from $\mathbf{V}$ to $\mathbf{W}$ is a function $T: \mathbf{V} \to \mathbf{W}$ that satisfies the following two conditions. For each $\mathbf{u}$ and $\mathbf{v}$ in $\mathbf{V}$ and scalar a,

   i. $T(a\mathbf{u}) = aT(\mathbf{u})$ (scalars factor out)

   ii. $T(\mathbf{u} + \mathbf{v}) = T(\mathbf{u}) + T(\mathbf{v})$ (T is additive).

230

EXAMPLE 6.1

Let $T: \mathbf{R}^2 \to \mathbf{R}^3$ be described by

$$T(x, y) = (x + y, x, 2x - y).$$

Show that T is a linear transformation from $\mathbf{R}^2$ to $\mathbf{R}^3$.

Solution

To check conditions (i) and (ii) of the definition, let $\mathbf{u} = (x_1, y_1)$, $\mathbf{v} = (x_2, y_2)$, and let a be a scalar.

i. $T(a\mathbf{u}) = T(a(x_1, y_1))$

$\qquad = T((ax_1, ay_1))$

$\qquad = (ax_1 + ay_1, ax_1, 2(ax_1) - ay_1)$

$\qquad = a(x_1 + y_1, x_1, 2x_1 - y_1)$

$\qquad = aT((x_1, y_1)) = aT(\mathbf{u}).$

ii. $T(\mathbf{u} + \mathbf{v}) = T((x_1, y_1) + (x_2, y_2))$

$\qquad = T((x_1 + x_2, y_1 + y_2))$

$\qquad = ((x_1 + x_2) + (y_1 + y_2), x_1 + x_2, 2(x_1 + x_2) - (y_1 + y_2))$

$\qquad = ((x_1 + y_1) + (x_2 + y_2), x_1 + x_2, (2x_1 - y_1) + (2x_2 - y_2))$

$\qquad = (x_1 + y_1, x_1, 2x_1 - y_1) + (x_2 + y_2, x_2, 2x_2 - y_2)$

$\qquad = T(\mathbf{u}) + T(\mathbf{v}).$

EXAMPLE 6.2

Let $T: \mathbf{V} \to \mathbf{W}$ be described by $T(\mathbf{u}) = \mathbf{0}$, for all $\mathbf{u}$ in $\mathbf{V}$. Show that T is a linear transformation.

Solution

i. If a is a scalar and $\mathbf{u}$ is in $\mathbf{V}$, $T(a\mathbf{u}) = \mathbf{0} = a\mathbf{0} = aT(\mathbf{u})$.

ii. If $\mathbf{u}$ and $\mathbf{v}$ are in $\mathbf{V}$, $T(\mathbf{u} + \mathbf{v}) = \mathbf{0} = \mathbf{0} + \mathbf{0} = T(\mathbf{u}) + T(\mathbf{v})$.

Thus T is a linear transformation.

DEFINITION

The transformation of Example 6.2 is called the **zero transformation** from **V** to **W** and will be denoted by 0.

Another simple linear transformation, this time from a vector space to *itself,* is the one that leaves every element fixed; that is, it does not change anything. This is just the *identity function* on the space. That this function is a linear transformation is straightforward and is to be verified in Exercise 26.

DEFINITION

Let **V** be a vector space. The linear transformation $I: \mathbf{V} \to \mathbf{V}$ described by $I(\mathbf{v}) = \mathbf{v}$, for every $\mathbf{v}$ in $\mathbf{V}$, is called the **identity transformation**.

EXAMPLE 6.3

Define $T: \mathbf{M}^{2,3} \to \mathbf{P}_2$ by

$$T(A) = T\begin{bmatrix} a_{11} & a_{12} & a_{13} \\ a_{21} & a_{22} & a_{23} \end{bmatrix} = (a_{11} + a_{13})x^2 + (a_{21} - a_{22})x + a_{23}.$$

Show that T is a linear transformation.

Solution

To verify condition (i) of the definition, let A be in $\mathbf{M}^{2,3}$, and let c be a scalar. Applying T to cA, we have

$$T(cA) = T\begin{bmatrix} ca_{11} & ca_{12} & ca_{13} \\ ca_{21} & ca_{22} & ca_{23} \end{bmatrix}$$

$$= (ca_{11} + ca_{13})x^2 + (ca_{21} - ca_{22})x + ca_{23}$$

$$= c[(a_{11} + a_{13})x^2 + (a_{21} - a_{22})x + a_{23}] = cT(A).$$

To verify condition (ii), let A and B be in $\mathbf{M}^{2,3}$. Applying T to $A + B$, we have

$$T(A + B) = T\begin{bmatrix} a_{11} + b_{11} & a_{12} + b_{12} & a_{13} + b_{13} \\ a_{21} + b_{21} & a_{22} + b_{22} & a_{23} + b_{23} \end{bmatrix}$$

$$= [(a_{11} + b_{11}) + (a_{13} + b_{13})]x^2 + [(a_{21} + b_{21})$$
$$\quad - (a_{22} + b_{22})]x + [a_{23} + b_{23}]$$

$$= (a_{11} + a_{13})x^2 + (a_{21} - a_{22})x + a_{23}$$
$$\quad + (b_{11} + b_{13})x^2 + (b_{21} - b_{22})x + b_{23}$$

$$= T(A) + T(B).$$

Thus T is a linear transformation.

The next example is very important in a more advanced study of linear algebra.

EXAMPLE 6.4

Let A be a square matrix of order n. Define $T: \mathbf{P} \to \mathbf{M}^{n,n}$ as follows. For each polynomial

$$\mathbf{p}(x) = a_m x^m + a_{m-1}x^{m-1} + \cdots + a_1 x + a_0,$$

define $T(\mathbf{p})$ to be the matrix

$$T(\mathbf{p}) = a_m A^m + a_{m-1}A^{m-1} + \cdots + a_1 A + a_0 I,$$

where I is the identity matrix of order n. Show that T is a linear transformation.

Solution

To show condition (i), let $\mathbf{p}$ be the polynomial

$$\mathbf{p}(x) = a_m x^m + \cdots + a_1 x + a_0$$

and c a scalar. Then $c\mathbf{p}$ is the polynomial

$$(c\mathbf{p})(x) = (ca_m)x^m + \cdots + (ca_1)x + ca_0.$$

Applying T to $c\mathbf{p}$, we have

$$T(c\mathbf{p}) = (ca_m)A^m + \cdots + (ca_1)A + (ca_0)I$$
$$= c(a_m A^m + \cdots + a_1 A + a_0 I) = cT(\mathbf{p}).$$

To show condition (ii), let $\mathbf{p}$ and $\mathbf{q}$ be the polynomials $\mathbf{p}(x) = a_m x^m + \cdots + a_1 x + a_0$ and $\mathbf{q}(x) = b_m x^m + \cdots + b_1 x + b_0$. (By using zero coefficients, if necessary, $\mathbf{p}$ and $\mathbf{q}$ can be made to appear as if they have the same degree.) Then $\mathbf{p} + \mathbf{q}$ is the polynomial

$$(\mathbf{p} + \mathbf{q})(x) = (a_m + b_m)x^m + \cdots + (a_1 + b_1)x + (a_0 + b_0).$$

Applying T to $\mathbf{p} + \mathbf{q}$, we have

$$T(\mathbf{p} + \mathbf{q}) = (a_m + b_m)A^m + \cdots + (a_1 + b_1)A + (a_0 + b_0)I$$
$$= (a_m A^m + \cdots + a_1 A + a_0 I) + (b_m A^m + \cdots + b_1 A + b_0 I)$$
$$= T(\mathbf{p}) + T(\mathbf{q}).$$

Thus T is a linear transformation.

To show that a given function is *not* a linear transformation, we need only show that *one* of the two conditions of the definition is violated in a *single* instance. This is illustrated in the next example.

EXAMPLE 6.5   Let $T: \mathbf{R}^2 \to \mathbf{R}^2$ be defined by $T(\mathbf{u}) = \mathbf{u} + (1, 2)$, for all $\mathbf{u}$ in $\mathbf{R}^2$. Show that T is *not* a linear transformation.

Solution   To show that condition (i) is violated by this function, we let $a = 0$ and $\mathbf{u} = (0, 0)$. Then

$$T(a\mathbf{u}) = T((0, 0)) = (0, 0) + (1, 2) = (1, 2),$$

but

$$aT(\mathbf{u}) = 0T((0, 0)) = 0((0, 0) + (1, 2)) = 0(1, 2) = (0, 0).$$

Thus it is not always true that $T(a\mathbf{u}) = aT(\mathbf{u})$, and condition (i) is violated: T is not linear. (It can be shown that condition (ii) does not hold either.)

Notice that the function described in Example 6.5 is a *translation* of the plane $\mathbf{R}^2$ by the vector $(1, 2)$: For each point $\mathbf{u}$, the line segment $\overrightarrow{\mathbf{u}T(\mathbf{u})}$ is equivalent to the vector $(1, 2)$, as shown in Figure 6.1. In particular, the origin $\mathbf{0}$ gets mapped by T to the point $(1, 2)$, $T(\mathbf{0}) = (1, 2)$. This fact, together with

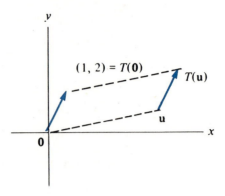

FIGURE 6.1

part (a) of the following theorem, provides another way to show that this function is not linear.

THEOREM I   Let $T: \mathbf{V} \to \mathbf{W}$ be a linear transformation. Then

   a. $T(\mathbf{0}) = \mathbf{0}$,

   b. $T(-\mathbf{u}) = -T(\mathbf{u})$ for each $\mathbf{u}$ in $\mathbf{V}$,

   c. $T(\mathbf{u} - \mathbf{v}) = T(\mathbf{u}) - T(\mathbf{v})$ for each $\mathbf{u}$ and $\mathbf{v}$ in $\mathbf{V}$.

NOTE   The vector $\mathbf{0}$ of $T(\mathbf{0})$ is an element of $\mathbf{V}$, and the vector $\mathbf{0}$ on the right side of the equation given in Theorem 1a is an element of $\mathbf{W}$. We use the same notation for each, but you should always be aware of where the vector is located.

....

Proofs

   1a. Since $\mathbf{0} = \mathbf{0} + \mathbf{0}$ (in $\mathbf{V}$),

$$T(\mathbf{0}) = T(\mathbf{0} + \mathbf{0}) \qquad \text{(in } \mathbf{W}\text{)},$$
$$T(\mathbf{0}) = T(\mathbf{0}) + T(\mathbf{0}) \qquad (T \text{ is additive)}.$$

Since $T(\mathbf{0})$ is in $\mathbf{W}$, using $\mathbf{0}$ in $\mathbf{W}$, we have

$$T(\mathbf{0}) + \mathbf{0} = T(\mathbf{0}) + T(\mathbf{0}).$$

Adding $-T(\mathbf{0})$ to both sides, we have

$$\mathbf{0} = T(\mathbf{0}).$$

1b. By Theorem 1a, and the fact that $0 = \mathbf{u} + (-\mathbf{u})$, we have

$$0 = T(0) = T(\mathbf{u} + -\mathbf{u}) = T(\mathbf{u}) + T(-\mathbf{u}).$$

That is,

$$0 = T(\mathbf{u}) + T(-\mathbf{u}).$$

Adding $-T(\mathbf{u})$ to both sides, we have

$$-T(\mathbf{u}) = T(-\mathbf{u}).$$

1c. The proof of 1c is Exercise 20 at the end of this section.

EXAMPLE 6.6   Let A be the matrix

$$A = \begin{bmatrix} 2 & 0 & 1 \\ 1 & 1 & -1 \end{bmatrix},$$

and let $\mathbf{x}$ be an element of $\mathbf{R}^3$. Let $T(\mathbf{x}) = A\mathbf{x}$, where $A\mathbf{x}$ denotes matrix multiplication. Show that T is a linear transformation.

Solution   We let $\mathbf{x} = (x, y, z)$ and compute $T(\mathbf{x})$:

$$T(\mathbf{x}) = A\mathbf{x} = \begin{bmatrix} 2 & 0 & 1 \\ 1 & 1 & -1 \end{bmatrix} \begin{bmatrix} x \\ y \\ z \end{bmatrix}$$

$$= \begin{bmatrix} 2x & + z \\ x + y & - z \end{bmatrix} = (2x + z, x + y - z).$$

Thus we have the rule $T(x, y, z) = (2x + z, x + y - z)$, and we could follow the procedure of Example 6.1 to verify that T is linear. (You should do it for the practice.)

Example 6.6 is a special case of the following theorem.

THEOREM 2   Let A be an $m \times n$ matrix. Then $T: \mathbf{R}^n \to \mathbf{R}^m$, defined by $T(\mathbf{x}) = A\mathbf{x}$, is a linear transformation.

Proof   First note that $\mathbf{x}$ is to be written as a column vector so that the matrix multiplication will be defined: A is $m \times n$, and $\mathbf{x}$ is $n \times 1$. Thus the product is defined, and the result is an $m \times 1$ matrix. The latter may be viewed as an element of $\mathbf{R}^m$.

To check property (i) of the definition of linear transformation, let a be a scalar, and let **x** be an element of $\mathbf{R}^n$. Then

$$T(a\mathbf{x}) = A(a\mathbf{x})$$
$$= aA\mathbf{x} \qquad \text{(Theorem 1n of Section 2.3)}$$
$$= aT(\mathbf{x}).$$

For property (ii), let **x** and **y** be elements of $\mathbf{R}^n$. Then

$$T(\mathbf{x} + \mathbf{y}) = A(\mathbf{x} + \mathbf{y})$$
$$= A\mathbf{x} + A\mathbf{y} \qquad \text{(Theorem 1g of Section 2.3)}$$
$$= T(\mathbf{x}) + T(\mathbf{y}).$$

Thus T is a linear transformation.

Examples of linear transformations given by Theorem 2 are of the utmost importance. In fact, it will be shown in Section 6.4 that in some sense all linear transformations are of this type. For the time being, such functions provide an infinite supply of linear transformations for which we need not check the definition; that has already been done in Theorem 2. We continue with some interesting special cases.

EXAMPLE 6.7

Let θ be any real number, and view θ as the radian measure of an angle. Let T_θ be given by $T_\theta(\mathbf{x}) = A_\theta \mathbf{x}$, where

$$A_\theta = \begin{bmatrix} \cos\theta & -\sin\theta \\ \sin\theta & \cos\theta \end{bmatrix}.$$

Then T_θ is the **rotation** by θ. In other words, T_θ is the linear transformation that rotates each vector of $\mathbf{R}^2$ through an angle of θ radians.

Solution

To see that this is so, we need only show that for all **x** in $\mathbf{R}^2$, **x** and $T_\theta(\mathbf{x})$ have the same length and that the angle between these vectors is θ (Figure 6.2).

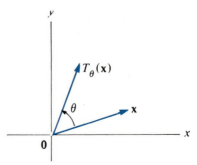

FIGURE 6.2

To show that $\mathbf{x}$ and $T_\theta(\mathbf{x})$ have the same length, let $\mathbf{x} = (x, y)$. Then $T_\theta(\mathbf{x}) = (x \cos \theta - y \sin \theta, x \sin \theta + y \cos \theta)$, so that

$\|T_\theta(\mathbf{x})\|$

$$= \|(x \cos \theta - y \sin \theta, x \sin \theta + y \cos \theta)\|$$

$$= \sqrt{(x \cos \theta - y \sin \theta)^2 + (x \sin \theta + y \cos \theta)^2}$$

$$= \sqrt{x^2 \cos^2 \theta - 2xy \cos \theta \sin \theta + y^2 \sin^2 \theta + x^2 \sin^2 \theta + 2xy \sin \theta \cos \theta + y^2 \cos^2 \theta}$$

$$= \sqrt{x^2(\cos^2 \theta + \sin^2 \theta) + y^2(\sin^2 \theta + \cos^2 \theta)}$$

$$= \sqrt{x^2 + y^2} \quad (\text{since } \sin^2 \theta + \cos^2 \theta = 1)$$

$$= \|\mathbf{x}\|.$$

By Theorem 2 of Section 1.2, the cosine of the angle ψ between x and $T_\theta(\mathbf{x})$ is given by

$$\cos \psi = \frac{\mathbf{x} \cdot T_\theta(\mathbf{x})}{\|\mathbf{x}\| \|T_\theta(\mathbf{x})\|} = \frac{x^2 \cos \theta - xy \sin \theta + xy \sin \theta + y^2 \cos \theta}{\|\mathbf{x}\| \|\mathbf{x}\|}$$

$$= \frac{(x^2 + y^2)}{\|\mathbf{x}\|^2} \cos \theta$$

$$= \frac{\|\mathbf{x}\|^2}{\|\mathbf{x}\|^2} \cos \theta = \cos \theta.$$

Thus $\cos \psi = \cos \theta$. If $0 \leq \theta \leq \pi$, this implies that $\psi = \theta$. If, on the other hand, $\pi < \theta < 2\pi$, then $\psi = 2\pi - \theta$. In either case, T_θ rotates $\mathbf{x}$ through an angle θ (see Figure 6.3).

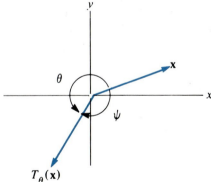

FIGURE 6.3

As a specific instance of Example 6.7, let $\theta = \pi/6$ radians. Then $\cos \pi/6 = \sqrt{3}/2$ and $\sin \pi/6 = 1/2$, so that

$$A = \begin{bmatrix} \sqrt{3}/2 & -1/2 \\ 1/2 & \sqrt{3}/2 \end{bmatrix}.$$

This matrix determines the linear transformation that rotates each vector in $\mathbf{R}^2$ by $\pi/6$ and leaves its length unchanged.

SPOTLIGHT

COMPUTER GRAPHICS

You have surely seen computer-generated graphics on television or in the movies. A spherical object spins and rockets through space, greenery grows at an amazing rate on a make-believe planet, or a spiraling picture of a television personality swoops (almost) into your living room. Each of these scenes is actually a sequence of discrete pictures, like the frames in movie film, mathematically generated by a computer.

How is this sequence of images generated? One way to produce them is to create in detail a single picture made up of many discrete points, or vectors, in $\mathbf{R}^3$; so many points that our minds visualize a solid object. This object is given motion (without changing its size or shape) by performing linear transformations of a special type on its vectors, by multiplying them by 3×3 matrices analogous to the rotation matrix of Example 6.7. (We will describe these linear transformations, which are represented by *orthogonal* matrices, in Section 7.2.)

To create images that make the object appear to move continuously before your eyes, a sequence of transformations is performed in which the multiplying matrices are changed very slightly to get each successive image. Although each multiplication is simple enough, an enormous number of them is required to make for an effective graphic. This necessitates the use of extremely fast computers containing very large amounts of memory.

EXAMPLE 6.8   Let c be a positive scalar, and let $A = cI$, where I is the $n \times n$ identity matrix. Then T_c given by $T_c(\mathbf{x}) = A\mathbf{x}$ is the **dilation of $\mathbf{R}^n$ by** c if $c > 1$ and is the **contraction by** c if $0 < c < 1$. In other words, T_c is the linear transformation that "stretches" or "shrinks" each vector of $\mathbf{R}^n$ by a factor of c but leaves its direction unchanged. Of course, if $c = 1$, T_c is the identity transformation. Figures 6.4 and 6.5 indicate the result when $c = 3$ and $c = \frac{1}{2}$, respectively.

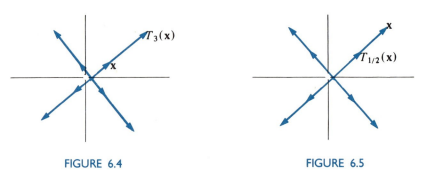

FIGURE 6.4   FIGURE 6.5

Solution

To verify the assertions, note that $T_c(\mathbf{x}) = c\mathbf{x}$. Thus T_c is just multiplication by the scalar c, and, by Section 1.4, both the length and the direction are as stated.

EXAMPLE 6.9

Let θ be any real number, and view θ as the radian measure of an angle. Let T^θ be given by $T^\theta(\mathbf{x}) = A\mathbf{x}$, where

$$A = \begin{bmatrix} \cos 2\theta & \sin 2\theta \\ \sin 2\theta & -\cos 2\theta \end{bmatrix}.$$

Then T^θ is the **reflection** in the line l through the origin that forms angle θ with the positive x-axis: For each $\mathbf{x}$ in $\mathbf{R}^2$, $T^\theta(\mathbf{x})$ is situated so that the perpendicular bisector of the segment from the endpoint of $\mathbf{x}$ to that of $T^\theta(\mathbf{x})$ is the line l (Figure 6.6). We leave the verification of this statement as Exercise 27 at the end of the section.

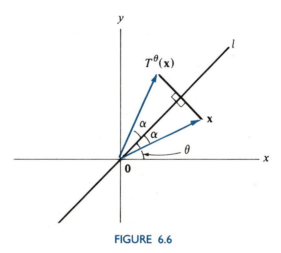

FIGURE 6.6

NOTE   The linear transformations of Examples 6.7, 6.8, and 6.9 are all **linear operators:** They are all linear transformations that map the space back into itself. They are in fact more special than that: They are **similarities;** they preserve the shape of geometric figures. Examples 6.7 and 6.9 are also **isometries,** which preserve not only shape but also size (distance between any two points).

In the preceding examples we checked the effect or "action" that a linear transformation $T: \mathbf{V} \to \mathbf{W}$ has on an arbitrary vector $\mathbf{x}$ in $\mathbf{V}$. Determining the effect of a linear transformation can also be done by checking its action on a specific, finite set of vectors. The next definition and theorem make this precise.

DEFINITION

Let $S: \mathbf{V} \to \mathbf{W}$ and $T: \mathbf{V} \to \mathbf{W}$ be linear transformations. Then S **equals** T if $S = T$ as functions; that is, if $S(\mathbf{v}) = T(\mathbf{v})$ for all $\mathbf{v}$ in $\mathbf{V}$.

THEOREM 3 If $S: \mathbf{V} \to \mathbf{W}$ and $T: \mathbf{V} \to \mathbf{W}$ are linear transformations, and $\mathcal{B} = \{\mathbf{b}_1, \mathbf{b}_2, \ldots, \mathbf{b}_n\}$ is a basis for $\mathbf{V}$ and $S(\mathbf{b}_i) = T(\mathbf{b}_i)$ for each i, then $S = T$.

Proof Since $\mathcal{B}$ is a basis for $\mathbf{V}$, for each $\mathbf{v}$ in $\mathbf{V}$ there exist scalars $c_1, \ldots, c_n$ with $\mathbf{v} = c_1\mathbf{b}_1 + c_2\mathbf{b}_2 + \cdots + c_n\mathbf{b}_n$. Then

$$S(\mathbf{v}) = S(c_1\mathbf{b}_1 + c_2\mathbf{b}_2 + \cdots + c_n\mathbf{b}_n)$$
$$= c_1 S(\mathbf{b}_1) + c_2 S(\mathbf{b}_2) + \cdots + c_n S(\mathbf{b}_n) \qquad \text{(Exercise 29)}$$
$$= c_1 T(\mathbf{b}_1) + c_2 T(\mathbf{b}_2) + \cdots + c_n T(\mathbf{b}_n) \qquad \text{(hypothesis)}$$
$$= T(c_1\mathbf{b}_1 + c_2\mathbf{b}_2 + \cdots + c_n\mathbf{b}_n) \qquad \text{(Exercise 29)}$$
$$= T(\mathbf{v}).$$

Thus $S = T$.

NOTE In words, Theorem 3 is stated as follows: A linear transformation from one vector space to another is determined by its *action* on a basis.

EXAMPLE 6.10

Describe the action of the linear transformation $T: \mathbf{R}^2 \to \mathbf{R}^2$ given by $T(x, y) = (2x - y, x + y)$.

Solution

Theorem 3 applies only to *linear transformations*, so we must verify that T is a linear transformation. This detail is Exercise 1 at the end of the section. We proceed to check the action of T on a basis for $\mathbf{R}^2$.

For the basis, we choose the most convenient one: the standard basis $\{\mathbf{e}_1, \mathbf{e}_2\}$. We have

$$T(\mathbf{e}_1) = T(1, 0) = (2, 1) \qquad \text{and} \qquad T(\mathbf{e}_2) = T(0, 1) = (-1, 1).$$

Next we sketch $\{\mathbf{e}_1, \mathbf{e}_2\}$ and $\{T(\mathbf{e}_1), T(\mathbf{e}_2)\}$ on separate copies of the space $\mathbf{R}^2$ (Figures 6.7 and 6.8).

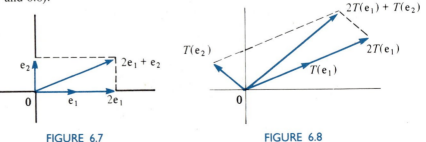

FIGURE 6.7   FIGURE 6.8

Here T is seen to rotate and stretch the vectors, but unevenly: e_1 is not rotated as much as e_2, and e_2 is not stretched as much as e_1. The importance of Theorem 3 is that it shows us that the effect of T on all other vectors is determined by its effect on these two. To be precise, for any $x = (x, y)$ in $\mathbf{R}^2$, since $x = xe_1 + ye_2$, we have $T(x) = xT(e_1) + yT(e_2)$. For example, consider the point $(2, 1)$. Since $(2, 1) = 2e_1 + e_2$, the point $(2, 1)$ completes the parallelogram determined by $2e_1$ and e_2. Then $T(2, 1)$ must be the point that completes the parallelogram determined by $2T(e_1)$ and $T(e_2)$ (Figure 6.8).

6.1 EXERCISES

In Exercises 1–8 show that each function is a linear transformation.

1. $S(x, y) = (2x - y, x + y)$ $\qquad$ $(S: \mathbf{R}^2 \rightarrow \mathbf{R}^2)$

2. $T(x, y) = (2x, x + y, x - 2y)$ $\qquad$ $(T: \mathbf{R}^2 \rightarrow \mathbf{R}^3)$

3. $U(x, y, z) = x + y + z$ $\qquad$ $(U: \mathbf{R}^3 \rightarrow \mathbf{R})$

4. $I(x_1, \ldots, x_n) = (x_1, \ldots, x_n)$ $\qquad$ $(I: \mathbf{R}^n \rightarrow \mathbf{R}^n)$

5. $T: \mathbf{R}^3 \rightarrow \mathbf{M}^{2,2}$ given by

$$T(x, y, z) = \begin{bmatrix} y & z \\ -x & 0 \end{bmatrix}$$

6. $T: \mathbf{P}_3 \rightarrow \mathbf{P}_3$ given by

$$T(a_0 + a_1x + a_2x^2 + a_3x^3) = (a_0 + a_2) - (a_1 + 2a_3)x^2$$

7. $T: \mathbf{P}_2 \rightarrow \mathbf{M}^{2,2}$ given by

$$T(a_0 + a_1x + a_2x^2) = \begin{bmatrix} a_0 & -a_2 \\ -a_2 & a_0 - a_1 \end{bmatrix}$$

8. $T: \mathbf{P}_2 \rightarrow \mathbf{P}_2$ given by

$$T(a_0 + a_1x + a_2x^2) = a_1 - (a_0 + a_2)x + (a_1 + a_2)x^2$$

In Exercises 9–12 determine whether or not the given function is a linear transformation, and justify your answer.

9. $F(x, y, z) = (0, 2x + y)$ $\qquad$ $(F: \mathbf{R}^3 \rightarrow \mathbf{R}^2)$

10. $G(x, y, z) = (xy, y, x - z)$ $\qquad$ $(G: \mathbf{R}^3 \rightarrow \mathbf{R}^3)$

11. $H(x, y) = \sqrt{x^2 + y^2}$ $\qquad$ $(H: \mathbf{R}^2 \rightarrow \mathbf{R})$

12. $K(x, y) = (x, \sin y, 2x + y)$ $\qquad$ $(K: \mathbf{R}^2 \rightarrow \mathbf{R}^3)$

In Exercises 13–16 let $T: \mathbf{R}^n \rightarrow \mathbf{R}^m$ be the linear transformation given by $T(x) = Ax$, multiplication by the matrix A. Find n and m and $T(x_0)$ for each given vector x_0.

13. $A = \begin{bmatrix} 2 & 3 & 1 \\ 1 & -1 & 0 \end{bmatrix}$, $\quad x_0 = (1, 4, 2)$ $\qquad$ **14.** $A = \begin{bmatrix} -1 & 1 & 2 \\ 2 & 4 & -1 \\ 1 & 0 & 1 \end{bmatrix}$, $\quad x_0 = (1, 4, 2)$

15. $A = \begin{bmatrix} 1 & 2 \\ 0 & 1 \\ 1 & 0 \end{bmatrix}$, $\mathbf{x}_0 = (2, -3)$ **16.** $A = \begin{bmatrix} 0 & 1 & 0 \\ 1 & 0 & 0 \\ 0 & 0 & 1 \end{bmatrix}$, $\mathbf{x}_0 = (x, y, z)$

17. The function $\Pi_i: \mathbf{R}^n \to \mathbf{R}$ defined by $\Pi_i(x_1, \ldots, x_n) = x_i$ is called the *i*th *projection map*. Prove that each projection map is a linear transformation.

18. If A is an $m \times n$ matrix, prove that the function $T: \mathbf{R}^m \to \mathbf{R}^n$ defined by $T(\mathbf{x}) = \mathbf{x}A$ is a linear transformation, where $\mathbf{x}$ is viewed as a matrix with one row. (*Hint:* Follow Theorem 2.)

19. For a linear transformation $T: \mathbf{R}^n \to \mathbf{R}^m$, the **negative** of T, $-T: \mathbf{R}^n \to \mathbf{R}^m$, is defined by $(-T)(\mathbf{x}) = -(T(\mathbf{x}))$. Prove that $-T$ is a linear transformation.

20. Prove Theorem 1c.

21. Find the matrix for the linear transformation $T_{3\pi/4}$, the rotation by $3\pi/4$.

22. Find the matrix for the linear transformation $T_{-\pi/2}$, the rotation by $-\pi/2$.

23. Find the matrix for the linear transformation $T^{\pi/4}$, the reflection in the line $x = y$.

24. Find the matrix for the linear transformation T^0, the reflection in the *x*-axis.

25. Let $D: \mathbf{P}_3 \to \mathbf{P}_2$ be defined as follows. For each

$$\mathbf{p}(x) = a_3x^3 + a_2x^2 + a_1x + a_0,$$

define

$$D(\mathbf{p}) = 3a_3x^2 + 2a_2x + a_1.$$

Prove that D is a linear transformation. (*Note from calculus*: This linear transformation gives the *derivative* of any polynomial in $\mathbf{P}_3$.)

26. Prove that the identity function on a vector space $\mathbf{V}$ is a linear transformation.

27. Complete the verification of Example 6.9.

28. Let $F: \mathbf{V} \to \mathbf{W}$ be a function. Prove that F is a linear transformation if and only if $F(a\mathbf{u} + \mathbf{v}) = aF(\mathbf{u}) + F(\mathbf{v})$ for each $\mathbf{u}$ and $\mathbf{v}$ in $\mathbf{V}$ and a in $\mathbf{R}$.

29. Let $T: \mathbf{V} \to \mathbf{W}$ be a linear transformation. Prove that for any linear combination $c_1\mathbf{v}_1 + \cdots + c_k\mathbf{v}_k$ in $\mathbf{V}$, $T(c_1\mathbf{v}_1 + \cdots + c_k\mathbf{v}_k) = c_1T(\mathbf{v}_1) + \cdots + c_kT(\mathbf{v}_k)$.

30. Let $T: \mathbf{R}^n \to \mathbf{R}^m$ be a linear transformation. If $\mathcal{S}$ is a subset of $\mathbf{R}^n$, then $T[\mathcal{S}]$ is the set of all images $T(\mathbf{s})$ for $\mathbf{s}$ in $\mathcal{S}$. If l is a line in $\mathbf{R}^n$, show that $T[l]$ is either a line or a point in $\mathbf{R}^m$. (*Hint:* Use Exercise 29 and the point-parallel or two-point form for a line.)

31. Extend Exercise 30 to include the case where the given set l is a plane.

6.2   ALGEBRA OF LINEAR TRANSFORMATIONS

In this section we will define operations on linear transformations that have the same properties as the analogous ones for matrices. Then, restricting consideration to Euclidean spaces, we will show that every linear transformation can be represented by a matrix and that there is a very simple correspondence between an operation on the transformation and on its matrix.

OPERATIONS ON LINEAR TRANSFORMATIONS

Sums, differences, negatives, and scalar multiples of linear transformations between vector spaces **V** and **W** can be defined in a natural way.

DEFINITION

Let S and T be linear transformations from **V** to **W**, and let c be a scalar.

i. The **sum** $S + T$ is the function given by

$$(S + T)(\mathbf{x}) = S(\mathbf{x}) + T(\mathbf{x})$$

for all **x** in **V**.

ii. The **difference** $S - T$ is the function given by

$$(S - T)(\mathbf{x}) = S(\mathbf{x}) - T(\mathbf{x})$$

for all **x** in **V**.

iii. The **negative** $-S$ is the function given by

$$(-S)(\mathbf{x}) = -S(\mathbf{x})$$

for all **x** in **V**.

iv. The **scalar multiple** cS is the function given by

$$(cS)(\mathbf{x}) = (c)(S(\mathbf{x}))$$

for all **x** in **V**.

EXAMPLE 6.11

Let S and T be linear transformations from $\mathbf{R}^3$ to $\mathbf{R}^2$ defined by $S(x, y, z) = (x + y - z, 3y + z)$ and $T(x, y, z) = (2x + z, y + z)$. Describe the transformation $3S - 2T$.

Solution

For any $\mathbf{x} = (x, y, z)$ we follow the definition:

$$
\begin{aligned}
(3S - 2T)(x, y, z) &= (3S)(x, y, z) - (2T)(x, y, z) \\
&= 3(S(x, y, z)) - 2(T(x, y, z)) \\
&= 3(x + y - z, 3y + z) - 2(2x + z, y + z) \\
&= (3x + 3y - 3z, 9y + 3z) - (4x + 2z, 2y + 2z) \\
&= (-x + 3y - 5z, 7y + z).
\end{aligned}
$$

It would be easy to verify directly that the resulting function $3S - 2T$ of Example 6.11 is a linear transformation, but the following theorem makes this unnecessary.

THEOREM 1   Let S and T be linear transformations from **V** to **W**, and let c be a scalar. Then $S + T$, $S - T$, $-S$, and cS are all linear transformations from **V** to **W**.

Proof   We prove only the case of $S + T$ and leave the remaining ones as an exercise. We must check the two defining properties of linear transformation. If **x** and **y** are vectors in **V**, then

$$
\begin{aligned}
(S + T)(\mathbf{x} + \mathbf{y}) &= S(\mathbf{x} + \mathbf{y}) + T(\mathbf{x} + \mathbf{y}) & \text{(definition of } S + T) \\
&= S(\mathbf{x}) + S(\mathbf{y}) + T(\mathbf{x}) + T(\mathbf{y}) & \text{(linearity of } S \text{ and } T) \\
&= S(\mathbf{x}) + T(\mathbf{x}) + S(\mathbf{y}) + T(\mathbf{y}) & \text{(vector algebra)} \\
&= (S + T)(\mathbf{x}) + (S + T)(\mathbf{y}) & \text{(definition of } S + T).
\end{aligned}
$$

If c is any scalar,

$$
\begin{aligned}
(S + T)(c\mathbf{x}) &= S(c\mathbf{x}) + T(c\mathbf{x}) & \text{(definition of } S + T) \\
&= cS(\mathbf{x}) + cT(\mathbf{x}) & \text{(linearity of } S \text{ and } T) \\
&= c(S(\mathbf{x}) + T(\mathbf{x})) & \text{(vector algebra)} \\
&= c((S + T)(\mathbf{x})) & \text{(definition of } S + T).
\end{aligned}
$$

Thus the sum of two linear transformations is a linear transformation.

In general, if $f\colon \mathscr{A} \to \mathscr{B}$ and $g\colon \mathscr{B} \to \mathscr{C}$ are functions, the **composition** of f and g is the function $g \circ f\colon \mathscr{A} \to \mathscr{C}$ defined by $(g \circ f)(a) = g(f(a))$ for all a in $\mathscr{A}$. That is, first apply f to obtain $f(a)$, and then apply g to the result. As usual, we will be concerned only with functions that are linear transformations between vector spaces.

EXAMPLE 6.12   Let $S\colon \mathbf{R}^2 \to \mathbf{R}^3$ be defined by $S(x, y) = (x - y, x + y, 2y)$, and let $T\colon \mathbf{R}^3 \to \mathbf{R}^3$ be defined by $T(x, y, z) = (x, 2x - y, x + y + z)$. Determine $T \circ S$.

Solution   Before computing the general result let us find $(T \circ S)(1, 2)$. First, $S(1, 2) = (-1, 3, 4)$. Now apply T to this result:

$$
(T \circ S)(1, 2) = T(S(1, 2)) = T(-1, 3, 4) = (-1, -5, 6).
$$

The general result is computed in exactly the same way, but unless we change variable names, it is easy to become confused. Let us rewrite T as $T(s, t, u) = (s, 2s - t, s + t + u)$. Now to evaluate T at $S(x, y) = (x - y, x + y, 2y)$, we will have $s = x - y$, $t = x + y$, and $u = 2y$:

$$
\begin{aligned}
(T \circ S)(x, y) &= T(S(x, y)) = T(x - y, x + y, 2y) \\
&= (x - y, 2(x - y) - (x + y), (x - y) + (x + y) + 2y) \\
&= (x - y, x - 3y, 2x + 2y).
\end{aligned}
$$

The resulting description of $T \circ S$ is thus

$$
(T \circ S)(x, y) = (x - y, x - 3y, 2x + 2y).
$$

In Example 6.12 you can check to see that S, T, and $T \circ S$ are all linear transformations. No mention was made of the composition in reverse order, $S \circ T$, since the image of any vector $T(\mathbf{x})$ is in $\mathbf{R}^3$, and S is only defined for vectors in $\mathbf{R}^2$. These observations are stated more generally in the following theorem.

THEOREM 2   Let $S: \mathbf{U} \to \mathbf{V}$ and $T: \mathbf{V} \to \mathbf{W}$ be linear transformations. Then the composition $T \circ S: \mathbf{U} \to \mathbf{W}$ is a linear transformation.

Proof   For any $\mathbf{u}$ in $\mathbf{V}$ and scalar a,

$$
\begin{aligned}
(T \circ S)(a\mathbf{u}) &= T(S(a\mathbf{u})) & \text{(definition of composition)} \\
&= T(aS(\mathbf{u})) & \text{(S is linear)} \\
&= aT(S(\mathbf{u})) & \text{(T is linear)} \\
&= a(T \circ S)(\mathbf{u}) & \text{(definition of composition).}
\end{aligned}
$$

That $(T \circ S)(\mathbf{u} + \mathbf{v}) = (T \circ S)(\mathbf{u}) + (T \circ S)(\mathbf{v})$ for all $\mathbf{u}$ and $\mathbf{v}$ is done in a similar manner; this is left as Exercise 22.

It is common to speak of the *product* of two linear transformations instead of their composition. In this vein, for $S: \mathbf{U} \to \mathbf{V}$ and $T: \mathbf{V} \to \mathbf{W}$, the composition $T \circ S$ is often abbreviated to TS. In general, a linear transformation cannot be composed with itself. In other words, for $T: \mathbf{V} \to \mathbf{W}$, $TT = T \circ T$ is not defined unless $\mathbf{V} = \mathbf{W}$. If $\mathbf{V} = \mathbf{W}$, the composition is defined and can be repeated as often as we wish. The convenient notation T^2, T^3, . . . symbolizes $T \circ T$, $T \circ (T \circ T)$, etc.

DEFINITION

Let $S: \mathbf{U} \to \mathbf{V}$, $T: \mathbf{V} \to \mathbf{W}$, and $L: \mathbf{V} \to \mathbf{V}$ be linear transformations.

    i. The **product** TS is the linear transformation $T \circ S$.

    ii. The **powers** of L are the linear transformations $L^0 = I$, $L^1 = L$, $L^2 = LL$, . . . , $L^{k+1} = L(L^k)$ for all $k = 2, 3, \ldots$.

We now present those properties of linear transformations that correspond to the analogous ones for matrices (Theorem 1 of Section 2.3).

THEOREM 3   Let S, T, and U be linear transformations and a and b scalars. Assume that the linear transformations are such that each of the following operations is defined. Then

    a. $S + T = T + S$ (commutative law for addition),

    b. $S + (T + U) = (S + T) + U$ (associative law for addition),

   c. $S + 0 = S,$

   d. $S + (-S) = 0,$

   e. $S(TU) = (ST)U$ (associative law for multiplication),

   f. $SI = S, IS = S,$

   g. $\left.\begin{array}{l} S(T + U) = ST + SU \\ (S + T)U = SU + TU \end{array}\right\}$ (distributive laws),

   h. $a(S + T) = aS + aT,$

   i. $(a + b)S = aS + bS,$

   j. $(ab)S = a(bS),$

   k. $1S = S,$

   l. $S0 = 0, 0S = 0,$

   m. $a0 = 0,$

   n. $a(ST) = (aS)T.$

SPOTLIGHT

DIFFERENTIATION (FROM CALCULUS)

In calculus you soon learn two of the basic rules for differentiating a function:

   i. "the derivative of a sum is the sum of the derivatives;"

   ii. "the derivative of a constant times a function is the constant times the derivative of the function."

If we denote the derivative of a function f by $D(f)$, then these rules can be more formally stated:

   i. $D(f + g) = D(f) + D(g)$;

   ii. $D(cf) = cD(f)$, where c is a constant.

But these are just the properties we need to claim that a derivative is a linear operator. (For simplicity's sake, we can take its domain and range spaces as $C^{\infty}(-\infty, \infty)$—see the note following Example 5.6.)

   Since the powers of a linear operator are just that transformation applied over and over again, we have:

   $D^2(f) = D(D(f))$, which is the derivative of the derivative—the *second derivative*;

   $D^3(f) = D(D^2(f))$, which is the derivative of the second derivative— the *third derivative*;

and in general, $D^n(f)$ is the nth derivative of the function f.

Proofs We prove parts (a), (e), (g), and (j) and leave the others as an exercise. To show that two linear transformations are equal, we show that they are equal as functions, that they agree at any vector for which they are defined. In these arguments, when we say "for each $\mathbf{x}$," we really mean "for each $\mathbf{x}$ for which the functions are defined."

3a. For each $\mathbf{x}$,

$$
\begin{aligned}
(S + T)(\mathbf{x}) &= S(\mathbf{x}) + T(\mathbf{x}) && \text{(definition of sum)} \\
&= T(\mathbf{y}) + S(\mathbf{x}) && \text{(definition of vector space)} \\
&= (T + S)(\mathbf{x}) && \text{(definition of sum)}.
\end{aligned}
$$

Therefore $S + T = T + S$.

3e. For each $\mathbf{x}$,

$$
\begin{aligned}
S(TU)(\mathbf{x}) &= S(TU(\mathbf{x})) && \text{(definition of product)} \\
&= S(T(U(\mathbf{x}))) && \text{(definition of product)} \\
&= ST(U(\mathbf{x})) = (ST)U(\mathbf{x}) && \text{(definition of product)}.
\end{aligned}
$$

Therefore $S(TU) = (ST)U$.

3g. For each $\mathbf{x}$,

$$
\begin{aligned}
(S(T + U))(\mathbf{x}) &= S((T + U)(\mathbf{x})) && \text{(definition of product)} \\
&= S(T(\mathbf{x}) + U(\mathbf{x})) && \text{(definition of sum)} \\
&= S(T(\mathbf{x})) + S(U(\mathbf{x})) && (S \text{ is a linear transformation)} \\
&= ST(\mathbf{x}) + SU(\mathbf{x}) && \text{(definition of product)} \\
&= (ST + SU)(\mathbf{x}) && \text{(definition of sum)}.
\end{aligned}
$$

Therefore $S(T + U) = ST + SU$. (The other case of 3g is similarly handled.)

3j. For each $\mathbf{x}$,

$$
\begin{aligned}
(ab)S(\mathbf{x}) &= (ab)(S(\mathbf{x})) && \text{(definition of multiplication by scalar)} \\
&= a(bS(\mathbf{x})) && \text{(definition of vector space)} \\
&= (a(bS))(\mathbf{x}) && \text{(definition of multiplication by scalar)}.
\end{aligned}
$$

Therefore $(ab)S = a(bS)$.

MATRIX OF A LINEAR TRANSFORMATION FROM $\mathbf{R}^n$ TO $\mathbf{R}^m$

In the first section of this chapter matrices were used to provide examples of linear transformations, while other linear transformations were presented by different means. In this section we will show that *every* linear transformation from $\mathbf{R}^n$ to $\mathbf{R}^m$ may be viewed as multiplication by an associated matrix. (In

Section 6.4 we will see that matrices can be used to represent linear transformations between general finite dimensional vector spaces.) We begin with a theorem that gives the procedure for finding the matrix just described.

THEOREM 4　Let $T: \mathbf{R}^n \to \mathbf{R}^m$ be a linear transformation, and let A be the $m \times n$ matrix with the ith column given by $\mathbf{a}^i = T(\mathbf{e}_i)$. Then A is the unique matrix for which $A\mathbf{x} = T(\mathbf{x})$ for all $\mathbf{x}$ in $\mathbf{R}^n$.

Proof　Since $\mathbf{a}^i = (a_{1i}, \ldots, a_{mi}) = T(\mathbf{e}_i)$ for each $i = 1, \ldots, n$, we have the following computation:

$$A\mathbf{e}_i = \begin{bmatrix} a_{11} & \cdots & a_{1i} & \cdots & a_{1n} \\ a_{21} & \cdots & a_{2i} & \cdots & a_{2n} \\ \vdots & & \vdots & & \vdots \\ a_{m1} & \cdots & a_{mi} & \cdots & a_{mn} \end{bmatrix} \begin{bmatrix} 0 \\ \vdots \\ 0 \\ 1 \\ 0 \\ \vdots \\ 0 \end{bmatrix}$$

$$= \begin{bmatrix} a_{1i} \\ a_{2i} \\ \vdots \\ a_{mi} \end{bmatrix} = (a_{1i}, \ldots, a_{mi}) = T(\mathbf{e}_i).$$

Thus multiplication on the left by A agrees with the linear transformation T on a basis for $\mathbf{R}^n$ (the standard one) and is itself a linear transformation (Theorem 2 of Section 6.1). By Theorem 3 of Section 6.1, $A\mathbf{x} = T(\mathbf{x})$ for all $\mathbf{x}$ in $\mathbf{R}^n$.

To show uniqueness, suppose B is a matrix such that $B\mathbf{x} = T(\mathbf{x})$ for all $\mathbf{x}$ in $\mathbf{R}^n$. Then, in particular, $B\mathbf{e}_i = T(\mathbf{e}_i)$ for $i = 1, \ldots, n$. But $B\mathbf{e}_i = \mathbf{b}^i$ and $T(\mathbf{e}_i) = \mathbf{a}^i$, which means the ith column of B is the ith column of A for each i. Thus $B = A$.

The matrix A of Theorem 4 is called the matrix **associated with** T or the matrix that **represents** T.

EXAMPLE 6.13

Let $T: \mathbf{R}^2 \to \mathbf{R}^3$ be the linear transformation of Example 6.1. For each $\mathbf{x} = (x, y)$ in $\mathbf{R}^2$, $T(x, y) = (x + y, x, 2x - y)$. Find the matrix that represents T.

Solution

We have already shown that T is linear, so we proceed with the computation of the associated matrix A. By Theorem 4, A is the 3×2 matrix with:

$$\mathbf{a}^1 = T(\mathbf{e}_1) = T(1, 0) = (1, 1, 2),$$
$$\mathbf{a}^2 = T(\mathbf{e}_2) = T(0, 1) = (1, 0, -1).$$

Thus the matrix A is given by

$$A = \begin{bmatrix} 1 & 1 \\ 1 & 0 \\ 2 & -1 \end{bmatrix}.$$

By Theorem 4, we do not need to verify that this matrix A represents T. To emphasize the point, however, we do it anyway. For any $\mathbf{x} = (x, y)$ in $\mathbf{R}^2$

$$A\mathbf{x} = \begin{bmatrix} 1 & 1 \\ 1 & 0 \\ 2 & -1 \end{bmatrix} \begin{bmatrix} x \\ y \end{bmatrix} = \begin{bmatrix} x + y \\ x \\ 2x - y \end{bmatrix}.$$

The resulting vector is exactly $T(x, y)$.

NOTE   Before applying Theorem 4, it is essential to check that the function T is indeed a linear transformation. Following the given procedure will always produce a matrix, even for nonlinear functions. However, unless the function is linear, the resulting matrix will not represent the function (see Exercise 28).

....

EXAMPLE 6.14   Find the matrix associated with the identity transformation $I: \mathbf{R}^n \rightarrow \mathbf{R}^n$.

Solution   By definition, $I(\mathbf{x}) = \mathbf{x}$ for each $\mathbf{x}$ in $\mathbf{R}^n$. Then, in particular, $I(\mathbf{e}_i) = \mathbf{e}_i = \mathbf{a}^i$ for each i, and thus the matrix A associated with I is the matrix with ith column equal to $\mathbf{e}_i$ for each i. Thus, $A = I$, the $n \times n$ identity matrix.

NOTE   The identity *transformation* $I: \mathbf{R}^n \rightarrow \mathbf{R}^n$ is represented by the $n \times n$ identity *matrix* I. Because of their close relationship, using the same symbol for both will present no difficulties. Similarly, the zero transformation $0: \mathbf{R}^n \rightarrow \mathbf{R}^m$ is represented by the $m \times n$ zero matrix, 0. Again, the slight ambiguity is no problem.

....

Since the sum, difference, negative, and scalar product functions are linear transformations, there is a matrix that represents each of them. These associated matrices are compatible with the corresponding operation in the following sense.

THEOREM 5   Let S and T be linear transformations from $\mathbf{R}^n$ to $\mathbf{R}^m$, and let their associated matrices be A and B, respectively. Then the associated matrix of the linear transformation

    i. $S + T$ is $A + B$,

    ii. $S - T$ is $A - B$,

    iii. $-S$ is $-A$,

    iv. cS is cA.

Proof   Again we prove only the case of $S + T$ and leave the remaining ones as an exercise. For any $\mathbf{x}$ in $\mathbf{R}^n$,

$$
\begin{aligned}
(A + B)\mathbf{x} &= A\mathbf{x} + B\mathbf{x} &&\text{(Theorem 1g of Section 2.3)} \\
&= S(\mathbf{x}) + T(\mathbf{x}) &&\text{(A and B represent S and T)} \\
&= (S + T)(\mathbf{x}) &&\text{(definition of $S + T$).}
\end{aligned}
$$

Therefore the matrix $A + B$ represents the linear transformation $S + T$.

EXAMPLE 6.15   Let S and T be the linear transformations of Example 6.11. Let A, B, and C be the matrices that represent S, T, and $3S - 2T$, respectively. Verify directly that $C = 3A - 2B$.

Solution   Following Theorem 4, we have from Example 6.11 that

$$
A = \begin{bmatrix} 1 & 1 & -1 \\ 0 & 3 & 1 \end{bmatrix}, \quad B = \begin{bmatrix} 2 & 0 & 1 \\ 0 & 1 & 1 \end{bmatrix}, \quad \text{and} \quad C = \begin{bmatrix} -1 & 3 & -5 \\ 0 & 7 & 1 \end{bmatrix}.
$$

We now compute $3A - 2B$ directly:

$$
3A - 2B = 3\begin{bmatrix} 1 & 1 & -1 \\ 0 & 3 & 1 \end{bmatrix} - 2\begin{bmatrix} 2 & 0 & 1 \\ 0 & 1 & 1 \end{bmatrix} = \begin{bmatrix} -1 & 3 & -5 \\ 0 & 7 & 1 \end{bmatrix}
$$

The resulting matrix is C, as required.

Since the composition of two linear transformations is a linear transformation, and since linear transformations can be represented by matrices (Theorem 4), it is natural to inquire about the nature of the matrix that represents the composition. The next theorem describes this very important situation.

THEOREM 6   Let $S: \mathbf{R}^n \rightarrow \mathbf{R}^m$ and $T: \mathbf{R}^m \rightarrow \mathbf{R}^l$ be linear transformations, and let A be the matrix that represents S and B the matrix that represents T. Then the matrix product BA represents the composition $T \circ S$.

Proof   Since B is $l \times m$, and A is $m \times n$, the matrix product BA is defined and is an $l \times n$ matrix as required ($T \circ S$ is a linear transformation from $\mathbf{R}^n$ to $\mathbf{R}^l$). For any $\mathbf{x}$ in $\mathbf{R}^n$,

$$
\begin{aligned}
(BA)\mathbf{x} &= (B)(A\mathbf{x}) & \text{(matrix multiplication is associative)} \\
&= (B)(S(\mathbf{x})) & (A \text{ represents } S) \\
&= T(S(\mathbf{x})) & (B \text{ represents } T) \\
&= (T \circ S)(\mathbf{x}) & \text{(definition of composition).}
\end{aligned}
$$

Thus multiplication on the left by BA has exactly the same effect as mapping by the linear transformation $T \circ S$, and BA represents $T \circ S$.

EXAMPLE 6.16

Let S and T be as in Example 6.12. Find the matrices A, B, and C that represent S, T, and $T \circ S$, respectively, and verify that $C = BA$.

Solution

From Example 6.12,

$$
\begin{aligned}
S(x, y) &= (x - y, x + y, 2y), \\
T(x, y, z) &= (x, 2x - y, x + y + z), \\
(T \circ S)(x, y) &= (x - y, x - 3y, 2x + 2y).
\end{aligned}
$$

Following the procedure described in Theorem 4 for each of these linear transformations,

$$
A = \begin{bmatrix} 1 & -1 \\ 1 & 1 \\ 0 & 2 \end{bmatrix}, \quad
B = \begin{bmatrix} 1 & 0 & 0 \\ 2 & -1 & 0 \\ 1 & 1 & 1 \end{bmatrix}, \quad \text{and} \quad
C = \begin{bmatrix} 1 & -1 \\ 1 & -3 \\ 2 & 2 \end{bmatrix}.
$$

The essence of Theorem 6 is that it is unnecessary to evaluate $T \circ S$ first and then compute its corresponding matrix; exactly the same result may be obtained by computing A and B and then $C = BA$. Performing this multiplication, we do indeed have

$$
BA = \begin{bmatrix} 1 & 0 & 0 \\ 2 & -1 & 0 \\ 1 & 1 & 1 \end{bmatrix} \begin{bmatrix} 1 & -1 \\ 1 & 1 \\ 0 & 2 \end{bmatrix} = \begin{bmatrix} 1 & -1 \\ 1 & -3 \\ 2 & 2 \end{bmatrix} = C.
$$

EXAMPLE 6.17

Use Theorem 6 to show that the composition of a rotation of $\mathbf{R}^2$ by an angle θ followed by a rotation by an angle ψ is a rotation by the angle $\theta + \psi$. (Obviously, this is true geometrically.)

Solution

From Example 6.7, we know that the rotation by θ is represented by the matrix

$$
A_\theta = \begin{bmatrix} \cos\theta & -\sin\theta \\ \sin\theta & \cos\theta \end{bmatrix},
$$

and the rotation by ψ is represented by the matrix

$$A_\psi = \begin{bmatrix} \cos \psi & -\sin \psi \\ \sin \psi & \cos \psi \end{bmatrix}.$$

By Theorem 6, the matrix that represents the composition is the product $A_\psi A_\theta$. Computing the product, we have

$$A_\psi A_\theta = \begin{bmatrix} \cos \psi \cos \theta - \sin \psi \sin \theta & -\cos \psi \sin \theta - \sin \psi \cos \theta \\ \sin \psi \cos \theta + \cos \psi \sin \theta & -\sin \psi \sin \theta + \cos \psi \cos \theta \end{bmatrix}$$

$$= \begin{bmatrix} \cos (\theta + \psi) & -\sin (\theta + \psi) \\ \sin (\theta + \psi) & \cos (\theta + \psi) \end{bmatrix} = A_{\theta + \psi}.$$

It is left as an exercise to show that the composition of two *reflections* is also a *rotation*.

EXAMPLE 6.18   Let $T: \mathbf{R}^3 \to \mathbf{R}^3$ be defined by $T(x, y, z) = (x + 2y - z, 3x - z, y + z)$. Describe the transformation $2T^2 - T + 3I$.

Solution   We use the algebra of the corresponding matrices to simplify the computation. Let A represent T, and note that the identity matrix I of order 3 represents I. Then $2A^2 - A + 3I$ represents $2T^2 - T + 3I$. By Theorem 4, A is given by

$$A = \begin{bmatrix} 1 & 2 & -1 \\ 3 & 0 & -1 \\ 0 & 1 & 1 \end{bmatrix},$$

and hence

$$2A^2 - A + 3I = 2\begin{bmatrix} 1 & 2 & -1 \\ 3 & 0 & -1 \\ 0 & 1 & 1 \end{bmatrix}^2 - \begin{bmatrix} 1 & 2 & -1 \\ 3 & 0 & -1 \\ 0 & 1 & 1 \end{bmatrix} + 3\begin{bmatrix} 1 & 0 & 0 \\ 0 & 1 & 0 \\ 0 & 0 & 1 \end{bmatrix}$$

$$= \begin{bmatrix} 14 & 2 & -8 \\ 6 & 10 & -8 \\ 6 & 2 & 0 \end{bmatrix} - \begin{bmatrix} 1 & 2 & -1 \\ 3 & 0 & -1 \\ 0 & 1 & 1 \end{bmatrix} + \begin{bmatrix} 3 & 0 & 0 \\ 0 & 3 & 0 \\ 0 & 0 & 3 \end{bmatrix}$$

$$= \begin{bmatrix} 16 & 0 & -7 \\ 3 & 13 & -7 \\ 6 & 1 & 2 \end{bmatrix}.$$

Finally,

$$(2T^2 - T + 3I)(x, y, z) = \begin{bmatrix} 16 & 0 & -7 \\ 3 & 13 & -7 \\ 6 & 1 & 2 \end{bmatrix}\begin{bmatrix} x \\ y \\ z \end{bmatrix} = \begin{bmatrix} 16x & - 7z \\ 3x + 13y - 7z \\ 6x + y + 2z \end{bmatrix},$$

so that $(2T^2 - T + 3I)(x, y, z) = (16x - 7z, 3x + 13y - 7z, 6x + y + 2z)$.

6.2 EXERCISES

The following functions are all linear transformations. Use them in Exercises 1–20.

$K(x, y, z) = (x, x + y, x + y + z)$

$L(x, y, z) = (2x - y, x + 2y)$

$S(x, y, z) = (z, y, x)$

$T(x, y) = (2x + y, x + y, x - y, x - 2y)$

$G: \mathbf{R}^3 \to \mathbf{M}^{2,2}$ defined by $G(x, y, z) = \begin{bmatrix} y & z \\ -x + y & -x \end{bmatrix}$

$H: \mathbf{M}^{2,2} \to \mathbf{P}_2$ defined by $H\left(\begin{bmatrix} a & b \\ c & d \end{bmatrix}\right) = (a + b) + (a - 2d)x + dx^2$

In Exercises 1–12 find the indicated linear transformation if it is defined. If it is not defined, explain why it is not.

1. $LK (= L \circ K)$

2. TL

3. S^2

4. $K + S$

5. T^2

6. GS

7. HG

8. $L + S$

9. $3K - 2S$

10. $2S^3 - S^2 + 3S - 4I$

11. $3H$

12. HGK

13–16. Find the matrix that represents each of the linear transformations K, L, S, and T.

17–20. Find the matrix that represents each of the linear transformations in Exercises 1–4 in two ways, first using the result of the exercise and then using the result of Exercises 13–16.

21. Show that the composition of one reflection T^θ followed by another T^ψ is the rotation by the angle $2(\psi - \theta)$.

22. Finish the proof of Theorem 2 by showing that if S and T are linear transformations such that $T \circ S$ is defined, then $(T \circ S)(\mathbf{u} + \mathbf{v}) = (T \circ S)\mathbf{u} + (T \circ S)\mathbf{v}$.

23. Complete the proof of Theorem 1.

24. Complete the proof of Theorem 3.

25. Complete the proof of Theorem 5.

26. Show that the composition of a contraction/dilation by a contraction/dilation is a contraction/dilation.

27. If $S: \mathbf{R}^n \to \mathbf{R}^n$ is a linear transformation, and $T_c: \mathbf{R}^n \to \mathbf{R}^n$ is a contraction/dilation, prove that $ST_c = T_cS$. In other words, show that a contraction/dilation *commutes* with any linear operator.

28. Let $f: \mathbf{R}^2 \to \mathbf{R}^2$ be the *nonlinear* function defined by $f(x, y) = (xy, x^2)$. Let A be the 2×2 matrix constructed as in Theorem 4. Show that A does *not* represent f. In other words, show that for some $\mathbf{x}$ in $\mathbf{R}^2$, $f(\mathbf{x}) \neq A\mathbf{x}$.

6.3 KERNEL AND IMAGE

In this section we will investigate further properties of linear transformations, as well as two very important subspaces associated with them.

KERNEL OF A LINEAR TRANSFORMATION

If T is a linear transformation, Theorem 1 of Section 6.1 showed us that $T(0) = 0$. There may or may not be other vectors x such that $T(x) = 0$.

DEFINITION

Let $T: V \rightarrow W$ be a linear transformation. The set of all vectors x in V that satisfy $T(x) = 0$ is called the **kernel** (or **nullspace**) of T and is denoted by Ker (T).

EXAMPLE 6.19

Computer Ker (T) for $T: \mathbf{R}^2 \rightarrow \mathbf{R}^3$ given by $T(x, y) = (x + y, x, 2x - y)$.

Solution

That the transformation T is linear has been shown in Example 6.1. To find Ker (T), we must solve the equation $T(x, y) = (0, 0, 0)$ for all possible ordered pairs (x, y). That is, we solve

$$(x + y, x, 2x - y) = (0, 0, 0).$$

Equating components, we have

$$x + y = 0$$
$$x \quad\;\; = 0$$
$$2x - y = 0.$$

This homogeneous system of linear equations can easily be solved by inspection. The second equation implies that $x = 0$, which in turn implies that $y = 0$, from the first equation. Thus Ker $(T) = \{0\}$, the set that consists solely of the zero vector.

EXAMPLE 6.20

Let 0 be the zero transformation from V to W, that is, $0(x) = 0$ for each x in V. Compute Ker (0).

Solution

That 0 is a linear transformation was verified in Example 6.2. Since every x in V gets mapped into 0, Ker $(0) = V$.

EXAMPLE 6.21

Describe Ker (T), where $T: \mathbf{R}^3 \rightarrow \mathbf{R}^2$ is the linear transformation given by $T(x) = Ax$ and

$$A = \begin{bmatrix} 2 & 0 & 1 \\ 1 & 1 & -1 \end{bmatrix}.$$

Solution

Since T is given by matrix multiplication, it is linear (by Theorem 2 of Section 6.1). To compute Ker (T), we let $\mathbf{x} = (x, y, z)$, determine $T(\mathbf{x})$, and set it equal to zero.

$$T(\mathbf{x}) = \begin{bmatrix} 2 & 0 & 1 \\ 1 & 1 & -1 \end{bmatrix} \begin{bmatrix} x \\ y \\ z \end{bmatrix} = \begin{bmatrix} 2x & + z \\ x + y & - z \end{bmatrix} = \begin{bmatrix} 0 \\ 0 \end{bmatrix}.$$

Equating components, we have

$$2x \qquad + z = 0$$
$$x + y - z = 0.$$

We solve the resulting homogeneous system by the methods of Chapter 2, forming the augmented matrix and transforming it to row-reduced echelon form:

$$\begin{bmatrix} 2 & 0 & 1 & | & 0 \\ 1 & 1 & -1 & | & 0 \end{bmatrix} \rightarrow \begin{bmatrix} 1 & 0 & \frac{1}{2} & | & 0 \\ 0 & 1 & -\frac{3}{2} & | & 0 \end{bmatrix}.$$

Thus

$$x = -\tfrac{1}{2}t,$$
$$y = \tfrac{3}{2}t,$$
$$z = t,$$

and we have a one-parameter family of solutions. Expressing this result in vector form,

$$\mathbf{x} = t(-\tfrac{1}{2}, \tfrac{3}{2}, 1),$$

we see that Ker (T) is the line through the origin determined by the vector $(-\tfrac{1}{2}, \tfrac{3}{2}, 1)$ (or more conveniently, by $(-1, 3, 2)$).

EXAMPLE 6.22

Compute Ker (T) for $T: \mathbf{P}_2 \rightarrow \mathbf{M}^{2,2}$ given by

$$T(a_0 + a_1x + a_2x^2) = \begin{bmatrix} a_1 & a_1 - a_0 \\ a_1 + a_0 & 2a_1 - a_2 \end{bmatrix}.$$

(That T is a linear transformation is left as an exercise.)

Solution

By definition of Ker (T), we must solve the equation

$$T(a_0 + a_1x + a_2x^2) = \begin{bmatrix} 0 & 0 \\ 0 & 0 \end{bmatrix}.$$

From the rule for T, this is equivalent to

$$\begin{bmatrix} a_1 & a_1 - a_0 \\ a_1 + a_0 & 2a_1 - a_2 \end{bmatrix} = \begin{bmatrix} 0 & 0 \\ 0 & 0 \end{bmatrix},$$

or equating coefficients,

$$
\begin{aligned}
a_1 &= 0 \\
-a_0 + a_1 &= 0 \\
a_0 + a_1 &= 0 \\
2a_1 - a_2 &= 0.
\end{aligned}
$$

Solving this system, we see that the only polynomials $\mathbf{p}(x) = a_0 + a_1 x + a_2 x^2$ in $\mathbf{P}_2$ for which $T(\mathbf{p}) = \mathbf{0}$ are those for which $a_0 = a_1 = a_2 = 0$. That is, $\mathbf{p} = \mathbf{0}$, so $\operatorname{Ker}(T) = \{\mathbf{0}\}$.

In each of these four examples, the kernel has been a subspace of $\mathbf{V}$. The following theorem shows that the kernel of a linear transformation is *always* a subspace. The dimension of the kernel is called the **nullity** of the linear transformation.

THEOREM I　　Let $T : \mathbf{V} \to \mathbf{W}$ be a linear transformation. Then the kernel of T is a subspace of $\mathbf{V}$.

Proof　　That $\operatorname{Ker}(T)$ is a *subset* of $\mathbf{V}$ is clear from the definition.
For $\mathbf{u}$ and $\mathbf{v}$ in $\operatorname{Ker}(T)$, we compute $T(\mathbf{u} + \mathbf{v})$:

$$T(\mathbf{u} + \mathbf{v}) = T(\mathbf{u}) + T(\mathbf{v}) = \mathbf{0} + \mathbf{0} = \mathbf{0}.$$

Thus $\mathbf{u} + \mathbf{v}$ is in $\operatorname{Ker}(T)$ as well.
For $\mathbf{u}$ in $\operatorname{Ker}(T)$ and c any scalar,

$$T(c\mathbf{u}) = cT(\mathbf{u}) = c\mathbf{0} = \mathbf{0}.$$

Thus $c\mathbf{u}$ is also in $\operatorname{Ker}(T)$.
These two computations confirm that the kernel of a linear transformation is a subspace of $\mathbf{V}$.

If the linear transformation is given by a matrix as in Example 6.21, computation of the kernel is particularly easy. Note that in this example we obtained the kernel by augmenting the matrix A by a column of zeros and row-reducing. That this is always the case is simply a statement of the fact that $T(\mathbf{x}) = \mathbf{0}$ if and only if $A\mathbf{x} = \mathbf{0}$. The latter equation may be solved by row-reducing the augmented matrix $[A \mid \mathbf{0}]$, and a basis for the kernel may be produced by following the procedure of Theorem 4 of Section 4.3. This observation is stated as the next theorem. Recall that the *rank* of a matrix is the number of nonzero rows in its row-reduced echelon form.

THEOREM 2   Let $T: \mathbf{R}^n \to \mathbf{R}^m$ be a linear transformation, and let A be the matrix that represents T. Then the kernel of T is the solution space of the vector equation $A\mathbf{x} = \mathbf{0}$. Thus dim (Ker (T)), the nullity of T, is $n - r$ where r is the rank of the matrix A.

PROCEDURE FOR COMPUTING A BASIS FOR THE KERNEL OF A TRANSFORMATION GIVEN BY A MATRIX A

Solve $A\mathbf{x} = \mathbf{0}$ by transforming the matrix A to row-reduced echelon form, and then follow the procedure of Theorem 4 of Section 4.3. The result is a basis for Ker (T).

EXAMPLE 6.23   Let $T: \mathbf{R}^5 \to \mathbf{R}^4$ be given by $T(\mathbf{x}) = A\mathbf{x}$, where A is the matrix

$$A = \begin{bmatrix} 1 & 2 & 0 & 1 & 0 \\ 2 & 4 & 1 & 0 & 0 \\ 0 & 0 & 1 & -2 & 1 \\ 1 & 2 & 1 & -1 & 1 \end{bmatrix}.$$

Find the nullity of T and a basis for Ker (T).

Solution   The row-reduced echelon form of A is the matrix

$$B = \begin{bmatrix} 1 & 2 & 0 & 1 & 0 \\ 0 & 0 & 1 & -2 & 0 \\ 0 & 0 & 0 & 0 & 1 \\ 0 & 0 & 0 & 0 & 0 \end{bmatrix}.$$

Since B has three nonzero rows, the rank of A is $r = 3$. Since $n = 5$ in this example, the nullity is $n - r = 2$. Following the procedure of Theorem 4 of Section 4.3, we determine that a basis for Ker (T) is $\{(-2, 1, 0, 0, 0), (-1, 0, 2, 1, 0)\}$.

A function $f: \mathscr{A} \to \mathscr{B}$ is **one-to-one** if, whenever a_1 and a_2 satisfy $f(a_1) = f(a_2)$, then $a_1 = a_2$. In other words, no two elements of $\mathscr{A}$ are mapped to the same element of $\mathscr{B}$. It is particularly easy to check to see if a *linear transformation* is one-to-one.

THEOREM 3   Let $T: \mathbf{V} \to \mathbf{W}$ be a linear transformation. Then T is one-to-one if and only if $\text{Ker}\,(T) = \{\mathbf{0}\}$.

Proof   If T is one-to-one, then $T(\mathbf{x}) = \mathbf{0}$ can have only one solution, namely $\mathbf{x} = \mathbf{0}$. That is, $\text{Ker}\,(T) = \{\mathbf{0}\}$. Conversely suppose $\text{Ker}\,(T) = \{\mathbf{0}\}$ and $T(\mathbf{u}) = T(\mathbf{v})$. We need to show that $\mathbf{u} = \mathbf{v}$. Since $T(\mathbf{u}) = T(\mathbf{v})$,
$$T(\mathbf{u}) - T(\mathbf{v}) = \mathbf{0},$$
and therefore $T(\mathbf{u} - \mathbf{v}) = \mathbf{0}$ (Theorem 1c of Section 6.1). Thus $\mathbf{u} - \mathbf{v}$ is in $\text{Ker}\,(T)$. But then $\mathbf{u} - \mathbf{v} = \mathbf{0}$ since $\mathbf{0}$ is the *only* element of $\text{Ker}\,(T)$. Hence $\mathbf{u} = \mathbf{v}$, and T is one-to-one.

From the computations given in Example 6.19, we see that the linear transformation $T(x, y) = (x + y, x, 2x - y)$ is one-to-one. On the other hand, the linear transformation $T(\mathbf{x}) = A\mathbf{x}$ given in Example 6.23 is not.

IMAGE OF A LINEAR TRANSFORMATION

Let $T: \mathbf{V} \to \mathbf{W}$ be a linear transformation. The set of all vectors $T(\mathbf{x})$ is called the **image** of T and is denoted by $\text{Im}\,(T)$.

EXAMPLE 6.24   Describe $\text{Im}\,(0)$, where 0 is the zero transformation from $\mathbf{V}$ to $\mathbf{W}$.

Solution   Since $0(\mathbf{x}) = \mathbf{0}$ for all $\mathbf{x}$ in $\mathbf{V}$, $\text{Im}\,(0) = \{\mathbf{0}\}$.

EXAMPLE 6.25   Describe $\text{Im}\,(T)$, where $T: \mathbf{R} \to \mathbf{R}^2$ is given by $T(x) = (x, 2x)$ for each x in $\mathbf{R}$.

Solution   Letting $T(x) = (x, y)$, we have $y = 2x$ for each x in the real numbers $\mathbf{R}$. Thus $\text{Im}\,(T)$ is the set of points on the line in $\mathbf{R}^2$ that passes through the origin and has slope 2.

EXAMPLE 6.26   Describe $\text{Im}\,(T)$ where $T: \mathbf{P}_2 \to \mathbf{M}^{2,2}$ is given by
$$T(a_0 + a_1 x + a_2 x^2) = \begin{bmatrix} a_1 & a_1 - a_0 \\ a_1 + a_0 & 2a_1 - a_2 \end{bmatrix},$$
which is the transformation of Example 6.22.

Solution   We express $T(a_0 + a_1 x + a_2 x^2)$ in the form
$$\begin{bmatrix} a_1 & a_1 - a_0 \\ a_1 + a_0 & 2a_1 - a_2 \end{bmatrix} = a_0 \begin{bmatrix} 0 & -1 \\ 1 & 0 \end{bmatrix} + a_1 \begin{bmatrix} 1 & 1 \\ 1 & 2 \end{bmatrix} + a_2 \begin{bmatrix} 0 & 0 \\ 0 & -1 \end{bmatrix}.$$

Since a_0, a_1, and a_2 are arbitrary, Im (T) is the set spanned by $\mathcal{S}$ where

$$\mathcal{S} = \left\{ \begin{bmatrix} 0 & -1 \\ 1 & 0 \end{bmatrix}, \begin{bmatrix} 1 & 2 \\ 1 & 2 \end{bmatrix}, \begin{bmatrix} 0 & 0 \\ 0 & -1 \end{bmatrix} \right\}.$$

Since $\mathcal{S}$ is linearly independent, Im (T) is a three-dimensional subspace of the four-dimensional vector space $\mathbf{M}^{2,2}$.

Just as the kernel of a linear transformation $T: \mathbf{V} \to \mathbf{W}$ is a subspace of $\mathbf{V}$ (Theorem 1), the image of T is a subspace of $\mathbf{W}$. The dimension of the image is called the **rank** of T.

THEOREM 4   Let $T: \mathbf{V} \to \mathbf{W}$ be a linear transformation. Then the image of T is a subspace of $\mathbf{W}$.

Proof   That Im (T) is a subset of $\mathbf{W}$ is clear from the definition.

For $\mathbf{u}$ and $\mathbf{v}$ in Im (T), we must show that $\mathbf{u} + \mathbf{v}$ is also in Im (T). Since $\mathbf{u}$ is in Im (T), there is some $\mathbf{x}$ in $\mathbf{V}$ such that $T(\mathbf{x}) = \mathbf{u}$. Similarly, there is some $\mathbf{y}$ in $\mathbf{V}$ such that $T(\mathbf{y}) = \mathbf{v}$. Let $\mathbf{z} = \mathbf{x} + \mathbf{y}$. Then

$$T(\mathbf{z}) = T(\mathbf{x} + \mathbf{y}) = T(\mathbf{x}) + T(\mathbf{y}) = \mathbf{u} + \mathbf{v},$$

and $\mathbf{u} + \mathbf{v}$ is the image of $\mathbf{z}$ under T. That is, $\mathbf{u} + \mathbf{v}$ is in Im (T).

For any scalar c, $T(c\mathbf{x}) = cT(\mathbf{x}) = c\mathbf{u}$. Thus $c\mathbf{u}$ is the image of $c\mathbf{x}$ under T. That is, $c\mathbf{u}$ is in Im (T).

These two computations confirm that the image of a linear transformation is a subspace of $\mathbf{W}$.

EXAMPLE 6.27

Describe Im (T), where $T: \mathbf{R}^3 \to \mathbf{R}^3$ is given by $T(\mathbf{x}) = A(\mathbf{x})$ and

$$A = \begin{bmatrix} 2 & 1 & 1 \\ 1 & -1 & 0 \\ 1 & 2 & 1 \end{bmatrix}.$$

Solution

Letting $\mathbf{x} = (x, y, z)$, we compute $T(\mathbf{x})$:

$$T(\mathbf{x}) = A\mathbf{x} = \begin{bmatrix} 2 & 1 & 1 \\ 1 & -1 & 0 \\ 1 & 2 & 1 \end{bmatrix} \begin{bmatrix} x \\ y \\ z \end{bmatrix} = \begin{bmatrix} 2x + y + z \\ x - y \\ x + 2y + z \end{bmatrix}$$

$$= x \begin{bmatrix} 2 \\ 1 \\ 1 \end{bmatrix} + y \begin{bmatrix} 1 \\ -1 \\ 2 \end{bmatrix} + z \begin{bmatrix} 1 \\ 0 \\ 1 \end{bmatrix}.$$

Thus we see that for any $\mathbf{x}$ in $\mathbf{R}^3$, $T(\mathbf{x})$ is a linear combination of the vectors $(2, 1, 1)$, $(1, -1, 2)$, and $(1, 0, 1)$. Furthermore, any given linear combination of these,

$$\mathbf{y} = a(2, 1, 1) + b(1, -1, 2) + c(1, 0, 1),$$

can be seen to be an image of some element $\mathbf{x}$ in $\mathbf{R}^3$ under T by letting $\mathbf{x} = (a, b, c)$:

$$T(a, b, c) = a(2, 1, 1) + b(1, -1, 2) + c(1, 0, 1).$$

Therefore the set of vectors $\{(2, 1, 1), (1, -1, 2), (1, 0, 1)\}$ generates $\text{Im}\,(T)$. Knowing a spanning set for $\text{Im}\,(T)$, we can produce a basis for it, following Theorem 1 of Section 4.3. We form the matrix with the spanning set as columns and row-reduce:

$$\begin{bmatrix} 2 & 1 & 1 \\ 1 & -1 & 0 \\ 1 & 2 & 1 \end{bmatrix} \rightarrow \begin{bmatrix} 1 & 0 & \frac{1}{3} \\ 0 & 1 & \frac{1}{3} \\ 0 & 0 & 0 \end{bmatrix}.$$

Thus a basis for $\text{Im}\,(T)$ is given by the first two columns, $\{(2, 1, 1), (1, -1, 2)\}$. From the geometric point of view, $\text{Im}\,(T)$ is a plane in $\mathbf{R}^3$.

For the remainder of this section, we will be making use of the matrix of a linear transformation, so the discussion will be restricted to Euclidean m-space. Then, in Section 6.4, we will extend all the results obtained here to general vector spaces.

You probably noticed that the matrix of columns we needed to row-reduce in Example 6.27 was the matrix A that defined the linear transformation T at the beginning of the example. The following theorem shows that this is no coincidence. Recall that the *column space* of a matrix A is the set of all linear combinations of its columns (see Section 4.4).

THEOREM 5   Let $T: \mathbf{R}^n \to \mathbf{R}^m$ be a linear transformation, and let A be the matrix that represents T. Then $\text{Im}\,(T)$ is the column space of A, and $\dim\,(\text{Im}\,(T))$, the rank of the transformation T, is equal to the rank of the matrix A.

PROCEDURE FOR COMPUTING A BASIS FOR THE IMAGE OF A TRANSFORMATION GIVEN BY A MATRIX A

Transform the matrix A to its row-reduced echelon matrix form, B. A basis for $\text{Im}\,(T)$ consists of those columns of A that are transformed into the distinct elementary columns of B.

Proof Let $\mathbf{x} = (x_1, \ldots, x_n)$. Then we have

$$T(\mathbf{x}) = A\mathbf{x} = \begin{bmatrix} a_{11} & a_{12} & \cdots & a_{1n} \\ a_{21} & a_{22} & \cdots & a_{2n} \\ \vdots & \vdots & & \vdots \\ a_{m1} & a_{m2} & \cdots & a_{mn} \end{bmatrix} \begin{bmatrix} x_1 \\ x_2 \\ \vdots \\ x_n \end{bmatrix}$$

$$= \begin{bmatrix} a_{11}x_1 + a_{12}x_2 + \cdots + a_{1n}x_n \\ a_{21}x_1 + a_{22}x_2 + \cdots + a_{2n}x_n \\ \vdots & \vdots & & \vdots \\ a_{m1}x_1 + a_{m2}x_2 + \cdots + a_{mn}x_n \end{bmatrix}$$

$$= x_1 \begin{bmatrix} a_{11} \\ a_{21} \\ \vdots \\ a_{m1} \end{bmatrix} + x_2 \begin{bmatrix} a_{12} \\ a_{22} \\ \vdots \\ a_{m2} \end{bmatrix} + \cdots + x_n \begin{bmatrix} a_{1n} \\ a_{2n} \\ \vdots \\ a_{mn} \end{bmatrix}$$

$$= x_1\mathbf{a}^1 + x_2\mathbf{a}^2 + \cdots + x_n\mathbf{a}^n,$$

where $\mathbf{a}^i$ denotes the ith column of A. In other words, for each $\mathbf{x}$ in $\mathbf{R}^n$, $T(\mathbf{x})$ is a linear combination of the columns of A. Conversely, if $\mathbf{y}$ in $\mathbf{R}^m$ is a linear combination of the columns of A,

$$\mathbf{y} = c_1\mathbf{a}^1 + c_2\mathbf{a}^2 + \cdots + c_n\mathbf{a}^n,$$

then $\mathbf{y} = T(\mathbf{c})$, where $\mathbf{c} = (c_1, \ldots, c_n)$. That is, $\mathbf{y}$ is in $\text{Im}(T)$ if and only if $\mathbf{y}$ is a linear combination of the columns of A.

The last procedure described is essentially that of Theorem 1 of Section 4.3. It is appropriate here as a consequence of Theorem 5.

EXAMPLE 6.28   For the matrix A, let $T: \mathbf{R}^5 \to \mathbf{R}^4$ be defined by $T(\mathbf{x}) = A\mathbf{x}$ for each $\mathbf{x}$ in $\mathbf{R}^5$. Find the rank of T and a basis for $\text{Im}(T)$.

$$A = \begin{bmatrix} 1 & 2 & 0 & 1 & 0 \\ 2 & 4 & 1 & 0 & 0 \\ 0 & 0 & 1 & -2 & 1 \\ 1 & 2 & 1 & -1 & 1 \end{bmatrix}.$$

Solution   We row-reduce A to obtain

$$B = \begin{bmatrix} 1 & 2 & 0 & 1 & 0 \\ 0 & 0 & 1 & -2 & 0 \\ 0 & 0 & 0 & 0 & 1 \\ 0 & 0 & 0 & 0 & 0 \end{bmatrix}.$$

Since B has three columns that contain leading ones (the first, third, and fifth), the rank of T is 3. A basis for $\text{Im}(T)$ is given by the columns of A that correspond to these, $\{\mathbf{a}^1, \mathbf{a}^3, \mathbf{a}^5\}$. That is, $\{(1, 2, 0, 1), (0, 1, 1, 1), (0, 0, 1, 1)\}$ is a basis for $\text{Im}(T)$.

A function $f: \mathscr{A} \to \mathscr{B}$ is **onto** if $\text{Im}(f)$, the set of all images $f(a)$, is the set $\mathscr{B}$. An immediate consequence of Theorem 7(iii) of Section 5.2 is that a *linear transformation* $T: \mathbf{V} \to \mathbf{W}$ is onto if and only if $\dim(\text{Im}(T)) = \dim \mathbf{W}$. In case $T: \mathbf{R}^n \to \mathbf{R}^m$ is represented by a matrix, the following theorem gives an easy test to see whether or not the linear transformation is onto.

THEOREM 6   Let $T: \mathbf{R}^n \to \mathbf{R}^m$ be a linear transformation, and let A be the matrix that represents T. Then T is onto if and only if the rank of A is equal to m.

Proof   Since $\text{Im}(T)$ is a subspace of $\mathbf{R}^m$, we have that $\text{Im}(T) = \mathbf{R}^m$ if and only if $\dim(\text{Im}(T)) = m$. By Theorem 5, $\dim(\text{Im}(T))$ is equal to the rank of A, so we have the desired result.

As examples of Theorem 6, note that the linear transformation of Example 6.21 is onto, since $\dim(\text{Im}(T)) = 2$, which is the rank of the matrix, but that of Example 6.23 is not, since $\dim(\text{Im}(T)) = 3$, which is less than 4.

Combining Theorem 2 with Theorem 5 yields an interesting result. Assume that A is the $m \times n$ matrix that represents the linear transformation T, and let r be the rank of A. As a consequence of Theorem 2, the dimension of the kernel (the nullity of A) is $n - r$. As a consequence of Theorem 5, the dimension of the image (the rank of A) is r. Combining these facts, we obtain the following theorem.

THEOREM 7   Let $T: \mathbf{R}^n \to \mathbf{R}^m$ be a linear transformation. Then the sum of the dimension of the kernel of T and the dimension of the image of T is equal to the dimension of the domain of T. That is,

$$\dim(\text{Ker}(T)) + \dim(\text{Im}(T)) = n,$$

or

$$\text{nullity} + \text{rank} = n.$$

In general, a *function* may be one-to-one without being onto, or it may be onto without being one-to-one. However, as the next theorem tells us, this cannot be the case for a *linear operator* on $\mathbf{R}^n$, a linear transformation from $\mathbf{R}^n$ into itself.

THEOREM 8 Let $T: \mathbf{R}^n \to \mathbf{R}^n$ be a linear transformation. Then T is one-to-one if and only if it is onto.

Proof From Theorem 7, we have

$$\dim (\mathrm{Ker}\,(T)) + \dim (\mathrm{Im}\,(T)) = n.$$

By Theorem 3, T is one-to-one if and only if $\mathrm{Ker}\,(T) = \{\mathbf{0}\}$; that is, if and only if $\dim (\mathrm{Ker}\,(T)) = 0$. But then the equation above implies that $\dim (\mathrm{Im}\,(T)) = n$; that is, T is onto $\mathbf{R}^n$ by Theorem 6.

Let A be the matrix that represents a linear transformation T from $\mathbf{R}^n$ into itself. Then A is an order n square matrix. Since T is one-to-one if and only if it is onto, by Theorems 3 and 6 it is one-to-one if and only if the columns of A are a basis for $\mathbf{R}^n$. By Theorem 6 of Section 4.3 and Theorem 5 of Section 3.2, this occurs if and only if A is *invertible*. In this situation A^{-1} also gives rise to a linear transformation T^{-1} on $\mathbf{R}^n$ defined by $T^{-1}(\mathbf{x}) = A^{-1}\mathbf{x}$, for each $\mathbf{x}$. The linear transformation T^{-1} is called the **inverse** of T, and we say that T is **invertible.** It is easy to confirm that $T^{-1}T = TT^{-1} = I$, since for each $\mathbf{x}$ in $\mathbf{R}^n$,

$$T^{-1}T(\mathbf{x}) = T^{-1}(T(\mathbf{x})) = T^{-1}(A\mathbf{x}) = A^{-1}A\mathbf{x} = I\mathbf{x} = I(\mathbf{x}),$$

and similarly, $TT^{-1}(\mathbf{x}) = I(\mathbf{x})$.

The following theorem is an immediate consequence of the preceding remarks and is left as an exercise.

THEOREM 9 Let $T: \mathbf{R}^n \to \mathbf{R}^n$ be a linear transformation. The following are equivalent:

    a. T is one-to-one,

    b. T is onto,

    c. T is invertible,

    d. $\det A \neq 0$, where A is the matrix that represents T.

EXAMPLE 6.29 Let $T: \mathbf{R}^3 \to \mathbf{R}^3$ be defined by $T(x, y, z) = (2x - y, x + y + z, 2y - z)$ for each $\mathbf{x} = (x, y, z)$ in $\mathbf{R}^3$. Show that T is invertible, and compute T^{-1}.

Solution   We first compute the matrix A that represents T,

$$A = \begin{bmatrix} 2 & -1 & 0 \\ 1 & 1 & 1 \\ 0 & 2 & -1 \end{bmatrix}.$$

Since $\det A = -7 \neq 0$, A is invertible. Its inverse is

$$A^{-1} = \frac{1}{7} \begin{bmatrix} 3 & 1 & 1 \\ -1 & 2 & 2 \\ -2 & 4 & -3 \end{bmatrix}.$$

Thus T^{-1} is given by

$$T^{-1}(\mathbf{x}) = A^{-1}\mathbf{x} = \frac{1}{7} \begin{bmatrix} 3 & 1 & 1 \\ -1 & 2 & 2 \\ -2 & 4 & -3 \end{bmatrix} \begin{bmatrix} x \\ y \\ z \end{bmatrix} = \begin{bmatrix} (3x + y + z)/7 \\ (-x + 2y + 2z)/7 \\ (-2x + 4y - 3z)/7 \end{bmatrix}$$

or

$$T^{-1}(x, y, z) = ((3x + y + z)/7, (-x + 2y + 2z)/7, (-2x + 4y - 3z)/7).$$

6.3 EXERCISES

In Exercises 1–6 determine whether or not $\mathbf{v}_1$ or $\mathbf{v}_2$ is in the kernel of the given linear transformation.

1. $T: \mathbf{R}^4 \to \mathbf{R}^3$ given by $T(\mathbf{x}) = A\mathbf{x}$, where

$$A = \begin{bmatrix} 1 & 2 & -1 & 1 \\ 1 & 0 & 1 & 1 \\ 2 & -4 & 6 & 2 \end{bmatrix} \quad \text{and} \quad \mathbf{v}_1 = (-2, 0, 0, 2), \quad \mathbf{v}_2 = (-2, 2, 2, 0).$$

2. $T: \mathbf{R}^3 \to \mathbf{R}^4$ given by $T(\mathbf{x}) = A\mathbf{x}$, where

$$A = \begin{bmatrix} 1 & 1 & 0 \\ 2 & 0 & 3 \\ -1 & 1 & 1 \\ 1 & 1 & 2 \end{bmatrix} \quad \text{and} \quad \mathbf{v}_1 = (-3, 3, 2), \quad \mathbf{v}_2 = (0, 0, 0).$$

3. $T: \mathbf{R}^3 \to \mathbf{R}^2$ given by $T(x, y, z) = (3x + y - z, x + 2y + z)$ and $\mathbf{v}_1 = (3, -4, 5)$, $\mathbf{v}_2 = (1, 4, 7)$.

4. $T: \mathbf{R}^4 \to \mathbf{R}$ given by $T(x_1, x_2, x_3, x_4) = x_1 + 2x_2 + 3x_3 + 4x_4$ and $\mathbf{v}_1 = (1, 2, 1, -2)$, $\mathbf{v}_2 = (2, 3, 0, -2)$.

5. $T: \mathbf{P}_2 \to \mathbf{M}^{2,2}$ of Example 6.22 and $\mathbf{v}_1 = 1 + x - 2x^2$, $\mathbf{v}_2 = \mathbf{0}$.

6. $T: \mathbf{M}^{2,3} \to \mathbf{P}_2$ of Example 6.3 and

$$\mathbf{v}_1 = \begin{bmatrix} -1 & 3 & 1 \\ 2 & 2 & 0 \end{bmatrix}, \qquad \mathbf{v}_2 = \begin{bmatrix} 2 & 1 & -2 \\ 1 & 1 & 0 \end{bmatrix}.$$

In Exercises 7–12 determine whether or not $\mathbf{w}_1$ or $\mathbf{w}_2$ is in the image of the corresponding linear transformation of Exercises 1–6.

7. $\mathbf{w}_1 = (1, 3, 1)$, $\mathbf{w}_2 = (-1, -1, -2)$

8. $\mathbf{w}_1 = (0, 0, -8, -4)$, $\mathbf{w}_2 = (-1, 13, 0, 7)$

9. $\mathbf{w}_1 = (1, 1)$, $\mathbf{w}_2 = (-3, 1)$

10. $\mathbf{w}_1 = -13$, $\mathbf{w}_2 = 0$

11. $\mathbf{w}_1 = \begin{bmatrix} 2 & 1 \\ 3 & 4 \end{bmatrix}$, $\mathbf{w}_2 = \begin{bmatrix} 3 & 2 \\ 3 & 4 \end{bmatrix}$

12. $\mathbf{w}_1 = x^2 + 2x + 1$, $\mathbf{w}_2 = x - 2$

13–18. For each of the linear transformations T in Exercises 1–6 find the nullity of T, and give a basis for Ker (T). Which of the transformations T are one-to-one?

19–24. For each of the linear transformations T of Exercises 1–6 find the rank of T, and give a basis for Im (T). Which of the transformations T are onto?

25. If $T: \mathbf{R}^n \to \mathbf{R}^m$ is a linear transformation, prove that $\dim (\text{Im}\,(T)) \leq n$.

26. If $T: \mathbf{R}^n \to \mathbf{R}^m$ is a linear transformation and $n < m$, prove that T is not onto.

27. If $T: \mathbf{R}^n \to \mathbf{R}^m$ is a linear transformation and $n > m$, prove that T is not one-to-one.

28. If $T: \mathbf{R}^n \to \mathbf{R}^m$ is a linear transformation that is one-to-one and $\mathcal{B}$ is a basis for $\mathbf{R}^n$, prove that $T[\mathcal{B}]$, the set of all images $T(\mathbf{b})$ for $\mathbf{b}$ in $\mathcal{B}$, is a basis for Im (T).

In Exercises 29–35 determine whether or not the given linear transformation is invertible. If it is invertible, compute its inverse.

29. $T: \mathbf{R}^3 \to \mathbf{R}^3$ given by $T(x, y, z) = (x + z, x - y + z, y + 2z)$

30. $T: \mathbf{R}^2 \to \mathbf{R}^2$ given by $T(x, y) = (3x + 2y, -6x - 4y)$

31. $T: \mathbf{R}^3 \to \mathbf{R}^3$ given by $T(\mathbf{x}) = A\mathbf{x}$, where

$$A = \begin{bmatrix} 3 & 1 & 2 \\ 5 & -3 & 4 \\ 1 & -2 & 1 \end{bmatrix}.$$

32. $T: \mathbf{R}^4 \to \mathbf{R}^4$ given by $T(\mathbf{x}) = A\mathbf{x}$, where

$$A = \begin{bmatrix} -1 & 2 & 1 & 0 \\ -1 & 1 & 0 & -1 \\ 2 & -1 & 0 & 4 \\ 1 & -2 & 0 & 0 \end{bmatrix}.$$

33. $T_\theta: \mathbf{R}^2 \to \mathbf{R}^2$, the rotation of the plane given by Example 6.7.

34. $T^\theta: \mathbf{R}^2 \to \mathbf{R}^2$, the reflection of the plane given by Example 6.9.

35. $T_c: \mathbf{R}^n \to \mathbf{R}^n$, the contraction/dilation given by Example 6.8.

6.4   MATRIX OF A GENERAL LINEAR TRANSFORMATION

In this section we show how the matrix representation of linear transformations from $\mathbf{R}^n$ to $\mathbf{R}^m$ (Section 6.2) can be extended to incorporate linear transformations between any two finite dimensional vector spaces.

Let $\mathbf{V}$ be a vector space with basis $\mathcal{B} = \{\mathbf{v}_1, \mathbf{v}_2, \ldots, \mathbf{v}_n\}$. Recall from Section 5.3 that the *coordinate vector* of a vector $\mathbf{v}$ in $\mathbf{V}$ is the unique $\mathbf{x}$ in $\mathbf{R}^n$ of coefficients needed to express $\mathbf{v}$ as a linear combination of elements of $\mathcal{B}$:

$$\mathbf{v} = x_1\mathbf{v}_1 + x_2\mathbf{v}_2 + \cdots + x_n\mathbf{v}_n.$$

The coordinate vector can be used to replace the original vector in many calculations. In the case of a linear transformation $T: \mathbf{V} \to \mathbf{W}$, there are two spaces involved, $\mathbf{V}$ and $\mathbf{W}$. If $\mathbf{V} = \mathbf{W}$, it is possible to start with the same basis for both spaces, but in general the two spaces might be completely different, so that a basis for one looks nothing like that for the other. It is still possible, however, to use a matrix to represent a linear transformation, but a fixed basis for *each* space must be chosen before the matrix can be constructed.

PROCEDURE FOR CONSTRUCTING A MATRIX THAT REPRESENTS A LINEAR TRANSFORMATION $T: \mathbf{V} \to \mathbf{W}$.

i. Choose $\mathcal{B} = \{\mathbf{b}_1, \ldots, \mathbf{b}_n\}$ and $\mathcal{B}' = \{\mathbf{b}'_1, \ldots, \mathbf{b}'_m\}$ as bases of $\mathbf{V}$ and $\mathbf{W}$, respectively.

ii. Express each $T(\mathbf{b}_i)$ as a linear combination of $\mathcal{B}'$.

iii. Let $\mathbf{a}^i$ be the coefficients of the $\mathbf{b}'_j$; that is, $\mathbf{a}^i$ is the coordinate vector of $T(\mathbf{b}_i)$ with respect to $\mathcal{B}'$.

iv. A is the matrix with ith column $\mathbf{a}^i$.

EXAMPLE 6.30   Let $T: \mathbf{M}^{2,3} \to \mathbf{P}_2$ be as in Example 6.3,

$$T\begin{bmatrix} a_{11} & a_{12} & a_{13} \\ a_{21} & a_{22} & a_{23} \end{bmatrix} = (a_{11} + a_{13})x^2 + (a_{21} - a_{22})x + a_{23}.$$

Find the matrix A in $\mathbf{M}^{3,6}$ that represents T with respect to the bases $\mathcal{B} = \{E^{ij} \mid i = 1, 2; j = 1, 2, 3\}$ and $\mathcal{B}' = \{1, x, x^2\}$.

Solution   Following Theorem 1, we compute $T(E^{ij})$ for each i, j and represent the result by means of the coordinate vector with respect to $\mathcal{B}'$:

$$T(E^{11}) = (1 + 0)x^2 + (0 - 0)x + 0 = 0(1) + 0(x) + 1(x^2) \leftrightarrow (0, 0, 1),$$
$$T(E^{12}) = (0 + 0)x^2 + (0 - 0)x + 0 = 0(1) + 0(x) + 0(x^2) \leftrightarrow (0, 0, 0),$$
$$T(E^{13}) = (0 + 1)x^2 + (0 - 0)x + 0 = 0(1) + 0(x) + 1(x^2) \leftrightarrow (0, 0, 1),$$

$$T(E^{21}) = (0 + 0)x^2 + (1 - 0)x + 0 = 0(1) + 1(x) + 0(x^2) \leftrightarrow (0, 1, 0),$$
$$T(E^{22}) = (0 + 0)x^2 + (0 - 1)x + 0 = 0(1) - 1(x) + 0(x^2) \leftrightarrow (0, -1, 0),$$
$$T(E^{23}) = (0 + 0)x^2 + (0 - 0)x + 1 = 1(1) + 0(x) + 0(x^2) \leftrightarrow (1, 0, 0).$$

The resulting matrix is then

$$A = \begin{bmatrix} 0 & 0 & 0 & 0 & 0 & 1 \\ 0 & 0 & 0 & 1 & -1 & 0 \\ 1 & 0 & 1 & 0 & 0 & 0 \end{bmatrix}.$$

EXAMPLE 6.31

Let the derivative transformation $D: \mathbf{P}_3 \to \mathbf{P}_2$ be defined as follows. For each

$$\mathbf{p}(x) = a_3x^3 + a_2x^2 + a_1x + a_0,$$

define

$$D(\mathbf{p}) = 3a_3x^2 + 2a_2x + a_1.$$

Find the matrix X that represents D with respect to the bases $\mathscr{B} = \{1, x, x^2, x^3\}$ and $\mathscr{B}' = \{1, x, x^2\}$.

Solution

We have shown in the Spotlight of Section 6.2 that D is a linear transformation. Now,

$$D(1) = 0 = 0(1) + 0(x) + 0(x^2) \leftrightarrow (0, 0, 0),$$
$$D(x) = 1 = 1(1) + 0(x) + 0(x^2) \leftrightarrow (1, 0, 0),$$
$$D(x^2) = 2x = 0(1) + 2(x) + 0(x^2) \leftrightarrow (0, 2, 0),$$
$$D(x^3) = 3x^2 = 0(1) + 0(x) + 3(x^2) \leftrightarrow (0, 0, 3).$$

The resulting matrix is then

$$A = \begin{bmatrix} 0 & 1 & 0 & 0 \\ 0 & 0 & 2 & 0 \\ 0 & 0 & 0 & 3 \end{bmatrix}.$$

NOTE (*From calculus*) Using the result of Example 6.31, we can give the derivative of any polynomial of degree less than or equal to 3 by multiplication by A. That is,

$$\frac{d}{dx}(a_3x^3 + a_2x^2 + a_1x + a_0) \leftrightarrow \begin{bmatrix} 0 & 1 & 0 & 0 \\ 0 & 0 & 2 & 0 \\ 0 & 0 & 0 & 3 \end{bmatrix} \begin{bmatrix} a_0 \\ a_1 \\ a_2 \\ a_3 \end{bmatrix}.$$

For example, if $\mathbf{p}(x) = 1 - 3x^2 + x^3$, then $(d/dx)(\mathbf{p}(x))$ can be obtained by computing

$$\begin{bmatrix} 0 & 1 & 0 & 0 \\ 0 & 0 & 2 & 0 \\ 0 & 0 & 0 & 3 \end{bmatrix} \begin{bmatrix} 1 \\ 0 \\ -3 \\ 1 \end{bmatrix} = \begin{bmatrix} 0 \\ -6 \\ 3 \end{bmatrix}.$$

Now $(0, -6, 3)$ is the coordinate vector of $(d/dx)(\mathbf{p}(x)) = -6x + 3x^2$.

The preceding note shows why we say that the matrix constructed *represents* the linear transformation. Instead of applying the original linear transformation to the original vector space, we multiply the constructed matrix times each coordinate vector to obtain a result that can be interpreted in the second vector space. In general, let $T: \mathbf{V} \to \mathbf{W}$ be a linear transformation from an n-dimensional space $\mathbf{V}$ to an m-dimensional space $\mathbf{W}$, and let A be the matrix that represents T with respect to a basis of each $\mathbf{V}$ and $\mathbf{W}$. Since A is an $m \times n$ matrix, multiplication by A is a linear transformation, $T_A: \mathbf{R}^n \to \mathbf{R}^m$, from $\mathbf{R}^n$ to $\mathbf{R}^m$. For each vector $\mathbf{v}$ in $\mathbf{V}$, the coordinate vector $\mathbf{x}$ with respect to the basis of $\mathbf{V}$ is in $\mathbf{R}^n$, so $T_A(\mathbf{x}) = A\mathbf{x}$ makes sense and is some element of $\mathbf{R}^m$. Each vector in $\mathbf{R}^m$ can be interpreted as the coordinate vector of some element of $\mathbf{W}$ with respect to the basis of $\mathbf{W}$. The next theorem asserts that the vector in $\mathbf{W}$ of which $A\mathbf{x}$ is the coordinate vector is exactly $T(\mathbf{v})$.

THEOREM I   Let $T: \mathbf{V} \to \mathbf{W}$ be a linear transformation and A the matrix that represents T with respect to the bases $\mathcal{B}$ of $\mathbf{V}$ and $\mathcal{B}'$ of $\mathbf{W}$. Let $\mathbf{v}$ be a vector in $\mathbf{V}$ with coordinate vector $\mathbf{x}$, and let $\mathbf{w}$ be a vector in $\mathbf{W}$ with coordinate vector $\mathbf{y}$. Then $T(\mathbf{v}) = \mathbf{w}$ if and only if $T_A(\mathbf{x}) = \mathbf{y}$; that is, if and only if $A\mathbf{x} = \mathbf{y}$. (See Figure 6.9.)

Proof   Since $T: \mathbf{V} \to \mathbf{W}$ and $T_A: \mathbf{R}^n \to \mathbf{R}^m$ are linear transformations, it suffices to show that each acts appropriately on a basis. The basis we choose

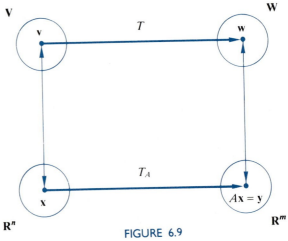

FIGURE 6.9

for **V** is $\mathcal{B} = \{\mathbf{b}_1, \ldots, \mathbf{b}_n\}$, and the corresponding basis for $\mathbf{R}^n$ is the standard basis for $\mathbf{R}^n$, $\{\mathbf{e}_1, \ldots, \mathbf{e}_n\}$, since $\mathbf{e}_i$ is the coordinate vector of $\mathbf{b}_i$ with respect to the basis $\mathcal{B}$. For each i, $T(\mathbf{b}_i)$ has as its coordinate vector $\mathbf{a}^i$, the ith column of A by the construction procedure for A. But $T_A(\mathbf{e}_i) = A\mathbf{e}_i = \mathbf{a}^i$, so T_A acting on the coordinate vector of $\mathbf{b}_i$ is the coordinate vector of $T(\mathbf{b}_i)$. Thus the two linear transformations agree in their actions on their respective bases, and the proof is complete.

NOTE Theorems 5 and 6 of Section 6.2, which demonstrate that the sum, scalar multiple, and composition of linear transformations correspond naturally to the analogous operations on their matrices, can easily be extended to this more general case. The proofs for the general case are almost identical to those given in Section 6.2.

··

The matrix that represents a linear transformation can be used to characterize the *kernel* and *image* of the linear transformation and thereby determine whether or not the transformation is one-to-one or onto. The following theorems explain this correspondence. They are almost immediate corollaries to Theorem 1, and their proofs are left as Exercises 18 and 19.

THEOREM 2 Let $T: \mathbf{V} \to \mathbf{W}$ be a linear transformation, and let A be a matrix that represents T. Then a vector $\mathbf{v}$ is in $\mathrm{Ker}\,(T)$ if and only if its coordinate vector $\mathbf{x}$ is in $\mathrm{Ker}\,(T_A)$; that is, if and only if $A\mathbf{x} = \mathbf{0}$.

THEOREM 3 Let $T: \mathbf{V} \to \mathbf{W}$ be a linear transformation, and let A be a matrix that represents T. Then a vector $\mathbf{w}$ is in $\mathrm{Im}\,(T)$ if and only if its coordinate vector $\mathbf{y}$ is in $\mathrm{Im}\,(T_A)$.

EXAMPLE 6.32   Let $T: \mathbf{M}^{2,3} \to \mathbf{P}_2$ be as in Example 6.30,

$$T\begin{bmatrix} a_{11} & a_{12} & a_{13} \\ a_{21} & a_{22} & a_{23} \end{bmatrix} = (a_{11} + a_{13})x^2 + (a_{21} - a_{22})x + a_{23}.$$

Find the kernel and image of T using the matrix that represents T with respect to the standard bases.

Solution   From Example 6.30 we have the appropriate matrix,

$$A = \begin{bmatrix} 0 & 0 & 0 & 0 & 0 & 1 \\ 0 & 0 & 0 & 1 & -1 & 0 \\ 1 & 0 & 1 & 0 & 0 & 0 \end{bmatrix}.$$

By inspection we see that the rank of A is 3. Since $T_A: \mathbf{R}^6 \to \mathbf{R}^3$, by Theorem 6 of Section 6.3 we have that T_A is onto. But then by Theorem 3, T must be onto as well. Thus, the image of T is $\mathbf{P}_2$.

To find $\text{Ker}(T)$, we also use the procedures of Section 6.3—this time the procedure to compute a basis for $\text{Ker}(T_A)$. Let B be the row-reduced echelon form for A. Then

$$B = \begin{bmatrix} 1 & 0 & 1 & 0 & 0 & 0 \\ 0 & 0 & 0 & 1 & -1 & 0 \\ 0 & 0 & 0 & 0 & 0 & 1 \end{bmatrix}.$$

Interpreting B as the matrix of coefficients of a homogeneous linear system in the variables $x_{11}, x_{12}, x_{13}, x_{21}, x_{22}$, and x_{23} (corresponding to the order of the E^{ij}), we have

$$
\begin{aligned}
x_{11} + x_{13} &= 0 \\
x_{21} - x_{22} &= 0 \\
x_{23} &= 0.
\end{aligned}
$$

Parameterizing x_{12}, x_{13}, and x_{22} with a, b, and c, we have

$$
\begin{aligned}
x_{11} &= 0a - 1b + 0c = -b \\
x_{21} &= 0a + 0b + 1c = c \\
x_{23} &= 0a + 0b + 0c = 0.
\end{aligned}
$$

Thus we have $\text{Ker}(T)$ as those matrices of the form

$$\begin{bmatrix} -b & a & b \\ c & c & 0 \end{bmatrix}.$$

With Theorems 1 through 3 it is a straightforward task to extend several theorems of Section 6.3 to the general case. These are summarized in the following theorem.

THEOREM 4   Let $T: \mathbf{V} \to \mathbf{W}$ be a linear transformation, and let A be a matrix that represents T with respect to bases for $\mathbf{V}$ and $\mathbf{W}$. Then

a. T is one-to-one if and only if $A\mathbf{x} = \mathbf{0}$ has only the trivial solution.

b. T is onto if and only if $\text{rank}(A) = \dim(\mathbf{W})$.

c. $\dim(\text{Ker}(T)) + \dim(\text{Im}(T)) = \dim(\mathbf{V})$.

d. If $\dim(\mathbf{V}) = \dim(\mathbf{W})$, then the following are equivalent:

   i. T is one-to-one.

   ii. T is onto.

   iii. $\det A \neq 0$.

EXAMPLE 6.33

Let $T: \mathbf{M}^{2,3} \to \mathbf{P}_2$ be as in Examples 6.30 and 6.32,

$$T\begin{bmatrix} a_{11} & a_{12} & a_{13} \\ a_{21} & a_{22} & a_{23} \end{bmatrix} = (a_{11} + a_{13})x^2 + (a_{21} - a_{22})x + a_{23}.$$

Is T one-to-one and/or onto?

Solution

That T is not one-to-one is immediate from the result of Example 6.32. For example, with $a = 1$, $b = 2$, and $c = 3$, the final matrix of that example yields the following nonzero matrix in Ker (T):

$$\begin{bmatrix} -2 & 1 & 2 \\ 3 & 3 & 0 \end{bmatrix}.$$

Theorem 4(a), however, gives an even easier solution to the question of whether or not T is one-to-one. The matrix A of T calculated in Example 6.30 has more columns than rows. Hence $A\mathbf{x} = \mathbf{0}$ has nontrivial solutions by Corollary 2 of Section 2.5, and therefore T is not one-to-one.

Since this same matrix A has rank 3 (its row-reduced echelon form is matrix B in Example 6.32), and since dim $(\mathbf{P}_2) = 3$, T is onto $\mathbf{P}_2$, by Theorem 4(b).

EXAMPLE 6.34

Let $D: \mathbf{P}_3 \to \mathbf{P}_3$ be given by the same rule as in Example 6.31. For each $\mathbf{p}(x) = a_3x^3 + a_2x^2 + a_1x + a_0$, define

$$D(\mathbf{p}) = 3a_3x^2 + 2a_2x + a_1.$$

Is D one-to-one and/or onto?

Solution

Since D is from $\mathbf{V} = \mathbf{P}_3$ to $\mathbf{W} = \mathbf{P}_3$, we have dim $(\mathbf{V}) = $ dim $(\mathbf{W})$, as required for Theorem 4(d). We do need a matrix to represent the transformation, and to construct one we use the same basis $\mathcal{B} = \{1, x, x^2, x^3\}$ for both $\mathbf{V}$ and $\mathbf{W}$. This gives almost the same matrix as was computed in Example 6.31; this time it has an extra zero row:

$$A = \begin{bmatrix} 0 & 1 & 0 & 0 \\ 0 & 0 & 2 & 0 \\ 0 & 0 & 0 & 3 \\ 0 & 0 & 0 & 0 \end{bmatrix}.$$

Because of the zero row, det $A = 0$. By Theorem 4(d), D is neither one-to-one nor onto.

6.4 EXERCISES

In Exercises 1–4 find the matrix A that represents the linear transformation T with respect to the bases $\mathcal{B}$ and $\mathcal{B}'$. (See Exercises 5–8 of Section 6.1.)

1. $T: \mathbf{R}^3 \to \mathbf{M}^{2,2}$ given by

$$T(x, y, z) = \begin{bmatrix} y & z \\ -x & 0 \end{bmatrix},$$

where $\mathcal{B} = \{\mathbf{e}_1, \mathbf{e}_2, \mathbf{e}_3\}$ and $\mathcal{B}' = \{E^{ij} \mid i = 1, 2; j = 1, 2\}$.

2. $T: \mathbf{P}_3 \to \mathbf{P}_3$ given by

$$T(a_0 + a_1 x + a_2 x^2 + a_3 x^3) = (a_0 + a_2) - (a_1 + 2a_3)x^2,$$

where $\mathcal{B} = \mathcal{B}' = \{1, x, x^2, x^3\}$.

3. $T: \mathbf{P}_2 \to \mathbf{M}^{2,2}$ given by

$$T(a_0 + a_1 x + a_2 x^2) = \begin{bmatrix} a_0 & -a_2 \\ -a_2 & a_0 - a_1 \end{bmatrix},$$

where $\mathcal{B} = \{1, x, x^2\}$ and $\mathcal{B}' = \{E^{ij} \mid i = 1, 2; j = 1, 2\}$.

4. $T: \mathbf{P}_2 \to \mathbf{P}_2$ given by

$$T(a_0 + a_1 x + a_2 x^2) = a_1 - (a_0 + a_2)x + (a_1 + a_2)x^2,$$

where $\mathcal{B} = \{1, x, x^2\}$ and $\mathcal{B}' = \{x + 1, x - 1, x^2 + 1\}$.

5–8. Use Theorem 2 to compute $\mathrm{Ker}\,(T)$ for each linear transformation in Exercises 1–4. Which ones are one-to-one?

9–12. Use Theorem 3 to compute $\mathrm{Im}\,(T)$ for each linear transformation in Exercises 1–4. Which ones are onto?

13. Find the 4×5 matrix associated with the derivative transformation

$$D: \mathbf{P}_4 \to \mathbf{P}_3$$

with respect to $\mathcal{B} = \{1, x, x^2, x^3, x^4\}$ and $\mathcal{B}' = \{1, x, x^2, x^3\}$ (see Example 6.31).

14. Use the matrix of Exercise 13 to find the derivative of

$$\mathbf{p}(x) = 1 + x^2 - 3x^3 + 2x^4.$$

15. Let $S: \mathbf{P}_2 \to \mathbf{P}_3$ be defined as follows. For each $\mathbf{p}(x) = a_2 x^2 + a_1 x + a_0$, define $S(\mathbf{p}) = a_2 x^3/3 + a_1 x^2/2 + a_0 x$. Find the matrix A that represents S with respect to the bases $\mathcal{B} = \{1, x, x^2\}$ and $\mathcal{B}' = \{1, x, x^2, x^3\}$. (The linear transformation S gives the *integral* of $\mathbf{p}(x)$, with the constant term equal to zero.)

16. Use the matrix of Exercise 15 to find the integral of $\mathbf{p}(x) = 1 - x + 2x^2$.

17. Let $T: \mathbf{V} \to \mathbf{W}$ be a linear transformation that is both one-to-one and onto. Prove that T is *invertible*. That is, prove that there exists a transformation $S: \mathbf{W} \to \mathbf{V}$ such that $S \circ T$ is the identity on $\mathbf{V}$ and $T \circ S$ is the identity on $\mathbf{W}$. (*Hint:* See Theorem 9 of Section 6.3.)

18. Prove Theorem 2.

19. Prove Theorem 3.

6.5   CHANGE OF BASIS

The matrix associated with a linear transformation $T: \mathbf{R}^n \to \mathbf{R}^m$ of Section 6.2 was computed using the standard basis vectors $\mathbf{e}_i$ (see Theorem 1 of Section 6.2). There are times when it is more convenient to compute the matrix of a linear transformation using a different basis (see Section 8.4). The procedure of Section 6.4 can be applied to construct this new matrix, but a more efficient procedure will be described in this section.

The components of any vector $\mathbf{x}$ in $\mathbf{R}^n$ are its coordinates with respect to the standard basis $\{\mathbf{e}_1, \ldots, \mathbf{e}_n\}$, since $\mathbf{x} = x_1\mathbf{e}_1 + \cdots + x_n\mathbf{e}_n$. When we find the coordinate vector for $\mathbf{x}$ with respect to some other basis, we are in effect making a *change of basis*. For example, let $n = 2$. Then any vector $\mathbf{x} = (x, y)$ can be written as $\mathbf{x} = x\mathbf{e}_1 + y\mathbf{e}_2$; the ordered pair (x, y) expresses $\mathbf{x}$ in terms of the standard basis. Now, if $\mathscr{B} = \{\mathbf{b}_1, \mathbf{b}_2\}$ is some other basis, there exist unique scalars s and t such that $\mathbf{x} = s\mathbf{b}_1 + t\mathbf{b}_2$ (Figure 6.10). The coordinate vector $(s, t)_{\mathscr{B}}$ of $\mathbf{x}$ with respect to $\mathscr{B}$ represents the same point as $\mathbf{x}$ (albeit with different coordinates), as long as it is understood that $(s, t)_{\mathscr{B}}$ means the point obtained by adding the vectors $s\mathbf{b}_1$ and $t\mathbf{b}_2$. When this convention is made, the usual coordinate axes are replaced by the lines determined by $\mathbf{b}_1$ and $\mathbf{b}_2$, and each point is expressed in terms of these. For instance, $(1, 0)_{\mathscr{B}}$ is the endpoint of $\mathbf{b}_1$ and $(1, 1)_{\mathscr{B}}$ is the endpoint of $\mathbf{b}_1 + \mathbf{b}_2$ in the new coordinates. The following theorem shows that the conversion of coordinates from the standard basis to some other basis can be effected by multiplication by a matrix, instead of by solving for the coordinate vector.

FIGURE 6.10

THEOREM I   Let $\mathscr{B} = \{\mathbf{b}_1, \ldots, \mathbf{b}_n\}$ be a basis for $\mathbf{R}^n$, and let P be the matrix with ith column $\mathbf{p}^i = \mathbf{b}_i$. Then P is invertible, and P^{-1} is the **change of basis** matrix. In other words, for each $\mathbf{x}$ in $\mathbf{R}^n$, $P^{-1}\mathbf{x}$ is the coordinate vector of $\mathbf{x}$ with respect to $\mathscr{B}$, $\mathbf{y}_{\mathscr{B}} = P^{-1}\mathbf{x}$.

Proof   The matrix P is invertible, since it is square of order n with n independent columns. Since a linear transformation is determined by its action on a basis (Theorem 3 of Section 6.1), we need only confirm that multiplication by P^{-1} "acts right" on a basis. Note that each basis element $\mathbf{b}_i$ has coordinate vector $\mathbf{e}_i$ with respect to $\mathscr{B}$, since

$$\mathbf{b}_i = 0\mathbf{b}_1 + \cdots + 0\mathbf{b}_{i-1} + 1\mathbf{b}_i + 0\mathbf{b}_{i+1} + \cdots + 0\mathbf{b}_n.$$

Thus the transformation must carry $\mathbf{b}_i$ into $\mathbf{e}_i$ for each i. But this is exactly the effect of multiplication on the left by P^{-1}, since $P^{-1}P = I$, or by columns, $P^{-1}\mathbf{p}^i = P^{-1}\mathbf{b}_i = \mathbf{e}_i$.

EXAMPLE 6.35

For $\mathscr{B} = \{(2, 1, 3), (1, 0, 1), (-1, 1, 1)\}$, a basis for $\mathbf{R}^3$, compute the change of basis matrix and use it to find the coordinate vector of $\mathbf{x} = (2, -1, 1)$.

Solution

Following Theorem 1, we let

$$P = \begin{bmatrix} 2 & 1 & -1 \\ 1 & 0 & 1 \\ 3 & 1 & 1 \end{bmatrix}.$$

Then the change of basis matrix is

$$P^{-1} = \begin{bmatrix} 1 & 2 & -1 \\ -2 & -5 & 3 \\ -1 & -1 & 1 \end{bmatrix}.$$

Finally, the required coordinate vector is

$$P^{-1}(2, -1, 1) = \begin{bmatrix} 1 & 2 & -1 \\ -2 & -5 & 3 \\ -1 & -1 & 1 \end{bmatrix}\begin{bmatrix} 2 \\ -1 \\ 1 \end{bmatrix} = \begin{bmatrix} -1 \\ 4 \\ 0 \end{bmatrix}_{\mathscr{B}} = (-1, 4, 0)_{\mathscr{B}}.$$

If P^{-1} is a change of basis matrix for $\mathbf{R}^n$, then any vector $\mathbf{x}$ is represented by $P^{-1}\mathbf{x}$ with respect to the new basis. In other words, $\mathbf{x}$ and $P^{-1}\mathbf{x}$ each represent the same point but express it in terms of different bases. If $T: \mathbf{R}^n \to \mathbf{R}^n$ is a linear transformation, the point represented by $T(\mathbf{x})$ can also be represented in terms of the new basis as $P^{-1}T(\mathbf{x})$.

EXAMPLE 6.36

Let $T: \mathbf{R}^3 \to \mathbf{R}^3$ be defined by $T(x, y, z) = (x + y, z, 2x - y)$. Find the coordinate vector of $T(2, -1, 1)$ with respect to the basis

$$\mathscr{B} = \{(2, 1, 3), (1, 0, 1), (-1, 1, 1)\}.$$

Solution

From Example 6.35 we have the change of basis matrix P^{-1}, so the result is

$$P^{-1}T(2, -1, 1) = \begin{bmatrix} 1 & 2 & -1 \\ -2 & -5 & 3 \\ -1 & -1 & 1 \end{bmatrix} \begin{bmatrix} 1 \\ 1 \\ 5 \end{bmatrix} = \begin{bmatrix} -2 \\ 8 \\ 3 \end{bmatrix}_{\mathcal{B}}.$$

That is, $P^{-1}T(2, -1, 1) = (-2, 8, 3)_{\mathcal{B}}$.

In Example 6.35 we computed the coordinates of the vector $(2, -1, 1)$ with respect to a basis $\mathcal{B}$, and in Example 6.36 we computed the coordinates of the vector $T(2, -1, 1)$ with respect to the same basis. However, to make the computation we essentially went back to the expression for the point in terms of the standard basis. Theorem 2 shows that it is possible to go directly from the coordinate vector of $\mathbf{x}$ to the coordinate vector of $T(\mathbf{x})$ by means of multiplication by the appropriate matrix.

DEFINITION

The matrix $A_{\mathcal{B}}$ **represents** T with **respect to the basis** $\mathcal{B}$ if

$$P^{-1}(T(\mathbf{x})) = A_{\mathcal{B}}(P^{-1}\mathbf{x})$$

for every $\mathbf{x}$ in $\mathbf{R}^n$, where P^{-1} is the change of basis matrix for $\mathcal{B}$. In other words, multiplication of the coordinate vector of $\mathbf{x}$ with respect to $\mathcal{B}$ by $A_{\mathcal{B}}$ gives the coordinate vector of $T(\mathbf{x})$ with respect to $\mathcal{B}$.

NOTE Let A be the matrix associated with a linear transformation $T: \mathbf{R}^n \to \mathbf{R}^n$. Then A is the matrix that represents T with respect to the *standard* basis, since $P^{-1} = P = I$ is the change of basis matrix for the standard basis.

· ·

THEOREM 2 Let $T: \mathbf{R}^n \to \mathbf{R}^n$ be a linear transformation with associated matrix A, and let $\mathcal{B}$ be a basis for $\mathbf{R}^n$. Let P^{-1} be the change of basis matrix for $\mathcal{B}$. Then $P^{-1}AP$ is the matrix that represents T with respect to the basis $\mathcal{B}$, $A_{\mathcal{B}} = P^{-1}AP$.

Proof For each $\mathbf{x}$ in $\mathbf{R}^n$, $P^{-1}\mathbf{x}$ is the coordinate vector of the same point with respect to the basis $\mathcal{B}$. Checking the action of $P^{-1}AP$ on any such vector, we have

$$(P^{-1}AP)(P^{-1}\mathbf{x}) = P^{-1}AI\mathbf{x} = P^{-1}A\mathbf{x} = P^{-1}T(\mathbf{x}),$$

since A is the matrix associated with T. But this last vector, $P^{-1}T(\mathbf{x})$, is the coordinate vector of $T(\mathbf{x})$ with respect to the basis $\mathcal{B}$.

SPOTLIGHT

SIMPLIFYING AN EQUATION

By making the proper change of basis in $\mathbf{R}^2$, certain equations in two variables, x and y, can be simplified considerably. As an example, consider the ellipse with equation $x^2 + xy + y^2 = 1$, pictured in the figure below. The axes of this ellipse are canted at a 45° angle with respect to the standard basis for $\mathbf{R}^2$, $\{(0, 1), (1, 0)\}$. But notice that the ellipse is symmetric with respect to the lines determined by the basis $\mathscr{B} = \{(1, 1), (-1, 1)\}$.

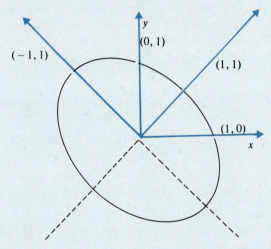

Let us see what effect a change to the basis $\mathscr{B}$ has on the given equation. The coordinate vector $\mathbf{x}' = (x', y')$ of $\mathbf{x} = (x, y)$ with respect to the basis $\mathscr{B}$ is given by $\mathbf{x}' = P^{-1}\mathbf{x}$, where

$$P = \begin{bmatrix} 1 & -1 \\ 1 & 1 \end{bmatrix}.$$

Thus $\mathbf{x} = P\mathbf{x}'$; that is,

$$\begin{bmatrix} x \\ y \end{bmatrix} = \begin{bmatrix} 1 & -1 \\ 1 & 1 \end{bmatrix}\begin{bmatrix} x' \\ y' \end{bmatrix} = \begin{bmatrix} x' - y' \\ x' + y' \end{bmatrix}.$$

Now replacing x with $x' - y'$ and y with $x' + y'$ in the given equation, we get

$$(x' - y')^2 + (x' - y')(x' + y') + (x' + y')^2 = 1,$$

which simplifies to $3(x')^2 + (y')^2 = 1$, clearly the equation of an ellipse. (For a much more detailed discussion of this process, see Section 8.6.)

EXAMPLE 6.37

Let $T: \mathbf{R}^3 \to \mathbf{R}^3$ be given by $T(x, y, z) = (x + y, z, 2x - y)$, and let $\mathcal{B} = \{(2, 1, 3), (1, 0, 1), (-1, 1, 1)\}$. Calculate $A_{\mathcal{B}}$ for T and use it to find the coordinate vector for the transformed point with coordinate vector $(-1, 4, 0)_{\mathcal{B}}$.

Solution

To use Theorem 2, we need the matrix A associated with T,

$$A = \begin{bmatrix} 1 & 1 & 0 \\ 0 & 0 & 1 \\ 2 & -1 & 0 \end{bmatrix}.$$

From Example 6.35 we have P and P^{-1}, so applying Theorem 2, we have

$$A_{\mathcal{B}} = P^{-1}AP = \begin{bmatrix} 1 & 2 & -1 \\ -2 & -5 & 3 \\ -1 & -1 & 1 \end{bmatrix} \begin{bmatrix} 1 & 1 & 0 \\ 0 & 0 & 1 \\ 2 & -1 & 0 \end{bmatrix} \begin{bmatrix} 2 & 1 & -1 \\ 1 & 0 & 1 \\ 3 & 1 & 1 \end{bmatrix}$$

$$= \begin{bmatrix} 6 & 1 & 5 \\ -12 & -1 & -14 \\ -3 & 0 & -4 \end{bmatrix}.$$

Finally, we use $A_{\mathcal{B}}$ to compute the coordinate vector for the transformed point with coordinate vector $(-1, 4, 0)_{\mathcal{B}}$,

$$\begin{bmatrix} 6 & 1 & 5 \\ -12 & -1 & -14 \\ -3 & 0 & -4 \end{bmatrix} \begin{bmatrix} -1 \\ 4 \\ 0 \end{bmatrix}_{\mathcal{B}} = \begin{bmatrix} -2 \\ 8 \\ 3 \end{bmatrix}_{\mathcal{B}}.$$

As it must, the result agrees with Example 6.36.

The relationship between the matrices $P^{-1}AP$ and A is important enough to warrant the following definition.

DEFINITION

Let A and B be square matrices of order n. We say that B is **similar** to A if there exists an invertible matrix P such that

$$B = P^{-1}AP.$$

Several remarks are in order. For one, P must also be of order n if the multiplication is to be defined. For another, $B = P^{-1}AP$ implies that $(P^{-1})^{-1}BP^{-1} = A$, so that if B is similar to A, then A is also similar to B (using P^{-1}). For a third, the matrix P is not unique. To see this, let $A = B = I$, the identity matrix of order n. Then *any* invertible matrix of order n will serve for P, since $I = P^{-1}IP$, no matter what P is.

By Theorem 2, if a matrix A is associated with a linear transformation $T: \mathbf{R}^n \to \mathbf{R}^n$, if $\mathcal{B}$ is a basis for $\mathbf{R}^n$, and if B is the matrix that represents T with respect to $\mathcal{B}$, then B is similar to A. It is natural to ask if the converse is true.

That is, let B be similar to A, and let T be the linear transformation $T(\mathbf{x}) = A\mathbf{x}$. Is there a basis $\mathcal{B}$ such that B represents T with respect to $\mathcal{B}$? The answer is yes.

THEOREM 3 Let A and B be square matrices of order n. Suppose B is similar to A, and let $T: \mathbf{R}^n \to \mathbf{R}^n$ be the linear transformation given by $T(\mathbf{x}) = A\mathbf{x}$, for each $\mathbf{x}$ in $\mathbf{R}^n$. Then if P is an invertible matrix such that $B = P^{-1}AP$, the set $\mathcal{B}$ of columns $\mathbf{b}_i = \mathbf{p}^i$, $i = 1, \ldots, n$, is a basis for $\mathbf{R}^n$ such that B represents T with respect to $\mathcal{B}$, $A_{\mathcal{B}} = B$.

Proof First note that $\mathcal{B}$ is a basis, since the columns of P are linearly independent (Theorem 5 of Section 2.5 and Theorem 4 of Section 4.1), and there are n of them.

Since P is the matrix with columns $\mathbf{b}_i = \mathbf{p}^i$, by Theorem 1, $P^{-1}\mathbf{x}$ is the coordinate vector of $\mathbf{x}$ with respect to $\mathcal{B}$.

Finally, by Theorem 2, $B = P^{-1}AP$ represents T with respect to $\mathcal{B}$.

EXAMPLE 6.38

Let $T: \mathbf{R}^3 \to \mathbf{R}^3$ be given by multiplication by the following matrix A. Using the given matrices P and B, show that B is similar to A, and find a basis for $\mathbf{R}^3$ with respect to which B represents the linear transformation T.

$$A = \begin{bmatrix} 2 & 1 & 3 \\ 1 & -2 & 1 \\ 0 & 2 & 1 \end{bmatrix}, \quad P = \begin{bmatrix} 1 & 0 & 1 \\ 2 & 1 & 2 \\ 0 & 2 & 1 \end{bmatrix}, \quad B = \begin{bmatrix} -22 & -25 & -30 \\ -11 & -14 & -16 \\ 26 & 32 & 37 \end{bmatrix}.$$

Solution

First compute P^{-1} to be

$$P^{-1} = \begin{bmatrix} -3 & 2 & -1 \\ -2 & 1 & 0 \\ 4 & -2 & 1 \end{bmatrix}.$$

Then direct multiplication confirms that $B = P^{-1}AP$, so B is similar to A. Finally, Theorem 3 assures us that the columns of P provide a basis for $\mathbf{R}^3$ with the required property,

$$\mathcal{B} = \{(1, 2, 0), (0, 1, 2), (1, 2, 1)\}.$$

6.5 EXERCISES

In Exercises 1–6 verify that the given set $\mathcal{B}$ is a basis for $\mathbf{R}^n$. Compute the change of basis matrix for each of the bases and use it to find the coordinate vector of $\mathbf{v}$ with respect to $\mathcal{B}$.

1. $\mathcal{B} = \{(1, 2), (1, -2)\}$, $\mathbf{v} = (-1, 3)$, $(n = 2)$

2. $\mathcal{B} = \{(2, 0), (1, 1)\}$, $\mathbf{v} = (3, -1)$, $(n = 2)$

3. $\mathcal{B} = \{(1, 1, 1), (1, 0, 1), (-1, 1, 0)\}$, $\mathbf{v} = (3, -1, 1)$, $(n = 3)$

4. $\mathcal{B} = \{(1, 2, 1), (-1, 1, 2), (1, 2, 3)\}$, $\mathbf{v} = (2, 1, 0)$, $(n = 3)$

5. $\mathcal{B} = \{(1, 2, 1, -1), (0, 2, 1, 0), (1, 1, 0, 1), (1, 3, 0, 0)\}$, $\mathbf{v} = (-1, 1, -1, 1)$, $(n = 4)$

6. $\mathcal{B} = \{(-1, 1, -1, 1), (2, 1, 0, 1), (1, 0, 0, 2), (0, 0, 3, 2)\}$, $\mathbf{v} = (2, 1, 2, 3)$, $(n = 4)$

In Exercises 7 and 8 let $T: \mathbf{R}^2 \rightarrow \mathbf{R}^2$ be given by $T(\mathbf{x}) = A\mathbf{x}$, where A is the matrix

$$A = \begin{bmatrix} 1 & 1 \\ 0 & -1 \end{bmatrix}.$$

7. Find the matrix $A_{\mathcal{B}}$ that represents T with respect to the basis $\mathcal{B}$ of Exercise 1.

8. Find the matrix $A_{\mathcal{B}}$ that represents T with respect to the basis $\mathcal{B}$ of Exercise 2.

In Exercises 9 and 10 let $T: \mathbf{R}^3 \rightarrow \mathbf{R}^3$ be given by $T(\mathbf{x}) = A\mathbf{x}$, where A is the matrix

$$A = \begin{bmatrix} 1 & 1 & 3 \\ 2 & -1 & 1 \\ 1 & 2 & 0 \end{bmatrix}.$$

9. Find the matrix $A_{\mathcal{B}}$ that represents T with respect to the basis $\mathcal{B}$ of Exercise 3.

10. Find the matrix $A_{\mathcal{B}}$ that represents T with respect to the basis $\mathcal{B}$ of Exercise 4.

In Exercises 11 and 12 let $T: \mathbf{R}^3 \rightarrow \mathbf{R}^3$ be given by $T(\mathbf{x}) = T(x, y, z) = (2x - y, x + y + z, y - z)$.

11. Find the matrix that represents T with respect to the basis $\mathcal{B}$ of Exercise 3.

12. Find the matrix that represents T with respect to the basis $\mathcal{B}$ of Exercise 4.

In Exercises 13 and 14 let $T: \mathbf{R}^4 \rightarrow \mathbf{R}^4$ be given by $T(\mathbf{x}) = T(w, x, y, z) = (w + y - z, 2w - y, w + 2x + y + z, y - z)$.

13. Find the matrix that represents T with respect to the basis $\mathcal{B}$ of Exercise 5.

14. Find the matrix that represents T with respect to the basis $\mathcal{B}$ of Exercise 6.

15. If A is similar to B, prove that $\det A = \det B$.

16. Let A and B be square matrices of the same order with $\det A = \det B$. Show that it is not necessarily true that A is similar to B.

17. If A is similar to B, prove that A^2 is similar to B^2.

18. If A is similar to B, prove that rank A = rank B. (*Hint:* Use Theorem 3.)

CHAPTER 6 REVIEW

This chapter dealt with *linear transformations* from one vector space to another. You should be able to determine:

1. whether or not a given function is a linear transformation;

2. the matrix that represents a linear transformation from $\mathbf{R}^n$ to $\mathbf{R}^m$;

3. the sum, difference, and product (composition) of linear transformations;

4. the kernel of a linear transformation and its nullity;

5. the image of a linear transformation and its rank;

6. whether or not a linear transformation is one-to-one and/or onto;

7. whether or not a linear transformation has an inverse;

8. the matrix that represents a linear transformation with respect to bases for the two vector spaces involved;

9. the change of bases matrix needed to express a given vector in $\mathbf{R}^n$ as the coordinate vector with respect to a given basis.

KEY WORDS

linear transformation

identity transformation

sum, difference, negative and scalar multiples of linear transformations

product and powers of linear transformations

nullity of a linear transformation

rank of a linear transformation

invertible linear transformation

similar matrices

zero transformation

matrix that represents a linear transformation

composition of linear transformations

kernel (nullspace) of a linear transformation

image of a linear transformation

inverse of a linear transformation

change of basis matrix

KEY THEOREMS

Theorem   A function from $\mathbf{R}^n$ to $\mathbf{R}^m$ given by multiplication by an $m \times n$ matrix A is a linear transformation, and any linear transformation can be so represented.

Theorem   The algebra of linear transformations is compatible with the algebra of the matrices that represent them (composition of linear transformations corresponds to multiplication of the corresponding matrices).

Theorem   The kernel of a linear transformation is a subspace. It is the zero subspace if and only if the linear transformation is one-to-one.

Theorem   The image of a linear transformation is a subspace. If $T: \mathbf{V} \to \mathbf{W}$ is a linear transformation represented by a matrix A, then T is onto if and only if rank A = dim $(\mathbf{W})$.

Theorem   If $T: \mathbf{V} \to \mathbf{W}$ is a linear transformation, then

$$\dim(\text{Ker}(T)) + \dim(\text{Im}(T)) = \dim(\mathbf{V}).$$

Theorem   If $T: \mathbf{V} \to \mathbf{V}$ is a linear transformation and A is a matrix that represents T, then the following are equivalent:

a. T is one-to-one;

b. T is onto;

c. $\det A \neq 0$.

SELF-TEST

Answers appear at the end of this chapter.

In Problems 1 and 2 determine whether or not the given function is a linear transformation. Justify your answer.

1. $L: \mathbf{R}^2 \to \mathbf{R}^3$ given by $L(x, y) = (x - y, x + y, 1)$

2. $K: \mathbf{P}_2 \to \mathbf{M}^{2,2}$ given by $K(a + bx + cx^2) = \begin{bmatrix} 0 & a - b \\ a + b & c \end{bmatrix}$ [Section 6.1]

In Problems 3–6 let $T: \mathbf{R}^3 \to \mathbf{R}^3$ be defined by

$$T(x, y, z) = (z, x + y, 0).$$

3. Find T^2 ($T \circ T$).   [Section 6.2]

4. Find the matrix that represents T.   [Section 6.2]

5. (a) Find a basis for the kernel of T.
   (b) Is T one-to-one?   [Section 6.3]

6. (a) Find a basis for the range of T.
   (b) Is T onto?   [Section 6.3]

7. Suppose that a linear transformation $L: \mathbf{R}^2 \to \mathbf{R}^2$ is given by $L(\mathbf{x}) = A\mathbf{x}$, where

$$A = \begin{bmatrix} 2 & 5 \\ 1 & 3 \end{bmatrix}.$$

   (a) Find $L(x, y)$ for arbitrary (x, y) in $\mathbf{R}^2$.   [Section 6.1]
   (b) Find $L^{-1}(x, y)$ for arbitrary (x, y) in $\mathbf{R}^2$.   [Section 6.3]

In Problems 8 and 9 let $S: \mathbf{P}_2 \to \mathbf{R}^4$ be the linear transformation given by

$$S(a + bx + cx^2) = (c, a - b, 0, 4a).$$ [Section 6.4]

8. Find the matrix that represents S with respect to the standard bases for $\mathbf{P}_2$ and $\mathbf{R}^4$.

9. (a) Find the dimension of the kernel of S. Is S one-to-one?
   (b) Find the dimension of the image of S. Is S onto?

10. Compute the change of basis matrix for the basis $\{(1, 0, 0), (1, 1, 0), (1, 1, 1)\}$ of $\mathbf{R}^3$.
                                                                                        [Section 6.5]

REVIEW EXERCISES

In Exercises 1 and 2 show that the given function is a linear transformation.

1. $S: \mathbf{R}^3 \to \mathbf{R}^2$ given by $S(x, y, z) = (x - 2y, x + y + z)$.

2. $T: \mathbf{P}_3 \to \mathbf{M}^{2,2}$ given by

$$T(a_0 + a_1 x + a_2 x^2 + a_3 x^3) = \begin{bmatrix} a_0 + a_2 & a_1 - a_3 \\ 2a_2 & 2a_0 + a_1 - a_3 \end{bmatrix}.$$

In Exercises 3 and 4 show that the given function is not a linear transformation.

3. $S: \mathbf{R}^2 \to \mathbf{R}^2$ given by $S(x, y) = (x + 2y, xy)$.

4. $T: \mathbf{R}^2 \to \mathbf{R}^3$ given by $T(x, y) = (x, x + y, x + |y|)$.

Assume that the following are linear transformations and use them in Exercises 5–16, 21 and 22.

$$D: \mathbf{R}^2 \to \mathbf{R}^3 \text{ given by } D(x, y) = (x + y, x - 2y, x)$$
$$E: \mathbf{R}^3 \to \mathbf{R}^3 \text{ given by } E(x, y, z) = (x, x + y + z, 2x + y + z)$$
$$F: \mathbf{R}^2 \to \mathbf{R}^3 \text{ given by } F(x, y) = (x, 2x + y, 2x - y).$$

In Exercises 5–8, find the indicated linear transformation if it is defined. If it is not, explain why it is not.

5. FD **6.** $D - 2F$

7. E^2 **8.** F^2

9. Find the matrix that represents D.   **10.** Find the matrix that represents E.

11. Is D one-to-one? Explain.   **12.** Is D onto? Explain.

13. Is E one-to-one? Explain.   **14.** Is E onto? Explain.

15. Find Ker (D).   **16.** Find Ker (E).

In Exercises 17–20, let T be the linear transformation of Exercise 2.

17. Find the matrix that represents T with respect to the standard bases for $\mathbf{P}_3$ and $\mathbf{M}^{2,2}$.

18. Is T one-to-one? Explain.   **19.** Is T onto? Explain.

20. Find Ker (T).

21. Directly find the matrix $A_{\mathcal{B}}$ that represents the linear transformation E with respect to the basis $\mathcal{B} = \{(1, 1, 0), (1, 0, 1), (0, 1, 1)\}$.

22. Again calculate $A_{\mathcal{B}}$ of Exercise 21, but this time use the result of Exercise 10 and the appropriate change of basis matrix.

ANSWERS TO SELF-TEST

1. L is not a linear transformation.

2. K is a linear transformation.

3. $T^2(x, y, z) = (0, x + y + z, 0)$

4. $\begin{bmatrix} 0 & 0 & 1 \\ 1 & 1 & 0 \\ 0 & 0 & 0 \end{bmatrix}$

5. (a) $\{(-1, 1, 0)\}$
   (b) T is not one-to-one.

6. (a) $\{(1, 0, 0), (0, 1, 0)\}$ (b) T is not onto.

7. (a) $L(x, y) = (2x + 5y, x + 3y)$
   (b) $L^{-1}(x, y) = (3x - 5y, -x + 2y)$

8. $\begin{bmatrix} 0 & 0 & 1 \\ 1 & -1 & 0 \\ 0 & 0 & 0 \\ 4 & 0 & 0 \end{bmatrix}$

9. (a) $\dim (\text{Ker} (S)) = 0; S$ is 1–1.
   (b) $\dim (\text{Im} (S)) = 3; S$ is not onto.

10. $\begin{bmatrix} 1 & -1 & 0 \\ 0 & 1 & -1 \\ 0 & 0 & 1 \end{bmatrix}$

EIGENVALUES AND EIGENVECTORS

In this chapter we will discuss the topic of *eigenvalues* and *eigenvectors* of linear transformations and matrices. These entities have many far-reaching applications in science and mathematics. One of these applications will be given in Section 7.2 and others in Chapter 8.

7.1 DEFINITIONS AND EXAMPLES

From a geometric point of view, a linear transformation T, of $\mathbf{R}^n$ into itself, maps one vector into another by altering the magnitude and/or direction of the former. The *eigenvectors* of T are those nonzero vectors $\mathbf{x}$ for which $T(\mathbf{x})$ and $\mathbf{x}$ are collinear. In other words, eigenvectors are characterized by the property that $T(\mathbf{x})$ is a scalar multiple of $\mathbf{x}$. The scalar involved is called the *eigenvalue* of T associated with the eigenvector $\mathbf{x}$. In more precise terms, we have the following definition.

DEFINITION

Let $\mathbf{V}$ be a vector space, and let $T: \mathbf{V} \to \mathbf{V}$ be a linear transformation. A scalar λ is called an **eigenvalue** of T if there is a *nonzero* vector $\mathbf{x}$ in $\mathbf{V}$ such that $T(\mathbf{x}) = \lambda \mathbf{x}$. The vector $\mathbf{x}$ is said to be an **eigenvector** corresponding to the eigenvalue λ.

Eigenvalues are sometimes referred to as *characteristic values, proper values,* or *latent values.* Similarly, other names for eigenvectors are *characteristic vectors, proper vectors,* and *latent vectors.* In the first two sections of this chapter we will consider only eigenvalues that are real numbers. However, in Section 7.3 we will allow eigenvalues and the components of eigenvectors to be complex numbers.

EXAMPLE 7.1

Let $T: \mathbf{R}^3 \to \mathbf{R}^3$ be the linear transformation given by

$$T(x, y, z) = (0, y + z, 3y - z).$$

Show that $\lambda_1 = 0$ and $\lambda_2 = 2$ are eigenvalues of T with corresponding eigenvectors $\mathbf{x}_1 = (1, 0, 0)$ and $\mathbf{x}_2 = (0, 1, 1)$, respectively.

Solution

We must verify that $T(\mathbf{x}_1) = \lambda_1 \mathbf{x}_1$ and $T(\mathbf{x}_2) = \lambda_2 \mathbf{x}_2$. Now

$$T(\mathbf{x}_1) = T(1, 0, 0) = (0, 0, 0) = 0(1, 0, 0) = \lambda_1 \mathbf{x}_1$$

and

$$T(\mathbf{x}_2) = T(0, 1, 1) = (0, 2, 2) = 2(0, 1, 1) = \lambda_2 \mathbf{x}_2,$$

as desired.

NOTE By definition, an eigen*vector* must be nonzero. A linear transformation may have a zero eigen*value*, however, as can be seen from Example 7.1.

....

It is easy to verify that for the linear transformation of Example 7.1 any nonzero vector of the form $(s, 0, 0)$ is an eigenvector corresponding to $\lambda_1 = 0$, whereas any nonzero vector of the form $(0, t, t)$ is an eigenvector for $\lambda_2 = 2$. In general, if T is any linear transformation with eigenvalue λ_0 and corresponding eigenvector $\mathbf{x}_0$, then all vectors of the form $c\mathbf{x}_0, c \neq 0$, are eigenvectors of T corresponding to λ_0. This statement follows from the fact that

$$T(c\mathbf{x}_0) = cT(\mathbf{x}_0) = c(\lambda\mathbf{x}_0) = \lambda(c\mathbf{x}_0).$$

EXAMPLE 7.2

Let $T: \mathbf{R}^2 \to \mathbf{R}^2$ be the linear transformation that maps every vector into its reflection in the x-axis (see Example 6.9). Find the eigenvalues and eigenvectors of T.

Solution

By the nature of the given transformation, the only nonzero vectors collinear with their images under T are those that lie along the coordinate axes. Thus all eigenvectors of T are of the form $c_1(1, 0)$ or $c_2(0, 1)$, where c_1 and c_2 are nonzero scalars. Moreover, for any vector $\mathbf{x}$ lying on the x-axis, $T(\mathbf{x}) = \mathbf{x}$, whereas if $\mathbf{x}$ lies on the y-axis, we have $T(\mathbf{x}) = -\mathbf{x}$. So the eigenvectors of the form $c_1(1, 0)$ satisfy $T(\mathbf{x}) = \mathbf{x}$, whereas those of the form $c_2(0, 1)$ satisfy $T(\mathbf{x}) = -\mathbf{x}$. Comparing these equations to the defining one for an eigenvalue, $T(\mathbf{x}) = \lambda\mathbf{x}$, we see that the eigenvalue corresponding to $c_1(1, 0)$ is $\lambda_1 = 1$ and that corresponding to $c_2(0, 1)$ is $\lambda_2 = -1$ (see Figure 7.1).

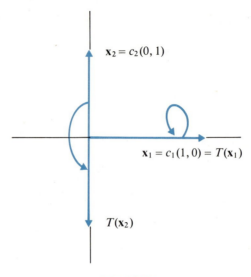

FIGURE 7.1

NOTE   Let $T: \mathbf{V} \to \mathbf{V}$ be a linear transformation. For a scalar λ and a vector $\mathbf{x}$, we may rewrite the equation $T(\mathbf{x}) = \lambda\mathbf{x}$ as $T(\mathbf{x}) = \lambda I(\mathbf{x})$ or $(T - \lambda I)(\mathbf{x}) = \mathbf{0}$, where I is the identity transformation. This last equation tells us that a nonzero vector $\mathbf{x}$ is an eigenvector of T if and only if $\mathbf{x}$ lies in the kernel of the linear transformation $T - \lambda I$. From Theorem 1 of Section 6.3, we know that the kernel of this linear transformation is a subspace of $\mathbf{V}$. Thus the set of all eigenvectors of T that correspond to λ, together with the zero vector, forms a subspace of $\mathbf{V}$, $\text{Ker}\,(T - \lambda I)$. This subspace is called the **eigenspace** of λ. For example, the eigenspaces of $\lambda_1 = 1$ and $\lambda_2 = -1$ of Example 7.2 are $\{c_1(1, 0)\}$ and $\{c_2(0, 1)\}$, respectively.

In Section 6.4 we demonstrated that any linear transformation $T: \mathbf{V} \to \mathbf{W}$ may be represented as multiplication by an $n \times m$ matrix A. (In dealing with eigenvalues and eigenvectors, $\mathbf{V} = \mathbf{W}$, so $n = m$, and the same basis can be used for each space.) This matrix A may then be used to calculate eigenvalues and eigenvectors of the transformation T. For this reason, we can restrict our attention to the matrix case. The following definition and theorem make this idea precise. The proof is an immediate consequence of Theorem 1 of Section 6.4 and the definition of eigenvalue and eigenvector.

DEFINITION   Let A be an $n \times n$ matrix. A scalar λ is called an **eigenvalue** of A if there is a *nonzero* vector $\mathbf{x}$ in $\mathbf{R}^n$ such that $A\mathbf{x} = \lambda\mathbf{x}$. The vector $\mathbf{x}$ is said to be an **eigenvector** corresponding to the eigenvalue λ.

THEOREM 1   Let $T: \mathbf{V} \to \mathbf{V}$ be a linear transformation, let $\mathscr{B}$ be a basis of $\mathbf{V}$, and let A be the matrix that represents T with respect to $\mathscr{B}$. Let λ be a real number, and let $\mathbf{v}$ be a vector in $\mathbf{V}$ with coordinate vector $\mathbf{x}$ with respect to $\mathscr{B}$. Then λ is an eigenvalue of T with corresponding eigenvector $\mathbf{v}$ if and only if λ is an eigenvalue of A with corresponding eigenvector $\mathbf{x}$.

We will now develop a systematic procedure for determining the eigenvalues and eigenvectors of an $n \times n$ matrix A. First suppose we know that λ_0 is an eigenvalue of A. How do we find corresponding eigenvectors $\mathbf{x}$? By definition, $A\mathbf{x} = \lambda_0\mathbf{x}$, which may be rewritten as $A\mathbf{x} = \lambda_0 I\mathbf{x}$ or $(A - \lambda_0 I)\mathbf{x} = \mathbf{0}$, where I is the $n \times n$ identity matrix. Thus $\mathbf{x}$ is an eivenvector corresponding to λ_0 if and only if it is a nonzero solution of the homogeneous linear system $(A - \lambda_0 I)\mathbf{x} = \mathbf{0}$.

EXAMPLE 7.3   The matrix

$$A = \begin{bmatrix} 1 & 1 \\ 6 & 2 \end{bmatrix}$$

has eigenvalues $\lambda_1 = -1$ and $\lambda_2 = 4$. Find the corresponding eigenvectors.

Solution   All eigenvectors corresponding to $\lambda_1 = -1$ will be nonzero solutions of the homogeneous linear system $(A - \lambda_1 I)\mathbf{x} = \mathbf{0}$ or $(A + I)\mathbf{x} = \mathbf{0}$, where

$$A + I = \begin{bmatrix} 2 & 1 \\ 6 & 3 \end{bmatrix}.$$

The solution of this system is found to be all vectors of the form $(s, -2s)$. So all eigenvectors corresponding to $\lambda_1 = -1$ are of the form $(s, -2s)$ where $s \neq 0$.

Similarly, to find the eigenvectors corresponding to $\lambda_2 = 4$, we solve the system $(A - 4I)\mathbf{x} = \mathbf{0}$. This yields eigenvectors of the form $(t, 3t)$, $t \neq 0$.

In Example 7.3 we saw how to find eigenvectors corresponding to a particular known eigenvalue. The next question is how to find the eigenvalues of an $n \times n$ matrix A in the first place.

Consider the equation $(A - \lambda I)\mathbf{x} = \mathbf{0}$. We seek scalars λ for which there exists *some nonzero* $\mathbf{x}$ satisfying the equation. By Theorem 5 of Section 2.5, there exists a nonzero solution $\mathbf{x}$ of the linear homogeneous system $(A - \lambda I)\mathbf{x} = \mathbf{0}$ if and only if the matrix $A - \lambda I$ is noninvertible. So the problem of finding the eigenvalues of A reduces to that of determining all scalars λ for which $A - \lambda I$ is noninvertible. But by Theorem 5 of Section 3.2, $A - \lambda I$ is noninvertible if and only if $\det(A - \lambda I) = 0$. Thus λ is an eigenvalue of A if and only if $\det(A - \lambda I) = 0$.

EXAMPLE 7.4   Find the eigenvalues of

$$A = \begin{bmatrix} 3 & -1 \\ 2 & 0 \end{bmatrix}.$$

Solution   We must determine all scalars λ for which $\det(A - \lambda I) = 0$. Now

$$A - \lambda I = \begin{bmatrix} 3 & -1 \\ 2 & 0 \end{bmatrix} - \lambda \begin{bmatrix} 1 & 0 \\ 0 & 1 \end{bmatrix} = \begin{bmatrix} 3 - \lambda & -1 \\ 2 & -\lambda \end{bmatrix},$$

so $\det(A - \lambda I) = (3 - \lambda)(-\lambda) + 2 = \lambda^2 - 3\lambda + 2$. Thus the eigenvalues of A are those scalars λ that satisfy $\lambda^2 - 3\lambda + 2 = 0$. Solving this quadratic equation, we obtain $\lambda_1 = 1$ and $\lambda_2 = 2$.

The discussions that preceded Examples 7.3 and 7.4 can be formalized to provide the proof of the following theorem.

THEOREM 2 Let A be an $n \times n$ matrix, and let I be the identity matrix of order n. Then

a. The eigenvalues of A are those scalars λ that satisfy the equation $\det(A - \lambda I) = 0$;

b. The eigenvectors of A corresponding to an eigenvalue λ_0 are the non-zero solutions $\mathbf{x}$ of the homogeneous linear system $(A - \lambda_0 I)\mathbf{x} = \mathbf{0}$.

To facilitate the implementation of Theorem 2, we restate it as the following procedure.

FINDING EIGENVALUES AND CORRESPONDING EIGENVECTORS OF A SQUARE MATRIX A

i. Form the matrix $A - \lambda I$ by subtracting λ from each diagonal entry of A.

ii. Solve the equation $\det(A - \lambda I) = 0$. The real solutions are the eigenvalues of A.

iii. For each eigenvalue λ_0, form the matrix $A - \lambda_0 I$.

iv. Solve the homogeneous linear system $(A - \lambda_0 I)\mathbf{x} = \mathbf{0}$. The nonzero solutions are the eigenvectors of A corresponding to λ_0 (procedure of Theorem 4 of Section 4.3).

DEFINITION

The **eigenspace** of an eigenvalue λ_0 of a matrix A is the set of *all* solutions of $(A - \lambda_0 I)\mathbf{x} = \mathbf{0}$.

From Theorem 4 of Section 4.3, the dimension of the eigenspace for λ_0 is $n - r$, where r is the rank of $A - \lambda_0 I$. The procedure associated with this theorem provides a basis for the eigenspace.

EXAMPLE 7.5

Find the eigenvalues and corresponding eigenvectors, as well as a basis for each eigenspace, of the matrix

$$A = \begin{bmatrix} 4 & 1 & -3 \\ 0 & 0 & 2 \\ 0 & 0 & -3 \end{bmatrix}.$$

Solution

Following the procedure for finding the eigenvalues of A, we compute

$$\det(A - \lambda I) = \det \begin{bmatrix} 4 - \lambda & 1 & -3 \\ 0 & -\lambda & 2 \\ 0 & 0 & -3 - \lambda \end{bmatrix} = (4 - \lambda)(-\lambda)(-3 - \lambda).$$

We now set this expression equal to zero, obtaining the equation

$$(4 - \lambda)(-\lambda)(-3 - \lambda) = 0.$$

The eigenvalues of A are the roots of this equation:

$$\lambda_1 = 4, \qquad \lambda_2 = 0, \qquad \text{and} \qquad \lambda_3 = -3.$$

To find the eigenvectors corresponding to $\lambda_1 = 4$, we form the matrix

$$A - 4I = \begin{bmatrix} 0 & 1 & -3 \\ 0 & -4 & 2 \\ 0 & 0 & -7 \end{bmatrix}$$

and solve the homogeneous linear system $(A - 4I)\mathbf{x} = \mathbf{0}$. Doing so, we see that all eigenvectors corresponding to $\lambda_1 = 4$ take the form $(t, 0, 0)$, where $t \neq 0$, and that a basis for the eigenspace of $\lambda_1 = 4$ is $\{(1, 0, 0)\}$.

To find the eigenspaces of $\lambda_2 = 0$ and $\lambda_3 = -3$, we work in a similar manner with the matrices $A - 0I \, (= A)$ and $A + 3I$, respectively. This yields eigenvectors of the form $(r, -4r, 0)$, $r \neq 0$, for λ_2 and $(11s, -14s, 21s)$, $s \neq 0$, for λ_3. The respective eigenspaces have bases given by $\{(1, -4, 0)\}$ and $\{(11, -14, 21)\}$.

The equation $\det(A - \lambda I) = 0$, whose roots are the eigenvalues of A, is called the **characteristic equation** of A. If A is $n \times n$, the expression $\det(A - \lambda I)$ is a polynomial of degree equal to n (see Exercise 26 of Section 3.1), known as the **characteristic polynomial** for A. For example, the characteristic polynomials of Examples 7.4 and 7.5 are, respectively, $\lambda^2 - 3\lambda + 2$ (of degree 2) and $(4 - \lambda)(-\lambda)(-3 - \lambda) = -\lambda^3 + \lambda^2 + 12\lambda$ (of degree 3). Since an nth degree polynomial equation has at most n roots, we see that an $n \times n$ matrix has *at most* n eigenvalues. As the next example illustrates, a matrix may have *fewer* than n eigenvalues.

EXAMPLE 7.6

Find the eigenvalues and bases for the corresponding eigenspaces of

$$A = \begin{bmatrix} -1 & 1 & 1 & 0 \\ 0 & 0 & 0 & -1 \\ 0 & 1 & 0 & 0 \\ 0 & 1 & 1 & -1 \end{bmatrix}.$$

Solution

Here the characteristic polynomial is of fourth degree and is computed (by expanding down the first column of $A - \lambda I$) to be

$$\det (A - \lambda I) = (-1 - \lambda) \det \begin{bmatrix} -\lambda & 0 & -1 \\ 1 & -\lambda & 0 \\ 1 & 1 & -1 - \lambda \end{bmatrix}$$

$$= (-1 - \lambda)[-\lambda^3 - \lambda^2 - \lambda - 1].$$

Thus one of the eigenvalues of A is $\lambda = -1$ (from the first factor), and the others are the real roots of the cubic equation $\lambda^3 + \lambda^2 + \lambda + 1 = 0$ (from the second factor). Since

$$\lambda^3 + \lambda^2 + \lambda + 1 = \lambda^2(\lambda + 1) + (\lambda + 1) = (\lambda^2 + 1)(\lambda + 1),$$

we see that -1 also satisfies this equation. We can find the remaining eigenvalues of A by solving the quadratic equation $\lambda^2 + 1 = 0$. Since the latter has no real roots (try, for example, the quadratic formula), A has only one eigenvalue, $\lambda_1 = -1$.

To find the eigenspace of this eigenvalue, we form the matrix

$$A - (-1)I = A + I = \begin{bmatrix} 0 & 1 & 1 & 0 \\ 0 & 1 & 0 & -1 \\ 0 & 1 & 1 & 0 \\ 0 & 1 & 1 & 0 \end{bmatrix}$$

and solve the system $(A + I)\mathbf{x} = \mathbf{0}$. Doing so, we see that the eigenspace of $\lambda_1 = -1$ is the set of all vectors of the form

$$(s, t, -t, t) = s(1, 0, 0, 0) + t(0, 1, -1, 1).$$

A basis for this eigenspace is $\{(1, 0, 0, 0), (0, 1, -1, 1)\}$.

 COMPUTATIONAL NOTE As you can see from Example 7.6, it is not very easy to determine the eigenvalues and eigenvectors of an $n \times n$ matrix by the procedure of Theorem 2, even when n is small. Moreover, this technique is not a very efficient one to use in computer applications and may lead to inaccurate answers, because it is somewhat susceptible to the accumulation of round-off error. For an outline of a better way to use a computer to approximate eigenvalues, see the Computational Note of Exercise 33 of Section 7.2. This matter is also discussed in the latter part of Section 8.5.

···

You may have noticed in Example 7.5 that the eigenvalues of A turned out to be identical to the diagonal entries of this matrix. However, as you can see from Examples 7.4 and 7.6, this is certainly not always true. The following theorem gives three special cases in which the diagonal entries of a matrix *are* its eigenvalues. Its proof follows from Theorem 4 of Section 3.1 and is left as an exercise.

THEOREM 3 The eigenvalues of an upper triangular, a lower triangular, or a diagonal matrix are identical to its diagonal entries.

EXAMPLE 7.7

Find the eigenvalues of the matrices

$$A = \begin{bmatrix} 8 & 0 & 0 \\ 0 & 0 & 0 \\ 0 & 0 & 7 \end{bmatrix} \quad \text{and} \quad B = \begin{bmatrix} 1 & 0 & 0 & 0 \\ 3 & -2 & 0 & 0 \\ 2 & -1 & 4 & 0 \\ 6 & 0 & 5 & -3 \end{bmatrix}.$$

Solution

The matrices A and B are diagonal and lower triangular, respectively. So applying Theorem 3, the eigenvalues of A are 8, 0, and 7, whereas those of B are 1, -2, 4, and -3.

7.1 EXERCISES

In Exercises 1–8 find the eigenvalues and eigenvectors of the given linear transformation.

1. The identity transformation, $I: \mathbf{R}^n \to \mathbf{R}^n$ defined by $I(\mathbf{x}) = \mathbf{x}$

2. The zero transformation, $0: \mathbf{R}^n \to \mathbf{R}^n$ defined by $0(\mathbf{x}) = \mathbf{0}$

3. The transformation $T_{\pi/2}: \mathbf{R}^2 \to \mathbf{R}^2$, which rotates every vector $\mathbf{x}$ in $\mathbf{R}^2$ by $\pi/2$

4. The transformation $T_\pi: \mathbf{R}^2 \to \mathbf{R}^2$, which rotates every vector $\mathbf{x}$ in $\mathbf{R}^2$ by π

5. The transformation $T^{\pi/2}: \mathbf{R}^2 \to \mathbf{R}^2$, which reflects every vector $\mathbf{x}$ in $\mathbf{R}^2$ in the y-axis

6. The transformation $T^{\pi/4}: \mathbf{R}^2 \to \mathbf{R}^2$, which reflects every vector $\mathbf{x}$ in $\mathbf{R}^2$ in the line $y = x$

7. The transformation $T: \mathbf{P}_3 \to \mathbf{P}_3$ given by

$$T(a_0 + a_1x + a_2x^2 + a_3x^3) = (a_0 + a_2) + (a_1 + 2a_3)x^3$$

8. The transformation $T: \mathbf{M}^{2,2} \to \mathbf{M}^{2,2}$ given by

$$T\left(\begin{bmatrix} a_{11} & a_{12} \\ a_{21} & a_{22} \end{bmatrix}\right) = \begin{bmatrix} a_{11} & -a_{21} \\ -a_{21} & a_{11} - a_{12} \end{bmatrix}$$

In Exercises 9–20 find the characteristic equation and eigenvalues of the given matrix.

9. $\begin{bmatrix} 1 & 4 \\ 2 & -1 \end{bmatrix}$ **10.** $\begin{bmatrix} 1 & 2 \\ -2 & -1 \end{bmatrix}$

11. $\begin{bmatrix} 0 & 1 \\ -1 & 2 \end{bmatrix}$ **12.** $\begin{bmatrix} 0 & 0 \\ 0 & 0 \end{bmatrix}$

13. $\begin{bmatrix} 0 & -1 & 0 \\ -1 & 0 & -1 \\ 0 & -1 & 0 \end{bmatrix}$ **14.** $\begin{bmatrix} 2 & -1 & 0 \\ -1 & 2 & -1 \\ 0 & -1 & 2 \end{bmatrix}$

15. $\begin{bmatrix} 1 & 1 & 1 \\ 1 & 1 & 1 \\ 1 & 1 & 1 \end{bmatrix}$ **16.** $\begin{bmatrix} 1 & 0 & -1 \\ 1 & 1 & 2 \\ 2 & 1 & 1 \end{bmatrix}$

17. $\begin{bmatrix} 1 & 0 & 3 \\ 0 & -2 & 1 \\ 0 & 0 & 1 \end{bmatrix}$ **18.** $\begin{bmatrix} 1 & 0 & 0 & 0 \\ 0 & 2 & 0 & 0 \\ 3 & 1 & -1 & 0 \\ -2 & 0 & 0 & 2 \end{bmatrix}$

19. $\begin{bmatrix} 1 & 1 & 1 & 1 \\ 1 & 1 & 1 & 1 \\ 1 & 1 & 1 & 1 \\ 1 & 1 & 1 & 1 \end{bmatrix}$ **20.** $\begin{bmatrix} 1 & 0 & 0 & 0 \\ 0 & 0 & 1 & 0 \\ 0 & 0 & 0 & 1 \\ 0 & 1 & 0 & 0 \end{bmatrix}$

21–32. Find all eigenvectors and bases for the eigenspaces of the matrices in Exercises 9–20.

In Exercises 33–36 find the eigenvalues of the given matrix by inspection.

33. $\begin{bmatrix} 1 & 2 & 0 \\ 0 & 3 & -1 \\ 0 & 0 & 4 \end{bmatrix}$ **34.** $\begin{bmatrix} 1 & 0 & 0 \\ 0 & 5 & 0 \\ 0 & 0 & -6 \end{bmatrix}$

33. $\begin{bmatrix} 1 & 0 & 0 & 0 \\ 2 & 3 & 0 & 0 \\ 4 & 5 & 6 & 0 \\ 7 & 8 & 9 & 10 \end{bmatrix}$ **36.** $\begin{bmatrix} -1 & 0 & 1 & 0 \\ 0 & 0 & 0 & -2 \\ 0 & 0 & 2 & 0 \\ 0 & 0 & 0 & 0 \end{bmatrix}$

37. Prove Theorem 3.

38. Prove that 0 is an eigenvalue of a square matrix A if and only if A is noninvertible. (*Hint*: Use Theorem 5 of Section 2.5 or Theorem 5 of Section 3.2.)

39. Let $T_\theta: \mathbf{R}^2 \to \mathbf{R}^2$ be the rotation by θ (see Example 6.7). Find the eigenvalues and eigenvectors of T_θ by (a) using the definition and (b) working with the matrix associated with T_θ.

40. Let $T^\theta: \mathbf{R}^2 \to \mathbf{R}^2$ be the reflection in the line through the origin that forms an angle θ with positive x-axis (see Example 6.9). Follow the instructions of Exercise 39 for T^θ.

41. Prove that A and A^T (the *transpose* of A) have the same eigenvalues.

42. Prove that λ is an eigenvalue of a nonsingular matrix A if and only if $1/\lambda$ is an eigenvalue of A^{-1}. What relationship holds between the eigenvectors of A and A^{-1}?

43. For the matrix

$$A = \begin{bmatrix} a & b \\ c & d \end{bmatrix},$$

show that A has (a) two eigenvalues if $(a - d)^2 + 4bc > 0$, (b) one eigenvalue if $(a - d)^2 + 4bc = 0$, and (c) no eigenvalues if $(a - d)^2 + 4bc < 0$.

44. Prove that the eigenvalues of a matrix are "invariant under similarity transformation." That is, if A is an $n \times n$ matrix, and P is an invertible matrix of order n, then the eigenvalues of A and $P^{-1}AP$ are identical.
(*Hint:* $\det [P^{-1}AP - \lambda I] = \det [P^{-1}(AP - \lambda PI)] = \det [P^{-1}(A - \lambda I)P].$)

45–48. If $P(x) = a_n x^n + \cdots + a_1 x + a_0$ is a polynomial with coefficients a_i, $i = 0, 1, \ldots, n$, from the real numbers, and A is an order n square matrix, then $P(A)$ is defined to be the matrix

$$P(A) = a_n A^n + \cdots + a_1 A + a_0 I.$$

The *Cayley–Hamilton theorem* states that if $P(x)$ is the characteristic polynomial of a square matrix A, then $P(A) = 0$. Verify the Cayley–Hamilton theorem for each matrix in Exercises 11–14.

7.2 DIAGONALIZATION

As we know from Section 6.4, the matrix of a linear transformation T from $\mathbf{V}$ into itself depends on the choice of basis for $\mathbf{V}$. Moreover (from Section 6.5), A and B are both matrices of a transformation T (corresponding to possibly different bases for $\mathbf{V}$) if and only if A and B are *similar*; that is, if and only if there is a nonsingular matrix P such that $B = P^{-1}AP$. Consequently the problem investigated in this section may be phrased in two equivalent ways:

i. If T is a linear transformation of $\mathbf{V}$ into itself, can we determine a basis for $\mathbf{V}$ for which the matrix associated with T is diagonal?

ii. If A is a square matrix, can we determine a nonsingular matrix P such that $P^{-1}AP$ is a diagonal matrix?

In either formulation we refer to this problem as **diagonalization** of the transformation T or the matrix A. As in Section 7.1, we will concentrate our efforts on the *matrix* rather than the transformation.

DEFINITION

A square matrix A is **diagonalizable** (or can be **diagonalized**) if it is similar to a diagonal matrix.

The following procedure will be justified by Theorems 1 through 3, given later in this section.

PROCEDURE FOR DIAGONALIZING A MATRIX

Let A be a square matrix of order n.

    i. Calculate the eigenvalues of A.

    ii. Calculate a basis for the eigenspace of each eigenvalue.

    iii. Let $\mathcal{B} = \{\mathbf{x}_1, \mathbf{x}_2, \ldots, \mathbf{x}_m\}$ be the union of all of the bases of (ii).

Then:

    iv. If $m < n$, A is not diagonalizable.

    v. If $m = n$, $\mathcal{B}$ is a basis for $\mathbf{R}^n$ and A is diagonalizable. A diagonalization of A is given by the $n \times n$ matrix D with ith diagonal element $d_{ii} = \lambda_i$, where λ_i is the eigenvalue to which $\mathbf{x}_i$ corresponds. (If P is the matrix with ith column $\mathbf{p}^i = \mathbf{x}_i$, then P is invertible and $D = P^{-1}AP$.)

EXAMPLE 7.8

Let

$$A = \begin{bmatrix} 1 & 0 & 0 \\ 0 & 0 & 1 \\ 0 & 1 & 0 \end{bmatrix}.$$

Show that A is diagonalizable, find a matrix P so that $P^{-1}AP$ is a diagonal matrix, and determine $P^{-1}AP$.

Solution

To diagonalize A, we first calculate the eigenvalues of A. To do so, we use the procedure of Theorem 2 of Section 7.1. The characteristic polynomial for A is

$$\det(A - \lambda I) = (1 - \lambda)(\lambda^2 - 1) = -(\lambda - 1)^2(\lambda + 1).$$

Thus the eigenvalues of A are $\lambda_1 = 1$ and $\lambda_2 = -1$. After some computation, we see that bases for their respective eigenspaces are $\{(1, 0, 0), (0, 1, 1)\}$ and $\{(0, 1, -1)\}$. Following the procedure above, the eigenvectors $(1, 0, 0)$, $(0, 1, 1)$, and $(0, 1, -1)$ form a basis for $\mathbf{R}^3$, and consequently A is diagonalizable. Moreover, the matrix P that we are seeking may be taken to be the one whose columns are these three eigenvectors, so that

$$P = \begin{bmatrix} 1 & 0 & 0 \\ 0 & 1 & 1 \\ 0 & 1 & -1 \end{bmatrix}.$$

Finally, to find the diagonal matrix $P^{-1}AP$, we could perform the indicated inversion and multiplications, but it is much easier to apply Step (v) of the procedure. Since the eigenvalues corresponding to the eigenvectors $(1, 0, 0)$, $(0, 1, 1)$, and $(0, 1, -1)$ are 1, 1, and -1, respectively, we have

$$P^{-1}AP = \begin{bmatrix} 1 & 0 & 0 \\ 0 & 1 & 0 \\ 0 & 0 & -1 \end{bmatrix}.$$

NOTE Reordering the eigenvectors would produce a different diagonal matrix, so if A is diagonalizable, it is usually similar to several diagonal matrices. Of course P changes accordingly.

EXAMPLE 7.9

Determine whether or not the matrix

$$A = \begin{bmatrix} 1 & 0 & 0 \\ 0 & 0 & -1 \\ 0 & 1 & 0 \end{bmatrix}$$

is diagonalizable.

Solution

The characteristic polynomial for A is $\det(A - \lambda I) = (1 - \lambda)(\lambda^2 + 1)$, so A has only one real number eigenvalue $\lambda = 1$. Moreover a basis for the eigenspace of $\lambda = 1$ is given by $\{(1, 0, 0)\}$ and is consequently one-dimensional. Thus A has no set of three independent eigenvectors and, by the procedure, it is not diagonalizable.

In Example 7.9, two of the roots of the characteristic equation were imaginary numbers. The next example illustrates that even if *all* its characteristic roots are real, a matrix still may not be diagonalizable.

EXAMPLE 7.10

Determine whether or not the matrix

$$A = \begin{bmatrix} 1 & 2 & -1 \\ 0 & -1 & 3 \\ 0 & 0 & 1 \end{bmatrix}$$

is diagonalizable.

Solution

Since A is upper triangular, by Theorem 3 of Section 7.1 its eigenvalues are given by the diagonal entries. Therefore A has two distinct eigenvalues, $\lambda_1 = 1$ and $\lambda_2 = -1$. Computing bases for the corresponding eigenspaces, we see that $\{(1, 0, 0)\}$ is a basis for the eigenspace of λ_1 and $\{(-1, 1, 0)\}$ is a basis for that of λ_2. Thus all eigenvectors are nonzero scalar multiples of these two vectors, and A has no set of three independent eigenvectors. Consequently, by the procedure, A is not diagonalizable.

SPOTLIGHT

DIAGONALIZATION AND POWERS OF A MATRIX

In some applications of linear algebra it is necessary to take the nth power of a given matrix A. (See Section 8.5 for a specific example.) If n is large, this process can be tedious and time-consuming. However, if we know a diagonalization, $D = P^{-1}AP$ of A, there is a simpler alternative. In this case, $A = PDP^{-1}$, so

$$A^n = (PDP^{-1})^n = (PDP^{-1})(PDP^{-1})(PDP^{-1}) \cdots (PDP^{-1}).$$

But $P^{-1}P = I$, so this simplifies to

$$A^n = PD^nP^{-1}.$$

Thus, instead of taking the nth power of A, we can perform the much simpler calculation of taking the nth power of the diagonal matrix D.

As an example, let

$$A = \begin{bmatrix} 1 & 2 \\ 2 & 4 \end{bmatrix}.$$

Then, $A = PDP^{-1}$, where

$$P = \begin{bmatrix} -2 & 1 \\ 1 & 2 \end{bmatrix}, \quad D = \begin{bmatrix} 0 & 0 \\ 0 & 5 \end{bmatrix}, \quad \text{and} \quad P^{-1} = \begin{bmatrix} -\frac{2}{5} & \frac{1}{5} \\ \frac{1}{5} & \frac{2}{5} \end{bmatrix}.$$

Now

$$A^n = PD^nP^{-1}$$

$$= \begin{bmatrix} -2 & 1 \\ 1 & 2 \end{bmatrix} \begin{bmatrix} 0 & 0 \\ 0 & 5^n \end{bmatrix} \begin{bmatrix} -\frac{2}{5} & \frac{1}{5} \\ \frac{1}{5} & \frac{2}{5} \end{bmatrix} = \begin{bmatrix} 5^{n-1} & 2 \cdot 5^{n-1} \\ 2 \cdot 5^{n-1} & 4 \cdot 5^{n-1} \end{bmatrix}.$$

We now state and prove the theorems that support the preceding procedure. The first theorem gives the reason why $\mathscr{B}$ is a basis when it has "enough" vectors in Step (v) of the procedure for diagonalizing a matrix.

THEOREM 1   For a linear transformation $T : \mathbf{V} \rightarrow \mathbf{V}$, nonzero eigenvectors that correspond to *distinct* eigenvalues are linearly independent. Moreover, if $\lambda_1 \neq \lambda_2$ are eigenvalues, if $\mathscr{T}_1$ is a linearly independent set of vectors that corresponds to λ_1, and if $\mathscr{T}_2$ is a linearly independent set of vectors that corresponds to λ_2, then $\mathscr{T} = \mathscr{T}_1 \cup \mathscr{T}_2$ is linearly independent.

Proof   Let $\mathcal{T}_1 = \{\mathbf{x}_1, \ldots, \mathbf{x}_k\}$ and $\mathcal{T}_2 = \{\mathbf{x}_{k+1}, \ldots, \mathbf{x}_m\}$, and suppose $\mathbf{0}$ is a linear combination of the vectors in $\mathcal{T} = \mathcal{T}_1 \cup \mathcal{T}_2$. Then

$$a_1\mathbf{x}_1 + \cdots + a_k\mathbf{x}_k + a_{k+1}\mathbf{x}_{k+1} + \cdots + a_m\mathbf{x}_m = \mathbf{0}. \qquad (1)$$

Applying T to both sides and using the fact that each $\mathbf{x}_i$ is an eigenvector of one of λ_1 or λ_2, we have

$$T(a_1\mathbf{x}_1 + \cdots + a_k\mathbf{x}_k + a_{k+1}\mathbf{x}_{k+1} + \cdots + a_m\mathbf{x}_m) = T(\mathbf{0}) = \mathbf{0},$$

so

$$a_1\lambda_1\mathbf{x}_1 + \cdots + a_k\lambda_1\mathbf{x}_k + a_{k+1}\lambda_2\mathbf{x}_{k+1} + \cdots + a_m\lambda_2\mathbf{x}_m = \mathbf{0}. \qquad (2)$$

Multiplying Equation (1) by λ_1, we have

$$a_1\lambda_1\mathbf{x}_1 + \cdots + a_k\lambda_1\mathbf{x}_k + a_{k+1}\lambda_1\mathbf{x}_{k+1} + \cdots + a_m\lambda_1\mathbf{x}_m = \mathbf{0}. \qquad (3)$$

Subtracting Equation (3) from Equation (2), we have

$$a_{k+1}(\lambda_2 - \lambda_1)\mathbf{x}_{k+1} + \cdots + a_m(\lambda_2 - \lambda_1)\mathbf{x}_m = \mathbf{0}. \qquad (4)$$

But the left side of Equation (4) is a linear combination of vectors in $\mathcal{T}_2$, a linearly independent set. Therefore the coefficients are each equal to 0; that is,

$$a_{k+1}(\lambda_2 - \lambda_1) = \cdots = a_m(\lambda_2 - \lambda_1) = 0.$$

Since the eigenvalues are distinct, $\lambda_2 - \lambda_1 \neq 0$, so we have $a_{k+1} = \cdots = a_m = 0$.

Returning to Equation (1), we now have

$$a_1\mathbf{x}_1 + \cdots + a_k\mathbf{x}_k = \mathbf{0}. \qquad (5)$$

But the left side of Equation (5) is a linear combination of vectors in $\mathcal{T}_1$, also linearly independent. Therefore the rest of the coefficients must also equal $0, a_1 = \cdots = a_k = 0$.

In conclusion, any linear combination of vectors in $\mathcal{T}$ can equal $\mathbf{0}$ only if the coefficients are all 0. That is, $\mathcal{T} = \mathcal{T}_1 \cup \mathcal{T}_2$ is linearly independent.

THEOREM 2   A square matrix of order n is diagonalizable if and only if A has n independent eigenvectors; that is, if and only if there exists a basis for $\mathbf{R}^n$ of eigenvectors of A.

Proof   Suppose A is diagonalizable, D is a diagonal matrix, and P is the matrix of similarity, so that

$$D = P^{-1}AP.$$

Then $AP = PD$, and letting $\lambda_1, \ldots, \lambda_n$ be the diagonal entries of $D(d_{ii} = \lambda_i)$, we have for each column $\mathbf{p}^i$ of P, $A\mathbf{p}^i = \lambda_i\mathbf{p}^i$. But this implies that each column of P is a eigenvector of A corresponding to the eigenvalue λ_i. By Theorem 3 of Section 4.1, these columns are independent (since P is invertible) and thus A has n independent eigenvectors.

Conversely, suppose A possesses n independent eigenvectors $\mathbf{x}_1, \ldots, \mathbf{x}_n$ that correspond respectively to (not necessarily distinct) eigenvalues $\lambda_1, \ldots, \lambda_n$. Following the procedure, we let P be the matrix of columns $\mathbf{p}^i = \mathbf{x}_i$ and D the diagonal matrix with corresponding λ_i for the ith entry, $d_{ii} = \lambda_i$. By Theorem 3 of Section 4.1, P is invertible, since its columns are linearly independent. Since λ_i is an eigenvalue of A with corresponding eigenvector $\mathbf{x}_i$, we have

$$A\mathbf{p}^i = A\mathbf{x}_i = \lambda_i \mathbf{x}_i = \lambda_i \mathbf{p}^i.$$

Thus the ith column of AP is $\lambda_i \mathbf{p}^i$, so that the ith column of $P^{-1}AP$ is

$$P^{-1}(\lambda_i \mathbf{p}^i) = \lambda_i(P^{-1}\mathbf{p}^i) = \lambda_i \mathbf{e}_i.$$

But $\lambda_i \mathbf{e}_i$ is just the ith column of the diagonal matrix D. So, taking all columns collectively, we conclude that

$$D = P^{-1}AP.$$

In one important case we can find a diagonalization D of a matrix (although not the matrix of similarity P) without calculating a basis of eigenvectors for each eigenspace, as the procedure generally requires. In this case, only steps (i) and (v) of the procedure are necessary. The following theorem describes this fortunate situation.

THEOREM 3   Let A be an $n \times n$ matrix. If A has n distinct eigenvalues, then A is diagonalizable.

Proof   Choosing a nonzero eigenvector for each eigenvalue, we have n linearly independent eigenvectors, by Theorem 1. By Theorem 2, A is diagonalizable.

EXAMPLE 7.11

Determine whether or not the matrix

$$A = \begin{bmatrix} 1 & 1 \\ 1 & 1 \end{bmatrix}$$

is diagonalizable.

Solution

The characteristic polynomial $\det(A - \lambda I)$ is given by $\lambda^2 - 2\lambda$, which has zeros $\lambda_1 = 0$ and $\lambda_2 = 2$. Thus the 2×2 matrix A has two distinct eigenvalues, and Theorem 3 tells us that A is diagonalizable.

Notice that Theorem 3 gives only a *sufficient* condition that a matrix be diagonalizable. Consequently, if an $n \times n$ matrix A has fewer than n eigenvalues, then A *may* or *may not* be diagonalizable. For example, the matrices of Examples 7.8 and 7.10 have fewer than three eigenvalues, yet the first of these is diagonalizable, whereas the second is not.

DIAGONALIZATION OF SYMMETRIC MATRICES

We will now consider the class of *symmetric* matrices. These matrices can easily be identified by inspection. As we will see, all symmetric matrices are diagonalizable.

DEFINITION

A square matrix A is **symmetric** if it is equal to its transpose A^T.

It is very easy to determine whether or not a given matrix is symmetric. Simply check for symmetry with respect to the main diagonal. For example, the matrices

$$\begin{bmatrix} 0 & 1 \\ 1 & 0 \end{bmatrix} \quad \text{and} \quad \begin{bmatrix} 1 & -1 & 3 \\ -1 & 2 & 0 \\ 3 & 0 & -5 \end{bmatrix}$$

are symmetric, but

$$\begin{bmatrix} 1 & 1 & -1 \\ 1 & 1 & -1 \\ -1 & 1 & 1 \end{bmatrix}$$

is not.

We now proceed to investigate the diagonalization of symmetric matrices. First, recall from Section 1.4 that two vectors $\mathbf{x}$ and $\mathbf{y}$ in $\mathbf{R}^n$ are *orthogonal* if and only if their dot product, $\mathbf{x} \cdot \mathbf{y}$, is equal to zero.

THEOREM 4 Let λ_1 and λ_2 be distinct eigenvalues of a symmetric matrix A with corresponding eigenvectors $\mathbf{x}_1$ and $\mathbf{x}_2$. Then $\mathbf{x}_1$ and $\mathbf{x}_2$ are orthogonal.

Proof We must show that $\mathbf{x}_1 \cdot \mathbf{x}_2 = 0$. By definition, $A\mathbf{x}_1 = \lambda_1\mathbf{x}_1$, so

$$\lambda_1(\mathbf{x}_1 \cdot \mathbf{x}_2) = (\lambda_1\mathbf{x}_1) \cdot \mathbf{x}_2 = (A\mathbf{x}_1) \cdot \mathbf{x}_2,$$

which by Exercise 28 is equal to $\mathbf{x}_1 \cdot (A^T\mathbf{x}_2)$. Since A is symmetric, the last expression is equal to $\mathbf{x}_1 \cdot (A\mathbf{x}_2)$, which is the same as $\mathbf{x}_1 \cdot (\lambda_2\mathbf{x}_2)$ or $\lambda_2(\mathbf{x}_1 \cdot \mathbf{x}_2)$. Thus $\lambda_1(\mathbf{x}_1 \cdot \mathbf{x}_2) = \lambda_2(\mathbf{x}_1 \cdot \mathbf{x}_2)$. Now, if $\mathbf{x}_1 \cdot \mathbf{x}_2$ were not 0, then dividing by this quantity, we would get $\lambda_1 = \lambda_2$, a contradiction. Therefore $\mathbf{x}_1 \cdot \mathbf{x}_2$ must be 0, as desired.

EXAMPLE 7.12   Verify Theorem 4 for the matrix

$$A = \begin{bmatrix} 3 & 2 \\ 2 & 0 \end{bmatrix}.$$

Solution   Since $\det (A - \lambda I) = \lambda^2 - 3\lambda - 4$, the eigenvalues of A are $\lambda_1 = 4$ and $\lambda_2 = -1$. Theorem 4 tells us that since A is symmetric and λ_1 and λ_2 are distinct, any pair of eigenvectors corresponding to these eigenvalues should be orthogonal. The eigenspaces of λ_1 and λ_2 are easily computed to be $\{(2s, s)\}$ and $\{(-t, 2t)\}$, respectively. To confirm that Theorem 4 holds in this case, we form the dot product

$$(2s, s) \cdot (-t, 2t) = -2st + 2st,$$

which is indeed equal to 0.

Since the matrix A of Example 7.12 has distinct eigenvalues, Theorem 3 implies that A is diagonalizable. In other words, there is a 2×2 invertible matrix P such that $P^{-1}AP = D$, a diagonal matrix. We can then apply the diagonalization procedure of this section to construct P by taking its columns to be linearly independent eigenvectors of A. By choosing these eigenvectors to have length 1 (that is, *normalizing* them), we obtain a matrix with particularly interesting properties. The matrix of normalized eigenvectors is

$$P = \begin{bmatrix} 2/\sqrt{5} & -1/\sqrt{5} \\ 1/\sqrt{5} & 2/\sqrt{5} \end{bmatrix}.$$

We not only have $\mathbf{p}^1 \cdot \mathbf{p}^2 = 0$ (reflecting the orthogonality of the eigenvectors) and $\mathbf{p}^1 \cdot \mathbf{p}^1 = \mathbf{p}^2 \cdot \mathbf{p}^2 = 1$ (reflecting the fact that we chose eigenvectors of length 1), but, computing P^{-1}, we see that

$$P^{-1} = \begin{bmatrix} 2/\sqrt{5} & 1/\sqrt{5} \\ -1/\sqrt{5} & 2/\sqrt{5} \end{bmatrix},$$

which is just P^T. As we shall see, *every* symmetric matrix A (whether or not it has distinct eigenvalues) is diagonalizable by a matrix P having these properties.

DEFINITION   A nonsingular matrix P is **orthogonal** if $P^{-1} = P^T$.

EXAMPLE 7.13   Let

$$P = \begin{bmatrix} 1/3 & 2\sqrt{2}/3 & 0 \\ 2/3 & -\sqrt{2}/6 & \sqrt{2}/2 \\ -2/3 & \sqrt{2}/6 & \sqrt{2}/2 \end{bmatrix}.$$

Show that P is orthogonal.

Solution

We could compute P^{-1} and then verify that $P^{-1} = P^T$, but there is an easier way. We simply compute the product $P^T P$. Since $P^T P = I$, by Theorem 2 of Section 2.4, $P^{-1} = P^T$, and thus P is orthogonal by definition.

NOTE It is easy to see that the columns of the matrix P of Example 7.13 have the following properties:

    i. $\mathbf{p}^i \cdot \mathbf{p}^j = 0$, if $i \neq j$ $(i, j = 1, 2, 3)$,

    ii. $\mathbf{p}^i \cdot \mathbf{p}^i = 1$ $(i = 1, 2, 3)$.

Recall that a set of vectors that possesses these two properties is called an *orthonormal set*. The proof of the next theorem is Exercise 29.

THEOREM 5 An $n \times n$ matrix P is orthogonal if and only if its columns

$$\{\mathbf{p}^1, \mathbf{p}^2, \ldots, \mathbf{p}^n\}$$

form an orthonormal set.

DEFINITION

A square matrix A is **orthogonally diagonalizable** if there exists an orthogonal matrix P such that $P^{-1}AP\,(=P^T AP)$ is a diagonal matrix.

The next theorem states that *every* symmetric matrix is orthogonally diagonalizable. The proof is omitted.

THEOREM 6 Let A be a symmetric matrix. Then there exists an orthogonal matrix P such that $D = P^{-1}AP$ is a diagonal matrix.

NOTE If A is a symmetric matrix, then by Theorem 6 it is diagonalizable, and for each i, $A\mathbf{p}^i = PP^{-1}A\mathbf{p}^i = P\mathbf{d}^i = d_{ii}\mathbf{p}^i$. Thus we have

    i. The columns of P are linearly independent eigenvectors of A, $\mathbf{x}_i = \mathbf{p}^i$, $i = 1, \ldots, n$,

    ii. The ith diagonal entry $\lambda_i = d_{ii}$ of D is the eigenvalue of A corresponding to the eigenvector $\mathbf{x}_i$.

Moreover, by Theorem 5 the columns of P form an orthonormal set. Consequently $\{\mathbf{x}_1, \mathbf{x}_2, \ldots, \mathbf{x}_n\}$ is an orthonormal set of eigenvectors.

PROCEDURE FOR ORTHOGONALLY DIAGONALIZING A SYMMETRIC MATRIX

Let A be a symmetric matrix of order n.

i. Calculate the eigenvalues of A.

ii. Calculate a basis for the eigenspace of each eigenvalue.

iii. With the aid of the Gram–Schmidt process (Section 5.4) replace each basis by an orthonormal basis. The union of all these bases is an orthonormal basis for $\mathbf{R}^n$.

iv. A diagonalization of A is the diagonal matrix D with $d_{ii} = \lambda_i$, the ith eigenvalue of A. The orthogonal matrix P has ith column $\mathbf{p}^i = \mathbf{x}_i$, the ith vector in the basis.

EXAMPLE 7.14   Let

$$A = \begin{bmatrix} -1 & 0 & 0 \\ 0 & 0 & 2 \\ 0 & 2 & 3 \end{bmatrix}.$$

Is the matrix A orthogonally diagonalizable? Is so, find an orthogonal matrix P and diagonal matrix D such that $D = P^{-1}AP$.

Solution   Since A is symmetric, by Theorem 6, A is orthogonally diagonalizable.

To find the matrices P and D, we apply the procedure just described. We first find the eigenvalues of A by solving the characteristic equation $\det(A - \lambda I) = 0$, where

$$\det(A - \lambda I) = \det \begin{bmatrix} -1 - \lambda & 0 & 0 \\ 0 & -\lambda & 2 \\ 0 & 2 & 3 - \lambda \end{bmatrix} = -(\lambda + 1)^2 (\lambda - 4).$$

Thus the eigenvalues of A are $\lambda_1 = -1$ and $\lambda_2 = 4$.

We now determine bases for the eigenspaces $\mathbf{S}_1$ and $\mathbf{S}_2$ of λ_1 and λ_2 by considering the linear homogeneous systems $(A + I)\mathbf{x} = \mathbf{0}$ and $(A - 4I)\mathbf{x} = \mathbf{0}$, respectively. Proceeding as in Section 7.1, we see that a basis for $\mathbf{S}_1$ is $\mathcal{T}_1 = \{(1, 0, 0), (0, -2, 1)\}$, whereas a basis for $\mathbf{S}_2$ is $\mathcal{T}_2 = \{(0, 1, 2)\}$.

Applying the Gram–Schmidt process to $\mathcal{T}_1$, we obtain the orthonormal basis $\mathcal{B}_1 = \{(1, 0, 0), (0, -2/\sqrt{5}, 1/\sqrt{5})\}$ for the eigenspace of λ_1. (Since $\mathcal{T}_1$ was orthogonal, all we actually needed to do was to normalize each vector in this set.) Then, normalizing the vector $(0, 1, 2)$ of $\mathcal{T}_2$, we obtain an orthonormal basis $\mathcal{B}_2 = \{(0, 1\sqrt{5}, 2/\sqrt{5})\}$ for $\mathbf{S}_2$. Finally, taking the union of $\mathcal{B}_1$ and $\mathcal{B}_2$, we obtain an orthonormal set consisting of the three eigenvectors $\mathbf{x}_1 = (1, 0, 0)$, $\mathbf{x}_2 = (0, -2/\sqrt{5}, 1/\sqrt{5})$, and $\mathbf{x}_3 = (0, 1/\sqrt{5}, 2/\sqrt{5})$.

Thus the matrix P is given by $\mathbf{p}^i = \mathbf{x}_i$ or

$$P = \begin{bmatrix} 1 & 0 & 0 \\ 0 & -2/\sqrt{5} & 1/\sqrt{5} \\ 0 & 1/\sqrt{5} & 2/\sqrt{5} \end{bmatrix}.$$

To construct D, we simply form the 3×3 diagonal matrix whose diagonal entries are the eigenvalues corresponding to the eigenvectors $\mathbf{x}_1$, $\mathbf{x}_2$, and $\mathbf{x}_3$. Therefore

$$D = \begin{bmatrix} -1 & 0 & 0 \\ 0 & -1 & 0 \\ 0 & 0 & 4 \end{bmatrix}.$$

We now know that every symmetric matrix is orthogonally diagonalizable. A natural question to ask at this point is whether any other types of matrices are orthogonally diagonalizable. The next theorem, which is the converse of Theorem 6, provides a negative answer to this question. Its proof is Exercise 30.

THEOREM 7   If a square matrix A is orthogonally diagonalizable, it is symmetric.

7.2 EXERCISES

In Exercises 1–12 determine whether or not the given matrix A is diagonalizable. If it is, find a matrix P such that $P^{-1}AP$ is diagonal, and find this diagonal matrix.

1. $\begin{bmatrix} 1 & 4 \\ 2 & -1 \end{bmatrix}$

2. $\begin{bmatrix} 0 & 1 \\ 1 & 0 \end{bmatrix}$

3. $\begin{bmatrix} 1 & 2 \\ -2 & -1 \end{bmatrix}$

4. $\begin{bmatrix} 0 & 1 \\ -1 & 2 \end{bmatrix}$

5. $\begin{bmatrix} 0 & -1 & 0 \\ -1 & 0 & -1 \\ 0 & -1 & 0 \end{bmatrix}$

6. $\begin{bmatrix} 2 & -1 & 0 \\ -1 & 2 & -1 \\ 0 & -1 & 2 \end{bmatrix}$

7. $\begin{bmatrix} 1 & 1 & 1 \\ 1 & 1 & 1 \\ 1 & 1 & 1 \end{bmatrix}$

8. $\begin{bmatrix} 1 & 1 & -1 \\ -1 & -1 & 1 \\ 2 & 2 & -2 \end{bmatrix}$

9. $\begin{bmatrix} 1 & 0 & 3 \\ 0 & -2 & 1 \\ 0 & 0 & 1 \end{bmatrix}$

10. $\begin{bmatrix} 1 & 0 & -1 \\ 1 & 1 & 2 \\ 2 & 1 & 1 \end{bmatrix}$

11. $\begin{bmatrix} 1 & 1 & 1 & 1 \\ 1 & 1 & 1 & 1 \\ 1 & 1 & 1 & 1 \\ 1 & 1 & 1 & 1 \end{bmatrix}$

12. $\begin{bmatrix} 1 & 0 & 0 & 0 \\ 0 & 2 & 0 & 0 \\ 0 & 1 & -1 & 0 \\ -2 & 0 & 0 & 2 \end{bmatrix}$

In Exercises 13–16 determine whether or not the given matrix is orthogonal.

13. $\begin{bmatrix} \frac{1}{3} & \frac{2}{3} & \frac{2}{3} \\ \frac{2}{3} & \frac{1}{3} & -\frac{2}{3} \\ \frac{2}{3} & -\frac{2}{3} & \frac{1}{3} \end{bmatrix}$

14. $\begin{bmatrix} 0 & 1 & 0 \\ 1 & 0 & 0 \\ 0 & 0 & 1 \end{bmatrix}$

15. $\begin{bmatrix} 1/\sqrt{5} & -2/\sqrt{5} & 1 \\ -2/\sqrt{5} & 1/\sqrt{5} & -1 \\ 0 & 0 & 1 \end{bmatrix}$

16. $\begin{bmatrix} 0 & 1 & 0 \\ 1 & 0 & 1 \\ 0 & 1 & 0 \end{bmatrix}$

In Exercises 17–24 find an orthogonal matrix P such that $P^{-1}AP$ is diagonal, where A is the given symmetric matrix.

17. $\begin{bmatrix} 0 & 2 \\ 2 & 0 \end{bmatrix}$

18. $\begin{bmatrix} 1 & 1 \\ 1 & 1 \end{bmatrix}$

19. $\begin{bmatrix} 0 & -1 & 0 \\ -1 & 0 & -1 \\ 0 & -1 & 0 \end{bmatrix}$

20. $\begin{bmatrix} 2 & -1 & 0 \\ -1 & 2 & -1 \\ 0 & -1 & 2 \end{bmatrix}$

21. $\begin{bmatrix} 1 & 1 & 1 \\ 1 & 1 & 1 \\ 1 & 1 & 1 \end{bmatrix}$

22. $\begin{bmatrix} 0 & 1 & 0 \\ 1 & 0 & 0 \\ 0 & 0 & 1 \end{bmatrix}$

23. $\begin{bmatrix} 0 & 1 & 0 & 0 \\ 1 & 0 & 0 & 0 \\ 0 & 0 & 0 & 1 \\ 0 & 0 & 1 & 0 \end{bmatrix}$

24. $\begin{bmatrix} 1 & 1 & 1 & 1 \\ 1 & 1 & 1 & 1 \\ 1 & 1 & 1 & 1 \\ 1 & 1 & 1 & 1 \end{bmatrix}$

25. Show that the rotation matrix

$$A = \begin{bmatrix} \cos\theta & -\sin\theta \\ \sin\theta & \cos\theta \end{bmatrix}$$

is orthogonal.

26. Let A be an orthogonal matrix. Show that $\det(A) = \pm 1$.
(*Hint*: $\det(A^TA) = \det(I) = 1$.)

27. For the matrix

$$A = \begin{bmatrix} a & b \\ c & d \end{bmatrix},$$

obtain conditions on a, b, c, and d such that (a) A is diagonalizable and (b) A is not diagonalizable. (*Hint*: See Exercise 43 of Section 7.1.)

28. Let A be an $n \times n$ matrix, and let $\mathbf{x}_1$ and $\mathbf{x}_2$ be in $\mathbf{R}^n$. Prove that $(A\mathbf{x}_1) \cdot \mathbf{x}_2 = \mathbf{x}_1 \cdot (A^T\mathbf{x}_2)$. (*Hint*: Viewing $\mathbf{u}$ and $\mathbf{v}$ as column vectors, the dot product is given by $\mathbf{u} \cdot \mathbf{v} = \mathbf{v}^T\mathbf{u}$.)

29. Prove (Theorem 5) that an $n \times n$ matrix P is orthogonal if and only if $\{\mathbf{p}^1, \mathbf{p}^2, \ldots, \mathbf{p}^n\}$ is an orthonormal set.

30. Prove (Theorem 7) that if a square matrix is orthogonally diagonalizable, then it is symmetric. (*Hint*: Consider $(P^{-1}AP)^T = D^T$.)

31. Let P and Q be orthogonal matrices of the same order. Show that PQ and QP are also orthogonal. (*Hint*: Show that $(PQ)(PQ)^T = I$.)

32. Let θ be a real number, and let s and t be integers with $1 \leq s < t \leq n$. Define an $n \times n$ matrix P by

$$p_{ss} = p_{tt} = \cos \theta, \qquad p_{ts} = -p_{st} = \sin \theta,$$

and

$$p_{ii} = 1, \quad p_{ij} = 0 \quad (i \neq j), \qquad \text{otherwise.}$$

Show that P is orthogonal. (The matrix P is called a *rotation matrix for the s,t-plane.*)

33. Let P be a third-order rotation matrix in the 1, 3-plane (see Exercise 32), that is

$$P = \begin{bmatrix} \cos \theta & 0 & -\sin \theta \\ 0 & 1 & 0 \\ \sin \theta & 0 & \cos \theta \end{bmatrix}.$$

Let A be an arbitrary 3×3 symmetric matrix. Show that if we choose θ so that $\tan 2\theta = 2a_{13}/(a_{11} - a_{33})$, then the matrix $B = P^T A P$ $(= P^{-1} A P)$ satisfies $b_{13} = b_{31} = 0$. (*Note*: In general, if P is a nth order rotation in the s,t-plane, and A is symmetric, then by choosing θ so that $\tan 2\theta = 2a_{st}/(a_{ss} - a_{tt})$, both the st and the ts entries of $P^T A P$ are equal to 0.)

 COMPUTATIONAL NOTE *Jacobi's method* approximates the eigenvalues of a symmetric matrix A by forming the sequence of products

$$(P_k^T P_{k-1}^T \cdots P_1^T) A (P_1 P_2 \cdots P_k), \qquad k = 1, 2, 3, \ldots.$$

In this expression each P_i is a rotation matrix (see Exercise 32) with θ chosen to "annihilate" (send to zero) the largest off-diagonal entry of

$$(P_{i-1}^T \cdots P_1^T) A (P_1 \cdots P_{i-1})$$

(see Exercise 33). Unfortunately, as one off-diagonal entry is annihilated, a previously annihilated one may become nonzero. Nevertheless, the size of all off-diagonal entries does approach zero as the process is continued, and the diagonal entries will in turn get closer and closer to the eigenvalues of A. Jacobi's method is perhaps the most straightforward way of approximating the eigenvalues of a symmetric matrix, but there are more sophisticated techniques that are much more efficient.

7.3 COMPLEX EIGENVALUES AND EIGENVECTORS

To this point in the text all scalars, including components of vectors and entries of matrices, have been real numbers. In this section we will allow scalars to be complex numbers. This generalization of the concept of scalar will alter very little of the material of the first six chapters and will make the subject of eigenvalues and eigenvectors more complete.

COMPLEX m-SPACE

Recall that a **complex number** is one of the form $a + bi$, where a and b are real numbers and $i = \sqrt{-1}$. We will denote the set of all complex numbers by **C**. The **absolute value (modulus)** of a complex number $z = a + bi$ is denoted $|z|$ and defined to be $|z| = \sqrt{a^2 + b^2}$. The **complex conjugate** of $z = a + bi$ is denoted $\bar{z}$ and is given by $\bar{z} = a - bi$.

Of course, the real numbers **R** form a subset of the set **C** of complex numbers, since any x in **R** may be written in the form $x + 0i$. Notice that the definition of absolute value for complex numbers is compatible with the one for real numbers, since $|x + 0i| = \sqrt{x^2 + 0^2} = \sqrt{x^2} = |x|$. Moreover, a necessary and sufficient condition that a given complex number z be real is that $z = \bar{z}$ (since, if $z = a + bi$, $z = \bar{z}$ is equivalent to $a = a$ and $b = -b$, and the latter is true if and only if $b = 0$).

The set of all m-tuples of complex numbers, denoted $\mathbf{C}^m$, is called **complex m-space.** The definitions of addition and subtraction of vectors in $\mathbf{C}^m$ and multiplication of a vector in $\mathbf{C}^m$ by a scalar (here, a complex number) are identical to the $\mathbf{R}^m$ case, but the definitions of dot product and norm in $\mathbf{C}^m$ are not.

DEFINITION

Let $\mathbf{w} = (w_1, w_2, \ldots, w_m)$ and $\mathbf{z} = (z_1, z_2, \ldots, z_m)$ be vectors in $\mathbf{C}^m$. Then

a. $\mathbf{w} \cdot \mathbf{z} = w_1 \bar{z}_1 + w_2 \bar{z}_2 + \cdots + w_m \bar{z}_m$ (dot product),

b. $\|\mathbf{w}\| = \sqrt{|w_1|^2 + |w_2|^2 + \cdots + |w_m|^2}$ (*norm* or *length*).

NOTE These definitions reduce to the corresponding ones for $\mathbf{R}^m$ in the case when $\mathbf{w}$ and $\mathbf{z}$ are in $\mathbf{R}^m$.

Notice that the dot product, $\mathbf{w} \cdot \mathbf{z}$, involves complex conjugates. It is defined in this manner to preserve the property that $\mathbf{w} \cdot \mathbf{w} = \|\mathbf{w}\|^2$. If we had defined the dot product as $\mathbf{w} \cdot \mathbf{z} = w_1 z_1 + w_2 z_2 + \cdots + w_m z_m$, then this would not always be the case (Exercise 42). However, with the definition just stated, we do have $\mathbf{w} \cdot \mathbf{w} = \|\mathbf{w}\|^2$ (Exercise 41).

EXAMPLE 7.15   Let $\mathbf{w} = (i, 1, 0)$ and $\mathbf{z} = (2i, 1 + i, 1 - i)$ be vectors in $\mathbf{C}^3$. Find $\mathbf{w} \cdot \mathbf{z}, \mathbf{z} \cdot \mathbf{w}$, and $\|\mathbf{w}\|$.

Solution

We have

i. $\mathbf{w} \cdot \mathbf{z} = i(\overline{2i}) + 1(\overline{1 + i}) + 0(\overline{1 - i}) = i(-2i) + 1(1 - i)$
$= -2i^2 + 1 - i = 3 - i,$

ii. $\mathbf{z} \cdot \mathbf{w} = 2i(\bar{i}) + (1 + i)\bar{1} + (1 - i)\bar{0} = 2i(-i) + (1 + i)1$
$= -2i^2 + 1 + i = 3 + i,$

iii. $\|\mathbf{w}\| = \sqrt{|i|^2 + 1^2 + 0^2} = \sqrt{1^2 + 1^2} = \sqrt{2}.$

Notice that for the $\mathbf{w}$ and $\mathbf{z}$ of Example 7.15, we have $\mathbf{w} \cdot \mathbf{z} = \overline{\mathbf{z} \cdot \mathbf{w}}$. In fact this equation holds for every $\mathbf{w}$ and $\mathbf{z}$ in $\mathbf{C}^m$. The next theorem presents some properties of dot product in $\mathbf{C}^m$, including this one.

THEOREM I Let $\mathbf{w}$ and $\mathbf{z}$ be vectors in $\mathbf{C}^m$, and let c be a scalar (a complex number). Then

a. $\mathbf{w} \cdot \mathbf{w} \geq 0$, $\mathbf{w} \cdot \mathbf{w} = 0$ if and only if $\mathbf{w} = \mathbf{0}$,

b. $\mathbf{w} \cdot \mathbf{z} = \overline{\mathbf{z} \cdot \mathbf{w}}$,

c. $(c\mathbf{w}) \cdot \mathbf{z} = c(\mathbf{w} \cdot \mathbf{z}), \mathbf{w} \cdot (c\mathbf{z}) = \bar{c}(\mathbf{w} \cdot \mathbf{z})$,

d. $\mathbf{w} \cdot \mathbf{w} = \|\mathbf{w}\|^2.$

Proof of (b) The proofs of (a), (c), and (d) are left as an exercise. Let $\mathbf{w} = (w_1, w_2, \ldots, w_m)$, $\mathbf{z} = (z_1, z_2, \ldots, z_m)$. Then

$$
\begin{aligned}
\overline{\mathbf{z} \cdot \mathbf{w}} &= \overline{z_1 \bar{w}_1 + z_2 \bar{w}_2 + \cdots + z_m \bar{w}_m} \\
&= \overline{z_1 \bar{w}_1} + \overline{z_2 \bar{w}_2} + \cdots + \overline{z_m \bar{w}_m} \quad \text{(Exercise 43a)} \\
&= \bar{z}_1 \bar{\bar{w}}_1 + \bar{z}_2 \bar{\bar{w}}_2 + \cdots + \bar{z}_m \bar{\bar{w}}_m \quad \text{(Exercise 43b)} \\
&= \bar{z}_1 w_1 + \bar{z}_2 w_2 + \cdots + \bar{z}_m w_m \quad \text{(Exercise 43c)} \\
&= w_1 \bar{z}_1 + w_2 \bar{z}_2 + \cdots + w_m \bar{z}_m \\
&= \mathbf{w} \cdot \mathbf{z}.
\end{aligned}
$$

MATRICES WITH ENTRIES IN C

In this section we will allow the entries of matrices to be complex numbers. All definitions and theorems concerning matrices and determinants given in Chapters 2 and 3 remain completely unchanged, with one minor exception. The product of an $m \times n$ matrix A and an $n \times p$ matrix B, whether the entries are complex or real, is defined to be the $m \times p$ matrix C in which

$$c_{ij} = a_{i1}b_{1j} + a_{i2}b_{2j} + \cdots + a_{in}b_{nj}.$$

However, because the complex dot product involves complex conjugates, this expression can no longer be given as simply $c_{ij} = \mathbf{a}_i \cdot \mathbf{b}^j$, as it was in Section 2.3.

EXAMPLE 7.16   Let

$$A = \begin{bmatrix} i & 1-i & 2 \\ 0 & -2i & 1+i \\ 1 & 0 & 3-i \end{bmatrix}, \qquad B = \begin{bmatrix} 1-i & i \\ -i & 0 \\ 0 & 2+i \end{bmatrix}.$$

Find $AB - 2B$.

Solution   We have

$$AB - 2B = \begin{bmatrix} i & 1-i & 2 \\ 0 & -2i & 1+i \\ 1 & 0 & 3-i \end{bmatrix} \begin{bmatrix} 1-i & i \\ -i & 0 \\ 0 & 2+i \end{bmatrix} - 2 \begin{bmatrix} 1-i & i \\ -i & 0 \\ 0 & 2+i \end{bmatrix}$$

$$= \begin{bmatrix} 0 & 3+2i \\ -2 & 1+3i \\ 1-i & 7+2i \end{bmatrix} - \begin{bmatrix} 2-2i & 2i \\ -2i & 0 \\ 0 & 4+2i \end{bmatrix}$$

$$= \begin{bmatrix} -2+2i & 3 \\ -2+2i & 1+3i \\ 1-i & 3 \end{bmatrix}.$$

Before proceeding to consider complex eigenvalues and eigenvectors, we should point out that all definitions and theorems of Chapters 4 through 6 (linear independence, basis, linear transformations, and so on) are exactly the same for the $\mathbf{C}^m$ case as they were in $\mathbf{R}^m$.

COMPLEX EIGENVALUES AND EIGENVECTORS IN $\mathbf{C}^n$

Although much of the material dealing with complex eigenvalues and eigenvectors is completely analogous to the real case (Section 7.1), we will also see some interesting differences.

DEFINITION   Let A be an $n \times n$ matrix with complex entries. A scalar λ is an **eigenvalue** of A if there is a nonzero vector $\mathbf{z}$ in $\mathbf{C}^n$ such that $A\mathbf{z} = \lambda\mathbf{z}$. The vector $\mathbf{z}$ is said to be an **eigenvector** corresponding to λ. The **eigenspace** of λ is the set of all solutions of the homogeneous linear system $(A - \lambda I)\mathbf{z} = \mathbf{0}$.

The computation of complex eigenvalues and their corresponding eigenvectors proceeds in a manner similar to that of the real case (see Theorem 2 of Section 7.1 and its accompanying procedure). However, since the characteristic equation, $\det(A - \lambda I) = 0$, is a polynomial equation in the *complex* numbers, it must have at least one complex root, by the *Fundamental Theorem of Algebra*. Consequently we now have the following theorem.

THEOREM 2   Every $n \times n$ matrix has at least one (complex) eigenvalue.

EXAMPLE 7.17   Find the eigenvalues and associated eigenspaces of the matrix

$$A = \begin{bmatrix} i & 0 & 0 \\ 0 & 0 & 1 \\ 0 & -1 & 0 \end{bmatrix}.$$

Solution   To find the eigenvalues of A, we compute the characteristic polynomial

$$\det(A - \lambda I) = \det \begin{bmatrix} i - \lambda & 0 & 0 \\ 0 & -\lambda & 1 \\ 0 & -1 & -\lambda \end{bmatrix} = (i - \lambda)(\lambda^2 + 1).$$

Thus the eigenvalues of A, the roots of the characteristic equation, are $\lambda_1 = i$ and $\lambda_2 = -i$ (since $\lambda^2 + 1 = 0$ has the solution $\lambda = \pm i$).

To compute the eigenspaces of λ_1 and λ_2, we solve the homogeneous linear systems $(A - iI)\mathbf{z} = \mathbf{0}$ and $(A + iI)\mathbf{z} = \mathbf{0}$, respectively. Doing so, we see that the eigenspace of $\lambda_1 = i$ is the set of all vectors of the form $(r, -is, s)$, whereas that of $\lambda_2 = -i$ consists of all vectors of the form $(0, it, t)$, where r, s, and t are scalars (complex numbers). (*Note:* Bases for the eigenspaces of λ_1 and λ_2 are, respectively, $\{(1, 0, 0), (0, -i, 1)\}$ and $\{(0, i, 1)\}$.)

DIAGONALIZATION OF COMPLEX MATRICES

The basic theory concerning diagonalization of matrices carries over to the case of matrices with complex entries. Specifically, Theorems 1, 2, and 3 of Section 7.2 continue to hold, and the proofs are essentially the same. We restate these theorems here in a more concise form.

THEOREM 3   Let A be an $n \times n$ matrix with complex entries.

a. Nonzero eigenvectors that correspond to distinct eigenvalues of A are linearly independent.

b. A is diagonalizable if and only if A has n linearly independent eigenvectors in $\mathbf{C}^n$.

c. If A has n distinct eigenvalues, then A is diagonalizable.

The next example illustrates the diagonalization procedure of Section 7.2 in the case of complex matrices.

EXAMPLE 7.18   Determine whether or not the matrix

$$A = \begin{bmatrix} i & 0 & 0 \\ 0 & 0 & 1 \\ 0 & -1 & 0 \end{bmatrix}$$

is diagonalizable. If it is, find an invertible matrix P such that $P^{-1}AP$ is diagonal. Also find this diagonal matrix.

Solution   Notice that the given matrix is the same as the one of Example 7.17. Since it has only two distinct eigenvalues, $\lambda_1 = i$ and $\lambda_2 = -i$, we cannot apply Theorem 3(c). However, the union of the bases for the eigenspaces of λ_1 and λ_2 (computed in Example 7.17) is $\{(1,0,0,), (0,-i,1), (0,i,1)\}$, a set of three linearly independent eigenvectors. Therefore, by Theorem 3(b), A is diagonalizable.

Moreover, following the procedure of Section 7.2, P is the matrix whose columns are these eigenvectors:

$$P = \begin{bmatrix} 1 & 0 & 0 \\ 0 & -i & i \\ 0 & 1 & 1 \end{bmatrix}.$$

Finally, the diagonal matrix $P^{-1}AP$ is the one with diagonal entries that are the eigenvalues of A. Thus

$$P^{-1}AP = \begin{bmatrix} i & 0 & 0 \\ 0 & i & 0 \\ 0 & 0 & -i \end{bmatrix}.$$

We next consider the complex analog of the orthogonal diagonalization of real symmetric matrices.

DEFINITION   The **conjugate transpose** of a matrix A, denoted A^H, is the transpose of the matrix obtained from A by replacing each entry with its complex conjugate. We say that a square matrix A is **hermitian** if $A = A^H$.

For example, the matrix

$$A = \begin{bmatrix} 1 & 1+i & 0 \\ 1-i & -3 & -2i \\ 0 & 2i & 0 \end{bmatrix}$$

is hermitian, since

$$A^H = \left(\begin{bmatrix} 1 & 1-i & 0 \\ 1+i & -3 & 2i \\ 0 & -2i & 0 \end{bmatrix}^T \right) = A.$$

NOTE The *diagonal* entries of a hermitian matrix must be real, since they are equal to their complex conjugates. Moreover, if *all* entries of a square matrix A are real, then $A^H = A^T$. Thus *hermitian* is the complex generalization of *symmetric*.

⋯⋯⋯⋯⋯⋯⋯⋯⋯⋯⋯⋯⋯⋯⋯⋯⋯⋯⋯⋯⋯⋯⋯⋯⋯⋯⋯⋯⋯⋯⋯⋯⋯⋯⋯⋯

The reason why we consider *conjugate transpose* instead of just *transpose* in the complex case is to preserve the validity of theorems such as Theorems 4, 6, and 7 of Section 7.2 when we replace *symmetric* with *hermitian*. For example, if a square matrix A has imaginary entries, it is *not* true that $A = A^T$ implies that A is diagonalizable (see Exercise 44). The underlying reason for the need for *conjugate* transpose in these situations is that complex dot products involve conjugates. Notice how this factor enters into the proofs of the following two theorems. (Theorem 5 is a complex analog of Theorem 4 of Section 7.2.)

THEOREM 4 Let A be a hermitian matrix. Then all eigenvalues of A are real numbers.

Proof Let λ be an eigenvalue of A with corresponding eigenvector $\mathbf{z}$. Then $A\mathbf{z} = \lambda\mathbf{z}$ and

$$\lambda(\mathbf{z} \cdot \mathbf{z}) = (\lambda\mathbf{z}) \cdot \mathbf{z} = (A\mathbf{z}) \cdot \mathbf{z} = \mathbf{z} \cdot (A^H\mathbf{z}),$$

by Exercise 45. Moreover, A is hermitian, so

$$\mathbf{z} \cdot (A^H\mathbf{z}) = \mathbf{z} \cdot (A\mathbf{z}) = \mathbf{z} \cdot (\lambda\mathbf{z}) = \bar{\lambda}(\mathbf{z} \cdot \mathbf{z}),$$

by Theorem 1(c). Hence $\lambda(\mathbf{z} \cdot \mathbf{z}) = \bar{\lambda}(\mathbf{z} \cdot \mathbf{z})$. But $\mathbf{z}$ is an eigenvector, and therefore $\mathbf{z} \neq \mathbf{0}$, which in turn implies that $\mathbf{z} \cdot \mathbf{z} \neq 0$ (Theorem 1(a)). Consequently $\lambda = \bar{\lambda}$, and we conclude that λ is real.

THEOREM 5 Let λ_1 and λ_2 be distinct eigenvalues of a hermitian matrix A with corresponding eigenvectors $\mathbf{z}_1$ and $\mathbf{z}_2$, respectively. Then $\mathbf{z}_1$ and $\mathbf{z}_2$ are *orthogonal*.

Proof The proof is virtually identical to that of Theorem 4 of Section 7.2. We have

$$\lambda_1(\mathbf{z}_1 \cdot \mathbf{z}_2) = (\lambda_1\mathbf{z}_1) \cdot \mathbf{z}_2 = (A\mathbf{z}_1) \cdot \mathbf{z}_2 = \mathbf{z}_1 \cdot (A^H\mathbf{z}_2),$$

by Exercise 45. But A is hermitian, so

$$\mathbf{z}_1 \cdot (A^H \mathbf{z}_2) = \mathbf{z}_1 \cdot (A\mathbf{z}_2) = \mathbf{z}_1 \cdot (\lambda_2 \mathbf{z}_2) = \bar{\lambda}_2(\mathbf{z}_1 \cdot \mathbf{z}_2),$$

by Theorem 1(c). Thus $\lambda_1(\mathbf{z}_1 \cdot \mathbf{z}_2) = \bar{\lambda}_2(\mathbf{z}_1 \cdot \mathbf{z}_2) = \lambda_2(\mathbf{z}_1 \cdot \mathbf{z}_2)$, by Theorem 4. Now if $\mathbf{z}_1 \cdot \mathbf{z}_2$ were not equal to zero, we would have $\lambda_1 = \lambda_2$, a contradiction. Thus $\mathbf{z}_1 \cdot \mathbf{z}_2 = 0$, as desired.

To state the main theorem concerning the diagonalization of hermitian matrices, we must first introduce the complex analog of an orthogonal matrix.

DEFINITION

A nonsingular matrix P with complex entries is **unitary** if $P^{-1} = P^H$.

For example, the matrix

$$P = \begin{bmatrix} i/\sqrt{2} & i/\sqrt{2} \\ i/\sqrt{2} & -i/\sqrt{2} \end{bmatrix}$$

is unitary. It is an easy matter to check that $P^H P = I$, which implies that $P^{-1} = P^H$.

The next theorem tells us that the columns of unitary matrices behave in exactly the same manner with respect to the complex dot product as do those of orthogonal matrices relative to the real dot product. Its proof is Exercise 46.

THEOREM 6   An $n \times n$ matrix P with complex entries is unitary if and only if its columns $\{\mathbf{p}^1, \mathbf{p}^2, \ldots, \mathbf{p}^n\}$ form an orthonormal set; that is, if and only if $\mathbf{p}^i \cdot \mathbf{p}^j = 0$ if $i \neq j$ while $\mathbf{p}^i \cdot \mathbf{p}^i = 1$.

Our last theorem for this section gives a generalization of Theorems 6 and 7 of Section 7.2 to the complex case. We omit the proof.

THEOREM 7   Let A be a square matrix with complex entries. Then there exists a unitary matrix P such that $P^{-1}AP$ $(= P^H AP)$ is diagonal with real entries if and only if A is hermitian.

EXAMPLE 7.19   Let

$$A = \begin{bmatrix} 0 & -i & 0 \\ i & 0 & 0 \\ 0 & 0 & 2 \end{bmatrix}.$$

Find a unitary matrix P such that $P^{-1}AP$ is diagonal.

Solution

To find the matrix P, we follow the procedure given in Section 7.2 for orthogonally diagonalizing a real symmetric matrix. Since the characteristic polynomial of A is $\det(A - \lambda I) = (2 - \lambda)(\lambda^2 - 1)$, the eigenvalues of A are $\lambda_1 = -1$, $\lambda_2 = 1$, and $\lambda_3 = 2$. Bases for the eigenspaces corresponding to these eigenvalues are then computed to be

$$\{(i, 1, 0)\} \quad \text{for} \quad \lambda_1 = -1,$$
$$\{(-i, 1, 0)\} \quad \text{for} \quad \lambda_2 = 1,$$

and

$$\{(0, 0, 1)\} \quad \text{for} \quad \lambda_3 = 2.$$

Finally, to obtain the columns of P, we normalize these three eigenvectors to get $(i/\sqrt{2}, 1/\sqrt{2}, 0)$, $(-i/\sqrt{2}, 1/\sqrt{2}, 0)$, and $(0, 0, 1)$. Hence

$$P = \begin{bmatrix} i/\sqrt{2} & -i/\sqrt{2} & 0 \\ 1/\sqrt{2} & 1/\sqrt{2} & 0 \\ 0 & 0 & 1 \end{bmatrix}.$$

NOTE   In general, to obtain a unitary matrix that diagonalizes a hermitian one, it may be necessary to employ the Gram–Schmidt process in the same manner as in Section 7.2.

··

7.3 EXERCISES

In Exercises 1–4, letting $\mathbf{w} = (i, 0, 1 - i, 1)$ and $\mathbf{z} = (-i, 2 + i, 0, -1)$, evaluate the expressions.

1. $\mathbf{w} \cdot \mathbf{z}$ **2.** $\mathbf{z} \cdot \mathbf{w}$ **3.** $\|\mathbf{w}\|$ **4.** $\mathbf{w} \cdot \mathbf{w}$

In Exercises 5–13 let

$$A = \begin{bmatrix} 1 & 0 & -i \\ 0 & 1+i & 1 \\ i & -1 & 2+i \end{bmatrix}, \qquad B = \begin{bmatrix} i & -i & 0 \\ i & i & 1 \\ 0 & 1 & i \end{bmatrix}.$$

5. Solve $A\mathbf{z} = \mathbf{b}$, where $\mathbf{b} = (i, 0, -1)$.

6. Solve $B\mathbf{z} = \mathbf{c}$, where $\mathbf{c} = (1 + i, 1 - i, 1)$.

7. Find AB.

8. Find BA.

9. Find A^{-1}.

10. Find B^{-1}.

11. Find det (A).

12. Find det (B).

13. Is either A or B hermitian?

14. Show that the dimension of $\mathbf{C}^m$ is equal to m.

In Exercises 15–18 determine whether or not the given set is a basis for $\mathbf{C}^3$.

15. $\{(i, 0, -1), (1, 1, 1), (0, -i, i)\}$

16. $\{(i, -i, 1), (1, 1 + i, 1 - i), (0, 2 + i, 1)\}$

17. $\{(i, 1, 0), (0, 0, 1)\}$

18. $\{(0, i, 1), (i, 0, 0), (1 + i, -1, 0), (1 - i, 0, 1)\}$

In Exercises 19–22 find bases for the given subspace of $\mathbf{C}^m$.

19. The subspace of $\mathbf{C}^3$ of all vectors of the form (z_1, z_2, z_3), where $z_1 + z_2 + z_3 = 0$.

20. The subspace of $\mathbf{C}^4$ of all vectors whose first and last components are zero.

21. The kernel of the transformation $T: \mathbf{C}^2 \to \mathbf{C}^2$ given by $T(z_1, z_2) = (z_1 - z_2, 0)$.

22. The kernel of the transformation $S: \mathbf{C}^3 \to \mathbf{C}^2$ given by $S(\mathbf{z}) = A\mathbf{z}$, where

$$A = \begin{bmatrix} i & 0 & 1 \\ 1 & -i & 0 \end{bmatrix}.$$

In Exercises 23–30 find the eigenvalues and associated eigenspaces for each matrix. Also find a basis for each eigenspace.

23. $\begin{bmatrix} 0 & 3 + 4i \\ 3 - 4i & 0 \end{bmatrix}$

24. $\begin{bmatrix} 0 & -2 \\ 2 & 0 \end{bmatrix}$

25. $\begin{bmatrix} 2 & 0 & -4 \\ 0 & 1 & 0 \\ 2 & 0 & -2 \end{bmatrix}$

26. $\begin{bmatrix} 1 & i & 0 \\ -i & 1 & -i \\ 0 & i & 1 \end{bmatrix}$

27. $\begin{bmatrix} i & 0 & 1 \\ 0 & i & 0 \\ 0 & 0 & i \end{bmatrix}$

28. $\begin{bmatrix} 1 + i & 0 & 0 \\ -1 & 1 - i & 0 \\ 0 & -1 & 0 \end{bmatrix}$

29. $\begin{bmatrix} 1 & 0 & 0 \\ 0 & 0 & -i \\ 0 & i & 0 \end{bmatrix}$

30. $\begin{bmatrix} 0 & 0 & -1 \\ 0 & 1 & 0 \\ 1 & 0 & 0 \end{bmatrix}$

31–38. Determine whether or not each matrix in Exercises 23–30 is diagonalizable. If it is diagonalizable, find an invertible matrix P that performs the diagonalization as well as the resultant diagonal matrix. If the given matrix is hermitian, find a unitary matrix P that performs the diagonalization as well as the resultant diagonal matrix.

39. Which, if any, of the matrices in Exercises 23–30 are unitary?

40. Let A be a unitary matrix. Show that det $(A) = \pm 1$. (See Exercise 26 of Section 7.2.)

41. Prove parts (a), (c), and (d) of Theorem 1.

42. Let $\mathbf{w}$ and $\mathbf{z}$ be in $\mathbf{C}^m$. Define a scalar-valued product, $\mathbf{w} \circ \mathbf{z}$, by

$$\mathbf{w} \circ \mathbf{z} = w_1 z_1 + w_2 z_2 + \cdots + w_m z_m.$$

Show that $\mathbf{w} \circ \mathbf{w}$ is not necessarily equal to $\|\mathbf{w}\|^2$.

43. Let z_1 and z_2 be complex numbers. Show that
    (a) $\overline{z_1 + z_2} = \bar{z}_1 + \bar{z}_2$ **(b)** $\overline{z_1 z_2} = \bar{z}_1 \bar{z}_2$
    (c) $\bar{\bar{z}}_1 = z_1$ **(d)** $z_1 \bar{z}_1 = |z_1|^2$

44. Let

$$A = \begin{bmatrix} 1 & 0 & 0 \\ 0 & 1 & i \\ 0 & i & 3 \end{bmatrix}.$$

Show that A is not diagonalizable. (Notice that $A = A^T$.)

45. Let A be an $n \times n$ matrix with complex entries, and let $\mathbf{w}$ and $\mathbf{z}$ be in $\mathbf{C}^n$. Show that $(A\mathbf{w}) \cdot \mathbf{z} = \mathbf{w} \cdot (A^H \mathbf{z})$. (*Hint:* See Exercise 28 of Section 7.2.)

46. Prove (Theorem 6) that an $n \times n$ matrix P with complex entries is unitary if and only if its columns form an orthonormal set (under the complex dot product).

CHAPTER 7 REVIEW

This chapter introduced the notions of eigenvalues and eigenvectors of a linear operator and discussed these concepts in greater depth for the equivalent case of a square matrix. Then the relation of eigenvalues and eigenvectors to the diagonalization of both real and complex matrices was discussed. You should be able to:

    1. calculate the eigenvalues of a square matrix;

    2. calculate the eigenvectors that correspond to a given eigenvalue and find a basis for each eigenspace;

    3. diagonalize a square matrix or demonstrate that it cannot be diagonalized;

    4. orthogonally diagonalize a symmetric matrix or, in the complex case, a hermitian matrix.

KEY WORDS

eigenvalue	eigenvector
eigenspace	characteristic equation
characteristic polynomial	diagonalize a matrix

symmetric matrix   orthogonal matrix

orthogonally diagonalize a matrix   conjugate transpose

hermitian matrix   unitary matrix

KEY THEOREMS

Theorem   A real number λ is an eigenvalue of a square matrix A if and only if $\det (A - \lambda I) = 0$.

Theorem   The eigenvalues of a triangular matrix are its diagonal entries.

Theorem   Eigenvectors that correspond to distinct eigenvalues are linearly independent.

Theorem   A square matrix of order n is diagonalizable if and only if there is a basis for $\mathbf{R}^n$ which consists of eigenvectors of A.

Theorem   If a matrix of order n has n distinct eigenvalues, then it is diagonalizable.

Theorem   A square matrix is orthogonal if and only if its columns form an orthonormal set.

Theorem   Eigenvectors that correspond to distinct eigenvalues of a symmetric matrix are orthogonal.

Theorem   A square matrix is orthogonally diagonalizable if and only if it is symmetric.

SELF-TEST

Answers appear at the end of this chapter.

1. Find the eigenvalues and corresponding eigenvectors of the linear transformation $T: \mathbf{R}^2 \to \mathbf{R}^2$ that projects every vector in $\mathbf{R}^2$ onto the x-axis; that is, for which $T(x, y) = (x, 0)$. [Section 7.1]

For each matrix in Problems 2–4 find:
   (a) its characteristic equation.
   (b) all real eigenvalues.
   (c) bases for the corresponding eigenspaces.   [Section 7.1]

2. $\begin{bmatrix} 1 & 1 \\ 0 & 1 \end{bmatrix}$
   3. $\begin{bmatrix} 2 & 2 & 2 \\ 2 & 2 & 2 \\ 2 & 2 & 2 \end{bmatrix}$
   4. $\begin{bmatrix} 2 & -1 & 0 \\ -1 & 2 & -1 \\ 0 & -1 & 2 \end{bmatrix}$

In Problems 5 and 6 determine whether or not the matrix A is diagonalizable. If it is, find matrices P and D such that $D = P^{-1}AP$ is a diagonal matrix.

5. $A = \begin{bmatrix} 2 & 2 & 2 \\ 2 & 2 & 2 \\ 2 & 2 & 2 \end{bmatrix}$
   6. $A = \begin{bmatrix} 1 & 0 & 0 \\ 0 & 0 & 1 \\ 0 & -1 & 2 \end{bmatrix}$ [Section 7.2]

7. Let $B = \begin{bmatrix} 0 & 0 & 1 \\ 0 & 1 & 0 \\ 1 & 0 & 1 \end{bmatrix}$.

(a) Is B a symmetric matrix?
(b) Is B an orthogonal matrix?   [Section 7.2]

8. Let $A = \begin{bmatrix} i & -i \\ i & i \end{bmatrix}$, $\mathbf{w} = (i, 1, 1 + i)$, and $\mathbf{z} = (-i, i, -i)$.

(a) Find $\mathbf{w} \cdot \mathbf{z}$.   (b) Find $\|\mathbf{w}\|$.
(c) Find A^2.   (d) Find A^{-1}.
(e) Is A hermitian?   (f) Is A unitary?   [Section 7.3]

9. Let $B = \begin{bmatrix} i & 0 & 0 \\ 0 & 0 & i \\ 0 & i & 0 \end{bmatrix}$.

Find all complex eigenvalues of B and bases for the corresponding eigenspaces.
[Section 7.3]

10. Is the matrix B of Problem 9 diagonalizable? If so, find matrices P and D such that $D = P^{-1}BP$ is a diagonal matrix.
[Section 7.3]

REVIEW EXERCISES

In Exercises 1–6 find the characteristic equation and eigenvalues of the given matrix.

1. $\begin{bmatrix} 2 & 1 \\ 1 & 2 \end{bmatrix}$ **2.** $\begin{bmatrix} 0 & 4 \\ 1 & 0 \end{bmatrix}$

3. $\begin{bmatrix} 1 & 0 & 0 \\ -1 & 2 & 1 \\ 0 & 0 & 1 \end{bmatrix}$ **4.** $\begin{bmatrix} -6 & 2 & 7 \\ 0 & 2 & 0 \\ -9 & 2 & 10 \end{bmatrix}$

5. $\begin{bmatrix} 1 & 1 & -1 & 2 \\ 0 & 3 & 0 & 1 \\ 0 & 0 & 5 & 1 \\ 0 & 0 & 0 & 7 \end{bmatrix}$ **6.** $\begin{bmatrix} 2 & 1 & 1 \\ 1 & 1 & 0 \\ 1 & 0 & 1 \end{bmatrix}$

7. Find a basis of eigenvectors for each eigenvalue in the matrix of Exercise 1.

8. Find a basis of eigenvectors for each eigenvalue in the matrix of Exercise 3.

9. Identify which matrices of Exercises 1–6 can be diagonalized. Explain each of your answers.

10. Find a matrix P such that $P^{-1}AP$ is a diagonal matrix where A is the matrix of Exercise 3.

11. Find a matrix P such that $P^{-1}AP$ is a diagonal matrix where A is the matrix of Exercise 6.

12. Identify which matrices of Exercises 1–6 are orthogonally diagonalizable. Explain each of your answers.

13. Is the following matrix orthogonal? Verify.

$$\begin{bmatrix} 1/\sqrt{5} & 0 & -2/\sqrt{5} \\ 0 & 1 & 0 \\ 2/\sqrt{5} & 0 & 1/\sqrt{5} \end{bmatrix}$$

14. Is the following matrix orthogonal? Verify.

$$\begin{bmatrix} -1 & 0 & 2 \\ 0 & 1 & 0 \\ 2 & 0 & 1 \end{bmatrix}$$

15. For each matrix A identified in Exercise 12, find an orthogonal matrix P such that $P^{-1}AP$ is a diagonal matrix.

In Exercises 16 and 17 determine whether or not the given set is a basis for $\mathbf{C}^3$.

16. $\{(1 + i, 1, 2 - i), (1, 1 + i, 1 - i), (1, 3 + 2i, 2 - i)\}$

17. $\{(1, i, -i), (1, 1 + i, 1 - i), (1, 1 - i, 1 + i)\}$

18. Find the kernel of the linear transformation $T : \mathbf{C}^3 \to \mathbf{C}^3$ given by $T(\mathbf{z}) = A\mathbf{z}$, where

$$A = \begin{bmatrix} i & -i & 1 \\ 1 & 1 + i & 1 - i \\ 0 & 2 + i & 1 \end{bmatrix}.$$

19. Is the linear transformation of Exercise 18 one-to-one? Explain.

20. Is the linear transformation of Exercise 18 onto? Explain.

In Exercises 21 and 22 find the eigenvalues and associated eigenspaces and a basis for each eigenspace.

21. $\begin{bmatrix} 1 & 0 & i \\ 0 & 1 & 0 \\ -i & 0 & 2 \end{bmatrix}$ **22.** $\begin{bmatrix} 1 & 2 & 1 \\ 0 & i & -i \\ 0 & 0 & 3 \end{bmatrix}$

23. Both of the matrices in Exercises 21 and 22 are diagonalizable. Explain why.

24. One of the matrices in Exercises 21 and 22 is orthogonally diagonalizable. Which one and why?

25. Find a unitary matrix P such that $P^{-1}AP$ is diagonal where A is the matrix of Exercise 24.

ANSWERS TO SELF-TEST

1. $\lambda_1 = 0$; corresponding eigenvalues have the form $c(0, 1)$, $c \neq 0$.
  $\lambda_2 = 1$; corresponding eigenvalues have the form $c(1, 0)$, $c \neq 0$.

2. (a) $1 - 2\lambda + \lambda^2 = 0$ **(b)** $\lambda = 1$ **(c)** $\{(1, 0)\}$

3. (a) $-\lambda^3 + 6\lambda^2 = 0$ **(b)** $\lambda = 0, 6$
   (c) For $\lambda = 0$, $\{(-1, 1, 0), (-1, 0, 1)\}$; for $\lambda = 6$, $\{(1, 1, 1)\}$.

4. (a) $(2 - \lambda)(\lambda^2 - 4\lambda + 2) = 0$ **(b)** $\lambda = 2, 2 \pm \sqrt{2}$
   (c) For $\lambda = 2$, $\{(-1, 0, 1)\}$; for $\lambda = 2 + \sqrt{2}$, $\{(1, -\sqrt{2}, 1)\}$; for $\lambda = 2 - \sqrt{2}$, $\{(1, \sqrt{2}, 1)\}$.

5. A is diagonalizable;

$$P = \begin{bmatrix} -1 & -1 & 1 \\ 1 & 0 & 1 \\ 0 & 1 & 1 \end{bmatrix}, \quad D = \begin{bmatrix} 0 & 0 & 0 \\ 0 & 0 & 0 \\ 0 & 0 & 6 \end{bmatrix}.$$

6. A is not diagonalizable.

7. B is symmetric; B is not orthogonal.

8. (a) -2 **(b)** 2 **(c)** $\begin{bmatrix} 0 & 2 \\ -2 & 0 \end{bmatrix}$ **(d)** $\begin{bmatrix} -\frac{i}{2} & -\frac{i}{2} \\ \frac{i}{2} & -\frac{i}{2} \end{bmatrix}$
   (e) No **(f)** No

9. $\lambda_1 = i$; basis for eigenspace is $\{(1, 0, 0), (0, 1, 1)\}$.
   $\lambda_2 = -i$; basis for eigenspace is $\{(0, -1, 1)\}$.

10. B is diagonalizable (but not unitarily diagonalizable);

$$P = \begin{bmatrix} 1 & 0 & 0 \\ 0 & 1 & -1 \\ 0 & 1 & 1 \end{bmatrix}, \quad D = \begin{bmatrix} i & 0 & 0 \\ 0 & i & 0 \\ 0 & 0 & -i \end{bmatrix}.$$

APPLICATIONS OF LINEAR ALGEBRA

In this chapter we will present some of the many applications of linear algebra. Each of these will make use of material introduced in previous chapters of this text. Specifically, the prerequisite material is as follows:

8.1   Interpolating Polynomials—Chapter 2

8.2   Least Square Polynomials—Chapter 2

8.3   Electrical Circuits and Pipe Networks—Sections 2.1 and 2.2

8.4   Input–Output Analysis: Leontief Models—Chapter 2

8.5   Markov Chains—Section 2.3 for the first part, Section 7.1 for the (optional) second part

8.6   Quadric Surfaces—Sections 7.1 and 7.2

8.7   Approximation of Continuous Functions; Fourier Series—Chapter 5, especially Section 5.4

8.8   Linear Differential Equations—Sections 5.1, 5.2, 6.1, 6.2, and 6.3

8.1   INTERPOLATING POLYNOMIALS

In this section we will develop a procedure to determine a polynomial that passes through a given set of points. Recall that a **polynomial** in the variable x is a function of the form

$$P(x) = a_0 + a_1x + a_2x^2 + \cdots + a_nx^n,$$

where the a_i are scalars and n is a nonnegative integer. If a_n is not equal to zero, the **degree** of the polynomial is equal to n. Polynomials of degree one, two, and three are called **linear**, **quadratic**, and **cubic**, respectively. The graph of a linear polynomial is a line in the plane, whereas that of a quadratic polynomial is a parabola.

We will first consider the simple case of determining an equation of the line (the polynomial of degree one) that passes through two distinct points. This is a standard problem in elementary algebra, but here we will solve it in a different way—one that illustrates the general process of passing a polynomial through a set of points.

Since a nonvertical line in the xy-plane may be represented by a linear polynomial $P(x) = a_0 + a_1 x$, we can find the equation of the line through the points (x_0, y_0) and (x_1, y_1) by determining the coefficients a_0 and a_1. Each point must satisfy the desired equation, so we have

$$a_0 + a_1 x_0 = y_0,$$

and

$$a_0 + a_1 x_1 = y_1.$$

But this is just a system of two linear equations in the two unknowns a_0 and a_1. Moreover, since the coefficient matrix

$$\begin{bmatrix} 1 & x_0 \\ 1 & x_1 \end{bmatrix}$$

row-reduces to the identity matrix if $x_0 \neq x_1$ (if the given points do not determine a vertical line), the system has a unique solution (Theorem 5 of Section 2.5). Solving it, we obtain a_0 and a_1 and hence the desired linear polynomial.

Following this lead, let us consider the situation in which we are given three points (x_0, y_0), (x_1, y_1), and (x_2, y_2), with x_0, x_1, and x_2 distinct. It is unlikely (although possible) that a line will pass through all three points. On the other hand, it is reasonable to believe that we might be able to construct a quadratic polynomial $P(x) = a_0 + a_1 x + a_2 x^2$ through these points, since P has *three* undetermined coefficients. Requiring the given points to satisfy this polynomial equation, we obtain the linear system

$$a_0 + a_1 x_0 + a_2 x_0^2 = y_0$$
$$a_0 + a_1 x_1 + a_2 x_1^2 = y_1$$
$$a_0 + a_1 x_2 + a_2 x_2^2 = y_2.$$

This linear system may be written in the matrix form $V\mathbf{a} = \mathbf{y}$, where

$$V = \begin{bmatrix} 1 & x_0 & x_0^2 \\ 1 & x_1 & x_1^2 \\ 1 & x_2 & x_2^2 \end{bmatrix}, \qquad \mathbf{a} = \begin{bmatrix} a_0 \\ a_1 \\ a_2 \end{bmatrix}, \qquad \text{and} \qquad \mathbf{y} = \begin{bmatrix} y_0 \\ y_1 \\ y_2 \end{bmatrix}.$$

As in the two-point case, we can show that this 3×3 linear system has a unique solution (Exercise 9). Solving it, we obtain $\mathbf{a} = (a_0, a_1, a_2)$ and hence the desired polynomial $P(x)$.

EXAMPLE 8.1

Find a parabola that passes through the points $(0, -5)$, $(1, -1)$, and $(2, 5)$.

Solution

We must find a quadratic polynomial $P(x) = a_0 + a_1 x + a_2 x^2$, where a_0, a_1, and a_2 satisfy

$$a_0 + 0a_1 + 0a_2 = -5$$
$$a_0 + 1a_1 + 1a_2 = -1$$
$$a_0 + 2a_1 + 4a_2 = 5.$$

Solving this linear system, we obtain $a_0 = -5$, $a_1 = 3$, and $a_2 = 1$. Thus the parabola is given by $P(x) = -5 + 3x + x^2$ (or $y = x^2 + 3x - 5$; see Figure 8.1.)

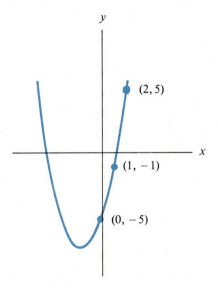

FIGURE 8.1

We now turn to the general case. We wish to find a polynomial of the form

$$P(x) = a_0 + a_1 x + a_2 x^2 + \cdots + a_n x^n$$

that passes through the given $n + 1$ points (x_0, y_0), (x_1, y_1), ..., (x_n, y_n), where all x_i are distinct. Since each point lies on the curve $y = P(x)$, we obtain the following system of $n + 1$ linear equations in the $n + 1$ unknowns $a_0, a_1, \ldots, a_n$:

$$a_0 + a_1 x_0 + a_2 x_0^2 + \cdots + a_n x_0^n = y_0$$
$$a_0 + a_1 x_1 + a_2 x_1^2 + \cdots + a_n x_1^n = y_1$$
$$\vdots \quad \vdots \quad \vdots \qquad \vdots \quad \vdots$$
$$a_0 + a_1 x_n + a_2 x_n^2 + \cdots + a_n x_n^n = y_n. \tag{1}$$

To determine the desired polynomial, we solve this system, obtaining the coefficients of $P(x)$ and hence the polynomial itself.

The following theorem summarizes this discussion and demonstrates the uniqueness of the solution of System (1). Notice that the same notation is used as in the 3×3 case, but we demonstrate the uniqueness in a different manner.

THEOREM 1   Let $(x_0, y_0), (x_1, y_1), \ldots, (x_n, y_n)$ be $n + 1$ points with all x_i distinct. Then there is a unique polynomial

$$P(x) = a_0 + a_1 x + \cdots + a_n x^n$$

such that $P(x_i) = y_i$ for $i = 0, 1, \ldots, n$. The vector of coefficients $\mathbf{a} = (a_0, a_1, \ldots, a_n)$ satisfies the linear system

$$V\mathbf{a} = \mathbf{y},$$

where $\mathbf{y} = (y_0, y_1, \ldots, y_n)$ and

$$V = \begin{bmatrix} 1 & x_0 & x_0^2 & \cdots & x_0^n \\ 1 & x_1 & x_1^2 & \cdots & x_1^n \\ \vdots & \vdots & \vdots & & \vdots \\ 1 & x_n & x_n^2 & \cdots & x_n^n \end{bmatrix}.$$

Proof   The only part of the theorem that still needs proof is that the system $V\mathbf{a} = \mathbf{y}$ has a solution for $\mathbf{a}$ and that the solution is unique. By Theorem 5 of Section 2.5, both of these will be demonstrated if we can show that the linear homogeneous system $V\mathbf{c} = \mathbf{0}$ has only the trivial solution $\mathbf{c} = \mathbf{0}$.

Suppose $\mathbf{c}$ is a solution of $V\mathbf{c} = \mathbf{0}$. Performing the matrix-vector multiplication and equating each resulting component to zero yields the $n + 1$ equations

$$c_0 + c_1 x_i + c_2 x_i^2 + \cdots + c_n x_i^n = 0, \qquad i = 0, 1, \ldots, n.$$

Letting $Q(x) = c_0 + c_1 x + \cdots + c_n x^n$, we have $Q(x_i) = 0$ for $i = 0, 1, \ldots, n$. Since the x_i are distinct, this implies that $Q(x) = 0$ has at least $n + 1$ roots. But the degree of Q is less than or equal to n, so by the Fundamental Theorem of Algebra, Q must be the zero polynomial; that is, $\mathbf{c} = (c_0, c_1, \ldots, c_n) = \mathbf{0}$, as desired. Thus $V\mathbf{a} = \mathbf{y}$ has a unique solution. But any polynomial, $P(x)$, of degree less than or equal to n that satisfies $P(x_i) = y_i$ must satisfy the equations of System (1). Consequently such a polynomial is unique.

DEFINITION

Let $(x_0, y_0), (x_1, y_1), \ldots, (x_n, y_n)$ be given points for which all x_i are distinct. The unique polynomial

$$P(x) = a_0 + a_1 x + a_2 x^2 + \cdots + a_n x^n$$

that passes through these points is called the **interpolating polynomial** for these points. The coefficient matrix V of System (1) is known as a **Vandermonde matrix**.

EXAMPLE 8.2

A table for e^x has the following entries

x	1.0	1.1	1.2	1.3
e^x	2.7183	3.0042	3.3201	3.6693

Use this data to find an approximation to e^x at $x = 1.15$ by means of a cubic interpolating polynomial.

Solution

Here (x_0, y_0), (x_1, y_1), (x_2, y_2), and (x_3, y_3) are $(1.0, 2.7183)$, $(1.1, 3.0042)$, $(1.2, 3.3201)$, and $(1.3, 3.6693)$, respectively. We shall find the cubic interpolating polynomial $P(x) = a_0 + a_1 x + a_2 x^2 + a_3 x^3$ for these points and then evaluate it at $x = 1.15$.

For this data, System (1) is

$$a_0 + (1.0)a_1 + (1.0)^2 a_2 + (1.0)^3 a_3 = 2.7183$$
$$a_0 + (1.1)a_1 + (1.1)^2 a_2 + (1.1)^3 a_3 = 3.0042$$
$$a_0 + (1.2)a_1 + (1.2)^2 a_2 + (1.2)^3 a_3 = 3.3201$$
$$a_0 + (1.3)a_1 + (1.3)^2 a_2 + (1.3)^3 a_3 = 3.6693$$

Solving this linear system for a_0, a_1, a_2, and a_3, we obtain $a_0 = 0.7832$, $a_1 = 1.7002$, $a_2 = 0.3152$, and $a_3 = 0.5501$. Therefore the desired approximation to $P(1.15)$ is 3.1582. (The correct value of $e^{1.15}$ to four decimal places is also 3.1582.)

NOTE The interpolating polynomial $P_n(x)$ for the $n + 1$ points $(x_0, y_0), (x_1, y_1), \ldots, (x_n, y_n)$ has degree less than or equal to n. Depending on the given data set, it may or may not have degree exactly equal to n. For example, in Examples 8.1 and 8.2 the interpolating polynomials for the given sets of three $(n = 2)$ and four $(n = 3)$ points did have degrees two and three, respectively. On the other hand, the unique polynomial of the form $P(x) = a_0 + a_1 x + a_2 x^2$ that passes through the three $(n = 2)$ points $(1, 10)$, $(3, 10)$, and $(5, 10)$ is just $P(x) = 10$, a constant function. Here the degree of the interpolating polynomial is zero, strictly less than n.

8.1 EXERCISES

In Exercises 1–6 find a polynomial of the specified degree that passes through the given points.

1. $\{(0, -3), (1, 0), (2, 5)\}$, $n = 2$
2. $\{(-1, 4), (1, 2), (3, -4)\}$, $n = 2$
3. $\{(0, -1), (1, -1), (2, 3), (3, 17)\}$, $n = 3$
4. $\{(-1, 4), (0, 1), (1, 0), (3, 4)\}$, $n = 3$
5. $\{(-2, -15), (-1, -2), (0, 1), (1, 0), (2, 1)\}$, $n = 4$
6. $\{(0, -3), (1, -9), (2, -3), (3, 87), (4, 381)\}$, $n = 4$

A table of logarithms contains the following entries:

x	1.0	2.0	3.0	4.0
$\log_{10} x$	0.0000	0.3010	0.4771	0.6021

In Exercises 7 and 8 approximate $\log_{10}(2.5)$ by constructing an interpolating polynomial.

7. Of degree two, using the entries at $x = 1.0$, 2.0, and 3.0

8. Of degree three, using all the entries

9. Let

$$V = \begin{bmatrix} 1 & x_0 & x_0^2 \\ 1 & x_1 & x_1^2 \\ 1 & x_2 & x_2^2 \end{bmatrix}.$$

where x_0, x_1, and x_2 are distinct numbers. Show that the row-reduced echelon form of V is the 3×3 identity matrix.

8.2 LEAST SQUARE POLYNOMIALS

In Section 8.1 we developed a procedure for determining the polynomial that passes through a given set of points. This technique works well if the number of points given is small. However, in most experimental situations in which we are trying to obtain an equation relating a dependent and independent variable (say, y and x), the experiment is performed a relatively large number of times, resulting in many data points (x_i, y_i). If we tried to determine the interpolating polynomial for these points, we would have to do quite a bit of computing. Moreover, data obtained experimentally is subject to measurement errors, and these errors would be passed on to the approximating interpolating polynomial. In this situation it would be better to construct a relatively low-degree polynomial that passes *near* the data points rather than through them. This is the idea behind *least square approximations*.

As an example, consider the simplest case—the *least square linear polynomial* or **least square line**. Let $(x_0, y_0), (x_1, y_1), \ldots, (x_m, y_m)$ be a given set of points with all x_i distinct, and let d_k denote the vertical distance from the

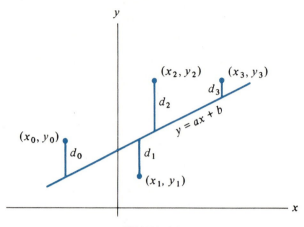

FIGURE 8.2

line $y = ax + b$ to the point (x_k, y_k) (see Figure 8.2). That is, let $d_k = |y_k - (ax_k + b)|$. We wish to determine a and b such that $S = d_0^2 + d_1^2 + \cdots + d_m^2$ is minimized. In other words, we seek a linear polynomial $Q(x) = ax + b$ that minimizes the expression

$$S = (y_0 - (ax_0 + b))^2 + (y_1 - (ax_1 + b))^2 + \cdots + (y_m - (ax_m + b))^2. \quad (1)$$

NOTE   It may seem more logical to seek a polynomial that minimizes the sum of the vertical distances, or perhaps the sum of the perpendicular distances from each point to the line. However, these alternatives result in a considerably more complicated analysis of the situation. Minimizing the expression S of Equation (1) has a relatively simple as well as unique solution and works well in practical applications.

....

Instead of demonstrating how one can determine the least square line, we will go directly to the more general case of the nth-degree least square polynomial. (The least square line is just the special case of the latter, when $n = 1$.)

Here we wish to determine a polynomial of the form

$$Q(x) = b_0 + b_1 x + b_2 x^2 + \cdots + b_n x^n$$

that minimizes the expression

$$S = (y_0 - Q(x_0))^2 + (y_1 - Q(x_1))^2 + \cdots + (y_m - Q(x_m))^2. \quad (2)$$

THEOREM I   Let $(x_0, y_0), (x_1, y_1), \ldots, (x_m, y_m)$ be $m + 1$ points with all x_i distinct, and let n be a positive integer with $n \le m$. Then there is a unique polynomial

$$Q(x) = b_0 + b_1 x + b_2 x^2 + \cdots + b_n x^n$$

that minimizes the expression S of Equation (2). The vector of coefficients $\mathbf{b} = (b_0, b_1, \ldots, b_n)$ satisfies the linear system

$$(U^T U)\mathbf{b} = U^T \mathbf{y}, \tag{3}$$

where $\mathbf{y} = (y_0, y_1, \ldots, y_m)$ and

$$U = \begin{bmatrix} 1 & x_0 & x_0^2 & \cdots & x_0^n \\ 1 & x_1 & x_1^2 & \cdots & x_1^n \\ \vdots & \vdots & \vdots & & \vdots \\ 1 & x_m & x_m^2 & \cdots & x_m^n \end{bmatrix}.$$

The proof of Theorem 1 is similar to that of Theorem 1 of Section 8.7, so we will omit the proof of the former.

DEFINITION

Given the points, $(x_0, y_0), (x_1, y_1), \ldots, (x_m, y_m)$, the unique polynomial, $Q(x)$, of degree less than or equal to n that minimizes the expression S of Equation (2) is called the nth-degree **least square polynomial** for these points. System (3) is called the set of **normal equations** for the given data.

EXAMPLE 8.3

A student performs an experiment in an attempt to obtain an equation relating two physical quantities x and y. He performs the experiment six times, letting $x = 1, 2, 4, 6, 8$, and 10 units. These experiments result in corresponding y values of $2, 2, 3, 4, 6$, and 8 units, respectively.

a. Find the least square linear polynomial (the least square line) for this data.

b. Find the least square quadratic polynomial (the least square parabola) for this data.

Solution

Using the notation of Theorem 1, the given data is the set of six points (x_i, y_i), $i = 0, 1, \ldots, 5$: $(1, 2), (2, 2), (4, 3), (6, 4), (8, 6), (10, 8)$. For part (a) we seek a polynomial of the form $Q_1(x) = b_0 + b_1 x$, whereas for part (b) we seek one of the form $Q_2(x) = b_0 + b_1 x + b_2 x^2$.

a. Here the vectors $\mathbf{b}$ and $\mathbf{y}$ are $\mathbf{b} = (b_0, b_1)$ and $\mathbf{y} = (2, 2, 3, 4, 6, 8)$, and the matrices U and U^T are

$$U = \begin{bmatrix} 1 & 1 \\ 1 & 2 \\ 1 & 4 \\ 1 & 6 \\ 1 & 8 \\ 1 & 10 \end{bmatrix} \qquad U^T = \begin{bmatrix} 1 & 1 & 1 & 1 & 1 & 1 \\ 1 & 2 & 4 & 6 & 8 & 10 \end{bmatrix}.$$

By computing $U^T U \mathbf{b}$ and $U^T \mathbf{y}$, we obtain the normal equations, System (3), $U^T U \mathbf{b} = U^T \mathbf{y}$, for part (a):

$$6b_0 + 31b_1 = 25$$
$$31b_0 + 221b_1 = 170.$$

Solving this linear system, we see that $b_0 = 0.6986$ and $b_1 = 0.6712$ (approximately), so that the least square line for the data is

$$Q_1(x) = 0.6986 + 0.6712x,$$

b. Here we have $\mathbf{b} = (b_0, b_1, b_2)$, $\mathbf{y} = (2, 2, 3, 4, 6, 8)$, and

$$U = \begin{bmatrix} 1 & 1 & 1 \\ 1 & 2 & 4 \\ 1 & 4 & 16 \\ 1 & 6 & 36 \\ 1 & 8 & 64 \\ 1 & 10 & 100 \end{bmatrix}, \qquad U^T = \begin{bmatrix} 1 & 1 & 1 & 1 & 1 & 1 \\ 1 & 2 & 4 & 6 & 8 & 10 \\ 1 & 4 & 16 & 36 & 64 & 100 \end{bmatrix}.$$

Therefore the normal equations for part (b) are

$$6b_0 + 31b_1 + 221b_2 = 25$$
$$31b_0 + 221b_1 + 1801b_2 = 170$$
$$221b_0 + 1801b_1 + 15665b_2 = 1386,$$

which has the approximate solution $b_0 = 1.836$, $b_1 = 0.0276$, and $b_2 = 0.0594$. Thus the least square parabola for this data is

$$Q_2(x) = 1.836 + 0.0276x + 0.0594x^2.$$

(In Figure 8.3 we have plotted the given data of this example and the least square line and parabola for this data.)

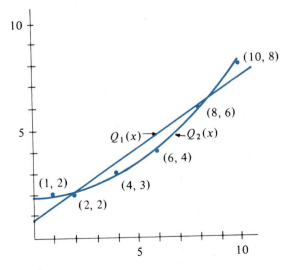

FIGURE 8.3

NOTE Theorem 1 is a generalization of Theorem 1 of Section 8.1. If in the former we have $m = n$, the matrix U is square and $U = V$, the Vandermonde matrix. Moreover, since V is invertible, the linear systems $V\mathbf{a} = \mathbf{y}$ and $V^TV\mathbf{a} = V^T\mathbf{y}$ are equivalent (just multiply both sides of the latter on the left by $(V^T)^{-1}$). Thus in the case that $m = n$, the interpolating polynomial and the least square polynomial are identical. Another way to see this last statement is to note that, when $m = n$, the interpolating polynomial will result in a minimum value (namely, 0) for Equation (2), since it passes *through* all the data points.

···

8.2 EXERCISES

In Exercises 1–4 find the least square polynomial of the specified degree for the given data points.

1. $\{(0,0), (1,0), (2,1), (3,3), (4,5)\}$, $n = 1$

2. $\{(x, y) \,|\, y = x^3$ and $x = 0, 1, 2, 3, 4, 5\}$, $n = 1$

3. $\{(x, y) \,|\, y = x^3$ and $x = 0, 1, 2, 3, 4, 5\}$, $n = 2$

4. $\{(-2, 2), (-1, 1), (0, -1), (1, 0), (2, 3)\}$, $n = 2$

In Exercises 5 and 6 determine the system of normal equations that occurs in the process of finding the least square polynomial of specified degree for the given data.

5. $\{(x, y) \,|\, y = x^4 - 8x^2, x = -3, -2, -1, 0, 1, 2, 3\}$, $n = 3$

6. $\{(0, 0), (1, 0), (2, 1), (3, 3), (4, 5), (5, 4), (6, 5)\}$, $n = 4$

A firm that manufactures widgets finds the daily consumer demand $d(x)$ for widgets as a function of their price x is as in the following table:

x	1	1.5	2	2.5	3
$d(x)$	200	180	150	100	25

In Exercises 7 and 8, using least square polynomials, approximate the daily consumer demand.

7. By a linear function

8. By a quadratic function

8.3   ELECTRICAL CIRCUITS AND PIPE NETWORKS

Linear algebra has many applications to physics. In this section we will discuss two of the simplest: determining the current flowing in each wire of an electrical circuit and finding the rate of flow of water in a network of pipes.

CURRENT FLOW IN AN ELECTRICAL CIRCUIT

We will consider direct current (dc) electrical circuits consisting of one or more *electromotive forces* (emf) and one or more *resistors*. Electromotive force, E, is measured in *volts* (denoted by V); *resistance, R,* is measured in *ohms* (Ω). The relationship between emf, resistance, and *current, I* (measured in *amperes* or *amps*), in a simple circuit is given by *Ohm's Law*:

$$E = IR.$$

For example, in a circuit containing a simple emf of 6 volts and a single resistance of 30 ohms, as in Figure 8.4, the current is calculated to be $\frac{6}{30} = \frac{1}{5}$ amp. The direction of the current is by convention assumed to be from $+$ to $-$, as indicated by the arrow in Figure 8.4.

6 V

30 Ω

FIGURE 8.4

Assuming that the wires in a circuit have negligible resistance, the *voltage drop* across the resistor of Figure 8.4 must equal the applied emf. Since the current across the 30 Ω resistor is $\frac{1}{5}$ amp, the voltage drop is $\frac{1}{5}(30) = 6$ V. In somewhat more complicated circuits, such as that of Figure 8.5, the sum of the voltage drops (across the resistors) must equal the *net* emf (the algebraic sum of the emfs). Here the net emf is 2 volts (8 V $-$ 6 V), since the two emfs are in opposition to one another. Equating total voltage drop to net emf, we obtain

$$6I + 4I + 2I = 2,$$

or $I = \frac{1}{6}$ amp (flowing in the direction indicated).

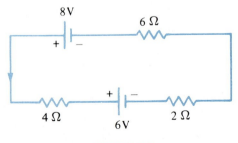

8 V

6 Ω

4 Ω

6 V

2 Ω

FIGURE 8.5

When a circuit consists of more than one *loop* (that is, *closed path*), we use Kirchhoff's laws to determine the current in any branch. In such circuits there are points, called *nodes*, where three or more wires come together. For example, the circuit of Figure 8.6 has three loops, *ABCD*, *ADEF*, and *ABCDEF*, and two nodes, *A* and *D*.

FIGURE 8.6

KIRCHHOFF'S LAWS

i. The algebraic sum of currents at any node in a circuit is zero.

ii. The algebraic sum of the voltage drops around any loop of the circuit is equal to the algebraic sum of the electromotive forces in that loop.

....

NOTE   It is not necessary to guess correctly the direction of the current in any given wire of a complicated circuit. An arbitrary assignment is made, and if the current turns out to be negative, it simply means that the guess was wrong—the current flows in the opposite direction.

....

EXAMPLE 8.4

Find the current in each wire of the circuit of Figure 8.6.

Solution

Let I_1 be the current from D to E to F to A, let I_2 be the current from A to B to C to D, and let I_3 be the current from A to D.

Since A is a node, and I_1 has been assigned to flow into A while I_2 and I_3 flow out of A, we have, by the first law,

$$I_1 - I_2 - I_3 = 0. \tag{1}$$

Since D is a node, and I_1 flows out of D while I_2 and I_3 flow into D, we have

$$I_2 + I_3 - I_1 = 0. \tag{2}$$

From loop *ABCD*, using the second law, we have

$$3I_2 + 4I_2 = 3 - 2,$$

or more simply,

$$7I_2 = 1. \tag{3}$$

From loop *ADEF*, we have

$$2I_1 + 6I_1 = 6 - 3,$$

or more simply,

$$8I_1 = 3. \tag{4}$$

From loop *ABCDEF*, we have

$$3I_2 + 4I_2 + 6I_1 + 2I_1 = 6 - 2,$$

or more simply,

$$8I_1 + 7I_2 = 4. \tag{5}$$

Equations (1) through (5) then provide the following system of linear equations:

$$
\begin{aligned}
I_1 - \ I_2 - I_3 &= 0 \\
-I_1 + \ I_2 + I_3 &= 0 \\
7I_2 &= 1 \\
8I_1 &= 3 \\
8I_1 + 7I_2 &= 4.
\end{aligned}
$$

Following the methods of Section 2.2, we obtain the unique solution $I_1 = \frac{3}{8}$, $I_2 = \frac{1}{7}$, $I_3 = \frac{13}{56}$.

WATER FLOW IN A NETWORK OF PIPES

A situation very similar to electrical circuits and Kirchhoff's laws is that of fluid flow through a plumbing network. In this case the equivalent of the voltage drop is the pressure drop along the length of a pipe due to internal friction, since the pipe does not move but the fluid does. The pressure drop d is assumed to obey the formula

$$d = \frac{vl}{\pi r^4} f = kf,$$

where f is the rate of flow in a pipe of length l and of radius r, and v is a constant based upon the viscosity of the fluid. The constant k is a combination of the other constants,

$$k = \frac{vl}{\pi r^4},$$

which can be calculated for each pipe in the system. These constants k will play the same role that resistances R did in a circuit analysis. Rather than list the pipe lengths, radii, and fluid viscosity, we will simply assume these calculations have been made and proceed with known values k. The pressure itself can be generated by pumps in the network, or by gravity if the pipes are not all level with each other. We will assume that the pressures and flows are represented by unspecified compatible units.

The rules that describe the flows are:

i. The algebraic sum of flows at any node in the network is zero.

ii. The algebraic sum of the pressure drops around any loop of the network is equal to pressures in the loop. (Equivalently, taking pressure sources to be negative pressure drops, the sum around any loop is zero.)

It is easily seen that the first equation, $d = kf$, and conditions (i) and (ii) are very similar to Ohm's law and Kirchhoff's laws of circuit analysis. It is therefore not surprising that the solution, *the rate of flow in each pipe*, can be obtained in a similar fashion. It is assumed that the system is operating in a *stable* manner; that is, the flows are all underway and there is enough capacity in the system to keep them flowing. Flows in and out of a lake or a large cistern would be examples of stability in this sense. Incidentally, the lake itself can be considered to be one of the pipes (with no pressure drop), to complete loops needed to solve the system.

EXAMPLE 8.5

In the network of pipes given in Figure 8.7, each node is given by a letter and each pipe by a number. Nodes B and C are pumps of pressure 200 and 300, respectively, that are driving the system, and E is a lake. Since the flows in 1 and 5 as well as in 3 and 4 must agree, they could be treated as one pipe. They do not look like one pipe, however, and could very well have different k values, so we will not make this assumption. Let the k values for the respective pipes be given by

Pipe i	1	2	3	4	5	6
k_i	5	4	2	4	2	1

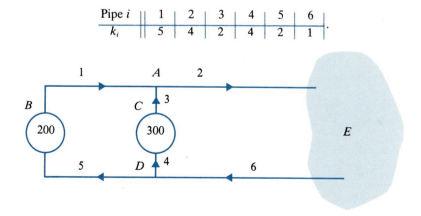

FIGURE 8.7

Let x_i represent the flow in pipe i. Then we have node equations at each of A through E, and loop equations corresponding to the loops $AEDBA$, $AEDCA$, and $ACDBA$. As was the case with Kirchhoff equations, there is some redundancy (dependence) in these equations, but that will show up in the solution process.

$$
\begin{array}{llll}
A: & x_1 - x_2 + x_3 & = & 0 \\
B: & x_1 \qquad\qquad\qquad - x_5 & = & 0 \\
C: & \qquad\quad x_3 - x_4 & = & 0 \\
D: & \qquad\qquad\quad x_4 + x_5 - x_6 & = & 0 \\
E: & \qquad x_2 \qquad\qquad\quad - x_6 & = & 0 \\
AEDBA: & 5x_1 + 4x_2 \qquad\qquad + 2x_5 + x_6 & = & 200 \\
AEDCA: & \qquad 4x_2 + 2x_3 + 4x_4 \qquad + x_6 & = & 300 \\
ACDBA: & 5x_1 \qquad - 2x_3 - 4x_4 + 2x_5 & = & -100
\end{array}
$$

Solving the system by Gauss–Jordan elimination, we have

$$
\begin{aligned}
x_1 &= 6.54 \\
x_2 &= 30.84 \\
x_3 &= 24.30 \\
x_4 &= 24.30 \\
x_5 &= 6.54 \\
x_6 &= 30.84.
\end{aligned}
$$

8.3 EXERCISES

In Exercises 1 and 2 let the circuit be given by Figure 8.8. Calculate the current through each resistor, using the given values of resistances and emfs.

1. $a = 3\,\Omega$, $b = 4\,\Omega$, $c = 3\,\Omega$, $d = 2\,\Omega$, $e = 1\,\Omega$, $f = 6\,V$, $g = 2\,V$

2. $a = 4\,\Omega$, $b = 4\,\Omega$, $c = 5\,\Omega$, $d = 6\,\Omega$, $e = 4\,\Omega$, $f = 8\,V$, $g = 10\,V$

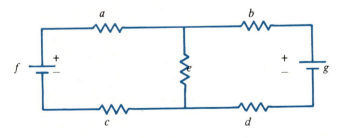

FIGURE 8.8

In Exercises 3 and 4 let the circuit be given by Figure 8.9. Calculate the current in each branch of the circuit, using the given values of resistances and emfs.

3. $a = 3\,\Omega$, $b = 4\,\Omega$, $c = 3\,\Omega$, $d = 2\,\Omega$, $e = 1\,\Omega$, $f = 6\,\text{V}$, $g = 2\,\text{V}$

4. $a = 4\,\Omega$, $b = 4\,\Omega$, $c = 5\,\Omega$, $d = 6\,\Omega$, $e = 4\,\Omega$, $f = 8\,\text{V}$, $g = 10\,\text{V}$

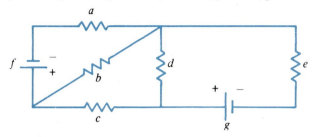

FIGURE 8.9

Induction ammeters can be used to measure the currents in live circuits. When the emf sources and these currents are known, Kirchhoff's laws can be used to calculate the resistances in a network. The system may have multiple solutions, since the resistances may not be completely determined, but additional information (for example, some known components in the system) can be used to complete the computations. When a parameterized family of solutions is all that is known, be sure to take into consideration the physical restriction that all resistances are nonnegative. In Exercises 5 and 6 use Figure 8.9 to calculate the resistances in each resistor if the current flow in amperes through each resistor is as indicated.

5. $a = 5$, $b = 3$, $c = 2$, $d = 4$, $e = 2$, $f = 6\,\text{V}$, $g = 2\,\text{V}$

6. $a = 6$, $b = 3$, $c = 3$, $d = 6$, $e = 3$, $f = 8\,\text{V}$, $g = 10\,\text{V}$

In Exercises 7 and 8 let a network of pipes be given by Figure 8.10. Calculate the flow in each branch, using the given pipe/viscosity constants k and pipe pressures.

7.

pipe i	1	2	3	4	5	6	7	$A = 200$
k_i	3	1	5	1	2	1	3	$B = 100$

8.

pipe i	1	2	3	4	5	6	7	$A = 200$
k_i	2	1	2	1	2	1	2	$B = 400$

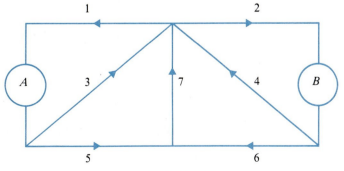

FIGURE 8.10

Analogously with Exercises 5 and 6, if we know the flow in each pipe of a network and the pressure at each source, the pipe/viscosity constants k can be computed. In Exercises 9 and 10 redraw Figure 8.9, interpreting the emf sources as pumps and the currents as flows. Then solve for the pipe constants, using the same numeric data as before.

9. $a = 5, b = 3, c = 2, d = 4, e = 2, f = 6, g = 2$

10. $a = 6, b = 3, c = 3, d = 6, e = 3, f = 8, g = 10$

8.4   INPUT–OUTPUT ANALYSIS: LEONTIEF MODELS

Linear algebra is useful in many fields other than mathematics and the physical sciences. In this section we will consider *input–output analysis*, an application to economics. **Input–output analysis** is the study of the economics of situations involving production and consumption of various goods and services. The method of investigation described here is due to the famous economist Wassily Leontief. We will consider two situations, first a closed model and then an open one.

CLOSED LEONTIEF MODEL

A **closed economic model** is one in which each item (or value of service) produced by one segment of the economy is consumed by another segment of the same economy. In other words, there are no stored surpluses (or monetary equivalents) and no exports from the economy. We further assume that each segment of the economy requires a fixed fraction of the production of the other segments. The problem is to determine whether or not it is possible to have such a system with no "leftovers" and, if so, how much each segment should produce in order to achieve this state.

If we number the segments of the economy, say from 1 to n, the *input–output matrix* of the economy is defined to be the $n \times n$ matrix A in which each entry a_{ij} is the fraction of the total production of segment j required to supply the needs of segment i. (Since all of any segment's production must be consumed, the sum of the entries in any column must be 1.)

EXAMPLE 8.6   Assume the economy has three segments—farming, manufacturing, and service (such as repairing or transporting)—and assume that $\frac{1}{3}$ of the food produced by the farming segment is consumed by farmers themselves, the manufacturers consume $\frac{1}{2}$ of it, and the remaining $\frac{1}{6}$ is consumed by those offering services. Further assume that farmers require $\frac{1}{2}$ of the manufactured goods, with the rest split evenly between the other two industries. Finally, assume that all three segments have equal need of services.

Solution

By numbering the segments in order, we have

$$A = \begin{bmatrix} \frac{1}{3} & \frac{1}{2} & \frac{1}{3} \\ \frac{1}{2} & \frac{1}{4} & \frac{1}{3} \\ \frac{1}{6} & \frac{1}{4} & \frac{1}{3} \end{bmatrix}.$$

Let $\mathbf{x} = (x_1, x_2, x_3)$ be the dollar values associated with the output of each segment. Since the value produced by the farming segment must equal the sum of the amounts spent by farmers to fulfill their needs from the other segments (including other farmers), we have

$$x_1 = \frac{1}{3}x_1 + \frac{1}{2}x_2 + \frac{1}{3}x_3.$$

Similar equations for the expenditures by manufacturing and by service are

$$x_2 = \frac{1}{2}x_1 + \frac{1}{4}x_2 + \frac{1}{3}x_3,$$

$$x_3 = \frac{1}{6}x_1 + \frac{1}{4}x_2 + \frac{1}{3}x_3.$$

In matrix form, these three equations can be written as

$$\mathbf{x} = A\mathbf{x},$$

and, since $\mathbf{x} = I\mathbf{x}$, we have the homogeneous linear system

$$(I - A)\mathbf{x} = \mathbf{0}.$$

Solving this homogeneous system, we obtain

$$x_1 = \frac{5}{3}s, \qquad x_2 = \frac{14}{9}s, \qquad x_3 = s.$$

Since these numbers are to represent the value of goods or services, only nonnegative solutions are realistic. Hence the system "balances" for any $s > 0$, and x_1, x_2, and x_3 as shown.

Returning to general considerations, let A be an input–output matrix, and let $\mathbf{x} = (x_1, \ldots, x_n)$ be the vector of values associated with the output of each segment. We wish to find the relative size of each segment of the economy.

Since the value produced by a given segment must equal the sum of the amounts spent by that segment to fulfill its needs from each of the other segments, we have the following system of equations:

$$x_1 = a_{11}x_1 + a_{12}x_2 + \cdots + a_{1n}x_n$$
$$x_2 = a_{21}x_1 + a_{22}x_2 + \cdots + a_{2n}x_n$$
$$\vdots \qquad \vdots \qquad \vdots \qquad \qquad \vdots$$
$$x_n = a_{n1}x_1 + a_{n2}x_2 + \cdots + a_{nn}x_n.$$

In matrix form, this becomes

$$\mathbf{x} = A\mathbf{x},$$

and as in Example 8.5, we have

$$(I - A)\mathbf{x} = \mathbf{0}.$$

This homogeneous system has the trivial solution (nobody produces anything!), but since each column of $I - A$ adds up to 0, the system has rank at most $n - 1$, so nontrivial solutions exist. To be realistic, we can only allow nonnegative values for the entries of $\mathbf{x}$, since they are dollar values of output. The following theorem, which we assume without proof, tells us that there are realistic, nontrival solutions.

THEOREM I   Let A be the input–output matrix of a closed model. Then $(I - A)\mathbf{x} = \mathbf{0}$ has a solution in which all of its entries are positive.

OPEN LEONTIEF MODEL

An **open economic model** is similar in structure to that of the closed model, except that one extra segment called *consumers* is added. The consumer segment is assumed to have a certain demand for each of the goods or services produced by the other segments. These demands are subject to change, however, so we wish to analyze the problem of what to do if the demands change. The other assumptions remain exactly as in the closed case—each segment requires a fixed fraction of the total output of each of the other segment's total production, and there are no surpluses (above consumer demand) produced.

Again we define the *input–output matrix A* of the economy in terms of fractions of production but with a slight change. We let a_{ij} be the fraction of the production of segment j required to supply the needs of segment i. Thus the matrix A is the same as that of a closed economy in the case when all consumer demands equal zero. Letting $\mathbf{x} = (x_1, \ldots, x_n)$ be the production vector, $A\mathbf{x}$ is the vector of internal consumption. For instance, the first row of $A\mathbf{x}$ is

$$a_{11}x_1 + a_{12}x_2 + \cdots + a_{1n}x_n.$$

This represents the fraction of product 1 consumed by segment 1, plus the fraction of product 1 consumed by segment 2, and so on. The resulting sum is the amount of product 1 consumed by all of the producing segments. Similarly, the ith row of $A\mathbf{x}$ is the amount *produced* by the ith segment which is *consumed* by the producing segments.

Since there is an assumption that no surplus exists, the internal consumption plus the demand for each product must equal the total production of that product. Letting **d** be the vector of demands, with d_i being the demand for the ith product, we have, for each i,

$$a_{i1}x_1 + a_{i2}x_2 + \cdots + a_{in}x_n + d_i = x_i.$$

In terms of matrices, this system becomes

$$A\mathbf{x} + \mathbf{d} = \mathbf{x}.$$

Transposing **x**, we have

$$(I - A)\mathbf{x} = \mathbf{d}$$

and, if $I - A$ is invertible,

$$\mathbf{x} = (I - A)^{-1}\mathbf{d}.$$

As in the case of the closed economy, we can only accept vectors **x** for which each entry is nonnegative. The following theorem (assumed without proof) gives fairly general conditions under which a solution can be obtained.

THEOREM 2 If a square matrix A has all positive entries, and the sum of the entries in each of its columns is less than one, then $(I - A)$ is invertible.

EXAMPLE 8.7

In the economy of farming, manufacturing, and service industries of Example 8.6, suppose there is also a consumer segment. Table 8.1 gives the dollar value of each product that is consumed by a given segment. That is, an entry in the ith row of Table 8.1 gives the value (in millions of dollars) of the product associated with that row which is consumed by the industry associated with that column. If the anticipated consumer demands for five years later are farming, 20, manufacturing, 30, and service, 20, find the amount each segment needs to produce in order to balance the system.

Industry	Farming	Manufacturing	Service	Consumer	Total
Farming	10	15	5	10	40
Manufacturing	14	7	7	21	49
Service	6	6	12	6	30

TABLE 8.1

Solution

The input–output matrix A for the economy is computed as follows: a_{ij} is that fraction of the total of the jth segment consumed by the ith segment (ignoring the consumer segment). For example the first column of A is $\mathbf{a}^1 = (10/40, 14/40, 6/40)$, since farming consumes 10 of the total 40, manufacturing 14, and service 6. Computing the other entries, we have

$$A = \begin{bmatrix} 10/40 & 15/49 & 5/30 \\ 14/40 & 7/49 & 7/30 \\ 6/40 & 6/49 & 12/30 \end{bmatrix}.$$

Letting $\mathbf{d} = (20, 30, 20)$ be the vector of demands, we determine the solution to be

$$\mathbf{x} = (I - A)^{-1}\mathbf{d} = \begin{bmatrix} 30/40 & -15/49 & -5/30 \\ -14/40 & 42/49 & -7/30 \\ -6/40 & -6/49 & 18/30 \end{bmatrix}^{-1} \begin{bmatrix} 20 \\ 30 \\ 20 \end{bmatrix} = \begin{bmatrix} 77.18 \\ 85.62 \\ 70.10 \end{bmatrix}.$$

Thus the approximate value of the farming segment should be $x_1 = 77.18$, manufacturing, $x_2 = 85.62$, and service, $x_3 = 70.10$.

8.4 EXERCISES

In Exercises 1 and 2 the given matrix M is the input–output matrix of a closed economy of three segments A, B, and C. Find the value of production for each segment.

1.
$$M = \begin{bmatrix} \frac{3}{4} & \frac{1}{5} & \frac{1}{3} \\ \frac{1}{8} & \frac{2}{5} & \frac{1}{6} \\ \frac{1}{8} & \frac{2}{5} & \frac{1}{2} \end{bmatrix}$$

2.
$$M = \begin{bmatrix} 0.2 & 0.1 & 0.3 \\ 0.4 & 0.8 & 0.3 \\ 0.4 & 0.1 & 0.4 \end{bmatrix}$$

3. Assume a closed economy has three segments A, B, and C, with A requiring $\frac{1}{2}$ of its own production and B and C splitting the rest evenly. All three segments require the same fraction of the production of B. Segment A requires $\frac{2}{3}$ of the production of C, while B and C require the same amount of C. Find the value of production for each segment.

4. The closed Leontief model problem is sometimes viewed as a problem in *pricing* rather than *production*. The assumption is that in a closed model each segment requires not only a fixed *fraction* of the production of the other segments but a fixed *amount* of the production. Viewed in this manner, the problem is to decide how each segment should price its production in order to spend exactly what it earns. Explain the result of Exercise 3 in this context.

In Exercises 5 and 6 the given matrix M is the input–output matrix of an open economy of three segments A, B, and C, and the vector $\mathbf{d}$ is the predicted demand vector of the consumer segment. Find the production level for each segment required to balance the economy.

5.
$$M = \begin{bmatrix} \frac{1}{3} & \frac{1}{5} & \frac{1}{6} \\ \frac{1}{4} & \frac{2}{5} & \frac{1}{3} \\ \frac{1}{4} & \frac{1}{5} & \frac{1}{6} \end{bmatrix}, \qquad \mathbf{d} = \begin{bmatrix} 3 \\ 5 \\ 1 \end{bmatrix}$$

6.
$$M = \begin{bmatrix} 0.2 & 0.1 & 0.5 \\ 0.3 & 0.2 & 0.3 \\ 0.1 & 0.1 & 0.1 \end{bmatrix}, \qquad \mathbf{d} = \begin{bmatrix} 20 \\ 15 \\ 5 \end{bmatrix}$$

7. Assume an open economy has three producing segments A, B, and C. Table 8.2 gives the current value of the production of the segment of each row that is consumed by the segment of each column. Calculate the production required of each industry to provide a balanced economy if the consumer demand rises to $\mathbf{d} = (15, 10, 6)$.

	A	B	C	Consumers
A	12	4	8	10
B	8	9	4	6
C	2	4	3	3

TABLE 8.2

8.5 MARKOV CHAINS

In this section we will demonstrate how multiplication of a vector by a matrix can be used to compute the probability of certain types of future events. Multiplication of matrices is the only knowledge required for most of the section. The latter part concerning equilibrium states presupposes a knowledge of eigenvalues and eigenvectors, Section 7.1.

MARKOV CHAINS

We first need to recall a few facts about probabilities in general. The probability of a random event is a number between 0 and 1 (inclusive) that measures the likelihood that the event will occur. For example, the probability of rolling the number 1 with a single die is $\frac{1}{6}$, since there is only one "successful" side out of the six sides of the die, and all outcomes are equally likely. The probability of rolling an even number, however, is $\frac{1}{2}$, since any of 2, 4, or 6 will be successful out of the six possible sides. Probability 0 implies that the event cannot occur, and probability 1 implies that the event must occur. As examples, the probability of rolling 7 with one die is 0, and the probability of rolling a number less than 10 with a single die is 1.

To compute probabilities of more complicated events, we use the following two facts. If two events are *independent,* the probability of both occurring is the *product* of the individual probabilities. For example, the probability of flipping a coin twice and obtaining heads both times is $(\frac{1}{2})(\frac{1}{2}) = \frac{1}{4}$. (The coin cannot remember what it did before, so the events are independent.) If two events are *disjoint* (that is, they cannot both occur), the probability of one *or* the other occurring is the *sum* of the individual probabilities. For example, the probability of rolling an even number on a die was shown to be $\frac{3}{6} = \frac{1}{2}$. This can be broken up into three disjoint events: roll a 2, roll a 4, or roll a 6. Each of these has probability $\frac{1}{6}$, and $\frac{1}{6} + \frac{1}{6} + \frac{1}{6} = \frac{1}{2}$.

We will now describe the setting for Markov chains. Repeated observations of a situation are made at regular intervals, and the current state of the situation is recorded. As an example, suppose the weather is observed every 6 hours, and the possible states of the weather are chosen to be fair (F), partly cloudy (P), cloudy (C), and rain (R). Based on past information, we assign to each state a probability of moving to some other state (or remaining the same). An important (and sometimes unrealistic) assumption is that these probabilities depend *only* on the current state, not on any previous ones. Continuing with the example, suppose the probabilities of moving from one weather state to another are as given in Table 8.3.

Next State

		F	P	C	R
	F	0.6	0.3	0.1	0.0
	P	0.2	0.4	0.2	0.2
Current State	*C*	0.1	0.4	0.3	0.2
	R	0.0	0.2	0.3	0.5

TABLE 8.3

For instance, if it is cloudy now, the probability that it will be raining at the next observation is 0.2. Notice that each entry is nonnegative, (probabilities must lie between 0 and 1) and that the sum of the entries in each row is 1 (no matter what the initial state is, *some* state must occur at the next observation, and the events are disjoint). Any such row is called a **probability vector**, a vector with nonnegative entries the sum of which is 1.

Our prediction for the weather will also be stated in terms of a probability vector, $\mathbf{x} = (x_1, x_2, x_3, x_4)$, where x_1 is the probability that the weather will be fair, x_2 is the probability of partly cloudy, and so on. Since *some* weather must occur, $x_1 + x_2 + x_3 + x_4 = 1$. For instance, if it is raining now, the prediction for 6 hours hence is $(0.0, 0.2, 0.3, 0.5)$. What is the prediction for the following time period? More generally, if $\mathbf{x}$ is the prediction for one time period, what is the prediction for the following one?

To answer this question, we reason as follows: let $\mathbf{y} = (y_1, y_2, y_3, y_4)$ be the prediction of probabilities in the next time period. That is, y_1 is the probability that the weather will be fair, y_2 is the probability that it will be partly cloudy, and so on. Looking just at y_1, there are four different ways of getting fair weather in the next time period. It could be fair now and stay fair, or it could change from one of three other states to fair. If it is fair, then the probability that it stays fair is 0.6. However, there is a probability of x_1 that it is fair, and these events are independent. Thus the probability that it is fair and stays fair (FF) is $(x_1)(0.6)$. Similarly, the probability that it is partly cloudy and changes to fair (PF) is $(x_2)(0.2)$. Listing all of the possibilities, we have

Probability of FF is $0.6x_1$,

Probability of PF is $0.2x_2$,

Probability of CF is $0.1x_3$,

Probability of RF is $0.0x_4$.

These four events are disjoint, so the probability that the weather is fair in the next time period is

$$y_1 = 0.6x_1 + 0.2x_2 + 0.1x_3 + 0.0x_4.$$

Note that y_1 is exactly the dot product of the first column of Table 8.3 with the vector $\mathbf{x}$. Carrying out a similar analysis for each of y_2, y_3, and y_4, we obtain y_i as the dot product of the ith column of the table with the vector $\mathbf{x}$. To make this conform to our usual convention for matrix multiplication, let T (for **transition matrix**) be the *transpose* of the entries in the table. Then this discussion yields the result that

$$\mathbf{y} = T\mathbf{x}. \tag{1}$$

Using Equation (1), we can answer the question posed earlier. If it is raining now, what is the probability vector for two periods in the future? Letting $\mathbf{x} = (0.0, 0.2, 0.3, 0.5)$, the prediction for the first time period, we have

$$\mathbf{y} = T\mathbf{x} = \begin{bmatrix} 0.6 & 0.2 & 0.1 & 0.0 \\ 0.3 & 0.4 & 0.4 & 0.2 \\ 0.1 & 0.2 & 0.3 & 0.3 \\ 0.0 & 0.2 & 0.2 & 0.5 \end{bmatrix} \begin{bmatrix} 0.0 \\ 0.2 \\ 0.3 \\ 0.5 \end{bmatrix} = \begin{bmatrix} 0.07 \\ 0.30 \\ 0.28 \\ 0.35 \end{bmatrix}.$$

Since $\mathbf{x}$ is equal to $T(0, 0, 0, 1)$, we have another way of computing $\mathbf{y}$, $\mathbf{y} = T^2(0, 0, 0, 1)$, Similarly, if we wanted the probability vector for the following time period, Equation (1) tells us that this prediction is given by

$$T\mathbf{y} = T(T^2(0, 0, 0, 1)) = T^3(0, 0, 0, 1).$$

Thus we see that to go from any given state to any later state, we simply multiply the given vector by a power of T. The results of this specific example are generalized as the following procedure.

MARKOV CHAIN PROCEDURE

Suppose there are n possible states:

    i. Prepare an $n \times n$ matrix where the ijth entry is the probability of moving from state i to state j.

    ii. Let T be the transpose of this matrix.

    iii. For any probability vector $\mathbf{x}$ in $\mathbf{R}^n$ and for any natural number k, the probability vector for k time periods in the future is given by $\mathbf{y} = T^k\mathbf{x}$.

....

NOTE   The vector $\mathbf{x}$ in the procedure is called an **initial vector** and is presented as a probability vector. If in fact we *know* the initial state (say, the ith state), then $\mathbf{x} = \mathbf{e}_i$, the ith elementary vector. Of course, $\mathbf{e}_i$ is itself a probability vector; the probability of the ith state is 1 and of any other is 0. As an example, in the preceding discussion, if the initial state is cloudy, the probability vector is $(0, 0, 1, 0)$.

....

EXAMPLE 8.8

Suppose that a particular telephone wire is capable of supporting up to four calls at any one time and that the number of calls that will be in progress one minute later than at a given time depends only on the number of calls in progress at the given time. Suppose further that the probability of moving from i calls at a given time to j calls one minute later is given in Table 8.4. If at some particular time all possible numbers of calls are equally likely, what is the probability that the line will be unavailable (four calls in progress) three minutes later?

		j				
		0	**1**	**2**	**3**	**4**
	0	0.3	0.3	0.2	0.2	0.0
	1	0.2	0.3	0.3	0.1	0.1
i	2	0.1	0.2	0.4	0.2	0.1
	3	0.1	0.1	0.2	0.3	0.3
	4	0.0	0.0	0.3	0.3	0.4

TABLE 8.4

Solution

We follow the Markov chain procedure stated earlier. Step (i) is already done. The matrix T is the transpose of the given array:

$$T = \begin{bmatrix} 0.3 & 0.2 & 0.1 & 0.1 & 0.0 \\ 0.3 & 0.3 & 0.2 & 0.1 & 0.0 \\ 0.2 & 0.3 & 0.4 & 0.2 & 0.3 \\ 0.2 & 0.1 & 0.2 & 0.3 & 0.3 \\ 0.0 & 0.1 & 0.1 & 0.3 & 0.4 \end{bmatrix}.$$

Since all numbers of calls are equally likely, the initial vector is

$$\mathbf{x} = (0.2, 0.2, 0.2, 0.2, 0.2).$$

The probability vector for three minutes hence is then

$$\mathbf{y} = T^3\mathbf{x} = \begin{bmatrix} 0.3 & 0.2 & 0.1 & 0.1 & 0.0 \\ 0.3 & 0.3 & 0.2 & 0.1 & 0.0 \\ 0.2 & 0.3 & 0.4 & 0.2 & 0.3 \\ 0.2 & 0.1 & 0.2 & 0.3 & 0.3 \\ 0.0 & 0.1 & 0.1 & 0.3 & 0.4 \end{bmatrix}^3 \begin{bmatrix} 0.2 \\ 0.2 \\ 0.2 \\ 0.2 \\ 0.2 \end{bmatrix} = \begin{bmatrix} 0.128 \\ 0.174 \\ 0.292 \\ 0.222 \\ 0.184 \end{bmatrix}.$$

The probability of having four calls in progress at that time is the last coordinate of $\mathbf{y}$, $y_5 = 0.184$.

EQUILIBRIUM STATE

DEFINITION

A probability vector $\mathbf{x}$ is an **equilibrium** or **steady state vector** if $\mathbf{x} = T\mathbf{x}$, where T is the transition matrix of a Markov chain.

In other words, an equilibrium vector is such that moving from one time period to the next results in no change. Every Markov chain has an equilibrium vector, and some have more than one. The following theorem shows how to determine them.

THEOREM 1 Let T be the transition matrix of a Markov chain. Then $\lambda = 1$ is an eigenvalue of T. If $\mathbf{x}$ is an eigenvector corresponding to $\lambda = 1$ with nonnegative components, and s is the sum of the components of $\mathbf{x}$, then $(1/s)\mathbf{x}$ is an equilibrium vector.

Proof Let there be n states so that T is an $n \times n$ matrix. Since the sum of the entries in each column is 1, the sum of the entries in each column of $T - I$ is 0. This implies that if each row is added to the last row, the result is a zero row, and hence the resulting matrix has determinant zero. But adding one row to another does not change the value of the determinant (Theorem 4 of Section 3.2), so the determinant of $T - I$ is 0. Since $T - I = T - 1I$, $\lambda = 1$ is an eigenvalue of T.

If $\mathbf{x}$ is an eigenvector corresponding to $\lambda = 1$, then $T\mathbf{x} = \mathbf{x}$. Assuming further that $\mathbf{x}$ has nonnegative components, and that $s = x_1 + \cdots + x_n$, the sum of the components of $\mathbf{x}$, we have $T((1/s)\mathbf{x}) = (1/s)T\mathbf{x} = (1/s)\mathbf{x}$. Now

$(1/s)\mathbf{x}$ has nonnegative components, and the sum of the components is $(1/s)(x_1 + \cdots + x_n) = (1/s)s = 1$. Therefore $(1/s)\mathbf{x}$ is a probability vector, and consequently it is an equilibrium vector.

EXAMPLE 8.9

Determine an equilibrium vector for a Markov chain with transition matrix

$$T = \begin{bmatrix} \frac{1}{3} & \frac{1}{3} & \frac{1}{3} \\ \frac{1}{2} & \frac{1}{6} & 0 \\ \frac{1}{6} & \frac{1}{2} & \frac{2}{3} \end{bmatrix}.$$

Solution

We first determine the eigenvectors of T corresponding to 1. To do so, we solve the homogeneous system $(T - I)\mathbf{x} = \mathbf{0}$ by row-reducing $T - I$,

$$T - I = \begin{bmatrix} -\frac{2}{3} & \frac{1}{3} & \frac{1}{3} \\ \frac{1}{2} & -\frac{5}{6} & 0 \\ \frac{1}{6} & \frac{1}{2} & -\frac{1}{3} \end{bmatrix} \rightarrow \begin{bmatrix} 1 & 0 & -\frac{5}{7} \\ 0 & 1 & -\frac{3}{7} \\ 0 & 0 & 0 \end{bmatrix}.$$

From the row-reduced echelon form we see that $(\frac{5}{7}, \frac{3}{7}, 1)$ is an eigenvector that corresponds to the eigenvalue 1. An (in fact, *the*) equilibrium vector is therefore

$$\left(1\Big/\left(\frac{5}{7} + \frac{3}{7} + 1\right)\right)\left(\frac{5}{7}, \frac{3}{7}, 1\right) = \left(\frac{1}{3}, \frac{1}{5}, \frac{7}{15}\right).$$

Often there is a *unique* equilibrium vector for a Markov chain, and succeeding stages of the chain give resulting vectors that are successively closer to this equilibrium vector and converge to it. The following theorem offers a very simple test for this situation. The proof is omitted.

THEOREM 2 Let T be the transition matrix of a Markov chain. If some power of T has strictly positive entries, then there is a unique equilibrium vector $\mathbf{x}$, and for any probability vector $\mathbf{y}$, $T^n\mathbf{y}$ converges to $\mathbf{x}$ as n becomes large.

EXAMPLE 8.10

Show that some power of the transition matrix T of Example 8.8,

$$T = \begin{bmatrix} 0.3 & 0.2 & 0.1 & 0.1 & 0.0 \\ 0.3 & 0.3 & 0.2 & 0.1 & 0.0 \\ 0.2 & 0.3 & 0.4 & 0.2 & 0.3 \\ 0.2 & 0.1 & 0.2 & 0.3 & 0.3 \\ 0.0 & 0.1 & 0.1 & 0.3 & 0.4 \end{bmatrix},$$

has strictly positive entries so that there is a unique equilibrium vector. Find an approximation to the equilibrium vector by computing values of $T^n\mathbf{y}$ for some $\mathbf{y}$ and various values of n.

Solution

The second power of T, T^2, has strictly positive entries, since

$$T^2 = \begin{bmatrix} 0.19 & 0.16 & 0.13 & 0.10 & 0.06 \\ 0.24 & 0.22 & 0.19 & 0.13 & 0.09 \\ 0.27 & 0.30 & 0.31 & 0.28 & 0.30 \\ 0.19 & 0.19 & 0.21 & 0.25 & 0.27 \\ 0.11 & 0.13 & 0.16 & 0.24 & 0.28 \end{bmatrix}.$$

Estimating the equilibrium vector by raising T to successively higher powers (with the aid of a computer), we see that T^7 and T^8 agree to the nearest hundredth.

$$T^7 = \begin{bmatrix} 0.125 & 0.123 & 0.123 & 0.122 & 0.121 \\ 0.170 & 0.170 & 0.169 & 0.167 & 0.170 \\ 0.295 & 0.295 & 0.295 & 0.295 & 0.295 \\ 0.224 & 0.224 & 0.224 & 0.225 & 0.226 \\ 0.188 & 0.188 & 0.189 & 0.190 & 0.192 \end{bmatrix}.$$

$$T^8 = \begin{bmatrix} 0.123 & 0.123 & 0.123 & 0.122 & 0.122 \\ 0.170 & 0.169 & 0.169 & 0.168 & 0.168 \\ 0.295 & 0.295 & 0.295 & 0.295 & 0.295 \\ 0.224 & 0.224 & 0.224 & 0.225 & 0.225 \\ 0.189 & 0.189 & 0.189 & 0.190 & 0.190 \end{bmatrix}.$$

Letting $\mathbf{y} = (1, 0, 0, 0, 0)$, we take $T^8\mathbf{y} = (0.123, 0.170, 0.295, 0.224, 0.189)$ as our approximation to the equilibrium vector.

 COMPUTATIONAL NOTE Estimating the equilibrium vector $\mathbf{x}$ by setting $\mathbf{x}$ approximately equal to $T^n\mathbf{y}$ for larger values of n and an arbitrary vector $\mathbf{y}$ (as in Example 8.10) is commonly used in place of computing the actual equilibrium vector when the conditions of Theorem 2 are met. By using the elementary vectors $\mathbf{e}_i$ for $\mathbf{y}$, we see that the columns of T^n must all equal $\mathbf{x}$ in the limit. Thus the procedure is to raise T to successively higher powers until all of the columns agree within the desired accuracy. Any column of this matrix is then a good approximation to the equilibrium vector.

8.5 EXERCISES

In Exercises 1–10 use the information given in the table of probabilities for a Markov chain, Table 8.5.

Next State

Current State		1	2	3
	1	0.2	0.5	0.3
	2	0.1	0.8	0.1
	3	0.2	0.4	0.4

TABLE 8.5

1. Find the probability of moving from state 3 to state 2.

2. Find the probability of moving from state 2 to state 3.

3. Find the transition matrix for the Markov chain.

4. Find the transition matrix for moving from the current state to two states in the future.

5. What is the initial vector if all states are equally likely?

6. What is the initial vector if state 1 is excluded but the other two are equally likely?

7. If all states are equally likely, find the probability vector for the next state.

8. If state 1 is excluded, but the other two are equally likely, find the probability vector for the next state.

9. If the initial state is state 1, find the probability of state 2 two time periods in the future.

10. If the initial state is state 1, find the probability of state 3 three time periods in the future.

In Exercises 11–15 use the following information. Suppose a man has three modes of transportation to work: He can walk, drive his car, or take the bus. He never walks two days in a row or drives two days in a row. If he walked the last time, he is twice as likely to drive as take the bus. In all other cases he shows no preference.

11. Prepare the table of probabilities for this Markov chain and find the transition matrix.

12. If he drives on Monday, what is the probability that he will walk on Tuesday? That he will drive on Tuesday?

13. If he is equally likely to choose any mode of transportation on Monday, what is the prediction for all modes of transportation for Tuesday?

14. If he takes the bus on Monday, what is the probability that he will take the bus on Wednesday?

15. If he takes the bus on Monday, what is the probability that he will take the bus all week (five days in succession)?

16. Find all equilibrium vectors of the transition matrix

$$T = \begin{bmatrix} \frac{1}{2} & \frac{1}{3} & 0 \\ \frac{1}{2} & \frac{1}{3} & 0 \\ 0 & \frac{1}{3} & 1 \end{bmatrix}.$$

17. Find all equilibrium vectors of the transition matrix

$$T = \begin{bmatrix} \frac{1}{2} & \frac{1}{2} & 0 \\ \frac{1}{2} & \frac{1}{2} & 0 \\ 0 & 0 & 1 \end{bmatrix}.$$

18. Find all equilibrium vectors of the transition matrix of Exercise 11.

19. If T is a transition matrix of a Markov chain, prove that the sum of the entries in each column of T^2 is also equal to 1.

20. Extend Exercise 19 to T^n.

 In Exercises 21 and 22 use a computer program to estimate the components of each equilibrium vector to the nearest hundredth.

21. Find the equilibrium vector for the transition matrix of Exercise 3.

22. Find the equilibrium vector for the transition matrix T obtained from Table 8.3.

8.6   QUADRIC SURFACES

A common topic in analytic geometry is the reduction of a quadratic equation in two variables—the equation of a *conic section*—to standard form. In this section we will discuss an extension of this idea. With the aid of the techniques of Section 7.2, we will reduce a quadratic equation in three variables—that of a *quadric surface*—to certain standard forms.

DEFINITION

A **second-degree polynomial** (or **quadratic**) **equation** in x, y, and z is one of the form

$$ax^2 + by^2 + cz^2 + dxy + exz + fyz + gx + hy + iz = k, \qquad (1)$$

where $a, b, \ldots, i, k$ are constants and $a, b, \ldots, f$ are not all zero. The graph of such an equation is called a **quadric** or **quadric surface**.

For example, the equation

$$x^2 + y^2 + z^2 - 2x + 4y = 4$$

is the special case of Equation (1), with $a = b = c = 1$, $d = e = f = 0$, $g = -2$, $h = 4$, $i = 0$, and $k = 4$. (Its graph is a *sphere* in 3-space, with center at the point $(1, -2, 0)$ and radius equal to 3.)

There are nine basic types of *nondegenerate* quadric surfaces. These are listed in *standard form* in the next subsection. We will develop a *reduction procedure* to transform Equation (1) to one of these standard forms. If it is impossible to reduce the equation to one of these forms, the corresponding surface is called a *degenerate quadric*. Graphs of the latter include planes, single points, and no points at all. It is not always clear from the original equation whether or not a quadric is nondegenerate, but as we will see, once the reduction is performed, the given surface can be easily identified.

STANDARD FORMS

i. Ellipsoid (Figure 8.11):

$$\frac{x^2}{p^2} + \frac{y^2}{q^2} + \frac{z^2}{r^2} = 1 \qquad (p, q, r > 0).$$

(A *sphere* is the special case of an ellipsoid for which $p = q = r$.)

ii. Elliptic paraboloid (Figure 8.12):

$$z = \frac{x^2}{p^2} + \frac{y^2}{q^2} \qquad (p, q > 0).$$

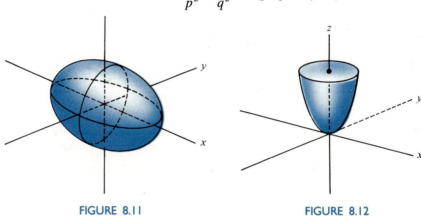

FIGURE 8.11   FIGURE 8.12

iii. Hyperbolic paraboloid (Figure 8.13):

$$z = \frac{x^2}{p^2} - \frac{y^2}{q^2} \qquad (p, q > 0).$$

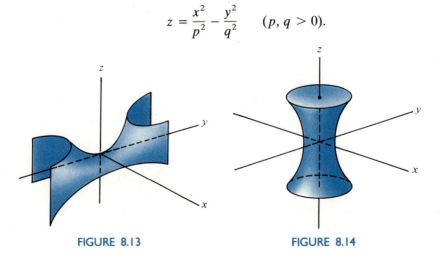

FIGURE 8.13   FIGURE 8.14

iv. Hyperboloid of one sheet (Figure 8.14):

$$\frac{x^2}{p^2} + \frac{y^2}{q^2} - \frac{z^2}{r^2} = 1 \qquad (p, q, r > 0).$$

v. Hyperboloid of two sheets (Figure 8.15):

$$\frac{x^2}{p^2} - \frac{y^2}{q^2} - \frac{z^2}{r^2} = 1 \qquad (p, q, r > 0).$$

vi. Cone (Figure 8.16):

$$\frac{x^2}{p^2} + \frac{y^2}{q^2} - \frac{z^2}{r^2} = 0 \qquad (p, q, r > 0).$$

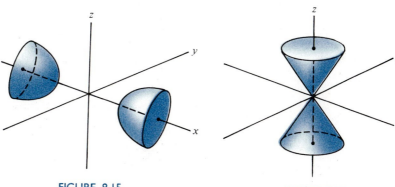

FIGURE 8.15

FIGURE 8.16

vii. Parabolic cylinder (Figure 8.17):

$$x^2 = py + qz \qquad (p, q \text{ not both zero}).$$

viii. Elliptic cylinder (Figure 8.18):

$$\frac{x^2}{p^2} + \frac{y^2}{q^2} = 1 \qquad (p, q > 0).$$

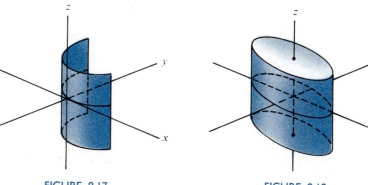

FIGURE 8.17

FIGURE 8.18

ix. Hyperbolic cylinder (Figure 8.19):

$$\frac{x^2}{p^2} - \frac{y^2}{q^2} = 1 \qquad (p, q > 0).$$

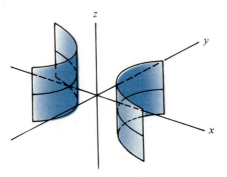

FIGURE 8.19

EXAMPLE 8.11   Identify the quadrics whose equations are

a. $\dfrac{x^2}{4} + \dfrac{y^2}{9} + z^2 = 1$,

b. $9x^2 - y^2 - 16z^2 = 144$,

c. $x^2 - y^2 - z = 0$.

Solution   Equation (a) fits the standard form (i), with $p = 2$, $q = 3$, and $r = 1$; it represents an ellipsoid. Dividing Equation (b) by 144, we obtain one of the standard form (v), with $p = 4$, $q = 12$, and $r = 3$; it represents a hyperboloid of two sheets. Finally, by transposing the z term of Equation (c), we see that it takes the standard form (iii), with $p = 1$ and $q = 1$; it represents a hyperbolic paraboloid.

THE REDUCTION PROCEDURE

We will now derive a procedure for transforming (or *reducing*) those special cases of Equation (1) that represent nondegenerate quadric surfaces to one of the given standard forms. We begin by writing Equation (1) in the matrix form

$$\mathbf{x}^T A \mathbf{x} + B \mathbf{x} = k, \tag{2}$$

where A, B, and $\mathbf{x}$ are the matrices

$$A = \begin{bmatrix} a & d/2 & e/2 \\ d/2 & b & f/2 \\ e/2 & f/2 & c \end{bmatrix}, \qquad B = [g \quad h \quad i], \qquad \text{and} \qquad \mathbf{x} = \begin{bmatrix} x \\ y \\ z \end{bmatrix}. \tag{3}$$

To see that Equations (1) and (2) are equivalent, simply perform the operations indicated in Equation (2), and Equation (1) will result. Notice that A is a *symmetric* matrix.

EXAMPLE 8.12

Find matrices A and B such that the equation $x^2 - 4y^2 - 2z^2 + 8xy - 6yz + 7x - 3y = 1$ takes the matrix form of Equation (2).

Solution

Comparing the given equation to Equation (1), we see that $a = 1$, $b = -4$, $c = -2$, $d = 8$, $e = 0$, $f = -6$, $g = 7$, $h = -3$, $i = 0$, and $k = 1$. Therefore

$$A = \begin{bmatrix} 1 & 4 & 0 \\ 4 & -4 & -3 \\ 0 & -3 & -2 \end{bmatrix} \quad \text{and} \quad B = \begin{bmatrix} 7 & -3 & 0 \end{bmatrix}.$$

The first step in the reduction procedure is the construction of a transformation that, when applied to Equation (1), results in the elimination of the *cross product* (the xy, xz, and yz) terms. In terms of Equation (2) this is equivalent to diagonalizing the symmetric matrix A, since the off-diagonal entries of A are the ones that produce the coefficients of the cross product terms.

Letting P be an orthogonal 3×3 matrix such that $D = P^T A P$ is diagonal (see Section 7.2), we make the change of variable

$$\mathbf{x} = P\mathbf{x}'$$

and substitute into Equation (2), obtaining

$$(P\mathbf{x}')^T A (P\mathbf{x}') + B(P\mathbf{x}') = k.$$

Now $(P\mathbf{x}')^T = (\mathbf{x}')^T P^T$, so this equation becomes

$$(\mathbf{x}')^T (P^T A P)\mathbf{x}' + BP\mathbf{x}' = k.$$

But $P^T A P = D$ (a diagonal matrix), so this last equation takes the form

$$(\mathbf{x}')^T D\mathbf{x}' + BP\mathbf{x}' = k. \tag{4}$$

Recalling from Section 7.2 that the diagonal entries of D are just the eigenvalues of A, λ_1, λ_2, and λ_3, we need not actually perform the matrix multiplications $P^T A P$ to obtain D; we simply set

$$D = \begin{bmatrix} \lambda_1 & 0 & 0 \\ 0 & \lambda_2 & 0 \\ 0 & 0 & \lambda_3 \end{bmatrix}.$$

Obtaining BP requires more effort, since the procedure for determining P involves finding an orthonormal basis of eigenvectors of A. Letting P be a matrix with these eigenvectors as columns, we form the product BP and obtain a matrix we denote by

$$BP = [a' \quad b' \quad c'].$$

Finally, letting $\mathbf{x}' = (x', y', z')$, we see that Equation (4) can be rewritten as the scalar equation

$$\lambda_1(x')^2 + \lambda_2(y')^2 + \lambda_3(z')^2 + a'x' + b'y' + c'z' = k. \tag{5}$$

Notice that Equation (5) contains no cross product terms.

NOTE   If Equation (1) contains no linear (x, y, or z) terms, then $B = 0$, and hence $BP = 0$. In this case we do not have to bother with P at all to identify the quadric surface, since Equation (5) becomes simply

$$\lambda_1(x')^2 + \lambda_2(y')^2 + \lambda_3(z')^2 = k$$

and depends only on the eigenvalues of A. This is the case in the following example.

··

EXAMPLE 8.13

Reduce the equation $5x^2 + 11y^2 + 2z^2 - 16xy + 20xz + 4yz = 18$ to standard form and identify the quadric.

Solution

We first write the given equation in the form

$$\mathbf{x}^T A\mathbf{x} = 18,$$

where $\mathbf{x} = (x, y, z)$ and A is the matrix

$$A = \begin{bmatrix} 5 & -8 & 10 \\ -8 & 11 & 2 \\ 10 & 2 & 2 \end{bmatrix}.$$

To diagonalize A, we compute its eigenvalues, $\lambda_1 = 9$, $\lambda_2 = 18$, $\lambda_3 = -9$, and set

$$D = \begin{bmatrix} 9 & 0 & 0 \\ 0 & 18 & 0 \\ 0 & 0 & -9 \end{bmatrix}.$$

Then the transformed equation, $(\mathbf{x}')^T D\mathbf{x}' = 18$, is simply

$$9(x')^2 + 18(y')^2 - 9(z')^2 = 18.$$

Finally, dividing both sides by 18, we obtain the standard form

$$\frac{(x')^2}{2} + \frac{(y')^2}{1} - \frac{(z')^2}{2} = 1.$$

Thus the quadric is a hyperboloid of one sheet (see Figure 8.20).

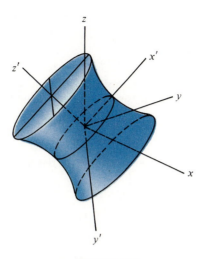

FIGURE 8.20

NOTE   Unless the eigenvalues of A (the diagonal entries of D) are arranged in the proper order, it may not be possible to obtain a final equation that is *exactly* the same as one of the listed standard forms. It may have some of the variables interchanged. This corresponds to a permutation of the coordinate axes relative to those of the given standard form. For instance, if in Example 8.12 we take

$$D = \begin{bmatrix} 9 & 0 & 0 \\ 0 & -9 & 0 \\ 0 & 0 & 18 \end{bmatrix},$$

the resulting equation will be

$$\frac{(x')^2}{2} - \frac{(y')^2}{2} + (z')^2 = 1,$$

a hyperboloid of one sheet whose axis of symmetry is the y'-axis.

···

We now know how to transform Equation (1) so that the cross product terms are eliminated. In general we will also have to eliminate certain linear terms. This is accomplished with the aid of the process of *completing the square*. Recall that to complete the square in the expression $pu^2 + qu$, we proceed as follows:

$$pu^2 + qu = p(u^2 + (q/p)u)$$
$$= p(u^2 + (q/p)u + q^2/4p^2) - q^2/4p$$
$$= p(u + q/2p)^2 - q^2/4p.$$

The manner in which completing the square is used in the reduction process is illustrated in the next example.

EXAMPLE 8.14

Reduce the equation $x^2 + 2y^2 - 2x + 8y - 4z = 3$ to standard form and identify the quadric.

Solution

Notice that no cross product terms appear in this equation. We complete the square in the quadratic variables, x and y, and transpose the z term to the right side:

$$(x^2 - 2x + 1) + 2(y^2 + 4y + 4) = 4z + 3 + 1 + 8,$$
$$(x - 1)^2 + 2(y + 2)^2 = 4z + 12.$$

We now divide both sides of this equation by the coefficient of z, obtaining

$$\frac{(x - 1)^2}{4} + \frac{(y + 2)^2}{2} = z + 3.$$

Finally, we make the change of variables $x' = x - 1$, $y' = y + 2$, and $z' = z + 3$. This yields

$$\frac{(x')^2}{4} + \frac{(y')^2}{2} = z',$$

which is the standard form of an equation (in variables x', y', and z') of an elliptic paraboloid (Figure 8.21).

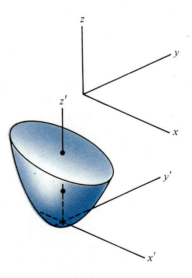

FIGURE 8.21

NOTE　Geometrically, the change of variables of the type used in Example 8.14, $x' = x - x_0$, $y' = y - y_0$, $z' = z - z_0$, *translates* the xyz-coordinate system into the $x'y'z'$-system. The origin of the $x'y'z'$-coordinate system will be the point (x_0, y_0, z_0) in the xyz-system. For example, the change of variables $x' = x - 1$, $y' = y + 2$, $z' = z + 3$ of Example 8.14 results in an $x'y'z'$ origin at $(1, -2, -3)$ of the xyz-system (see Figure 8.21).

Before giving another example, we will present this technique in the form of a procedure.

PROCEDURE TO REDUCE A QUADRATIC EQUATION IN THREE VARIABLES TO STANDARD FORM

Given the equation

$$ax^2 + by^2 + cz^2 + dxy + exz + fyz + gx + hy + iz = k,$$

i. Determine the matrices A and B described by (3).

ii. Find an orthogonal matrix P and a diagonal matrix D such that $P^TAP = D$.

iii. Make the change of variable $\mathbf{x} = P\mathbf{x}'$ by performing the indicated matrix operations in

$$(\mathbf{x}')^TD\mathbf{x}' + BP\mathbf{x}' = k$$

to eliminate the cross product terms, and convert it to a scalar equation.

iv. Complete the square, translate the coordinate system by making a change of variables, and rearrange terms in this scalar equation so that it is in standard form.

EXAMPLE 8.15

Reduce the equation $2x^2 + 2y^2 + 2z^2 + 2xz + 4\sqrt{2}x + 12y = -10$ to standard form and identify the quadric.

Solution

The given equation has the matrix form $\mathbf{x}^TA\mathbf{x} + B\mathbf{x} = -10$, where

$$A = \begin{bmatrix} 2 & 0 & 1 \\ 0 & 2 & 0 \\ 1 & 0 & 2 \end{bmatrix} \quad \text{and} \quad B = [4\sqrt{2} \quad 12 \quad 0].$$

Since the eigenvalues and corresponding normalized eigenvectors of A are

$$\lambda_1 = 1, \quad \lambda_2 = 2, \quad \lambda_3 = 3,$$

and

$$\mathbf{x}_1 = \left(\frac{-1}{\sqrt{2}}, 0, \frac{1}{\sqrt{2}}\right), \quad \mathbf{x}_2 = (0, 1, 0), \quad \mathbf{x}_3 = \left(\frac{1}{\sqrt{2}}, 0, \frac{1}{\sqrt{2}}\right),$$

we may take

$$P = \begin{bmatrix} -1/\sqrt{2} & 0 & 1/\sqrt{2} \\ 0 & 1 & 0 \\ 1/\sqrt{2} & 0 & 1/\sqrt{2} \end{bmatrix} \quad \text{and} \quad D = \begin{bmatrix} 1 & 0 & 0 \\ 0 & 2 & 0 \\ 0 & 0 & 3 \end{bmatrix}.$$

Thus, applying the transformation $\mathbf{x} = P\mathbf{x}'$, we obtain the equation

$$(\mathbf{x}')^T D\mathbf{x}' + BP\mathbf{x}' = -10,$$

or

$$(x')^2 + 2(y')^2 + 3(z')^2 - 4x' + 12y' + 4z' = -10.$$

We now complete the square in x', y', and z' and obtain

$$(x' - 2)^2 + 2(y' + 3)^2 + 3\left(z' + \frac{2}{3}\right)^2 = \frac{40}{3}.$$

Finally, letting $x'' = x' - 2$, $y'' = y' + 3$, and $z'' = z' + \frac{2}{3}$ and dividing both sides of the last equation by $\frac{40}{3}$ yields the standard form

$$\frac{(x'')^2}{40/3} + \frac{(y'')^2}{20/3} + \frac{(z'')^2}{40/9} = 1.$$

The graph is an ellipsoid.

NOTE   The techniques of this section can also be applied to the simpler case of a quadratic equation in two variables

$$ax^2 + bxy + cy^2 + dx + ey = f,$$

which is the equation of a *conic section*. The reduction of a conic section to standard form is explored in Exercises 11 through 18.

··

8.6 EXERCISES

In Exercises 1–4 the given equation contains no linear terms. Reduce it to standard form and identify the quadric surface it represents.

1. $x^2 + y^2 + z^2 - 2xy - 2xz - 2yz = 2$

2. $3x^2 + 3y^2 + 3z^2 - 3xy - 4yz = 66$

3. $x^2 + y^2 + z^2 + 6xy + 8yz = 0$

4. $x^2 + 2\sqrt{2}xy = 4$

In Exercises 5 and 6 the given equation contains no cross product terms. Reduce it to standard form and identify the quadric surface it represents.

5. $2x^2 + y^2 + z^2 - 4x + 8y + 6z = 9$

6. $2x^2 - 2y^2 - z^2 - 8x + 4y - 4z = 2$

In Exercises 7–10 reduce the given equation to standard form and identify the quadric surface it represents.

7. $2x^2 + y^2 + z^2 + 2\sqrt{6}xz + 2y + 2\sqrt{5}z = 2$

8. $x^2 + y^2 + z^2 + 2xy + 2xz + 2yz + \sqrt{6}y = 0$

9. $x^2 + y^2 - z^2 + 2xy - 2z = 3$

10. $3x^2 + 3y^2 + z^2 - 2xy + 12z = 7$

In Exercises 11–14 a method for reducing quadratic equations in two variables—those of *conic sections*—to standard form is investigated.

11. Show that the equation $ax^2 + bxy + cy^2 + dx + ey = f$ can be written in the matrix form $\mathbf{x}^T A \mathbf{x} + B\mathbf{x} = f$, where

$$A = \begin{bmatrix} a & b/2 \\ b/2 & c \end{bmatrix}, \qquad B = [d \quad e], \qquad \text{and} \qquad \mathbf{x} = \begin{bmatrix} x \\ y \end{bmatrix}.$$

12. Use the result of Exercise 11 to write the equation

$$3x^2 + 4xy - y^2 + 5x - 2y = 1$$

in matrix form.

13. Given the matrix equation $\mathbf{x}^T A \mathbf{x} + B\mathbf{x} = f$, let P be an orthogonal matrix such that $P^T A P = D$, a diagonal matrix. Show that the change of variable $\mathbf{x} = P\mathbf{x}'$ transforms the given equation into one of the form

$$(\mathbf{x}')^T D\mathbf{x}' + BP\mathbf{x}' = f.$$

14. Show that the matrix equation $(\mathbf{x}')^T D\mathbf{x}' + BP\mathbf{x}' = f$ of Exercise 13 is equivalent to the scalar equation $a'(x')^2 + c'(y')^2 + d'x' + e'y' = f$, where a' and c' are the diagonal entries of D and d' and e' are the entries of BP. (Notice that there is no *cross product* (that is, $x'y'$) term in this scalar equation.)

In Exercises 15 and 16 use the results of Exercises 11, 13, and 14 to transform the given equation into one in which no cross product terms appear.

15. $x^2 + 4xy + y^2 + 5\sqrt{2}x + \sqrt{2}y = 0$

16. $x^2 - 4xy + 4y^2 + 10\sqrt{5}x = 15$

17–18. With the aid of the *completing the square* process, reduce the transformed equations obtained in Exercises 15 and 16 to one of the following *standard forms* for a *nondegenerate conic section*:

i. *ellipse*, $\dfrac{x^2}{p^2} + \dfrac{y^2}{q^2} = 1 \qquad (p, q > 0)$,

ii. *parabola*, $y^2 = px \qquad (p \neq 0)$,

iii. *hyperbola*, $\dfrac{x^2}{p^2} - \dfrac{y^2}{q^2} = 1 \qquad (p, q > 0)$.

8.7 APPROXIMATION OF CONTINUOUS FUNCTIONS; FOURIER SERIES

In Sections 8.1 and 8.2 we considered the problem of obtaining a polynomial approximation to a discrete set of data. Now we will turn to the related idea of approximation of continuous functions by other, generally simpler, functions. The key to the development of this material is the topic of inner product spaces (Section 5.4).

LEAST SQUARE APPROXIMATIONS TO CONTINUOUS FUNCTIONS

There are many ways of approximating a continuous function on an interval by a polynomial. If $\mathbf{f}(x)$ is the function, and $[a, b]$ is the interval, we could evaluate $\mathbf{f}$ at n distinct points in $[a, b]$ to obtain the data set $(x_i, \mathbf{f}(x_i))$, $i = 1, 2, \ldots, n$, and then construct the interpolating or least square polynomial for this data (Sections 8.1 and 8.2). However, this approach makes use of only a small part of the given information. To involve all our knowledge of $\mathbf{f}(x)$, we employ the following approach.

Recall that the least square polynomial, $\mathbf{p}(x)$, for a finite set of data, $\{(x_i, y_i) \mid i = 1, 2, \ldots, n\}$, is the one that minimizes

$$S = (y_1 - \mathbf{p}(x_1))^2 + (y_2 - \mathbf{p}(x_2))^2 + \cdots + (y_n - \mathbf{p}(x_n))^2.$$

By letting $\mathbf{y} = (y_1, \ldots, y_n)$ and $\mathbf{z} = (\mathbf{p}(x_1), \ldots, \mathbf{p}(x_n))$, and making use of the dot product for $\mathbf{R}^n$, we may write this equation as

$$S = (\mathbf{y} - \mathbf{z}) \cdot (\mathbf{y} - \mathbf{z}) = \|\mathbf{y} - \mathbf{z}\|^2.$$

This suggests one way to extend the idea of the least square approximation to a continuous function $\mathbf{f}(x)$: Find a polynomial $\mathbf{p}(x)$ that minimizes

$$(\mathbf{f} - \mathbf{p}, \mathbf{f} - \mathbf{p}) = \|\mathbf{f} - \mathbf{p}\|^2,$$

where this inner product is the one on $\mathbf{C}[a, b]$ (the vector space of functions continuous on the interval $[a, b]$) given by

$$(\mathbf{g}, \mathbf{h}) = \int_a^b \mathbf{g}(x)\mathbf{h}(x) \, dx \tag{1}$$

(see Section 5.4).

The following theorem provides the basis for the approximation techniques of this section. It will be proved at the end of the section.

THEOREM I Let $\mathbf{f}(x)$ be a member of $\mathbf{C}[a, b]$, and let $\mathbf{S}$ be a finite dimensional subspace of $\mathbf{C}[a, b]$. Then there is a unique function $\mathbf{p}(x)$ in $\mathbf{S}$ that minimizes

$$\|\mathbf{f} - \mathbf{p}\|^2 = \int_a^b (\mathbf{f}(x) - \mathbf{p}(x))^2 \, dx.$$

NOTE   The function $\mathbf{p}$ of Theorem 1 is called the **least square approximation** to $\mathbf{f}$ from $\mathbf{S}$. From a practical point of view the subspace $\mathbf{S}$ usually consists of polynomials of some sort.

PROCEDURE FOR DETERMINING THE LEAST SQUARE APPROXIMATION TO A CONTINUOUS FUNCTION

Let $\mathbf{f}$ be a function in $\mathbf{C}[a, b]$, and let $\mathbf{S}$ be a finite dimensional subspace of approximating functions. To construct the least square approximation to $\mathbf{f}$ from $\mathbf{S}$,

i.   find an orthonormal basis, $\{\mathbf{p}_1, \mathbf{p}_2, \ldots, \mathbf{p}_n\}$ for $\mathbf{S}$,

ii.   let $\mathbf{p} = c_1\mathbf{p}_1 + c_2\mathbf{p}_2 + \cdots + c_n\mathbf{p}_n$, where

$$c_i = (\mathbf{f}, \mathbf{p}_i) = \int_a^b \mathbf{f}(x)\mathbf{p}_i(x) \, dx.$$

Then $\mathbf{p}$ is the desired least square approximation.

LEAST SQUARE POLYNOMIAL APPROXIMATION

Perhaps the simplest choice of the subspace $\mathbf{S}$ of Theorem 1 is $\mathbf{P}_n[a, b]$, the vector space of all polynomials of degree less than or equal to n with domain restricted to $[a, b]$. In this case the desired approximating function is called the **least square polynomial** for $\mathbf{f}$ on $[a, b]$. To apply the procedure of Theorem 1 to construct this polynomial, we first need an orthonormal basis for $\mathbf{P}_n[a, b]$. Since $\mathscr{B} = \{1, x, x^2, \ldots, x^n\}$ is a basis for this subspace, we can find an orthonormal basis by applying the Gram–Schmidt process of Section 5.4 to $\mathscr{B}$.

EXAMPLE 8.16

Find the least square quadratic polynomial approximating $\mathbf{f}(x) = x^3$ on $[-1, 1]$.

Solution

We apply the procedure of Theorem 1. Here $\mathbf{S} = \mathbf{P}_2[-1, 1]$, which has a basis $\mathscr{B} = \{1, x, x^2\}$. By applying the Gram–Schmidt process to $\mathscr{B}$, we obtain the orthonormal basis (see Example 5.35) $\{\mathbf{p}_0, \mathbf{p}_1, \mathbf{p}_2\}$, where

$$\mathbf{p}_0(x) = \sqrt{1/2},$$

$$\mathbf{p}_1(x) = (\sqrt{3/2})x,$$

and

$$\mathbf{p}_2(x) = (1/2)(\sqrt{5/2})(3x^2 - 1).$$

Therefore the quadratic least square polynomial, $\mathbf{p}(x)$, for $\mathbf{f}(x) = x^3$ on $[-1, 1]$ is given by

$$\mathbf{p}(x) = c_0\mathbf{p}_0(x) + c_1\mathbf{p}_1(x) + c_2\mathbf{p}_2(x),$$

where

$$c_0 = (\mathbf{f}, \mathbf{p}_0) = \int_{-1}^{1} x^3(\sqrt{1/2})\, dx = 0$$

$$c_1 = (\mathbf{f}, \mathbf{p}_1) = \int_{-1}^{1} x^3(\sqrt{3/2})x\, dx = \sqrt{3/2}\int_{-1}^{1} x^4\, dx = (2/5)\sqrt{3/2}$$

and

$$c_2 = (\mathbf{f}, \mathbf{p}_2) = \int_{-1}^{1} x^3(1/2)(\sqrt{5/2})(3x^2 - 1)\, dx$$

$$= (1/2)\sqrt{5/2}\int_{-1}^{1} (3x^5 - x^3)\, dx = 0.$$

Thus

$$\mathbf{p}(x) = [(2/5)\sqrt{3/2}](\sqrt{3/2})x = (3/5)x.$$

(The given $\mathbf{f}(x)$ and approximating $\mathbf{p}(x)$ are shown in Figure 8.22.)

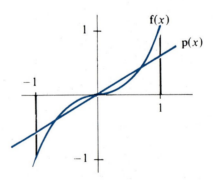

FIGURE 8.22

NOTE   As we can see from Example 8.16, the least square approximation to $\mathbf{f}$ from $\mathbf{P}_2[a, b]$ does not need to have its degree exactly equal to 2.

The polynomials that form an orthonormal basis for $\mathbf{P}_n[-1, 1]$ are called *orthonormal Legendre* polynomials and are often listed in mathematical reference books. The next theorem demonstrates how we can construct a set of polynomials that are orthonormal on any interval $[a, b]$ in terms of the orthonormal Legendre polynomials.

THEOREM 2 Let $\{q_0(t), q_1(t), \ldots, q_n(t)\}$ be an orthonormal basis for $P_n[-1, 1]$. Then $\mathcal{B} = \{p_0(x), p_1(x), \ldots, p_n(x)\}$ is an orthonormal basis for $P_n[a, b]$, where

$$p_i(x) = \left(\sqrt{\frac{2}{b-a}} \right) q_i \left(\frac{2x - (a+b)}{b-a} \right).$$

Proof All we need do is to verify that $\mathcal{B}$ is an orthonormal *set* on $[a, b]$; that is,

$$\int_a^b (p_i(x))^2 \, dx = 1$$

and

$$\int_a^b p_i(x)p_j(x) \, dx = 0, \qquad \text{if } i \neq j.$$

Then by Theorem 3 of Section 5.4 and Theorem 7 of Section 5.2, $\mathcal{B}$ is an orthonormal *basis* for $P_n[a, b]$. We leave the details of this verification as an exercise.

EXAMPLE 8.17 Find the least square linear polynomial, $p(x)$, approximating

$$f(x) = \sqrt{x}$$

on $[0, 1]$.

Solution We again apply the procedure of Theorem 1. The approximation we seek is of the form

$$p = c_0 p_0 + c_1 p_1,$$

where $\{p_0, p_1\}$ is an orthonormal basis for $P_1[0, 1]$,

$$c_0 = \int_0^1 f(x)p_0(x) \, dx$$

and

$$c_1 = \int_0^1 f(x)p_1(x) \, dx.$$

To find p_0 and p_1, we use Theorem 2. Since an orthonormal basis for $P_1[-1, 1]$ is

$\{1/\sqrt{2}, (\sqrt{3/2})t\}$, and for this example $a = 0$ and $b = 1$, we obtain

$$\mathbf{p}_0(x) = \sqrt{2}(1/\sqrt{2}) = 1$$

and

$$\mathbf{p}_1(x) = \sqrt{2}(\sqrt{3/2})(2x - 1) = \sqrt{3}(2x - 1).$$

Consequently

$$c_0 = \int_0^1 \sqrt{x}\, dx = 2/3,$$

$$c_1 = \int_0^1 \sqrt{x}\sqrt{3}(2x - 1)\, dx$$

$$= \sqrt{3} \int_0^1 (2x^{3/2} - x^{1/2})\, dx = 2\sqrt{3}/15,$$

yielding

$$\mathbf{p}(x) = c_0\mathbf{p}_0(x) + c_1\mathbf{p}_1(x) = (4/5)x + 4/15.$$

(The given $\mathbf{f}(x)$ and approximating $\mathbf{p}(x)$ are shown in Figure 8.23.)

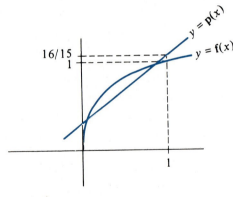

FIGURE 8.23

FOURIER SERIES

Another type of relatively simple function used for approximation purposes is the *trigonometric polynomial*.

DEFINITION

A **trigonometric polynomial** is a function of the form

$$\mathbf{p}(x) = a_0 + a_1 \cos x + a_2 \cos 2x + \cdots + a_n \cos nx$$
$$+ b_1 \sin x + b_2 \sin 2x + \cdots + b_n \sin nx.$$

If a_n and b_n are not both zero, we say that $\mathbf{p}(x)$ has **order** n.

It can be shown that $\mathbf{T}_n[a, b]$, the set of all trigonometric polynomials on $[a, b]$ of order less than or equal to n, is a subspace of $\mathbf{C}[a, b]$. If we choose the subspace $\mathbf{S}$ of Theorem 1 to be $\mathbf{T}_n[-\pi, \pi]$, the resulting approximation is called the **Fourier series of order** n for the given function.

To obtain the Fourier series of order n for a continuous function $\mathbf{f}(x)$, we apply the procedure of Theorem 1 with $\mathbf{S} = \mathbf{T}_n[-\pi, \pi]$. This subspace has the orthogonal basis (Exercise 17)

$$\mathcal{B} = \{1, \cos x, \cos 2x, \ldots, \cos nx, \sin x, \sin 2x, \ldots, \sin nx\}.$$

Thus, to obtain an orthonormal basis for $\mathbf{T}_n[-\pi, \pi]$, we need only normalize the vectors in $\mathcal{B}$. In other words, we seek constants $c_0, c_1, \ldots, c_n, d_1, \ldots, d_n$ so that

$$(c_0, c_0) = \int_{-\pi}^{\pi} c_0^2 \, dx = 1,$$

$$(c_k \cos kx, c_k \cos kx) = \int_{-\pi}^{\pi} c_k^2 \cos^2 kx \, dx = 1 \qquad (k = 1, 2, \ldots, n),$$

and

$$(d_k \sin kx, d_k \sin kx) = \int_{-\pi}^{\pi} d_k^2 \sin^2 kx \, dx = 1 \qquad (k = 1, 2, \ldots, n).$$

Performing the indicated integrations and solving for the c_k and d_k yields

$$c_0 = 1/\sqrt{2\pi},$$
$$c_k = d_k = 1/\sqrt{\pi} \qquad \text{for } k = 1, 2, \ldots, n.$$

Now that we have an orthonormal basis for $\mathbf{T}_n[-\pi, \pi]$, we can complete the procedure of Theorem 1 and obtain the following theorem.

THEOREM 3   Let $\mathbf{f}(x)$ be a continuous function on $[-\pi, \pi]$. Then the Fourier series for $\mathbf{f}$ (the least square trigonometric polynomial approximating $\mathbf{f}$ on $[-\pi, \pi]$) of order n is

$$\mathbf{p}(x) = a_0 + a_1 \cos x + a_2 \cos 2x + \cdots + a_n \cos nx$$
$$+ b_1 \sin x + b_2 \sin 2x + \cdots + b_n \sin nx,$$

where

$$a_0 = (1/2\pi) \int_{-\pi}^{\pi} \mathbf{f}(x) \, dx,$$

$$a_k = (1/\pi) \int_{-\pi}^{\pi} \mathbf{f}(x) \cos kx \, dx \qquad \text{for } k = 1, 2, \ldots, n,$$

and

$$b_k = (1/\pi) \int_{-\pi}^{\pi} \mathbf{f}(x) \sin kx \, dx \qquad \text{for } k = 1, 2, \ldots, n.$$

EXAMPLE 8.18   Find the Fourier series of order 2 for $\mathbf{f}(x) = x$.

Solution   Applying Theorem 3, the desired function has the form

$$\mathbf{p}(x) = a_0 + a_1 \cos x + a_2 \cos 2x + b_1 \sin x + b_2 \sin 2x,$$

where

$$a_0 = (1/2\pi) \int_{-\pi}^{\pi} x \, dx = 0,$$

$$a_1 = (1/\pi) \int_{-\pi}^{\pi} x \cos x \, dx = 0,$$

$$a_2 = (1/\pi) \int_{-\pi}^{\pi} x \cos 2x \, dx = 0,$$

$$b_1 = (1/\pi) \int_{-\pi}^{\pi} x \sin x \, dx = 2,$$

$$b_2 = (1/\pi) \int_{-\pi}^{\pi} x \sin 2x \, dx = -1.$$

Therefore $\mathbf{p}(x) = 2 \sin x - \sin 2x$. (The given $\mathbf{f}(x)$ and the approximating $\mathbf{p}(x)$ are shown in Figure 8.24.)

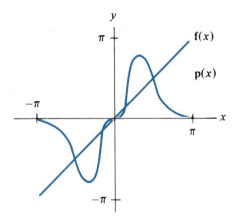

FIGURE 8.24

NOTE   If a function $\mathbf{f}$ is continuous on $[-\pi, \pi]$, there is a Fourier series of order n for $\mathbf{f}$ for every natural number n. If this sequence of Fourier series is denoted by $\{\mathbf{p}_n\}$, then $\|\mathbf{f} - \mathbf{p}_n\| \to 0$ as $n \to \infty$. Therefore the infinite series

$$a_0 + \sum_{k=1}^{\infty} (a_k \cos kx + b_k \sin kx)$$

converges to the function **f**. This series is called the **Fourier series** (without the qualifier *of order n*) for **f**.

···

We close this section with the proof of Theorem 1. This proof will also justify the procedure immediately following the statement of the theorem.

Proof of Theorem 1   We first show existence of the least square approximation. Let $\mathscr{B} = \{\mathbf{p}_1, \mathbf{p}_2, \ldots, \mathbf{p}_n\}$ be an orthonormal basis for **S**. Then any vector in **S** can be written

$$\mathbf{p} = c_1\mathbf{p}_1 + c_2\mathbf{p}_2 + \cdots + c_n\mathbf{p}_n.$$

We wish to find **p** in **S** that minimizes

$$\|\mathbf{f} - \mathbf{p}\|^2 = (\mathbf{f} - \mathbf{p}, \mathbf{f} - \mathbf{p})$$
$$= (\mathbf{f} - (c_1\mathbf{p}_1 + c_2\mathbf{p}_2 + \cdots + c_n\mathbf{p}_n),$$
$$\mathbf{f} - (c_1\mathbf{p}_1 + c_2\mathbf{p}_2 + \cdots + c_n\mathbf{p}_n)).$$

But the $\mathbf{p}_i$ form an orthonormal set, so $(\mathbf{p}_i, \mathbf{p}_i) = 1$ and $(\mathbf{p}_i, \mathbf{p}_j) = 0$ if $i \neq j$. This implies that

$$\|\mathbf{f} - \mathbf{p}\|^2 = (c_1^2) + (c_2^2) + \cdots + (c_n^n) - 2c_1(\mathbf{f}, \mathbf{p}_1)$$
$$- 2c_2(\mathbf{f}, \mathbf{p}_2) - \cdots - 2c_n(\mathbf{f}, \mathbf{p}_n) + (\mathbf{f}, \mathbf{f}).$$

We now let $\mathbf{c} = (c_1, c_2, \ldots, c_n)$, $\mathbf{d} = (d_1, d_2, \ldots, d_n)$, where $d_i = (\mathbf{f}, \mathbf{p}_i)$ and $(\mathbf{f}, \mathbf{f}) = e$. Then, using the dot product in $\mathbf{R}^n$, we can write

$$\|\mathbf{f} - \mathbf{p}\|^2 = \mathbf{c} \cdot \mathbf{c} - 2\mathbf{c} \cdot \mathbf{d} + e. \tag{2}$$

Since the desired function **p** is completely determined by the n-vector **c**, we seek **c** that minimizes $\|\mathbf{f} - \mathbf{p}\|^2$. If we can find a vector **r** so that the change of variable

$$\mathbf{c} = \mathbf{c}' + \mathbf{r} \tag{3}$$

results in an equation of the form

$$\|\mathbf{f} - \mathbf{p}\|^2 = \mathbf{c}' \cdot \mathbf{c}' + e', \tag{4}$$

and e' depends only on the fixed choice of **r**, not on the variable **c**', the resulting function **p** will minimize $\|\mathbf{f} - \mathbf{p}\|^2$. This is because $\mathbf{c}' \cdot \mathbf{c}' \geq 0$ is minimized if $\mathbf{c}' = \mathbf{0}$. Thus, if we can determine such an **r**, we see from Equation (4) that $\mathbf{c} = \mathbf{0} + \mathbf{r} = \mathbf{r}$ will minimize $\|\mathbf{f} - \mathbf{p}\|^2$.

To find the desired **r**, we substitute Equation (3) into Equation (2). This results in

$$\|\mathbf{f} - \mathbf{p}\|^2 = (\mathbf{c}' + \mathbf{r}) \cdot (\mathbf{c}' + \mathbf{r}) - 2(\mathbf{c}' + \mathbf{r}) \cdot \mathbf{d} + e$$
$$= \mathbf{c}' \cdot \mathbf{c}' + 2\mathbf{c}' \cdot (\mathbf{r} - \mathbf{d}) + (\mathbf{r} \cdot \mathbf{r} - 2\mathbf{r} \cdot \mathbf{d} + e).$$

Choosing $\mathbf{r} = \mathbf{d}$, we obtain Equation (4), with

$$e' = \mathbf{r} \cdot \mathbf{r} - 2\mathbf{r} \cdot \mathbf{d} + e = -\mathbf{d} \cdot \mathbf{d} + e.$$

Thus taking $\mathbf{c} = \mathbf{d}$, that is,

$$c_i = (\mathbf{f}, \mathbf{p}_i),$$

results in a minimum for $\|\mathbf{f} - \mathbf{p}\|^2$.

To show uniqueness, suppose $\mathbf{q}$ is a vector in $\mathbf{S}$, with

$$\|\mathbf{f} - \mathbf{q}\|^2 = \|\mathbf{f} - \mathbf{p}\|^2,$$

where

$$\mathbf{p} = (\mathbf{f}, \mathbf{p}_1)\mathbf{p}_1 + (\mathbf{f}, \mathbf{p}_2)\mathbf{p}_2 + \cdots + (\mathbf{f}, \mathbf{p}_n)\mathbf{p}_n.$$

Writing $\mathbf{q}$ as a linear combination of the basis vectors

$$\mathbf{q} = a_1\mathbf{p}_1 + a_2\mathbf{p}_2 + \cdots + a_n\mathbf{p}_n,$$

we let $\mathbf{a} = (a_1, \ldots, a_n)$ and make the change of variable of Equation (3) (now with $\mathbf{r} = \mathbf{d}$),

$$\mathbf{a} = \mathbf{a}' + \mathbf{d}.$$

Using Equation (4) and the fact that the minimum $\mathbf{p}$ was obtained with $\mathbf{c}' = \mathbf{0}$, we have

$$e' = \|\mathbf{f} - \mathbf{p}\|^2 = \|\mathbf{f} - \mathbf{q}\|^2 = \mathbf{a}' \cdot \mathbf{a}' + e'.$$

This implies that $\mathbf{a}' = \mathbf{0}$, and hence $\mathbf{a} = \mathbf{d}$. Thus $\mathbf{q} = \mathbf{p}$, as desired.

8.7 EXERCISES

In Exercises 1–4 find the *linear* least square polynomial approximating the given function on the indicated interval.

1. $f(x) = x^4$ on $[-1, 1]$ **2.** $f(x) = x^3 + 3x$ on $[-1, 1]$

3. $f(x) = 1/x$ on $[1, 2]$ **4.** $f(x) = 1/\sqrt{x}$ on $[1, 4]$

5–8. Find the *quadratic* least square polynomial approximating the functions of Exercises 1–4 on the indicated intervals.

9–10. Find the *cubic* least square polynomial approximating the functions of Exercises 1 and 2 on the indicated intervals.

In Exercises 11–16 find the Fourier series of order n (for the indicated n) for the given function.

11. $f(x) = x + 1$ $(n = 1)$ **12.** $f(x) = x^2$ $(n = 1)$

13. $f(x) = x^3 + 1$ $(n = 1)$ **14.** $f(x) = e^x$ $(n = 1)$

15. $f(x) = x - 1$ $(n = 2)$ **16.** $f(x) = \sin 3x$ $(n = 2)$

17. Show that $\{1, \cos x, \cos 2x, \ldots, \cos nx, \sin x, \sin 2x, \ldots, \sin nx\}$ is an orthogonal basis for $\mathbf{T}_n[-\pi, \pi]$, the set of all trigonometric polynomials on $[-\pi, \pi]$ of order less than or equal to n.

8.8 LINEAR DIFFERENTIAL EQUATIONS

In this section we will use linear algebra in discussing the topic of linear differential equations. Prerequisites for a complete understanding of this material include Sections 5.1, 5.2, and 6.1 through 6.3.

We begin by giving some definitions basic to the study of differential equations. A **differential equation** is one involving an "unknown" function and at least one of its derivatives. The **order** of a differential equation is that of its highest-order derivative. For example,

$$\mathbf{y}'' + 3\mathbf{y}' - 5\mathbf{y} = x^3$$

is a differential equation of second order. This particular equation symbolizes the problem of determining a function $\mathbf{y} = \mathbf{f}(x)$ such that $\mathbf{f}''(x) + 3\mathbf{f}'(x) - 5\mathbf{f}(x) = x^3$. Such a function is called a **solution** of the differential equation.

We will consider only *ordinary, linear* differential equations. An **ordinary differential equation** is one in which the unknown is a function of a single variable. (If it is a function of several variables, we have a *partial* differential equation.) An (ordinary) **linear differential equation** of order n is one of the form

$$a_n\mathbf{y}^{(n)} + a_{n-1}\mathbf{y}^{(n-1)} + \cdots + a_1\mathbf{y}' + a_0\mathbf{y} = \mathbf{b}, \tag{1}$$

where, in general, the a_i and $\mathbf{b}$ are functions of the independent variable. We will consider only the case for which the a_i are constants—the *constant coefficient* case—and $a_n \neq 0$.

EXAMPLE 8.19   The differential equation

$$\mathbf{y}''' - 4\mathbf{y}' = \mathbf{0}$$

is a linear one of order 3. Comparing it to the general form (1), we have $n = 3$, $a_3 = 1$, $a_2 = 0$, $a_1 = -4$, $a_0 = 0$, and $\mathbf{b} = \mathbf{0}$. One solution of this differential equation is $\mathbf{y} = e^{2x}$. This is easily verified by differentiating to get $\mathbf{y}' = 2e^{2x}$ and $\mathbf{y}''' = 8e^{2x}$ so that

$$\mathbf{y}''' - 4\mathbf{y}' = 8e^{2x} - 4(2e^{2x}) = \mathbf{0},$$

as desired. (We will soon see how to determine *all* solutions of this differential equation.)

We can further divide the study of linear differential equations into two cases—*homogeneous* and *nonhomogeneous* equations. A **homogeneous** (ordinary) linear differential equation has the form of Equation (1) with $\mathbf{b} = \mathbf{0}$; it is **nonhomogeneous** if the function $\mathbf{b}$ (which we will assume to be infinitely differentiable) takes on nonzero values. For example, the equation $\mathbf{y}'' - \mathbf{y} = \mathbf{0}$ is a linear homogeneous one, whereas $\mathbf{y}'' - \mathbf{y}' = \sin x$ is nonhomogeneous.

HOMOGENEOUS LINEAR DIFFERENTIAL EQUATIONS

We will first consider the homogeneous version of Equation (1),

$$a_n\mathbf{y}^{(n)} + a_{n-1}\mathbf{y}^{(n-1)} + \cdots + a_1\mathbf{y}' + a_0\mathbf{y} = \mathbf{0}, \tag{2}$$

with the a_i constant and $a_n \neq 0$. To help analyze the solutions of this differential equation, we will restate the problem in terms of linear operators (linear transformations from a vector space into itself).

Let $\mathbf{C}^\infty$ ($= \mathbf{C}^\infty(-\infty, \infty)$) denote the vector space of all functions $\mathbf{y}(x)$ that have continuous derivatives of all orders for all values of x (see the Note following Example 5.6). Also, let $D: \mathbf{C}^\infty \to \mathbf{C}^\infty$ be the linear operator that maps each function into its derivative (see the Spotlight in Section 6.2). Notice that since $D(\mathbf{y}) = \mathbf{y}'$, we have $D^2\mathbf{y} = D(D(\mathbf{y})) = D\mathbf{y}' = \mathbf{y}''$, and in general $D^k(\mathbf{y})$ gives the kth derivative of $\mathbf{y}$. Finally, we define the linear operator $L: \mathbf{C}^\infty \to \mathbf{C}^\infty$ to be

$$L = a_nD^n + a_{n-1}D^{n-1} + \cdots + a_1D + a_0I,$$

so that Equation (2) may be written as simply $L(\mathbf{y}) = \mathbf{0}$ (and Equation (1) as $L(\mathbf{y}) = \mathbf{b}$). The verification that L is indeed a linear operator on $\mathbf{C}^\infty$ is left as an exercise.

Using this operator notation, we see that the problem of solving the homogeneous equation (2) may be restated as that of determining the *kernel* of the operator L—those functions $\mathbf{y}$ for which $L(\mathbf{y}) = \mathbf{0}$ (Section 6.3). Before discussing how to find the kernel of L—the solutions of Equation (2)—we will present several theorems that give *properties* of these solutions.

THEOREM 1   Let $\mathbf{y}_1$ and $\mathbf{y}_2$ be solutions of $L(\mathbf{y}) = \mathbf{0}$ and let c be a constant. Then

   i. $\mathbf{y}_1 + \mathbf{y}_2$ is a solution of $L(\mathbf{y}) = \mathbf{0}$;

   ii. $c\mathbf{y}_1$ (and $c\mathbf{y}_2$) is a solution of $L(\mathbf{y}) = \mathbf{0}$.

Proof   Since $\mathbf{y}_1$ and $\mathbf{y}_2$ are solutions of $L(\mathbf{y}) = \mathbf{0}$, they are both in the kernel of L. But by Theorem 1 of Section 6.3, the kernel of L is a subspace of $\mathbf{C}^\infty$, and hence $\mathbf{y}_1 + \mathbf{y}_2$ and $c\mathbf{y}_i$ are also in the kernel. Thus they are also solutions of $L(\mathbf{y}) = \mathbf{0}$.

We will prove the following two theorems at the end of this section.

THEOREM 2 Let $y_1, y_2, \ldots, y_n$ be solutions of $L(y) = 0$. Then $y_1, y_2, \ldots, y_n$ form a linearly dependent set if and only if

$$\det \begin{bmatrix} y_1 & y_2 & \cdots & y_n \\ y_1' & y_2' & \cdots & y_n' \\ \vdots & \vdots & & \vdots \\ y_1^{(n-1)} & y_2^{(n-1)} & \cdots & y_n^{(n-1)} \end{bmatrix} = 0;$$

that is, the determinant is identically zero (zero for all values of the variable).

NOTE The determinant of Theorem 2 is called the **Wronskian** of $y_1, y_2, \ldots, y_n$ and is denoted by $W(y_1, y_2, \ldots, y_n)$. It can also be shown that if $y_1, y_2, \ldots, y_n$ are solutions of $L(y) = 0$, then $W(y_1, y_2, \ldots, y_n)$ is either identically zero or else it is never zero.

···

EXAMPLE 8.20

Show that the solutions $y_1 = e^{2x}$, $y_2 = e^{-2x}$, and $y_3 = 3$ of the differential equation $y''' - 4y' = 0$ are linearly independent.

Solution

We compute the Wronskian

$$W(y_1, y_2, y_3) = \det \begin{bmatrix} e^{2x} & e^{-2x} & 3 \\ 2e^{2x} & -2e^{-2x} & 0 \\ 4e^{2x} & 4e^{-2x} & 0 \end{bmatrix} = 48.$$

Thus by Theorem 2, $\{y_1, y_2, y_3\}$ is a linearly independent set.

THEOREM 3 The dimension of the kernel of the linear operator L is n, the order of the differential equation $L(y) = 0$ (or, in other words, there exist exactly n linearly independent solutions of this differential equation).

EXAMPLE 8.21

Determine the general form for the solution of $u''' - 4u' = 0$.

Solution

This is the differential equation of Example 8.19, although here the unknown function is called "u". Thus we know that e^{2x}, e^{-2x}, and 3 are three linearly independent solutions of the given equation. By Theorem 3, this set of functions provides a basis for the kernel of $L = D^3 - 4D$, the subspace of all solutions of the given differential equation. Hence these three functions span this subspace, and every solution can be written as a linear combination of them. Thus,

$$\mathbf{u}(x) = c_1 e^{2x} + c_2 e^{-2x} + c_3(3).$$

NOTE   The general form of the solution of $L(\mathbf{y}) = \mathbf{0}$ (or $L(\mathbf{y}) = \mathbf{b}$) is called the **general solution** of this equation. Thus the general solution of $\mathbf{u}''' - 4\mathbf{u}' = \mathbf{0}$ is (from Example 8.20) $\mathbf{u}(x) = c_1 e^{2x} + c_2 e^{-2x} + c_3(3)$.

···

SOLVING LINEAR HOMOGENEOUS DIFFERENTIAL EQUATIONS

The linear operator

$$L = a_n D^n + a_{n-1} D^{n-1} + \cdots + a_1 D + a_0 I$$

can be viewed as a polynomial of degree n in the "variable" D. This polynomial,

$$P(\lambda) = a_n \lambda^n + a_{n-1} \lambda^{n-1} + \cdots + a_1 \lambda + a_0, \tag{3}$$

is called the **characteristic polynomial** for L (or for the associated differential equation (1)). The algebraic equation $P(\lambda) = 0$ is called the **characteristic equation** of (1). As we will see, the roots of the characteristic equation play a key role in the process of solving linear differential equations.

EXAMPLE 8.22   Find the roots of the characteristic equation for the differential equation $\mathbf{y}''' - 4\mathbf{y}' = \mathbf{0}$.

Solution   The characteristic equation is easily obtained by replacing the derivatives of $\mathbf{y}$ with the corresponding powers of λ. Doing this, we get $\lambda^3 - 4\lambda = 0$. The left side of this equation factors as $\lambda^3 - 4\lambda = \lambda(\lambda - 2)(\lambda + 2)$, and so the roots of the characteristic equations are 0, 2, and -2.

To simplify matters, we will first consider the special case of the second-order equation

$$\mathbf{y}'' + p\mathbf{y}' + q\mathbf{y} = \mathbf{0}. \tag{4}$$

Letting $L: \mathbf{C}^\infty \to \mathbf{C}^\infty$ be the linear operator

$$L = D^2 + pD + qI,$$

Equation (4) takes the form $L(\mathbf{y}) = \mathbf{0}$. Here the characteristic polynomial is a quadratic one,

$$P(\lambda) = \lambda^2 + p\lambda + q. \tag{5}$$

The roots of a quadratic equation can always be found by the quadratic formula, which, using the notation in Equation (5), is

$$\lambda = \frac{-p \pm \sqrt{p^2 - 4q}}{2}.$$

If we denote the two roots of the characteristic polynomial (5) by α and β, we can restrict ourselves to three cases:

i. α and β are real numbers with $\alpha \neq \beta$;

ii. α and β are real numbers with $\alpha = \beta$;

iii. α and β are complex (but nonreal) numbers and are complex conjugates of one another—if $\alpha = \alpha_1 + \alpha_2 i$, then $\beta = \alpha_1 - \alpha_2 i$.

By Theorem 3 we know that the differential equation (4) has exactly two linearly independent solutions. The next theorem tells us how to find these solutions once we know the roots of the characteristic polynomial (5).

THEOREM 4 Let the two roots of the characteristic equation for $\mathbf{y}'' + p\mathbf{y}' + q\mathbf{y} = \mathbf{0}$ (p, q are constant) be denoted by α and β.

i. If α and β are real numbers with $\alpha \neq \beta$, then

$$e^{\alpha x} \quad \text{and} \quad e^{\beta x}$$

are linearly independent solutions.

ii. If α and β are real numbers with $\alpha = \beta$, then

$$e^{\alpha x} \quad \text{and} \quad xe^{\alpha x}$$

are linearly independent solutions.

iii. If α and β are complex conjugates with $\alpha = \alpha_1 + \alpha_2 i$, then

$$e^{\alpha_1 x} \sin \alpha_2 x \quad \text{and} \quad e^{\alpha_1 x} \cos \alpha_2 x$$

are linearly independent solutions.

Proof We will prove only part (ii) and leave the others as exercises. First we must show that $e^{\alpha x}$ and $xe^{\alpha x}$ are solutions of the differential equation $\mathbf{y}'' + p\mathbf{y}' + q\mathbf{y} = \mathbf{0}$. Since the characteristic polynomial is $P(\lambda) = \lambda^2 + p\lambda + q = (\lambda - \alpha)(\lambda - \alpha) = \lambda^2 - 2\alpha\lambda + \alpha^2$, we may rewrite the differential equation in terms of α and β as

$$(D^2 - 2\alpha D + \alpha^2 I)(\mathbf{y}) = \mathbf{0} \quad \text{or} \quad (D - \alpha I)(D - \alpha I)(\mathbf{y}) = \mathbf{0}.$$

Now,

$$(D - \alpha I)(D - \alpha I)(e^{\alpha x}) = (D - \alpha I)(De^{\alpha x} - \alpha e^{\alpha x})$$
$$= (D - \alpha I)(\alpha e^{\alpha x} - \alpha e^{\alpha x})$$
$$= (D - \alpha I)(0) = 0,$$

so $e^{\alpha x}$ is a solution. Similarly,

$$(D - \alpha I)(D - \alpha I)(xe^{\alpha x}) = (D - \alpha I)(D(xe^{\alpha x}) - \alpha xe^{\alpha x})$$
$$= (D - \alpha I)(\alpha xe^{\alpha x} + e^{\alpha x} - \alpha xe^{\alpha x})$$
$$= (D - \alpha I)(e^{\alpha x}) = 0,$$

and $xe^{\alpha x}$ is a solution as well.

To show that $e^{\alpha x}$ and $xe^{\alpha x}$ are linearly independent, we need only show that their Wronskian is nonzero (by Theorem 2). Now,

$$W(e^{\alpha x}, xe^{\alpha x}) = \det \begin{bmatrix} e^{\alpha x} & xe^{\alpha x} \\ \alpha e^{\alpha x} & \alpha xe^{\alpha x} + e^{\alpha x} \end{bmatrix}$$
$$= e^{\alpha x}(\alpha xe^{\alpha x} + e^{\alpha x}) - \alpha xe^{\alpha x}e^{\alpha x}$$
$$= e^{\alpha x}e^{\alpha x} = e^{2\alpha x},$$

which is nonzero regardless of the values of α and x.

EXAMPLE 8.23   Find the general solution of $\mathbf{y}'' - 4\mathbf{y}' + 4\mathbf{y} = \mathbf{0}$.

Solution   The characteristic equation is $\lambda^2 - 4\lambda + 4 = 0$, which has the double root $\lambda = 2$ (case (ii) of Theorem 4). Hence two linear independent solutions are e^{2x} and xe^{2x}, and by Theorem 3, the general solution of the given differential equation is $\mathbf{y} = c_1 e^{2x} + c_2 xe^{2x}$.

EXAMPLE 8.24   Find the general solution of $(d^2\mathbf{s}/dt^2) + \mathbf{s} = \mathbf{0}$.

Solution   Here the characteristic equation is $\lambda^2 + 1 = 0$, which has roots $\lambda_1 = i (= 0 + 1i)$ and $\lambda_2 = -i (= 0 - 1i)$. Thus, by Theorem 4(iii), two linearly independent solutions of the given differential equation are $e^{0t} \sin t = \sin t$ and $e^{0t} \cos t = \cos t$. By Theorem 3, its general solution is $\mathbf{s} = c_1 \sin t + c_2 \cos t$.

HIGHER-ORDER HOMOGENEOUS EQUATIONS

For the more general case of Equation (2),

$$a_n\mathbf{y}^{(n)} + a_{n-1}\mathbf{y}^{(n-1)} + \cdots + a_1\mathbf{y}' + a_0\mathbf{y} = \mathbf{0},$$

Theorem 3 tells us that there will be n linearly independent solutions. To find them, we turn again to the characteristic equation. The problem here is that the latter now has degree n, and it is difficult (sometimes impossible) to find the roots of a general nth-degree polynomial equation. However, if we *can* determine all the roots of this equation, then the following procedure, which generalizes Theorem 4, will provide the needed n linearly independent solutions.

PROCEDURE FOR SOLVING A LINEAR HOMOGENEOUS DIFFERENTIAL EQUATION

To solve the differential equation

$$a_n y^{(n)} + a_{n-1} y^{(n-1)} + \cdots + a_1 y' + a_0 y = 0,$$

where the a_i are constant and $a_n \neq 0$:

i. Find the roots $\alpha_1, \alpha_2, \ldots, \alpha_n$ of the characteristic equation

$$a_n \lambda^n + a_{n-1} \lambda^{n-1} + \cdots + a_1 \lambda + a_0 = 0.$$

ii. Form the following collection of solutions:

a. if α is a real root, $e^{\alpha x}$ is a solution;

b. if $\alpha = \alpha_1 + \alpha_2 i$ is a complex root, then $\bar{\alpha} = \alpha_1 - \alpha_2 i$ is also a root and $e^{\alpha_1 x} \sin \alpha_2 x$ and $e^{\alpha_1 x} \cos \alpha_2 x$ are solutions;

c. if α is any root of multiplicity k, and $\bar{y}$ is a solution constructed from α as described above, then so are $x\bar{y}, x^2\bar{y}, \ldots, x^{k-1}\bar{y}$.

iii. The collection of distinct solutions $y_1, y_2, \ldots, y_n$ described in (ii) forms a linearly independent set of n solutions and the general solution is given by

$$y = c_1 y_1 + c_2 y_2 + \cdots + c_n y_n.$$

EXAMPLE 8.25

Find the general solution of $y^{(v)} - 2y^{(iv)} + 2y''' = 0$.

Solution

Following the preceding procedure, we first must solve the characteristic equation

$$\lambda^5 - 2\lambda^4 + 2\lambda^3 = 0,$$

or

$$\lambda^3(\lambda^2 - 2\lambda + 2) = 0.$$

The first factor gives the triple root (a root of multiplicity 3) $\lambda = 0$, whereas the roots corresponding to the second factor can be determined by the quadratic formula to be $\lambda = 1 + i$ and $\lambda = 1 - i$.

Now, the solution of the differential equation corresponding to the real root $\lambda = 0$ is $y = e^{0x} = 1$, and since 0 is a triple root, $y = x$ and $y = x^2$ are also solutions. Corresponding to the complex roots $1 + i$ (and $1 - i$) are the solutions $y = e^x \sin x$ and

$\mathbf{y} = e^x \cos x$. We have thus found the five linearly independent solutions of the given fifth-order differential equation; its general solution is

$$\mathbf{y} = c_1 + c_2 x + c_3 x^2 + c_4 e^x \cos x + c_5 e^x \sin x.$$

THE NONHOMOGENEOUS EQUATION

Although we will not go into the details of finding *particular* solutions of the nonhomogeneous linear differential equation (1), we will give a theorem showing how its general solution can be found.

THEOREM 5   Let $\mathbf{y}_p$ be a given solution of the differential equation (1), $L(\mathbf{y}) = \mathbf{b}$. Then the general solution of $L(\mathbf{y}) = \mathbf{b}$ is

$$\mathbf{y} = \mathbf{y}_p + \mathbf{y}_c,$$

where $\mathbf{y}_c$ is the general solution of the *associated homogeneous equation* $L(\mathbf{y}) = \mathbf{0}$.

NOTE   The functions $\mathbf{y}_p$ and $\mathbf{y}_c$ of Theorem 5 are often called a **particular solution** and the **complementary function** for $L(\mathbf{y}) = \mathbf{b}$, respectively. The function $\mathbf{y}_p + \mathbf{y}_c$ is said to be the **general solution** of this equation.

Proof   First we will show that $\mathbf{y} = \mathbf{y}_p + \mathbf{y}_c$ is a solution of $L(\mathbf{y}) = \mathbf{b}$. We have

$$L(\mathbf{y}) = L(\mathbf{y}_p + \mathbf{y}_c) = L(\mathbf{y}_p) + L(\mathbf{y}_c),$$

since L is linear. Now, $L(\mathbf{y}_p) = \mathbf{b}$ and $L(\mathbf{y}_c) = \mathbf{0}$, so $L(\mathbf{y}) = \mathbf{b} + \mathbf{0} = \mathbf{b}$, as desired.

Now we will show that if $\mathbf{v}$ is an arbitrary solution of $L(\mathbf{y}) = \mathbf{b}$, then it can be written as $\mathbf{y}_p + \mathbf{y}_c$ for some particular solution $\mathbf{y}_p$ of this equation. Let $\mathbf{y}_k = \mathbf{v} - \mathbf{y}$. Then $L(\mathbf{y}_k) = L(\mathbf{v} - \mathbf{y}_p) = L(\mathbf{v}) - L(\mathbf{y}_p) = \mathbf{b} - \mathbf{b} = \mathbf{0}$. Therefore $\mathbf{y}_k$ is in the kernel of L, so it is a specific member of the general solution $\mathbf{y}_c$. Transposing, we have $\mathbf{v} = \mathbf{y}_p + \mathbf{y}_k$, and $\mathbf{v}$ is expressed in the required form.

As we said, we will not delve into the various techniques for finding a particular solution of $L(\mathbf{y}) = \mathbf{b}$. However, if we do know one particular solution, we can then use Theorem 5 to determine its general solution.

EXAMPLE 8.26

Given that $y_p = e^x$ is a particular solution of $y'' + y = 2e^x$, find the general solution of this differential equation.

Solution

Applying Theorem 5, we see that we must find the general solution y_c of the associated homogeneous equation $y'' + y = 0$. Since the characteristic equation of the latter is $\lambda^2 + 1 = 0$, which has roots $\lambda = i$ and $\lambda = -i$, we have $y_c = c_1 \cos x + c_2 \sin x$. Therefore the general solution of $y'' + y = 2e^x$ is $y = y_p + y_c = e^x + c_1 \cos x + c_2 \sin x$.

INITIAL VALUE PROBLEMS

The general solution of an nth-order linear differential equation will contain n arbitrary constants (which we have been calling $c_1, c_2, \ldots, c_n$). In applications, additional information about the solution is usually known, and this information can be used to determine the arbitrary constants. The case in which we know the value of the unknown function as well as its first $n - 1$ derivatives at a single point leads to an *initial value problem*.

DEFINITION

Given a differential equation $L(y) = b$, a point x_0 and numbers $y_0, y_1, \ldots, y_{n-1}$, the problem of determining a function $y(x)$ that satisfies

$$L(y) = b$$

and

$$y(x_0) = y_0, y'(x_0) = y_1, \ldots, y^{(n-1)}(x_0) = y_{n-1} \tag{6}$$

is called an **initial value problem**. The equations (6) are called the **initial conditions** for the problem.

EXAMPLE 8.27

Solve the initial value problem

$$y''' - y' = 0$$

$$y(0) = 1, \quad y'(0) = 3, \quad y''(0) = -1.$$

Solution

We first solve the differential equation, ignoring the initial conditions for the time being. Since its characteristic equation is $\lambda^3 - \lambda = 0$ with roots $\lambda = 0$, $\lambda = 1$, and $\lambda = -1$, its general solution is given by

$$y = c_1 + c_2 e^x + c_3 e^{-x}.$$

Differentiating this solution, we have

$$y' = c_2 e^x - c_3 e^{-x}$$

and

$$\mathbf{y}'' = c_2 e^x + c_3 e^{-x}.$$

Thus the initial conditions imply the following:

$$\mathbf{y}(0) = \quad 1 \rightarrow c_1 + c_2 + c_3 = \quad 1$$
$$\mathbf{y}'(0) = \quad 3 \rightarrow \quad\quad c_2 - c_3 = \quad 3$$
$$\mathbf{y}''(0) = -1 \rightarrow \quad\quad c_2 + c_3 = -1.$$

The solution of this system of linear (algebraic) equations is $c_1 = 2$, $c_2 = 1$, $c_3 = -2$. Hence, substituting these values into the general solution, we arrive at the solution of the initial value problem,

$$\mathbf{y} = 2 + e^x - 2e^{-x}.$$

EXAMPLE 8.28   A weight attached to a spring suspended from the ceiling moves up and down in such a way that its displacement **s** (in inches) from its equilibrium position (position while "at rest") is given by $(d^2s/dt^2) + 4\mathbf{s} = \mathbf{0}$, where t represents time in seconds (see Figure 8.25). If the initial position of the spring is $\mathbf{s} = \mathbf{0}$ (its equilibrium position) and its initial velocity, ds/dt, is 2 inches per second in the upward direction, find the position $\mathbf{s}(t)$ of the spring at any time t.

FIGURE 8.25

Solution

Phrasing it as an initial value problem, we have the differential equation

$$\frac{d^2s}{dt^2} + 4\mathbf{s} = \mathbf{0},$$

with initial conditions

$$\mathbf{s}(0) = 0, \quad \mathbf{s}'(0) = 2.$$

Solving the differential equation, we obtain the general solution

$$\mathbf{s}(t) = c_1 \cos 2t + c_2 \sin 2t,$$

so

$$\mathbf{s}'(t) = -2c_1 \sin 2t + 2c_2 \cos 2t.$$

Thus the initial conditions yield:

$$s(0) = 0 \rightarrow c_1(1) + c_2(0) = 0,$$

$$s'(0) = 2 \rightarrow -2c_1(0) + 2c_2(1) = 2.$$

So, $c_1 = 0$ and $c_2 = 1$, resulting in the solution of the initial value problem:

$$s(t) = \sin 2t.$$

We will now state (without proof) a basic theorem in the study of initial value problems. This theorem is used in the proofs of Theorems 2 and 3 which follow.

THEOREM 6   The initial value problem

$$L(\mathbf{y}) = \mathbf{b} \qquad \text{(Equation (1))}$$

$$\mathbf{y}(x_0) = y_0, \qquad \mathbf{y}'(x_0) = y_1, \ldots, \mathbf{y}^{(n-1)}(x_0) = y_{n-1}$$

has a solution, and this solution is unique.

Proof of Theorem 2   We first assume that $\mathbf{y}_1, \mathbf{y}_2, \ldots, \mathbf{y}_n$ are linearly dependent and show that $W(\mathbf{y}_1, \mathbf{y}_2, \ldots, \mathbf{y}_n) = \mathbf{0}$. By definition there exist n scalars, not all zero, such that

$$c_1 \mathbf{y}_1 + c_2 \mathbf{y}_2 + \cdots + c_n \mathbf{y}_n = \mathbf{0}.$$

This equation taken together with the ones obtained by differentiating both sides of it $n - 1$ times results in:

$$
\begin{aligned}
c_1 \mathbf{y}_1(x) \quad &+ c_2 \mathbf{y}_2(x) \quad + \cdots + c_n \mathbf{y}_n(x) \quad = 0 \\
c_1 \mathbf{y}_1'(x) \quad &+ c_2 \mathbf{y}_2'(x) \quad + \cdots + c_n \mathbf{y}_n'(x) \quad = 0 \\
&\vdots \\
c_1 \mathbf{y}_1^{(n-1)}(x) &+ c_2 \mathbf{y}_2^{(n-1)}(x) + \cdots + c_n \mathbf{y}_n^{(n-1)}(x) = 0,
\end{aligned}
\tag{7}
$$

for each x. Thus the vector $(c_1, c_2, \ldots, c_n)$ is a solution of a system of homogeneous linear equations. Since not all c_i are zero, this linear system has a nonzero solution, so its determinant, $W(\mathbf{y}_1(x), \mathbf{y}_2(x), \ldots, \mathbf{y}_n(x))$ is zero (by Theorem 6 of Section 3.2). Since this is true for each x, $W(\mathbf{y}_1, \mathbf{y}_2, \ldots, \mathbf{y}_n) = \mathbf{0}$, as desired.

We now assume that the Wronskian is zero for some fixed real number a and demonstrate the linear dependence of the $\mathbf{y}_i$. Since the Wronskian is zero, there exist scalars $c_1, c_2, \ldots, c_n$, not all zero, such that the linear system (7) is satisfied at $x = a$. Now, let $\mathbf{y}$ be the function

$$\mathbf{y} = c_1 \mathbf{y}_1 + c_2 \mathbf{y}_2 + \cdots + c_n \mathbf{y}_n.$$

By Theorem 1, $\mathbf{y}$ is a solution of $L(\mathbf{y}) = \mathbf{0}$, since each of the $\mathbf{y}_i$ are. Moreover, $\mathbf{y}(x)$ satisfies the initial conditions

$$\mathbf{y}(a) = \mathbf{y}'(a) = \cdots = \mathbf{y}^{(n-1)}(a) = 0.$$

But this implies (Exercise 32) that $\mathbf{y} = \mathbf{0}$. Hence we have found scalars, $c_1, c_2, \ldots, c_n$, not all zero, such that

$$c_1\mathbf{y}_1 + c_2\mathbf{y}_2 + \cdots + c_n\mathbf{y}_n = \mathbf{0},$$

and therefore the $\mathbf{y}_i$ are linearly dependent.

Proof of Theorem 3   We will explicitly give a basis for the kernel of L, Ker (L), consisting of n functions, $\mathbf{y}_1, \mathbf{y}_2, \ldots, \mathbf{y}_n$. Let a be any fixed real number and let $\mathbf{y}_i$ be the solutions of the n initial value problems

$$L(\mathbf{y}) = \mathbf{0}$$
$$\mathbf{y}_i(a) = \mathbf{y}_i'(a) = \cdots = \mathbf{y}_i^{(i-2)}(a) = \mathbf{y}_i^{(i)}(a) = \cdots = \mathbf{y}_i^{(n-1)}(a) = 0$$
$$\mathbf{y}_i^{(i-1)}(a) = 1,$$

for $i = 1, 2, \ldots, n$. The fact that these $\mathbf{y}_i$ exist (and are unique) is a consequence of Theorem 6.

We now show that the set $\mathcal{S} = \{\mathbf{y}_1, \mathbf{y}_2, \ldots, \mathbf{y}_n\}$ is linearly independent and that it spans Ker (L). For these functions $\mathbf{y}_i$, the Wronskian evaluated at $x = a$ is equal to $\det (I) = 1 \neq 0$. Hence by Theorem 2, $\mathcal{S}$ is a linearly independent set.

To show that $\mathcal{S}$ spans Ker (L), let $\mathbf{v}$ be any function in this space, and define the function $\mathbf{u}$ by

$$\mathbf{u} = \mathbf{v} - \mathbf{v}(a)\mathbf{y}_1 - \mathbf{v}'(a)\mathbf{y}_2 - \cdots - \mathbf{v}^{(n-1)}(a)\mathbf{y}_n.$$

Since $\mathbf{v}$ and each $\mathbf{y}_i$ is in Ker (L), $\mathbf{u}$ is in Ker (L). Differentiating both sides, we obtain

$$\mathbf{u}^{(k)} = \mathbf{v}^{(k)} - \mathbf{v}(a)\mathbf{y}_1^{(k)} - \mathbf{v}'(a)\mathbf{y}_2^{(k)} - \cdots - \mathbf{v}^{(n-1)}(a)\mathbf{y}_n^{(k)},$$

for $k = 1, 2, \ldots, n$. But direct substitution shows that

$$\mathbf{u}(a) = \mathbf{u}'(a) = \cdots = \mathbf{u}^{(n-1)}(a) = \mathbf{0},$$

and since $L(\mathbf{u}) = \mathbf{0}$, we have (by Exercise 32) $\mathbf{u} = \mathbf{0}$. Thus $\mathbf{v}$ can be written as a linear combination of the $\mathbf{y}_i$, and $\mathcal{S}$ is a spanning set.

8.8 EXERCISES

In Exercises 1–16 find the general solution of the given differential equation. (Assume that x is the independent variable unless otherwise specified.)

1. $y'' - 3y' + 2y = 0$ **2.** $y'' + 2y' - 3y = 0$

3. $y'' + 2y' - y = 0$

4. $y'' - 4y' + y = 0$

5. $u'' + 4u' + 4u = 0$

6. $4u'' - 4u' + u = 0$

7. $4\dfrac{d^2s}{dt^2} + s = 0$

8. $\dfrac{d^2s}{dt^2} + 9s = 0$

9. $y'' + 2y' + 5y = 0$

10. $y'' - 2y' + 4y = 0$

11. $u''' + 2u'' + u' = 0$

12. $u''' - 4u'' + 4u' = 0$

13. $\dfrac{d^4s}{dt^4} - s = 0$

14. $\dfrac{d^4s}{dt^4} + \dfrac{d^2s}{dt^2} = 0$

15. $\dfrac{d^4s}{dt^4} - \dfrac{2d^3s}{dt^3} + \dfrac{4d^2s}{dt^2} = 0$

16. $\dfrac{d^4s}{dt^4} + \dfrac{2d^2s}{dt^2} + s = 0$

In Exercises 17–20 a particular solution of the differential equation is given. Find its general solution.

17. $y'' - 2y' - 3y = 10 \sin x, \qquad y_p = -2 \sin x + \cos x$

18. $y'' - y = x^2 + 1, \qquad y_p = -x^2 - 3$

19. $u'' + u = 2e^x + 2e^{-x}, \qquad u_p = e^x + e^{-x}$

20. $u'' + u' = 2 \cos x, \qquad u_p = \sin x - \cos x$

In Exercises 21–26 solve the given initial value problems.

21. $y'' + 2y' - 3y = 0, \quad y(0) = 1, \quad y'(0) = 2$

22. $y'' - y' = 0, \quad y(1) = 0, \quad y'(1) = 1$

23. $\dfrac{d^2s}{dt^2} + s = 0, \quad s(\pi/2) = 1, \quad \dfrac{ds}{dt}(\pi/2) = 0$

24. $\dfrac{d^2s}{dt^2} = 0, \quad s(2) = 0, \quad \dfrac{ds}{dt}(2) = 1$

25. $u^{(\mathrm{iv})} - u'' = 0, \quad u(0) = 0, \quad u'(0) = 1, \quad u''(0) = 2, \quad u'''(0) = 4$

26. $u''' + u' = 0, \quad u(0) = 2, \quad u'(0) = 1, \quad u''(0) = 0$

In Exercises 27 and 28 a weight is placed on a spring suspended from the ceiling. It is then disturbed as follows. If its motion is described by the differential equation

$$\frac{d^2s}{dt^2} + 9s = 0,$$

where s is the displacement from its equilibrium position in inches and t is time in seconds, in each case find the formula s(t) for the motion of the spring.

27. The weight is pulled down 2 inches below its equilibrium position and released with an upward velocity of 6 inches per second.

28. The weight is pulled 2 inches below its equilibrium position and let go (initial velocity = 0).

In Exercises 29 and 30 a pendulum swings in such a way that its motion is described by the differential equation

$$\frac{d^2\theta}{dt^2} + 16\theta = 0,$$

FIGURE 8.26

where t is time in seconds and θ is the angle in radians pictured in Figure 8.26 (θ is positive if the pendulum bob is to the right of the vertical). In each case find the formula, $\theta(t)$, that describes the motion of the pendulum.

29. The pendulum bob, hanging vertically, is hit so that its initial angular velocity is 1 radian per second.

30. The pendulum bob is raised to an angle of $\pi/6$ radians (to the right of vertical) and released (zero initial velocity).

31. Prove parts (i) and (iii) of Theorem 4.

32. Use Theorem 6 to prove that if $\mathbf{f}$ is a solution of the initial value problem

$$a_n \mathbf{y}^{(n)} + a_{n-1} \mathbf{y}^{(n-1)} + \cdots + a_1 \mathbf{y}' + a_0 \mathbf{y} = 0$$

$$\mathbf{y}(x_0) = \mathbf{y}'(x_0) = \cdots = \mathbf{y}^{(n-1)}(x_0) = 0,$$

then $\mathbf{f}$ must be the zero function.

In Exercises 33 and 34 use Theorem 2 to show that the following sets of solutions of the indicated differential equation are linearly independent.

33. $\{1, x, e^x, e^{-x}\}, \ \mathbf{y}^{(iv)} - \mathbf{y}'' = \mathbf{0}$

34. $\{e^x, \cos x, \sin x\}, \ \mathbf{y}''' - \mathbf{y}'' + \mathbf{y}' - \mathbf{y} = \mathbf{0}$

LINEAR PROGRAMMING

....

The subject of linear programming deals with the problem of optimizing (maximizing or minimizing) a linear function, subject to the constraints imposed by a system of linear inequalities. After introducing the topic in the first section, we will demonstrate the basic solution technique in Section 9.2 and then extend this technique to a more general situation in Section 9.3. The proofs of the major theorems will be delayed until Section 9.4. Adequate preparation for this chapter is given by Chapter 2.

9.1 INTRODUCTION AND TERMINOLOGY

In this section you will see how linear programming problems might arise and how very simple ones can be solved geometrically. We will also introduce some of the basic terminology and a standard form for linear programming problems.

A SIMPLE EXAMPLE

Suppose a manufacturer produces two types of liquids, X and Y. Because of past sales experience the market researcher estimates that at least twice as much Y as X is needed. The manufacturing capacity of the plant allows for a total of 9 units to be manufactured. If each unit of liquid X results in a profit of \$2 and the profit for each unit of Y is \$1, how much of each should be produced to maximize the profit?

Let x be the amount of X to be produced, and let y be the amount of Y. Then the condition that there be at least twice as much Y as X becomes

$$2x \leq y.$$

The manufacturing capacity of 9 units implies that

$$x + y \le 9.$$

Let $z = z(x, y)$ be the profit associated with the production schedule of producing x units of X and y units of Y. Then the *profit function* or *objective function* is given by

$$z = 2x + y.$$

There is also a further condition implied but not stated in the problem. The variables must be nonnegative, since we cannot produce negative amounts of a product. Thus we have

$$x \ge 0, \qquad y \ge 0.$$

Collecting all of this information together, we may state the given problem mathematically as follows: maximize $z = 2x + y$ subject to the constraints

$$2x - y \le 0$$
$$x + y \le 9$$
$$x \ge 0, \qquad y \ge 0.$$

In the next section we develop a technique (the *simplex algorithm*) for solving such problems in general, but we can solve this particular one geometrically. First we sketch the set of points in $\mathbf{R}^2$ that satisfy the set of constraints (shaded part of Figure 9.1). This region is called the *feasible region*.

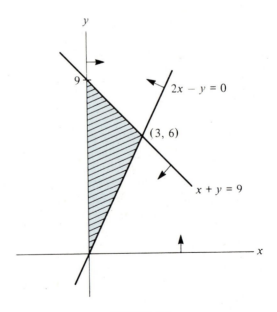

FIGURE 9.1

If a point is in the region, it satisfies all of the constraints and is called *feasible*; if it is not in the region, it violates at least one of them and is called *infeasible*.

Different points in the feasible region give different values of the objective function. For example, the origin $(0,0)$ is feasible and gives a profit of $2(0) + (0) = 0$. Similarly, $(1,5)$ is feasible and gives a profit of $2(1) + (5) = 7$, which is better than 0. We seek the point or points of the feasible region that yield the maximal profit.

Corresponding to a *fixed* value of z, the set of solutions to the equation $z = 2x + y$ is the line (called the *objective line*) with slope -2 and y-intercept z. In other words, all points along the line $y = -2x + z$ correspond to the same z value. Remember that we are interested only in those points that lie in the feasible region, and we want z to be as large as possible (Figure 9.2).

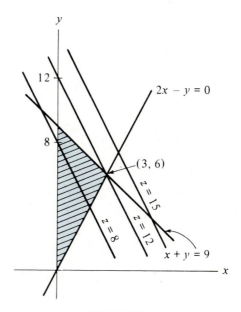

FIGURE 9.2

It is clear from Figure 9.2 that the optimal value occurs when $z = 12$, since moving the objective line up increases its y-intercept (the value of z), but moving beyond the point $(3, 6)$ gives a line that does not intersect the feasible region at all.

NOTE In the preceding discussion we obtained a unique solution; the maximum value $z = 12$ occurs only at the point $(3,6)$. If, however, the objective function is changed to $z = 2x + 2y$, all of the points in the feasible region that lie on the edge determined by $x + y = 9$ will be optimal, with

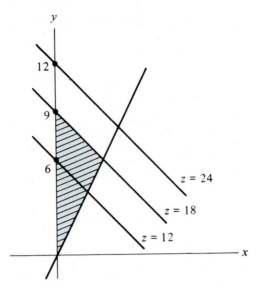

FIGURE 9.3

value $z = 18$ (Figure 9.3). Thus the point at which the optimal value occurs need not be unique.

Moreover, if the feasible region is *unbounded*, as would be the case in the preceding discussion if the constraint $x + y \le 9$ were removed, an optimal value may not exist. That is, it is possible that no matter which point is chosen, a "better" one always exists.

THE GENERAL PROBLEM

The previous example is a simple case of the general *linear programming problem*. We will describe this problem mathematically after introducing the following notation for comparing certain vectors.

DEFINITION

Let $\mathbf{u}$ and $\mathbf{v}$ be vectors in $\mathbf{R}^m$. We say $\mathbf{u}$ **is less than** $\mathbf{v}$ and write $\mathbf{u} < \mathbf{v}$ if $u_i < v_i$ for each $i = 1, \ldots, m$. Likewise, $\mathbf{u} \le \mathbf{v}$ means $u_i \le v_i$ for each i. Similar definitions hold for $\mathbf{u} > \mathbf{v}$ and $\mathbf{u} \ge \mathbf{v}$. The vector $\mathbf{v}$ is **positive** if $\mathbf{v} > \mathbf{0}$ and **nonnegative** if $\mathbf{v} \ge \mathbf{0}$.

EXAMPLE 9.1

Let $\mathbf{u} = (1, 2, -3)$, $\mathbf{v} = (2, 4, 1)$, and $\mathbf{w} = (2, 0, 1)$. Then $\mathbf{u} < \mathbf{v}$ and $\mathbf{w} \le \mathbf{v}$, but no special relationship exists between $\mathbf{u}$ and $\mathbf{w}$. It is also true that $\mathbf{v} > \mathbf{0}$ and $\mathbf{w} \ge \mathbf{0}$.

Given a system of inequalities

$$
\begin{aligned}
a_{11}x_1 + a_{12}x_2 + \cdots + a_{1n}x_n &\le b_1 \\
a_{21}x_1 + a_{22}x_2 + \cdots + a_{2n}x_n &\le b_2 \\
\vdots \qquad \vdots \qquad\qquad \vdots \qquad \vdots& \\
a_{m1}x_1 + a_{m2}x_2 + \cdots + a_{mn}x_n &\le b_m,
\end{aligned}
$$

we can use the vector inequality notation to write the system as

$$
Ax \le \mathbf{b},
$$

where, as was the case with linear equations,

$$
A = \begin{bmatrix} a_{11} \cdots a_{1n} \\ \vdots \qquad \vdots \\ a_{m1} \cdots a_{mn} \end{bmatrix}, \qquad \mathbf{b} = \begin{bmatrix} b_1 \\ \vdots \\ b_m \end{bmatrix}, \qquad \text{and} \qquad \mathbf{x} = \begin{bmatrix} x_1 \\ \vdots \\ x_n \end{bmatrix}.
$$

EXAMPLE 9.2

Graph the set of solutions of the system of inequalities

$$
Ax \le \mathbf{b}, \qquad \mathbf{x} \ge \mathbf{0},
$$

where

$$
A = \begin{bmatrix} 2 & -1 \\ 1 & 1 \end{bmatrix} \qquad \text{and} \qquad \mathbf{b} = \begin{bmatrix} 0 \\ 9 \end{bmatrix}.
$$

Solution

The matrix inequality,

$$
Ax = \begin{bmatrix} 2 & -1 \\ 1 & 1 \end{bmatrix}\begin{bmatrix} x \\ y \end{bmatrix} \le \begin{bmatrix} 0 \\ 9 \end{bmatrix}, \qquad \mathbf{x} \ge \mathbf{0},
$$

is equivalent to the system of inequalities

$$
\begin{aligned}
2x - y &\le 0 \\
x + y &\le 9 \\
x \ge 0, \qquad y &\ge 0.
\end{aligned}
$$

This is the same set of inequalities as was derived at the beginning of this section. Thus the set of solutions is just that of Figure 9.1.

Consider the system of linear inequalities

$$
Ax \le \mathbf{b}, \qquad \mathbf{b} \ge \mathbf{0}, \qquad \mathbf{x} \ge \mathbf{0}.
$$

The set of all solutions to this system is called the **feasible region.** Notice that **0** is always in the feasible region of this system, since $A\mathbf{0} = \mathbf{0} \le \mathbf{b}$, and usually there are infinitely many feasible points. As in the example presented earlier, we will be seeking solutions that *optimize (maximize or minimize)* the value z

of a linear function

$$z = \mathbf{c} \cdot \mathbf{x} = c_1 x_1 + c_2 x_2 + \cdots + c_n x_n,$$

where the c_i are scalars. The function $z = \mathbf{c} \cdot \mathbf{x}$ is called the **objective function.**

DEFINITION

A linear programming problem is in **standard form** if it is stated in the following way: Optimize the function $z = \mathbf{c} \cdot \mathbf{x}$, subject to the conditions

$$A\mathbf{x} \leq \mathbf{b}, \quad \mathbf{b} \geq \mathbf{0}, \quad \text{and} \quad \mathbf{x} \geq \mathbf{0}.$$

To apply our knowledge of linear *equations* to the study of linear *inequalities*, we convert the system of linear inequalities to a system of linear equations by adding a nonnegative variable, called a **slack variable,** to each constraint. For example, the inequality

$$2x + 3y \leq 5$$

is equivalent to the equation

$$2x + 3y + z = 5,$$

with the restriction that z be nonnegative. Notice that $z = 5 - (2x + 3y)$ is the "slack" or difference between $2x + 3y$ and 5. There is a tradeoff in this process—in order to work with equations instead of inequalities, the number of variables increases by one for each inequality.

The mathematical description of this conversion process is as follows. Let

$$A\mathbf{x} \leq \mathbf{b}, \quad \mathbf{b} \geq \mathbf{0}, \quad \mathbf{x} \geq \mathbf{0}$$

be a system of m linear inequalities in n unknowns. The corresponding system of equations is

$$A'\mathbf{x}' = \mathbf{b}, \quad \mathbf{b} \geq \mathbf{0}, \quad \mathbf{x}' \geq \mathbf{0},$$

where A' is the $m \times (n + m)$ matrix whose first n columns are those of A and last m columns are those of the identity matrix of order m. The first n components of the vector $\mathbf{x}'$ correspond to those of $\mathbf{x}$, and the last m components are the slack variables of the system.

EXAMPLE 9.3

Convert the following system of linear inequalities to the corresponding system of linear equations.

$$2x_1 - x_2 + x_3 \leq 5$$
$$x_2 - 2x_3 \leq 2$$
$$x_1 - x_2 + 2x_3 \leq 4$$
$$x_i \geq 0, \quad i = 1, 2, 3.$$

Solution The resulting system in equation form is

$$
\begin{aligned}
2x_1 - x_2 + x_3 + x_4 \qquad\qquad\qquad &= 5 \\
x_2 - 2x_3 \qquad\quad + x_5 \qquad\quad &= 2 \\
x_1 - x_2 + 2x_3 \qquad\qquad\quad + x_6 &= 4 \\
x_i \geq 0, \qquad i = 1, \ldots, 6.
\end{aligned}
$$

The matrices A and A' are given by

$$
A = \begin{bmatrix} 2 & -1 & 1 \\ 0 & 1 & -2 \\ 1 & -1 & 2 \end{bmatrix}, \qquad
A' = \begin{bmatrix} 2 & -1 & 1 & 1 & 0 & 0 \\ 0 & 1 & -2 & 0 & 1 & 0 \\ 1 & -1 & 2 & 0 & 0 & 1 \end{bmatrix}.
$$

NOTE We are not interested in the *values* of the slack variables. Thus in a linear programming problem with objective function $z = \mathbf{c} \cdot \mathbf{x}$, in this conversion process we take as the new objective function $z' = \mathbf{c}' \cdot \mathbf{x}'$, where $c_j' = c_j$ for $j = 1, \ldots, n$ and $c_j' = 0$ for $j = n + 1, \ldots, n + m$.

· ·

A system of linear equations that corresponds to the linear inequalities of a linear programming problem in standard form is always *row-reduced* in the following sense.

DEFINITION A matrix of rank r is **row-reduced** if it contains r nonzero rows, the rest are zero, and there exists a subset of r columns that are distinct elementary columns.

NOTE A matrix in row-reduced echelon form is of course row-reduced in this sense. The columns that contain the leading ones provide the set of elementary columns.

· ·

EXAMPLE 9.4 Which of the following are row-reduced?

$$
B = \begin{bmatrix} 1 & 3 & 1 & 0 \\ -1 & 2 & 0 & 1 \\ 0 & 0 & 0 & 0 \end{bmatrix}, \qquad
C = \begin{bmatrix} 2 & 0 & 3 & 0 \\ 1 & 1 & 2 & 0 \\ 3 & 0 & 4 & 1 \end{bmatrix},
$$

$$
D = \begin{bmatrix} 2 & 0 & 0 & 2 & 1 \\ 1 & 1 & 0 & 1 & 0 \\ -1 & 0 & 1 & 4 & 0 \end{bmatrix}.
$$

Solution The matrix B is row-reduced, since $r = 2$ and $\mathbf{b}^3 = \mathbf{e}_1$ and $\mathbf{b}^4 = \mathbf{e}_2$. D is also row-reduced, since $r = 3$ and $\mathbf{d}^5 = \mathbf{e}_1$, $\mathbf{d}^2 = \mathbf{e}_2$, and $\mathbf{d}^3 = \mathbf{e}_3$. However, C is not row-reduced, since $r = 3$, but there is no column equal to $\mathbf{e}_1$.

For an $m \times n$ matrix of rank m, being row-reduced is tantamount to possessing a subset of m columns that may be rearranged to form an identity matrix. It is sometimes convenient to speak of an *embedded identity matrix* when referring to the set of elementary columns. For example, in the matrix A' of a linear system of equations that corresponds to a system of linear inequalities, the last m columns form an embedded identity matrix.

As in the case of a system in row-reduced echelon form, it is easy to identify a particular solution, called a **basic solution.** To obtain the corresponding basic solution, let the variables that correspond to the elementary columns assume the corresponding **b** column value (these are called the **basic variables** of the solution), and let all others be zero (these are the **nonbasic variables**). Basic solutions correspond to the vertices of the feasible region in $\mathbf{R}^n$.

EXAMPLE 9.5

Find a basic solution to the following system of equations and identify the basic and nonbasic variables.

$$
\begin{aligned}
2x_2 + x_3 \quad - \quad x_5 &= 2 \\
x_1 - x_2 \quad\quad\quad + 3x_5 &= 3 \\
x_2 \quad\quad + x_4 \quad\quad &= 4.
\end{aligned}
$$

Solution

Forming the corresponding augmented matrix, we see that it is already in row-reduced form:

$$
\begin{bmatrix}
0 & 2 & 1 & 0 & -1 & | & 2 \\
1 & -1 & 0 & 0 & 3 & | & 3 \\
0 & 1 & 0 & 1 & 0 & | & 4
\end{bmatrix}.
$$

A particular solution is then obtained by letting $x_3 = 2$ (because $\mathbf{a}^3 = \mathbf{e}_1$), $x_1 = 3$ (because $\mathbf{a}^1 = \mathbf{e}_2$), $x_4 = 4$ (because $\mathbf{a}^4 = \mathbf{e}_3$), and $x_2 = x_5 = 0$. The basic variables are $x_1, x_3,$ and x_4, and the nonbasic variables are x_2 and x_5.

9.1 EXERCISES

In Exercises 1–4 sketch the feasible region in $\mathbf{R}^2$ described by the constraints.

1. $x - y \le 1$
$x + 2y \le 8$
$x, y \ge 0$

2. $x - y \le 2$
$3x - y \le 6$
$x, y \ge 0$

3. $x + y \le 2$
$-x + 3y \le 9$
$2x - y \le 6$
$x, y \ge 0$

4. $x - y \le 1$
$2x - 2y \le 5$
$2x - y \ge 6$
$x \ge 1, y \le 2$

5. Maximize the objective function $z = 2x + y$ on the feasible region of Exercise 1, or show that no maximum exists.

6. Maximize the objective function $z = 3x - y$ on the feasible region of Exercise 2, or show that no maximum exists.

7. Maximize the function $z = x - 3y$ on the feasible region of Exercise 3, or show that no minimum exists.

8–10. Convert each system of constraints in Exercises 1–3 into equation form.

In Exercises 11–14 decide whether or not the system of linear equations is row-reduced. If it is row-reduced, identify the columns that yield an embedded identity matrix and find the corresponding basic solution.

11. $\begin{aligned} 2x_1 \quad\quad + x_3 - 3x_4 \quad &= 2 \\ x_1 + x_2 \quad\quad - x_4 \quad &= 1 \\ 3x_1 \quad\quad\quad + 2x_4 + x_5 &= 4 \end{aligned}$

12. $\begin{aligned} x_1 \quad\quad\quad - x_4 &= 1 \\ x_2 \quad + x_4 &= 3 \\ x_3 + 2x_4 &= 5 \end{aligned}$

13. $\begin{aligned} x_1 \quad\quad + x_3 \quad\quad &= 2 \\ -x_1 + x_2 \quad\quad + x_4 \quad &= 6 \\ 2x_1 \quad\quad\quad\quad + x_5 &= 1 \end{aligned}$
(*Note*: two possible answers)

14. $\begin{aligned} 2x_2 - x_3 + 2x_4 + x_5 + 2x_6 &= 2 \\ x_1 - x_2 \quad\quad\quad\quad - 3x_6 &= 2 \end{aligned}$

In Exercises 15 and 16 convert each to a linear programming problem in standard form and then to equation form. You do *not* need to try to solve them.

15. A man wishes to invest a portion of \$50,000 in savings, stocks, and property. He feels that no more than $\frac{1}{5}$ of the total amount should be in savings and no more than $\frac{2}{5}$ in savings and stocks together. If he estimates an 8% return from savings, 12% from real estate, and 15% from stocks, how should he invest his money?

16. A toy manufacturer has available 200 lb of wood, 400 lb of moldable plastic, and 50 lb of metal to produce toys A, B, C, and D, each of which requires some or all of the materials for component parts. Table 9.1 gives the requirements for each

	Wood	Plastic	Metal	\$/Toy
A	0.2	0.3	0.1	5
B	0	0.5	0.2	4
C	0	0.9	0.1	3
D	1	0	0.1	6
\$/lb	1	0.5	2	

TABLE 9.1

toy, the price of each toy, and the price of the materials used in production of the toys. How many of each toy should the manufacturer produce to maximize profit?

9.2   THE SIMPLEX ALGORITHM

In this section we introduce a general procedure for solving linear programming problems. Now that the translation from inequalities to equations has been described (Section 9.1), we will assume that the linear programming problem is in equation form and row-reduced.

Let the linear programming problem be given as follows: Optimize the linear function

$$z = \mathbf{c} \cdot \mathbf{x},$$

subject to the system of constraints

$$A\mathbf{x} = \mathbf{b}, \quad \mathbf{b} \geq 0, \quad \mathbf{x} \geq 0, \tag{1}$$

where A has an embedded identity matrix.

Let $\mathbf{c}^*$ be the $1 \times m$ matrix that consists of the components of $\mathbf{c}$ associated with the basic variables in the current basic solution. Then the value associated with the current solution is $\mathbf{c}^*\mathbf{b}$ (the nonbasic variables are zero). The m-vector $\mathbf{c}^*A - \mathbf{c}$ is called the vector of **simplex indicators.** (Here $\mathbf{c}$ is viewed as a $1 \times n$ matrix so that the operations are defined.) We call the jth component of this vector, $\mathbf{c}^*\mathbf{a}^j - c^j$, the jth **simplex indicator.**

Associated with System (1), we prepare the array shown in Figure 9.4, called the **tableau,** associated with the current solution. For convenience, tableaux will sometimes be presented with the row of variables at the top and the basic variables and the vector $\mathbf{c}^*$ on the left (see, for example, Figure 9.5).

A	$\mathbf{b}$
$\mathbf{c}^*A - \mathbf{c}$	$\mathbf{c}^*\mathbf{b}$

FIGURE 9.4

EXAMPLE 9.6   Let $A\mathbf{x} = \mathbf{b}, \mathbf{b} \geq 0, \mathbf{x} \geq 0$ be the system of Example 9.5,

$$\begin{aligned}
2x_2 + x_3 \quad\quad - x_5 &= 2 \\
x_1 - x_2 \quad\quad\quad + 3x_5 &= 3 \\
x_2 \quad + x_4 \quad\quad &= 4,
\end{aligned}$$

with objective function $z = 2x_1 - x_2 + 4x_3 - 2x_5$. Prepare the corresponding tableau.

Solution

Since the basic solution is $x_3 = 2$, $x_1 = 3$, and $x_4 = 4$, we have $\mathbf{c}^* = (c_3, c_1, c_4) = (4, 2, 0)$. The tableau is given in Figure 9.5.

	$\mathbf{c}^*$	x_1	x_2	x_3	x_4	x_5	$\mathbf{b}$
x_3	4	0	2	1	0	−1	2
x_1	2	1	−1	0	0	3	3
x_4	0	0	1	0	1	0	4
		0	7	0	0	4	14

FIGURE 9.5

Notice in Figure 9.5 that the second simplex indicator has been computed $\mathbf{c}^*\mathbf{a}^2 - c_2 = (4, 2, 0) \cdot (2, -1, 1) - (-1) = 7$. The rightmost bottom row entry is $\mathbf{c}^*\mathbf{b} = 14$, the value associated with the basic solution $x_3 = 2$, $x_1 = 3$, and $x_4 = 4$.

CONDITIONS FOR TERMINATING THE ALGORITHM

The next theorem implies that the solution of Example 9.6 is maximal. In other words, there is no solution with value greater than 14, which occurs when $x_3 = 2$, $x_1 = 3$, $x_4 = 4$, and $x_2 = x_5 = 0$. The proof of this and subsequent theorems will be delayed until Section 9.4.

THEOREM I Let a linear programming problem be given as follows: Optimize $z = \mathbf{c} \cdot \mathbf{x}$, subject to the constraints

$$A\mathbf{x} = \mathbf{b}, \quad \mathbf{b} \geq \mathbf{0}, \quad \mathbf{x} \geq \mathbf{0},$$

where A has an embedded identity matrix. If no simplex indicator of the associated tableau is negative (that is, $\mathbf{c}^*A - \mathbf{c} \geq \mathbf{0}$), then the corresponding basic solution is maximal. If no simplex indicator is positive, the corresponding basic solution is minimal.

Theorem 1 gives us conditions under which we have solved the problem and can terminate the procedure. The next theorem also gives us conditions under which we can cease our search, for a different reason—no optimal solution exists.

THEOREM 2 Let a linear programming problem be given as follows: Optimize $z = \mathbf{c} \cdot \mathbf{x}$, subject to the constraints

$$A\mathbf{x} = \mathbf{b}, \qquad \mathbf{b} \geq \mathbf{0}, \qquad \mathbf{x} \geq \mathbf{0},$$

where A has an embedded identity matrix. Let j be a column number such that the simplex indicator is negative ($\mathbf{c}^* \mathbf{a}^j - c_j < 0$). If every element of $\mathbf{a}^j$ is less than or equal to zero ($\mathbf{a}^j \leq \mathbf{0}$), then no maximal solution exists. If everything is the same except that the simplex indicator is positive, then no minimal solution exists.

EXAMPLE 9.7

Show that the following linear programming problem has no maximal solution. Subject to the constraints

$$
\begin{aligned}
x_1 \quad &+ 2x_3 - x_4 \quad\quad + x_6 = 2 \\
&- x_3 \quad\quad + x_5 + 3x_6 = 1 \\
x_2 + \; &x_3 - 2x_4 \quad\quad + x_6 = 4 \\
&x_j \geq 0, \qquad j = 1, \ldots, 6,
\end{aligned}
$$

maximize the function

$$z = x_1 + 3x_2 - x_3 + 3x_4 - 2x_5.$$

Solution

We prepare the corresponding tableau, Figure 9.6, using $\mathbf{c} = (1, 3, -1, 3, -2, 0)$ and $\mathbf{c}^* = (1, -2, 3)$, since the basic variables are x_1, x_5, and x_2.

	$\mathbf{c}^*$	x_1	x_2	x_3	x_4	x_5	x_6	$\mathbf{b}$
x_1	1	1	0	2	-1	0	1	2
x_2	-2	0	0	-1	0	1	3	1
x_2	3	0	1	1	-2	0	1	4
		0	0	8	-10	0	-2	12

FIGURE 9.6

We have $\mathbf{c}^* \mathbf{a}^4 - c_4 = -10$ and $\mathbf{a}^4 = (-1, 0, -2) \leq \mathbf{0}$, so by Theorem 2, no maximal solution exists.

FINDING A BETTER SOLUTION

We now have a criterion for optimality (Theorem 1) and a criterion under which no optimal solution exists (Theorem 2). Next we will introduce a procedure that takes us from a basic solution that meets neither criterion to a new one that might meet one of them. This process is called **pivoting**; it is just the row-reduction technique of Chapter 2. Once the pivot is chosen, divide that row by the pivot value (to obtain a 1 in the pivot position), and use the pivot row to obtain zeros in the rest of the pivot column.

EXAMPLE 9.8

In the following tableau (that of Example 9.7 without the bottom row), pivot on the 1,3 position:

1	0	②	−1	0	1	2
0	0	−1	0	1	3	1
0	1	1	−2	0	1	4

Solution

We first divide the top row by the circled entry in the 1, 3 position to obtain

$\frac{1}{2}$	0	1	$-\frac{1}{2}$	0	$\frac{1}{2}$	1
0	0	−1	0	1	3	1
0	1	1	−2	0	1	4

Now add the top row to the second and the negative of the top row to the third, to create zeros in the rest of the third column. This yields the new tableau:

$\frac{1}{2}$	0	1	$-\frac{1}{2}$	0	$\frac{1}{2}$	1
$\frac{1}{2}$	0	0	$-\frac{1}{2}$	1	$\frac{7}{2}$	2
$-\frac{1}{2}$	1	0	$-\frac{3}{2}$	0	$\frac{1}{2}$	3

The new tableau is row-reduced and corresponds to the basic solution $x_3 = 1$, $x_5 = 2$, $x_2 = 3$, and $x_1 = x_4 = x_6 = 0$. The new $\mathbf{c}^*$ vector is $\mathbf{c}^* = (-1, -2, 3)$. Completing the tableau to check for optimality, we have the tableau of Figure 9.7.

	$\mathbf{c}^*$	x_1	x_2	x_3	x_4	x_5	x_6	$\mathbf{b}$
x_3	−1	$\frac{1}{2}$	0	1	$-\frac{1}{2}$	0	$\frac{1}{2}$	1
x_5	−2	$\frac{1}{2}$	0	0	$-\frac{1}{2}$	1	$\frac{7}{2}$	2
x_2	3	$-\frac{1}{2}$	1	0	$-\frac{3}{2}$	0	$\frac{1}{2}$	3
		−4	0	0	−6	0	−6	4

FIGURE 9.7

By Theorem 1, this new basic solution is minimal.

Notice that in Example 9.8 we could have computed the bottom row of the new tableau (Figure 9.7) by continuing the pivot procedure to the bottom row. In other words, the bottom row of the tableau of Figure 9.6 was

$$[0 \quad 0 \quad 8 \quad -10 \quad 0 \quad -2 \mid 12].$$

If we add -8 times the pivot row to it, we obtain

$$[-4 \quad 0 \quad 0 \quad -6 \quad 0 \quad -6 \mid 4],$$

which is the bottom row of the new tableau of Figure 9.7. This technique will always work, and we state it as the next theorem.

THEOREM 3   The bottom row of a new tableau may be computed from the preceding tableau by using the pivot row and row operations to obtain a zero in the pivot column position of the bottom row.

In this pivoting process we must check two conditions. The simplex indicator (bottom row entry) determines whether the value of the solution will increase or decrease in the next tableau. That is, if the simplex indicator is *negative,* that column can be chosen to *increase* the value in the next tableau; if it is *positive,* that column will cause the value to *decrease.* However, care must also be taken to assure that the next solution remains *feasible*; that is, that all entries in the new **b** column are nonnegative.

Once an appropriate column has been chosen, a row is chosen as follows. Compute the quotients of the **b** column entries divided by the corresponding *positive* entries in the desired column. Choose a row that gives the smallest such quotient, because it will lead to a new solution that is feasible. In Example 9.8 we chose to pivot on the 1,3 position by the following procedure. To minimize, we first chose the third column, since the simplex indicator of this column, 8, is greater than zero. Then we computed the quotients $b_1/a_{13} = 2/2 = 1$ and $b_3/a_{33} = 4/1 = 4$ (b_2/a_{23} was ignored because $a_{23} \leq 0$), using the smaller of the two quotients (the first one) as the basis for choosing the first row. The justification for the process just described is the following theorem.

THEOREM 4   Let a linear programming problem be given as follows: Optimize $z = \mathbf{c} \cdot \mathbf{x}$, subject to the constraints

$$A\mathbf{x} = \mathbf{b}, \quad \mathbf{b} \geq \mathbf{0}, \quad \mathbf{x} \geq \mathbf{0},$$

where A has an embedded identity matrix. Let j be a column number such that the simplex indicator is negative (that is, $\mathbf{c}^*\mathbf{a}^j - c_j < 0$). If a_{ij} is positive for at least one row number $i = 1, \ldots, m$, let k be a row number such that $a_{kj} > 0$

and $b_k/a_{kj} \leq b_i/a_{ij}$ for all $i = 1, \ldots, m$ with $a_{ij} > 0$. Then pivoting on the k, j position leads to a new basic solution with value greater than or equal to the former. If everything is the same except that the simplex indicator is positive, the new solution will have value less than or equal to the former.

THE SIMPLEX ALGORITHM

Combining Theorems 1 through 4, we have the following procedure for *maximizing* a linear programming problem. To *minimize*, consideration of negative simplex indicators should be replaced by that of *positive* simplex indicators.

  i. Convert the given linear programming problem to the form: Optimize $z = \mathbf{c} \cdot \mathbf{x}$, subject to the constraints

$$A\mathbf{x} = \mathbf{b}, \qquad \mathbf{x} \geq \mathbf{0}, \qquad \mathbf{b} \geq \mathbf{0},$$

  where A has an embedded identity matrix.

  ii. Prepare the first tableau.

  iii. If no simplex indicator is negative, the solution is maximal, and the algorithm terminates.

  iv. If some simplex indicator is negative, and no entry in that column is positive, there is no maximal solution, and the algorithm terminates.

  v. If at least one simplex indicator is negative, and each such column has a positive entry, choose any such column j and pivot on the k, j position, where $b_k/a_{kj} = \min \{b_i/a_{ij} \mid a_{ij} > 0\}$.

  vi. Repeat step (v) until either step (iii) or step (iv) is satisfied.

Because of Theorem 3, we only need to construct $\mathbf{c}^*$ for the first tableau. For each succeeding tableau, the vector of simplex indicators $\mathbf{c}^*A - \mathbf{c}$ is obtained by pivoting. Furthermore, we do not need to worry about which variables are basic and which are nonbasic until the final tableau. For these reasons, tableaux subsequent to the first will not include the columns that indicate the basic variables or $\mathbf{c}^*$.

EXAMPLE 9.9

Use the simplex algorithm to solve the motivating example of Section 9.1. Maximize the function $z = 2x + y$, subject to the constraints

$$2x - y \leq 0$$
$$x + y \leq 9$$
$$x, y \geq 0.$$

Solution

We first convert it to equation form by use of slack variables s and t, as follows:

Maximize the function $z = 2x + y$, subject to the constraints

$$2x - y + s \quad\quad = 0$$
$$x + y \quad\quad + t = 9,$$
$$x, y, s, t \geq 0.$$

We now have the problem in the desired form, and we proceed to prepare the first tableau with $\mathbf{c} = (2, 1, 0, 0)$ and $\mathbf{c}^* = (0, 0)$, since the embedded identity matrix columns correspond to the slack variables s and t:

	$\mathbf{c}^*$	x	y	s	t	$\mathbf{b}$
s	0	2	−1	1	0	0
t	0	1	1	0	1	9
		−2	−1	0	0	0

Here we could choose either column 1 or column 2 as the pivot column, since the simplex indicators are negative for these columns (-2 and -1, respectively). We choose column 1. Both entries in this column are positive, so to determine the pivot row, we need to form both quotients $0/2 = 0$ and $9/1 = 9$. Since $0 < 9$, the pivot row is row 1. Pivoting on the 1, 1 position, we have

1	$-\frac{1}{2}$	$\frac{1}{2}$	0	0
0	$\frac{3}{2}$	$-\frac{1}{2}$	1	9
0	−2	1	0	0

Now the only negative simplex indicator is in the second column, so this will be the pivot column. There is only one positive entry in the second column ($\frac{3}{2} > 0$), so the new pivot row is the second row. Pivoting on the 2, 2 position yields

1	0	$\frac{5}{6}$	$\frac{1}{3}$	3
0	1	$-\frac{1}{3}$	$\frac{2}{3}$	6
0	0	$\frac{1}{3}$	$\frac{4}{3}$	12

Since all of the simplex indicators are nonnegative, the resulting basic solution $x = 3$, $y = 6$, and $s = t = 0$ is maximal, and the value of the objective function at this point is 12. Of course, this is the result that was obtained geometrically in Section 9.1.

NOTE From now on we will proceed directly from one tableau to the next without explanation, only circling the pivot entry at each stage. For example, the simplex algorithm for Example 9.9 would be presented as shown in Figure 9.8.

	$\mathbf{c}^*$	x	y	s	t	$\mathbf{b}$
s	0	②	-1	1	0	0
t	0	1	1	0	1	9
		-2	-1	0	0	0
	1	1	$-\frac{1}{2}$	$\frac{1}{2}$	0	0
	0	0	③⁄②	$-\frac{1}{2}$	1	9
		0	-2	1	0	0
	1	1	0	$\frac{5}{6}$	$\frac{1}{3}$	3
	0	0	1	$-\frac{1}{3}$	$\frac{2}{3}$	6
		0	0	$\frac{1}{3}$	$\frac{4}{3}$	12

FIGURE 9.8

EXAMPLE 9.10

Minimize $z = 3x_1 - x_2 - 2x_3 + x_4$, subject to the constraints

$$2x_1 - 4x_2 - x_3 + x_4 \leq 8$$
$$x_1 + x_2 + 2x_3 - 3x_4 \leq 7$$
$$x_1 - x_2 - x_3 + x_4 \leq 3$$
$$x_j \geq 0, \quad j = 1, \ldots, 4.$$

Solution

We first convert the problem to equation form: Minimize $z = 3x_1 - x_2 - 2x_3 + x_4 + 0x_5 + 0x_6 + 0x_7$, subject to

$$2x_1 - 4x_2 - x_3 + x_4 + x_5 = 8$$
$$x_1 + x_2 + 2x_3 - 3x_4 + x_6 = 7$$
$$x_1 - x_2 - x_3 + x_4 + x_7 = 3$$
$$x_j \geq 0, \quad j = 1, \ldots, 7.$$

We then prepare the first tableau, Figure 9.9, and use it to start the simplex algorithm.

	c^*	x_1	x_2	x_3	x_4	x_5	x_6	x_7	b
x_5	0	2	−4	−1	1	1	0	0	8
x_6	0	1	①	2	−3	0	1	0	7
x_7	0	1	−1	−1	1	0	0	1	3
		−3	1	2	−1	0	0	0	0
		6	0	7	−11	1	4	0	36
		1	1	2	−3	0	1	0	7
		2	0	1	−2	0	1	1	10
		−4	0	0	2	0	−1	0	−7

FIGURE 9.9

The pivot used is circled in the tableau. From the fourth column of the last tableau, we see that no minimal solution exists.

EXAMPLE 9.11   A factory makes three products, A, B, and C. The company has available each week 200 hours of lathe time, 100 hours of milling time, 100 hours of grinding time, and 80 hours of packing time. The time requirements for production of one unit of each product are given in Table 9.2.

	Lathe	Mill	Grind	Pack
A	3	1	2	1
B	1	2	1	2
C	4	2	1	2

TABLE 9.2

If each unit of A yields a profit of \$20, each unit of B yields a profit of \$16, and each unit of C yields a profit of \$24, find a week's production schedule that maximizes profit.

Solution   Let x_1 be the number of units of A to be produced, x_2 the number of B, and x_3 the number of C. Knowing that the total lathe time cannot exceed 200, we have

$$3x_1 + x_2 + 4x_3 \le 200.$$

Similarly, the milling, grinding, and packing limitations yield the constraints

$$x_1 + 2x_2 + 2x_3 \le 100$$
$$2x_1 + x_2 + x_3 \le 100$$
$$x_1 + 2x_2 + 2x_3 \le 80.$$

The profit function is $z = 20x_1 + 16x_2 + 24x_3$, and obviously each variable must be nonnegative. Putting all these conditions together, we have the mathematical problem: Maximize $z = 20x_1 + 16x_2 + 24x_3$, subject to the constraints

$$3x_1 + x_2 + 4x_3 \leq 200$$
$$x_1 + 2x_2 + 2x_3 \leq 100$$
$$2x_1 + x_2 + x_3 \leq 100$$
$$x_1 + 2x_2 + 2x_3 \leq 80$$
$$x_j \geq 0, \qquad j = 1, 2, 3.$$

We restate the problem in equation form: Maximize

$$z = 20x_1 + 16x_2 + 24x_3 + 0x_4 + 0x_5 + 0x_6 + 0x_7,$$

subject to the constraints

$$3x_1 + x_2 + 4x_3 + x_4 \qquad\qquad = 200$$
$$x_1 + 2x_2 + 2x_3 \qquad + x_5 \qquad\qquad = 100$$
$$2x_1 + x_2 + x_3 \qquad\qquad + x_6 \qquad = 100$$
$$x_1 + 2x_2 + 2x_3 \qquad\qquad\qquad + x_7 = 80$$
$$x_j \geq 0, \qquad j = 1, \dots, 7.$$

We prepare the first tableau (Figure 9.10) and use it to start the simplex algorithm. The pivot used at each iteration is circled.

	c^*	x_1	x_2	x_3	x_4	x_5	x_6	x_7	b
x_4	0	3	1	4	1	0	0	0	200
x_5	0	1	2	2	0	1	0	0	100
x_6	0	2	1	1	0	0	1	0	100
x_7	0	1	2	②	0	0	0	1	80
		-20	-16	-24	0	0	0	0	0
		1	-3	0	1	0	0	-2	40
		0	0	0	0	1	0	-1	20
		③⁄₂	0	0	0	0	1	$-\frac{1}{2}$	60
		$\frac{1}{2}$	1	1	0	0	0	$\frac{1}{2}$	40
		-8	8	0	0	0	0	12	960
		0	-3	0	1	0	$-\frac{2}{3}$	$-\frac{5}{3}$	0
		0	0	0	0	1	0	-1	20
		1	0	0	0	0	$\frac{2}{3}$	$-\frac{1}{3}$	40
		0	1	1	0	0	$-\frac{1}{3}$	$\frac{2}{3}$	20
		0	8	0	0	0	$\frac{16}{3}$	$\frac{28}{3}$	1280

FIGURE 9.10

Since no simplex indicator of the last tableau is negative, the solution is maximal. The conclusion is that the company should produce 40 units of A, 20 units of C, and 0 units of B. The associated profit is $1280 per week.

9.2 EXERCISES

In Exercises 1 and 2 complete the first tableau for the given linear programming problem.

1. Objective function: $z = 3x_1 - 2x_2 + 2x_4 - x_5$

	c^*	x_1	x_2	x_3	x_4	x_5	b
x_3	0	3	1	1	0	0	1
x_4	2	-1	2	0	1	0	5
x_5	-1	4	5	0	0	1	9

2. Objective function: $z = 2x_1 + x_2 - 3x_3 + x_4$

c^*	x_1	x_2	x_3	x_4	x_5	x_6	x_7	x_8	b
	2	1	1	0	3	0	-2	0	10
	1	0	-1	0	1	0	1	1	9
	1	0	1	0	2	1	2	0	7
	3	0	2	1	0	0	-1	0	2

In Exercises 3–6 prepare the first tableau for each linear programming problem.

3. Maximize $z = x_1 - 2x_2 + x_3 + 2x_5$, subject to

$$3x_1 + x_2 + 2x_3 \qquad\qquad = 2$$
$$x_1 \qquad - 3x_3 \qquad + x_5 = 5$$
$$2x_1 \qquad - x_3 + x_4 \qquad = 8$$
$$x_j \geq 0, \qquad j = 1, \ldots, 5.$$

4. Minimize $z = 2x_1 - x_2 + x_3$, subject to

$$x_1 + x_2 + x_3 \leq 10$$
$$2x_1 - x_2 - x_3 \leq 5$$
$$x_1 + 3x_2 - 4x_3 \leq 1$$
$$x_j \geq 0, \qquad j = 1, 2, 3.$$

5. Minimize $z = x_1 - 2x_2 + x_3 + 4x_4$, subject to

$$2x_1 + x_2 + x_3 + 3x_4 \leq 12$$
$$x_1 - 3x_2 - x_3 + x_4 \leq 8$$
$$x_j \geq 0, \quad j = 1, \ldots, 4.$$

6. Maximize $z = x_1 - x_2 + 2x_3$, subject to

$$x_1 - 3x_2 - x_3 \leq 6$$
$$x_1 + 3x_2 - 4x_3 \leq 8$$
$$x_1 - x_2 - 3x_3 \leq 10$$
$$x_j \geq 0, \quad j = 1, 2, 3.$$

In Exercises 7–10 explain why each tableau is the final tableau for some linear programming problem. If the solution is optimal, give the optimal solution and corresponding value of the objective function.

7. Maximize:

x_1	x_2	x_3	x_4	x_5	b
-3	1	0	-2	0	2
-1	0	0	3	1	4
-2	0	1	1	0	1
2	0	0	3	0	5

8. Same as Exercise 7, only *minimize*.

9. Minimize:

x_1	x_2	x_3	x_4	x_5	x_6	b
0	1	1	-3	1	0	8
1	2	0	-2	-3	4	11
0	-2	0	-1	-2	-5	-8

10. Same as Exercise 9, only *maximize*.

11–14. Solve completely each of Exercises 3–6.

15–16. Solve Exercises 15 and 16 of Section 9.1.

9.3 THE TWO-PHASE PROBLEM

In this section we show how other types of problems can be put in the form needed for the simplex algorithm.

THE TWO-PHASE PROBLEM

So far we have considered only problems where the constraints are of the type

$$A\mathbf{x} \le \mathbf{b}, \quad \mathbf{b} \ge \mathbf{0}, \quad \text{and} \quad \mathbf{x} \ge \mathbf{0}.$$

What if we have constraints of the type $\mathbf{a}_i \cdot \mathbf{x} \ge b_i$ for $b_i \ge 0$? To handle this situation, we introduce a slack variable, as before, but this time we *subtract* it. That is, $\mathbf{a}_i \cdot \mathbf{x} \ge b_i$ is equivalent to the equation $\mathbf{a}_i \cdot \mathbf{x} - x_{n+1} = b_i$, where $x_{n+1} \ge 0$ (and $\mathbf{x} = (x_1, \ldots, x_n)$). Unfortunately, a problem arises here. We no longer have an embedded identity matrix to start the solution process. The same problem arises if we have equations as well as inequalities in the system.

EXAMPLE 9.12

Convert the following problem to equation form:

$$3x_1 - 2x_2 + 5x_3 \le 14$$
$$x_1 - x_2 - 2x_3 \ge 1$$
$$x_2 - x_3 = 5$$
$$x_j \ge 0, \quad j = 1, 2, 3.$$

Solution

We add a slack variable to the first equation and subtract one from the second to obtain the system:

$$3x_1 - 2x_2 + 5x_3 + x_4 \qquad = 14$$
$$x_1 - x_2 - 2x_3 \qquad - x_5 = 1$$
$$x_2 - x_3 \qquad = 5$$
$$x_j \ge 0, \quad j = 1, \ldots, 5.$$

In Example 9.12 we have only one elementary column (the fourth), and the simplex algorithm cannot be initiated. There is, however, a device that allows us to use the same method. As needed, we add new variables, called **artificial variables,** to create an embedded identity matrix. In this case we add one artificial variable to each of the second and third constraints, to obtain

$$3x_1 - 2x_2 + 5x_3 + x_4 \qquad = 14$$
$$x_1 - x_2 - 2x_3 \qquad - x_5 + x_6 \qquad = 1$$
$$x_2 - x_3 \qquad + x_7 = 5$$
$$x_j \ge 0, \quad j = 1, \ldots, 7.$$

Now we have an embedded identity matrix: the entries of the fourth, sixth, and seventh columns.

Unfortunately, the new system is *not* equivalent to the original. In the last equation, for example, if x_7 is strictly positive, then $x_2 - x_3 < 5$ instead of

$x_2 - x_3 = 5$. If $x_7 = 0$, however, we have $x_2 - x_3 + 0 = 5$, which is equivalent to the original. The idea then is to use the basic solution that has been "rigged" ($x_4 = 14$, $x_6 = 1$, $x_7 = 5$) to start the simplex algorithm, and then try to move to a basic solution in which the artificial variables are *not* basic. To this end, we introduce an objective function that is zero if the artificial variables are all zero and positive otherwise. This objective function is taken to be the sum of the artificial variables. In the case of Example 9.12, this is $z = x_6 + x_7$. We then use the simplex method to *minimize* this function. If the minimum value is zero, we have a solution—in fact a *basic* solution—to the original problem. If the minimum value is positive, the original system can have *no* solutions; that is, the constraints are inconsistent.

EXAMPLE 9.13

Find a basic solution to the following system, or show that none exists.

$$3x_1 - 2x_2 + 5x_3 \le 14$$
$$x_1 - x_2 - 2x_3 \ge 1$$
$$x_2 - x_3 = 5$$
$$x_j \ge 0, \qquad j = 1, 2, 3.$$

Solution

Including the slack variables x_4 and x_5 and artificial variables x_6 and x_7, we minimize the function $z = x_6 + x_7$. The tableaux are given in Figure 9.11.

	c^*	x_1	x_2	x_3	x_4	x_5	x_6	x_7	b
x_4	0	3	-2	5	1	0	0	0	14
x_6	1	①	-1	-2	0	-1	1	0	1
x_7	1	0	1	-1	0	0	0	1	5
	1	0	-3	0	-1	0	0.	6
	0	1	11	1	3	-3	0	11
	1	-1	-2	0	-1	1	0	1
	0	①	-1	0	0	0	1	5
	0	1	-1	0	0	-1	0	5
	0	0	12	1	3	-3	-1	6
	1	0	-3	0	-1	1	1	6
	0	1	-1	0	0	0	1	5
	0	0	0	0	0	-1	-1	0

FIGURE 9.11

From the final tableau of Figure 9.11, we see that x_6 and x_7 are nonbasic variables, so the basic solution that corresponds to the final tableau is a solution to the original problem. That is, $x_4 = 6$, $x_1 = 6$, $x_2 = 5$, and $x_3 = x_5 = 0$ is a basic solution to the original problem. Ignoring variables x_6 and x_7, the final tableau corresponds to the following system of equations,

$$12x_3 + x_4 + 3x_5 = 6$$
$$x_1 \quad - \quad 3x_3 \quad - \quad x_5 = 6 \tag{1}$$
$$x_2 - \quad x_3 \quad = 5.$$

The process just described is called the **first phase** of the linear programming problem: locating a basic feasible solution. The **second phase** is taking the result of the first phase to construct the first tableau with the *original* objective function. From then on, the problem is optimized in the usual manner.

EXAMPLE 9.14

Maximize the function $z = x_1 - 2x_2 + 2x_3$, subject to the constraints

$$3x_1 - 2x_2 + 5x_3 \le 14$$
$$x_1 - \quad x_2 - 2x_3 \ge 1$$
$$x_2 - \quad x_3 = 5$$
$$x_j \ge 0, \qquad j = 1, 2, 3.$$

Solution

Since the constraints are exactly those of Example 9.12, the first phase has already been completed, and we start from there. The first tableau of the second phase is constructed from system (1) of Example 9.13 and the objective function $z = x_1 - 2x_2 + 2x_3 + 0x_4 + 0x_5$. Notice that except for the bottom row, the columns of the first tableau will be exactly those of the final tableau of the first stage, without the columns that correspond to the artificial variables. The tableaux of the second phase are shown in Figure 9.12.

	c^*	x_1	x_2	x_3	x_4	x_5	b
x_4	0	0	0	12	1	③	6
x_1	1	1	0	-3	0	-1	6
x_2	-2	0	1	-1	0	0	5
		0	0	-3	0	-1	-4
		0	0	4	$\frac{1}{3}$	1	2
		1	0	1	$\frac{1}{3}$	0	8
		0	1	-1	0	0	5
		0	0	1	$\frac{1}{3}$	0	-2

FIGURE 9.12

Since no simplex indicators in the final tableau are negative, we have the maximal solution $x_1 = 8$, $x_2 = 5$, $x_5 = 2$, and $x_3 = x_4 = 0$.

We summarize the preceding discussion with the following procedure.

THE TWO-PHASE PROCEDURE

    i. Add artificial variables to create an embedded identity matrix.

    ii. Use the simplex algorithm to minimize this linear programming problem where the objective function is the sum of the artificial variables.

    iii. If the minimum is not zero, there is no solution to the system of constraints.

    iv. If the minimum is zero, use the resulting basic solution, and delete the columns that correspond to artificial variables to create a first tableau with the original objective function.

    v. Use the simplex algorithm to optimize the resulting linear programming problem.

....

NOTE In some kinds of problems, knowing whether or not a set of constraints has *any* solution is the only important question. In this case only the first phase is necessary.

....

EXAMPLE 9.15

Show that there is no solution to the set of constraints

$$4x_1 + 2x_2 - x_3 \leq 5$$
$$2x_1 + x_2 + x_3 \geq 8$$
$$-2x_1 - x_2 + 2x_3 \leq 2$$
$$x_j \geq 0, \quad j = 1, 2, 3.$$

Solution

Adding slack variables to the first and third constraints and subtracting one from the second, we obtain the equivalent set of equations,

$$4x_1 + 2x_2 - x_3 + x_4 \qquad\qquad = 5$$
$$2x_1 + x_2 + x_3 \qquad - x_5 \qquad\quad = 8$$
$$-2x_1 - x_2 + 2x_3 \qquad\qquad + x_6 = 2$$
$$x_j \geq 0, \quad j = 1, \ldots, 6.$$

Adding an artificial variable to the second constraint, we obtain the (nonequivalent) first-phase problem: Minimize $z = x_7$, subject to the constraints

$$4x_1 + 2x_2 - x_3 + x_4 \qquad\qquad\qquad = 5$$
$$2x_1 + x_2 + x_3 \qquad - x_5 \qquad + x_7 = 8$$
$$-2x_1 - x_2 + 2x_3 \qquad\qquad + x_6 \qquad = 2$$
$$x_j \geq 0, \quad j = 1, \ldots, 7.$$

	c^*	x_1	x_2	x_3	x_4	x_5	x_6	x_7	b
x_4	0	④	2	-1	1	0	0	0	5
x_7	1	2	1	1	0	-1	0	1	8
x_6	0	-2	-1	2	0	0	1	0	2
		2	1	1	0	-1	0	0	8
		1	$\frac{1}{2}$	$-\frac{1}{4}$	$\frac{1}{4}$	0	0	0	$\frac{5}{4}$
		0	0	$\frac{3}{2}$	$-\frac{1}{2}$	-1	0	1	$\frac{11}{2}$
		0	0	$\circled{\frac{3}{2}}$	$\frac{1}{2}$	0	1	0	$\frac{9}{2}$
		0	0	$\frac{3}{2}$	$-\frac{1}{2}$	-1	0	0	$\frac{11}{2}$
		1	$\frac{1}{2}$	0	$\frac{1}{3}$	0	$\frac{1}{6}$	0	2
		0	0	0	-1	-1	-1	1	1
		0	0	1	$\frac{1}{3}$	0	$\frac{2}{3}$	0	3
		0	0	0	-1	-1	-1	0	1

FIGURE 9.13

The tableaux for the simplex algorithm are shown in Figure 9.13. Since this solution is minimal, and the artificial variable is still in the corresponding basic solution $(x_1 = 2, x_3 = 3, x_7 = 1)$, there is no solution to the original problem.

9.3 EXERCISES

In Exercises 1–4 convert each problem to the form $A\mathbf{x} = \mathbf{b}, \mathbf{b} \geq \mathbf{0}, \mathbf{x} \geq \mathbf{0}$, where A has an embedded identity matrix, by use of slack and/or artificial variables.

1. $2x - 3y - z \leq 5$
$\quad\quad y + 2z \geq 8$
$\quad x \quad\quad - 2z = 3$
$\quad x, y, z \geq 0$

2. $x_1 + 2x_2 \quad\quad + x_4 \geq 6$
$\quad\quad x_2 + x_3 + 4x_4 \geq 12$
$\quad x_1 \quad\quad - 2x_3 + x_4 \leq 4$
$\quad x_j \geq 0, \quad j = 1, \ldots, 4.$

3. $2x - y = 7$
$\quad x + 2y \leq 18$
$\quad 3x - 5y \leq -3$
$\quad x, y \geq 0$

4. $x_1 + 2x_2 - x_3 \quad\quad = 5$
$\quad\quad 4x_2 - x_3 + x_4 \geq 8$
$\quad\quad x_2 - 3x_3 + x_4 \leq -2$
$\quad x_j \geq 0, \quad j = 1, \ldots, 4$

5–8. Find a basic solution to each of Exercises 1–4, or show that none exists by using the simplex algorithm with objective function given by the sum of the artificial variables.

In Exercises 9–12 use the results of Exercises 5–8 and the simplex algorithm to optimize the given objective function, subject to the constraints imposed in Exercises 1–4.

9. Maximize $z' = x + 3y - z$

10. Minimize $z = x_1 + 2x_3 + x_4$

11. Minimize $z = 2x - 3y$

12. Minimize $z = x_1 + x_2 + x_3 + x_4$

In Exercises 13 and 14 use the two-phase method to solve each linear programming problem.

13. Maximize the function $z = 2x_1 + x_2 + 2x_4$, subject to

$$\begin{align}
x_1 + 2x_2 - x_3 &\geq 5 \\
x_2 + x_3 + x_4 &\leq 12 \\
2x_1 - x_2 - 3x_3 - 2x_4 &\leq -10 \\
x_j \geq 0, \quad j &= 1, \ldots, 4.
\end{align}$$

14. Maximize the function $z = x_1 - x_2 + x_3 - x_4$, subject to

$$\begin{align}
x_2 + 3x_3 + x_4 &\leq 20 \\
x_1 - 3x_2 + x_3 - 2x_4 &\geq 5 \\
x_2 - x_3 + 3x_4 &\geq 6 \\
x_j \geq 0, \quad j &= 1, \ldots, 4.
\end{align}$$

15. *The diet problem:* A company makes dog food from a mixture of four different ingredients (A, B, C, and D) that contain three basic nutrients (P, Q, and R) in various quantities. The chart of Figure 9.14 gives the amount of each nutrient in one pound of each ingredient. The column at the right gives the price per pound (in cents) of each ingredient, and the row at the bottom gives the minimum daily requirement of each nutrient per pound of final product. Find the dog food mix that supplies all of the needs at the cheapest price.

	P	Q	R	
A	3	2	3	40
B	3	1	1	20
C	0	4	4	10
D	1	2	3	20
	2	2	3	

FIGURE 9.14

16. *The transportation problem:* A company has a particular raw material, stored at warehouse W_1, W_2, and W_3, that is required by factories F_1 and F_2. The chart of Figure 9.15 gives the number of miles (in hundreds) between each warehouse and each factory. The column at the right gives the quantity of material available at each warehouse, and the row at the bottom gives the amount needed by each factory. Find a shipping formula that supplies the needs of each factory and minimizes the total shipping distance.

	F_1	F_2	
W_1	4	8	6
W_2	2	1	8
W_3	0	3	5
	7	4	

FIGURE 9.15

9.4 THEORY OF LINEAR PROGRAMMING

We now present the proofs of Theorems 1–4 of Section 9.2, describing the simplex algorithm. We consider only the problem of *maximizing*. The problem of minimizing requires only minor modifications and will therefore be omitted.

THEOREM I   Let a linear programming problem be given as follows: Maximize $z = \mathbf{c} \cdot \mathbf{x}$, subject to the constraints

$$A\mathbf{x} = \mathbf{b}, \qquad \mathbf{b} \geq \mathbf{0}, \qquad \mathbf{x} \geq \mathbf{0},$$

where A has an embedded identity matrix. If no simplex indicator is negative (that is, $\mathbf{c}^*A - \mathbf{c} \geq \mathbf{0}$), then the corresponding basic solution is maximal.

Proof   As usual, assume that A is an $m \times n$ matrix. By renumbering the variables if necessary, we may assume that the embedded identity matrix occurs in the first m columns and in their natural order. $A\mathbf{x} = \mathbf{b}$ has the form

$$
\begin{aligned}
x_1 & & + a_{1,m+1}x_{m+1} + \cdots + a_{1,n}x_n &= b_1 \\
& x_2 & + a_{2,m+1}x_{m+1} + \cdots + a_{2,n}x_n &= b_2 \\
& \ddots & \vdots \qquad\qquad \vdots \qquad\quad \vdots & \\
& x_m & + a_{m,m+1}x_{m+1} + \cdots + a_{m,n}x_n &= b_m.
\end{aligned}
\tag{1}
$$

The corresponding basic solution is obtained by letting $x_i = b_i$ for $i = 1, \ldots, m$ and $x_{m+1} = x_{m+2} = \cdots = x_n = 0$. With the truncated cost vector $\mathbf{c}^* = (c_1, \ldots, c_m)$, the value of the current basic solution is

$$
\begin{aligned}
z &= c_1b_1 + c_2b_2 + \cdots + c_mb_m + c_{m+1}(0) + \cdots + c_n(0) \\
&= c_1b_1 + c_2b_2 + \cdots + c_mb_m = \mathbf{c}^*\mathbf{b}.
\end{aligned}
$$

From the theory of linear equations of Chapter 2, we have the *general* solution to the system of linear equations (1) as a parameterized family by letting

$x_{m+1} = s_1, x_{m+2} = s_2, \ldots, x_n = s_{n-m}$, and then

$$
\begin{aligned}
x_1 &= b_1 - a_{1,m+1}s_1 - \cdots - a_{1,n}s_{n-m} \\
x_2 &= b_2 - a_{2,m+1}s_1 - \cdots - a_{2,n}s_{n-m} \\
&\ \vdots \qquad \vdots \qquad \vdots \qquad\qquad \vdots \\
x_m &= b_m - a_{m,m+1}s_1 - \cdots - a_{m,n}s_{n-m}. \\
x_{m+1} &= \qquad\qquad s_1 \\
&\ \vdots \\
x_n &= \qquad\qquad\qquad\qquad\qquad s_{n-m}.
\end{aligned}
\tag{2}
$$

The value associated with this general solution $\mathbf{x}$ is given by

$$
\begin{aligned}
z = \mathbf{c} \cdot \mathbf{x} &= c_1 x_1 + c_2 x_2 + \cdots + c_n x_n \\
&= \quad c_1 b_1 - c_1 a_{1,m+1}s_1 - \cdots - c_1 a_{1,n}s_{n-m} \\
&\quad + c_2 b_2 - c_2 a_{2,m+1}s_1 - \cdots - c_2 a_{2,n}s_{n-m} \\
&\qquad \vdots \qquad\qquad \vdots \qquad\qquad\qquad \vdots \\
&\quad + c_m b_m - c_m a_{m,m+1}s_1 - \cdots - c_m a_{m,n}s_{n-m} \\
&\qquad\qquad\quad + c_{m+1}s_1 + \cdots + \qquad c_n s_{n-m}.
\end{aligned}
$$

Simplifying by columns, we have

$$
z = \mathbf{c}^* \mathbf{b} - (\mathbf{c}^* \mathbf{a}^{m+1} - c_{m+1})s_1 - (\mathbf{c}^* \mathbf{a}^{m+2} - c_{m+2})s_2 - \cdots - (\mathbf{c}^* \mathbf{a}^n - c_n)s_{n-m}.
$$

Since $\mathbf{c}^* \mathbf{a}^{m+j} - c_{m+j} \geq 0$ and $s_j \geq 0$ for each $j = 1, \ldots, n - m$, the value of z is maximal when $s_1 = s_2 = \cdots = s_{n-m} = 0$. Thus the corresponding basic solution is maximal.

An important special case of the final expression for z is the following.

COROLLARY Let $A\mathbf{x} = \mathbf{b}$, $\mathbf{b} \geq \mathbf{0}$, $\mathbf{x} \geq \mathbf{0}$, and $z = \mathbf{c} \cdot \mathbf{x}$ be as in Theorem 1. Let j be the number of a nonbasic variable, and let $\mathbf{x}$ be the solution obtained from the basic solution by letting $x_j = s$ and the other nonbasic variables be zero. Then the value of the solution $\mathbf{x}$ is given by

$$
z = \mathbf{c}^* \mathbf{b} - (\mathbf{c}^* \mathbf{a}^j - c_j)s.
$$

Thus we see that if $\mathbf{c}^* \mathbf{a}^j - c_j < 0$, a larger value of z results in the new solution (if $s > 0$). That is, if the simplex indicator is negative, the new solution will be better for the purpose of maximizing. In fact, the equation $z = \mathbf{c}^* \mathbf{b} - (\mathbf{c}^* \mathbf{a}^j - c_j)s$ expresses the objective function value z as a *linear* function of s. In other words, $-(\mathbf{c}^*\mathbf{a}^j - c_j)$ measures the *rate of increase* (or *decrease* if $\mathbf{c}^*\mathbf{a}^j - c_j$ is positive) in the value of z as a function of s. If there are no restrictions on the size of s, the value of z is unbounded, and no maximum occurs. This is the situation of Theorem 2. If s is restricted, we make s as large as the restriction will allow. This yields a new basic solution—the same situation as Theorem 4. We restate both theorems and prove them simultaneously.

Let a linear programming problem be given as follows: Optimize $z = \mathbf{c} \cdot \mathbf{x}$, subject to the constraints

$$A\mathbf{x} = \mathbf{b}, \qquad \mathbf{b} \geq \mathbf{0}, \qquad \mathbf{x} \geq \mathbf{0},$$

where A has an embedded identity matrix. Let j be a column number such that the simplex indicator is negative (that is, $\mathbf{c}^*\mathbf{a}^j - c_j < 0$).

THEOREM 2   If no element of $\mathbf{a}^j$ is positive $(\mathbf{a}^j \leq \mathbf{0})$, then no maximal solution exists.

THEOREM 4   If a_{ij} is positive for at least one row number $i = 1, \ldots, m$, let k be a row number such that $a_{kj} > 0$ and $b_k/a_{kj} \leq b_i/a_{ij}$ for all $i = 1, \ldots, m$ such that $a_{ij} > 0$. Then pivoting on the k, j position leads to a new basic solution with value greater than or equal to the former.

Proof   From System (2), the solution $\mathbf{x}$ obtained from the basic solution $x_1 = b_1, \ldots, x_m = b_m, x_{m+1} = \cdots = x_n = 0$, by letting nonbasic variable $x_j = s$, is given by

$$
\begin{aligned}
x_1 &= b_1 - a_{1j}s \\
x_2 &= b_2 - a_{2j}s \\
&\vdots \quad \vdots \quad \vdots \\
x_m &= b_m - a_{mj}s \\
x_j &= \qquad s \\
x_i &= 0 \qquad \text{if } i \neq 1, \ldots, m \qquad \text{and} \qquad i \neq j.
\end{aligned}
\tag{3}
$$

From this we see that $\mathbf{x}$ is a solution only if $s \geq 0$ (since $x_j = s$), and, for $i = 1, \ldots, m, b_i - a_{ij}s \geq 0$. If $a_{ij} \leq 0$, then $b_i - a_{ij}s \geq b_i \geq 0$, so this condition is satisfied for any value of $s \geq 0$. If $\mathbf{a}^j \leq \mathbf{0}$ (that is, if *every* $a_{ij} \leq 0$), we have no constraints on s (except $s \geq 0$), and by the corollary to Theorem 1, Theorem 2 is proved. If, however, $a_{ij} > 0$ for some i, the condition $b_i - a_{ij}s \geq 0$ implies that $s \leq b_i/a_{ij}$. Thus s must be chosen less than or equal to b_i/a_{ij} for each $i = 1, \ldots, m$ such that $a_{ij} > 0$. By choosing $s = \min\{b_i/a_{ij} \mid a_{ij} > 0\}$, we obtain a new solution. Theorem 4 will be proved when it is shown that this new solution is the result of pivoting on the k, j position, where $s = b_k/a_{kj}$. To this end, we put System (1) in augmented matrix form (emphasizing the pivot column j and pivot row k):

$$\begin{bmatrix} 1 & & & \cdots & a_{1j} & \cdots & b_1 \\ & \ddots & & & \vdots & & \vdots \\ & & 1 & \cdots & \boxed{a_{kj}} & \cdots & b_k \\ & & & \ddots & \vdots & & \vdots \\ & & 1 & \cdots & a_{mj} & \cdots & b_m \end{bmatrix}. \qquad (4)$$

Pivoting on the k, j position, we obtain the new matrix

$$\begin{bmatrix} 1 & -a_{1j}/a_{kj} & & \cdots & 0 & \cdots & b_1 - a_{1j}(b_k/a_{kj}) \\ & \ddots & \vdots & & \vdots & & \vdots \\ & & 1/a_{kj} & \cdots & 1 & \cdots & b_k/a_{kj} \\ & & \vdots & \ddots & \vdots & & \vdots \\ & & -a_{mj}/a_{kj} & 1 & \cdots & 0 & \cdots & b_m - a_{mj}(b_k/a_{kj}) \end{bmatrix}. \qquad (5)$$

Since $s = b_k/a_{kj}$, the corresponding basic solution is that of System (3), as desired.

The final theorem needed for the simplex algorithm states that a new bottom row may be computed by continuing row operations to the bottom row.

THEOREM 3 The bottom row of a new tableau may be computed from the preceding tableau by using the pivot row and row operations to obtain a zero in the pivot column position of the bottom row.

Proof Again we assume that the current basic solution is in terms of the first m variables and in their natural order. Then matrix (4) of the preceding proof is the "heart" of the simplex tableau for the current solution, and matrix (5) is that of the new tableau. The bottom row of the first tableau is given by $\mathbf{c}^*A - \mathbf{c}$ for the first n entries and $\mathbf{c}^*\mathbf{b}$ for the last, where $\mathbf{c}^* = (c_1, \ldots, c_m)$. Then the entry in the jth (the pivot column) position is $\mathbf{c}^*\mathbf{a}^j - c_j$. To obtain a zero in that position by use of the pivot row $\mathbf{a}_k$, we must add $-[(\mathbf{c}^*\mathbf{a}^j - c_j)/a_{kj}]\mathbf{a}_k$ to the bottom row. The result is that the first n positions in the bottom row will be given by

$$(\mathbf{c}^*A - \mathbf{c}) - [(\mathbf{c}^*\mathbf{a}^j - c_j)/a_{kj}]\mathbf{a}_k,$$

and the last position by

$$\mathbf{c}^*\mathbf{b} - [(\mathbf{c}^*\mathbf{a}^j - c_j)/a_{kj}]b_k.$$

To complete the proof, we must show that this is also the bottom row if we construct the bottom row from matrix (5), $\mathbf{c}^{*\prime}A' - \mathbf{c}$, where $\mathbf{c}^{*\prime}$ is the $1 \times m$ matrix of components of $\mathbf{c}$ that corresponds to basic variables in the new solution and A' is the new tableau. In the new solution, x_j has replaced x_k, so $\mathbf{c}^{*\prime} = (c_1, \ldots, c_{k-1}, c_j, c_{k+1}, \ldots, c_m)$. The matrix A' from matrix (5) is given by its rows, as

$$\mathbf{a}_i' = \mathbf{a}_i - \frac{a_{ij}}{a_{kj}}\mathbf{a}_k \qquad \text{for } i \neq k$$

and

$$\mathbf{a}_k' = \frac{1}{a_{kj}}\mathbf{a}_k.$$

We can then compute $\mathbf{c}^{*\prime}A' - \mathbf{c}$ by rows as follows:

$$\mathbf{c}^{*\prime}A' - \mathbf{c} = c_1\left(\mathbf{a}_1 - \frac{a_{1j}}{a_{kj}}\mathbf{a}_k\right) + \cdots + c_j\left(\frac{1}{a_{kj}}\mathbf{a}_k\right)$$

$$+ \cdots + c_m\left(\mathbf{a}_m - \frac{a_{mj}}{a_{kj}}\mathbf{a}_k\right) - \mathbf{c}$$

$$= (c_1\mathbf{a}_1 + \cdots + c_m\mathbf{a}_m - c_k\mathbf{a}_k)$$

$$- \left(\frac{c_1 a_{1j}}{a_{kj}} + \cdots - \frac{c_j}{a_{kj}} + \cdots + \frac{c_m a_{mj}}{a_{kj}}\right)\mathbf{a}_k - \mathbf{c}$$

$$= \mathbf{c}^*A - [(c_1 a_{1j} + \cdots + c_m a_{mj} - c_j)/a_{kj}]\mathbf{a}_k - \mathbf{c}$$

$$= \mathbf{c}^*A - [(\mathbf{c}^*\mathbf{a}^j - c_j)/a_{kj}]\mathbf{a}_k - \mathbf{c}$$

$$= (\mathbf{c}^*A - \mathbf{c}) - [(\mathbf{c}^*\mathbf{a}^j - c_j)/a_{kj}]\mathbf{a}_k.$$

Thus the first n positions in the bottom row are correct, and we need only check the last position. This entry is just the value of the new basic solution, and by the corollary to Theorem 1 with $s = b_k/a_{kj}$, this position must be

$$\mathbf{c}^*\mathbf{b} - (\mathbf{c}^*\mathbf{a}^j - c_j)(b_k/a_{kj}).$$

Since this agrees with the row operation result, the proof is complete.

One natural question that remains is whether or not the process must terminate. In other words, if we are maximizing the objective function, are we guaranteed that eventually some tableau must satisfy the condition of Theorem 1 or Theorem 2? The answer is, "Not quite, but close enough." Although examples have been constructed that *cycle* (in other words, repeat the same sequence of basic solutions ad infinitum without locating the optimal value), none has arisen by accident. In practical terms, it is not necessary to take such situations into account. In one important case (the most common situation), it is easy to prove that the process must terminate, as the next theorem shows.

THEOREM 5   If in each tableau of the simplex algorithm the basic solution $\mathbf{b}$ is strictly positive, then the process terminates in a finite number of steps.

Proof   Assume the problem is to maximize the objective function. Then at any stage in the algorithm, we choose a column j for which $\mathbf{c}^*\mathbf{a}^j - c_j < 0$ and let $s = \min \{b_i/a_{ij} \mid a_{ij} > 0\}$. Since $\mathbf{b} > \mathbf{0}$, each $b_i > 0$, and thus $s > 0$. By the corollary to Theorem 1, the value of the next solution is $\mathbf{c}^*\mathbf{b} - (\mathbf{c}^*\mathbf{a}^j - c_j)s$, which is strictly greater than the value of the current solution $\mathbf{c}^*\mathbf{b}$. That is, the value associated with each basic solution *strictly increases* with each iteration of the method. Thus no solution can be repeated, or else the value would drop back to its former level—a contradiction. The proof is then complete if we know that there are only finitely many basic solutions. This follows from the fact that any basic solution arises as a transformation of the original system $A\mathbf{x} = \mathbf{b}$ into a new row-equivalent one with elementary columns that correspond to the basic solution variables. Since A has only finitely many columns, and since any such row-reduced form is unique, there are only finitely many such matrices. Thus there are only finitely many basic solutions.

9.4 EXERCISES

1. Each of the theorems in this section was stated as a maximizing problem. Prove the analogous results for minimizing problems by proving the following statement. A solution $\mathbf{x}_0$ maximizes a linear function $z = \mathbf{c} \cdot \mathbf{x}$, subject to some set of constraints, if and only if it minimizes the linear function $z' = (-\mathbf{c}) \cdot \mathbf{x}$, subject to the same set of constraints.

2. At some stage in the operation of the simplex algorithm, there may be several simplex indicators that will improve the value of the objective function if chosen for the pivot column. Show that the best choice is the simplex indicator

$$\mathbf{c}^*\mathbf{a}^j - c_j \qquad \text{such that} \qquad |(\mathbf{c}^*\mathbf{a}^j - c_j)q_j|$$

is greatest where

$$q_j = \min \{b_i/a_{ij} \mid a_{ij} > 0\}.$$

3. Theorem 5 guarantees that if $\mathbf{b} > \mathbf{0}$ at each stage, the simplex algorithm will terminate in a finite number of iterations. If $\mathbf{b}$ has some zero entries, this need not be the case. Demonstrate this fact by showing that the following problem cycles if the simplex indicator of proper sign and maximal absolute value are always chosen. If a choice between pivot rows occurs, always choose the first one. (The example is due to Beale.)
   Minimize the linear function

$$z = -0.75x_1 + 150x_2 - 0.02x_3 + 6x_4,$$

subject to the constraints

$$0.25x_1 - 60x_2 - 0.04x_3 + 9x_4 \leq 0$$
$$0.50x_1 - 90x_2 - 0.02x_3 + 3x_4 \leq 0$$
$$x_3 \qquad \leq 1$$
$$x_j \geq 0, \qquad j = 1, \ldots, 4.$$

4. Find the optimal solution to the problem in Exercise 3 by row-reducing the equational form and obtaining a matrix with embedded identity matrix in columns 1, 3, and 4. Form the corresponding tableau and proceed in the usual manner.

APPENDIX: COMPUTER PROGRAMS

The interactive programs in this appendix can be used to reduce or even eliminate the computations inherent in various linear algebra problems. Diskettes containing these programs are available directly from the publisher for several popular microcomputers, including the IBM PC, Apple II, Apple Macintosh, Commodore 64, and Atari 800. If you are using a time-sharing system, they can be entered into the system from the listings given in this appendix. The programs themselves require no computer experience and are easy to use.

The programs given here have one distinct advantage over many similar software packages. They allow coefficients that are fractions to be entered as quotients of integers rather than as decimal equivalents. Moreover, answers are given in the same form, provided the denominators are not too large. That is, if $x = 1/7$, that is the way the answer will be given rather than $x = 0.142857$. Also included are sample runs of some of the programs, showing how they operate. If the response to some inquiry (?) is omitted, it means that the program defaults to the option indicated. For example, to enter ordinary fractions you must enter YES when asked, but to enter integer or decimal coefficients, a simple $\langle$RETURN$\rangle$ is sufficient.

Note that these programs are designed for instructional purposes rather than for efficiency. For example, they are written in BASIC, which, despite its lack of speed and elegance, is universally available and easy to understand. As another example, the programs assume that a coefficient smaller than 0.00001 is zero. This is no limitation for the kinds of problems contained in this book, although it would be disastrous if a problem were carefully written to demonstrate the shortcoming.

CROSS PRODUCT OF TWO VECTORS

This program computes the cross product of two vectors in $\mathbf{R}^2$.

```
1000 DIM B(2,3),C(3),A$(3)
1010 DIM E(2,3),F(2,3)
1020 PRINT " "
1022 PRINT "CROSS PRODUCT OF TWO VECTORS"
1024 PRINT " "
1060 PRINT "DO YOU WANT TO ENTER FRACTIONS?  YES OR NO"
1070 INPUT A$
1080 PRINT " "
1090 IF A$="YES" THEN GO TO 1120
1100 PRINT "ENTER THE VECTORS BY ROWS (RETURN BETWEEN ENTRIES)"
1110 GO TO 1130
1120 PRINT "ENTER THE NUMERATORS BY ROWS"
1130 FOR I=1 TO 2
1140 PRINT " "
1150 PRINT "VECTOR";I
1160 FOR J=1 TO 3
1170 INPUT B
1172 B(I,J)=B
1180 LET F(I,J)=1
1190 NEXT J
1200 NEXT I
1210 PRINT " "
1220 IF A$<>"YES" THEN GO TO 1310
1230 PRINT "ENTER DENOMINATORS BY ROWS"
1240 FOR I=1 TO 2
1250 PRINT " "
1260 PRINT "VECTOR ";I
1270 FOR J=1 TO 3
1280 INPUT F
1282 F(I,J)=F
1290 NEXT J
1300 NEXT I
1310 PRINT " "
1320 PRINT "THE VECTORS ARE:"
1330 FOR I=1 TO 2
1340 FOR J=1 TO 3
1350 PRINT B(I,J);
1360 IF F(I,J)<>1 THEN GO TO 1390
1370 PRINT ,
1380 GO TO 1400
1390 PRINT "/";F(I,J),
1400 NEXT J
1410 PRINT " "
1420 NEXT I
1430 FOR I=1 TO 2
1440 FOR J=1 TO 3
1450 IF ABS(F(I,J))>1.0E-05 THEN GO TO 1480
1460 PRINT "INVALID DENOMINATOR"
1470 GO TO 1210
1480 LET B(I,J)=B(I,J)/F(I,J)
1490 NEXT J
1500 NEXT I
1510 LET C(1)=B(1,2)*B(2,3)-B(1,3)*B(2,2)
1520 LET C(2)=B(1,3)*B(2,1)-B(1,1)*B(2,3)
1530 LET C(3)=B(1,1)*B(2,2)-B(1,2)*B(2,1)
1790 FOR J=1 TO 3
1800 LET K=1
1810 LET G=ABS(C(J))
1820 IF G*K-INT(G*K)<1.0E-05 THEN GO TO 1870
1830 IF G*K-INT(G*K)>0.99999 THEN GO TO 1870
1840 LET K=K+1
1850 IF K>1000 THEN GO TO 1910
1860 GO TO 1820
1870 LET E(1,J)=INT(C(J)*K+1.0E-05)
1880 LET F(1,J)=K
1890 NEXT J
1910 PRINT " "
1920 PRINT "THE CROSS PRODUCT IS:"
1930 IF K<=1000 THEN GO TO 2020
1950 FOR J=1 TO 3
```

```
1960 PRINT C(J),
1970 NEXT J
2000 GO TO 2110
2020 FOR J=1 TO 3
2030 PRINT E(1,J);
2040 IF F(1,J)<>1 THEN GO TO 2070
2050 PRINT ,
2060 GO TO 2080
2070 PRINT "/";F(1,J),
2080 NEXT J
2110 PRINT " "
2120 PRINT "RUN PROGRAM AGAIN? YES OR NO"
2130 INPUT A$
2140 IF A$="YES" THEN GOTO 1020
2160 END
```

```
CROSS PRODUCT OF TWO VECTORS

DO YOU WANT TO ENTER FRACTIONS?  YES OR NO
? no

ENTER THE VECTORS BY ROWS (RETURN BETWEEN ENTRIES)

VECTOR 1
? 2
? 3
? 1

VECTOR 2
? 2
? -1
? 4


THE VECTORS ARE:
  2             3                 1
  2            -1                 4

THE CROSS PRODUCT IS:
 13            -6                -8
RUN PROGRAM AGAIN? YES OR NO
? no
```

ROW-REDUCTION TO ROW-REDUCED ECHELON FORM

This program is by far the most important, and there is pedagogical justification for allowing only this one throughout the course. Except for matrix arithmetic problems, nearly every question that requires computation can be completely answered or substantially simplified by row-reducing an appropriate matrix.

```
1000 DIM B(10,15),C(15),A$(3)
1010 DIM E(10,15),F(10,15)
1020 PRINT " "
1030 PRINT "ENTER THE DIMENSIONS OF THE MATRIX AS M,N"
1040 INPUT M,N
1050 PRINT " "
1060 PRINT "DO YOU WANT TO ENTER FRACTIONS?  YES OR NO"
1070 INPUT A$
1080 PRINT " "
```

```
1090 IF A$="YES" THEN GO TO 1120
1100 PRINT "ENTER THE MATRIX BY ROWS (RETURN BETWEEN ENTRIES)"
1110 GO TO 1130
1120 PRINT "ENTER MATRIX OF NUMERATORS BY ROWS"
1130 FOR I=1 TO M
1140 PRINT " "
1150 PRINT "ROW ";I
1160 FOR J=1 TO N
1170 INPUT B
1180 LET  B(I,J)=B
1190 LET F(I,J)=1
1200 NEXT J
1210 NEXT I
1220 PRINT " "
1230 IF A$<>"YES" THEN GO TO 1330
1240 PRINT "ENTER MATRIX OF DENOMINATORS BY ROWS"
1250 FOR I=1 TO M
1260 PRINT " "
1270 PRINT "ROW ";I
1280 FOR J=1 TO N
1290 INPUT F
1300 LET  F(I,J)=F
1310 NEXT J
1320 NEXT I
1330 PRINT " "
1340 PRINT "THE MATRIX IS:"
1350 FOR I=1 TO M
1360 FOR J=1 TO N
1370 PRINT B(I,J);
1380 IF F(I,J)<>1 THEN GO TO 1410
1390 PRINT ,
1400 GO TO 1420
1410 PRINT "/";F(I,J),
1420 NEXT J
1430 PRINT " "
1440 NEXT I
1450 FOR I=1 TO M
1460 FOR J=1 TO N
1470 IF ABS(F(I,J))>.00001 THEN GO TO 1500
1480 PRINT "INVALID DENOMINATOR
1490 GO TO 1220
1500 LET B(I,J)=B(I,J)/F(I,J)
1510 NEXT J
1520 NEXT I
1530 LET J=1
1540 FOR I=1 TO M
1550 LET K=I-1
1560 LET K=K+1
1570 IF ABS(B(K,J))>.00001 THEN GO TO 1620
1580 IF K < M THEN GO TO 1560
1590 LET J=J+1
1600 IF J>N THEN GO TO 1800
1610 GO TO 1550
1620 LET D=B(K,J)
1630 FOR J1=J TO N
1640 LET C(J1)=B(I,J1)
1650 LET B(I,J1)=B(K,J1)
```

```
1660 LET B(K,J1)=C(J1)
1670 LET B(I,J1)=B(I,J1)/D
1680 NEXT J1
1690 FOR I1=1 TO M
1700 IF I1=I THEN GO TO 1770
1710 LET D=B(I1,J)
1720 FOR J1=J TO N
1730 LET B(I1,J1)=B(I1,J1)-B(I,J1)*D
1740 IF ABS(B(I1,J1))>.00001 THEN GO TO 1760
1750 LET B(I1,J1)=0
1760 NEXT J1
1770 NEXT I1
1780 LET J=J+1
1790 NEXT I
1800 FOR I=1 TO M
1810 FOR J=1 TO N
1820 LET K=1
1830 LET G=ABS(B(I,J))
1840 IF G*K-INT(G*K)<.00001 THEN GO TO 1890
1850 IF G*K-INT(G*K)>.99999 THEN GO TO 1890
1860 LET K=K+1
1870 IF K>1000 THEN GO TO 1930
1880 GO TO 1840
1890 LET E(I,J)=INT(B(I,J)*K+.00001)
1900 LET F(I,J)=K
1910 NEXT J
1920 NEXT I
1930 PRINT " "
1940 PRINT "THE ROW-REDUCED ECHELON FORM IS:"
1950 IF K<=1000 THEN GO TO 2030
1960 FOR I=1 TO M
1970 FOR J=1 TO N
1980 PRINT B(I,J),
1990 NEXT J
2000 PRINT " "
2010 NEXT I
2020 GO TO 2130
2030 FOR I=1 TO M
2040 FOR J=1 TO N
2050 PRINT E(I,J);
2060 IF F(I,J)<>1 THEN GO TO 2090
2070 PRINT ,
2080 GO TO 2100
2090 PRINT "/";F(I,J),
2100 NEXT J
2110 PRINT " "
2120 NEXT I
2130 PRINT " "
2140 END


ENTER THE DIMENSIONS OF THE MATRIX AS M,N
? 2,4


DO YOU WANT TO ENTER FRACTIONS?   YES OR NO
? NO
```

```
ENTER THE MATRIX BY ROWS (RETURN BETWEEN ENTRIES)

ROW  1
? 2
? 3
? 1
? 2

ROW  2
? 3
? 1
? 0
? -1


THE MATRIX IS:
  2              3              1              2
  3              1              0             -1

THE ROW-REDUCED ECHELON FORM IS:
  1              0             -1 / 7         -5 / 7
  0              1              3 / 7          8 / 7
```

MATRIX MULTIPLICATION

This program multiplies two matrices (if the multiplication is defined) and allows a new matrix to be entered for products of more than two, once the first operation is completed.

```
1000 DIM A(15,15),B(15,15),A$(3),F$(2),G$(3)
1010 DIM C(15,15),E(15,15),F(15,15)
1020 PRINT " "
1030 PRINT "GIVE THE DIMENSIONS OF A AS M,N"
1040 INPUT M,N
1050 PRINT "GIVE THE DIMENSIONS OF B AS M,N"
1060 INPUT M1,N1
1070 IF N=M1 THEN GO TO 1110
1080 PRINT " "
1090 PRINT "THE PRODUCT IS NOT DEFINED"
1100 GO TO 1020
1110 PRINT "DO YOU WANT TO ENTER FRACTIONS?   YES OR NO"
1120 INPUT A$
1130 IF A$="YES" THEN GO TO 1160
1140 LET F$="N"
1150 GO TO 1210
1160 PRINT " "
1170 PRINT "IF ONLY  A  HAS FRACTIONS, TYPE  A , IF ONLY  B  , TYPE  B"
1180 PRINT "IF BOTH HAVE FRACTIONS, TYPE  AB ,  IF NEITHER, TYPE  N"
1190 INPUT F$
1200 IF F$="A" OR F$="AB" THEN GO TO 1240
1210 PRINT " "
1220 PRINT "ENTER A BY ROWS"
1230 GO TO 1260
```

```
1240 PRINT " "
1250 PRINT "ENTER MATRIX OF NUMERATORS OF A BY ROWS"
1260 FOR I=1 TO M
1270 PRINT " "
1280 PRINT "ROW ";I
1290 FOR J=1 TO N
1300 INPUT A
1310 LET A(I,J)=A
1320 LET C(I,J)=1
1330 NEXT J
1340 NEXT I
1350 IF F$="B" OR F$="N" THEN GO TO 1460
1360 PRINT " "
1370 PRINT "ENTER MATRIX OF DENOMINATORS OF A BY ROWS"
1380 FOR I=1 TO M
1390 PRINT " "
1400 PRINT "ROW ";I
1410 FOR J=1 TO N
1420 INPUT C
1430 LET C(I,J)=C
1440 NEXT J
1450 NEXT I
1460 PRINT " "
1470 PRINT "THE MATRIX  A  IS:"
1480 FOR I=1 TO M
1490 FOR J=1 TO N
1500 PRINT A(I,J);
1510 IF C(I,J)<>1 THEN GO TO 1540
1520 PRINT ,
1530 GO TO 1550
1540 PRINT "/";C(I,J),
1550 NEXT J
1560 PRINT " "
1570 NEXT I
1580 FOR I=1 TO M
1590 FOR J=1 TO N
1600 IF ABS(C(I,J))>.00001  THEN GO TO 1630
1610 PRINT "INVALID DENOMINATOR"
1620 GO TO 1360
1630 LET A(I,J)=A(I,J)/C(I,J)
1640 NEXT J
1650 NEXT I
1660 IF F$="B" OR F$="AB" THEN GO TO 1700
1670 PRINT " "
1680 PRINT "ENTER B BY ROWS"
1690 GO TO 1720
1700 PRINT" "
1710 PRINT "ENTER MATRIX OF NUMERATORS OF  B BY ROWS"
1720 FOR I=1 TO M1
1730 PRINT " "
1740 PRINT "ROW ";I
1750 FOR J=1 TO N1
1760 INPUT B
1770 LET B(I,J)=B
1780 LET C(I,J)=1
1790 NEXT J
1800 NEXT I
```

```
1810 IF F$="A" OR F$="N" THEN GO TO 1920
1820 PRINT " "
1830 PRINT "ENTER MATRIX OF DENOMINATORS OF  B BY ROWS"
1840 FOR I=1 TO M1
1850 PRINT " "
1860 PRINT "ROW ";I
1870 FOR J=1 TO N1
1880 INPUT C
1890 LET C(I,J)=C
1900 NEXT J
1910 NEXT I
1920 PRINT " "
1930 PRINT "THE MATRIX  B   IS:"
1940 FOR I=1 TO M1
1950 FOR J=1 TO N1
1960 PRINT B(I,J);
1970 IF C(I,J)<>1 THEN GO TO 2000
1980 PRINT ,
1990 GO TO 2010
2000 PRINT "/";C(I,J),
2010 NEXT J
2020 PRINT " "
2030 NEXT I
2040 FOR I=1 TO M1
2050 FOR J=1 TO N1
2060 IF ABS(C(I,J))>.00001 THEN GO TO 2090
2070 PRINT "INVALID DENOMINATOR"
2080 GO TO 1820
2090 LET B(I,J)=B(I,J)/C(I,J)
2100 NEXT J
2110 NEXT I
2120 FOR I=1TO M
2130 FOR J=1 TO N1
2140 LET C(I,J)=0
2150 FOR K=1 TO N
2160 LET C(I,J)=C(I,J)+A(I,K)*B(K,J)
2170 NEXT K
2180 NEXT J
2190 NEXT I
2200 FOR I=1 TO M
2210 FOR J=1 TO N1
2220 LET A(I,J)=C(I,J)
2230 NEXT J
2240 NEXT I
2250 FOR I=1 TO M
2260 FOR J=1 TO N1
2270 LET K=1
2280 LET G =ABS(C(I,J))
2290 IF G*K-INT(G*K)<.00001 THEN GO TO 2350
2300 IF G*K-INT(G*K)>.99999 THEN GO TO 2350
2310 LET K=K+1
2320 IF K>1000 THEN GO TO 2390
2330 LET T=2
2340 GO TO 2290
2350 LET E(I,J)=INT(C(I,J)*K+.00001)
```

```
2360 LET F(I,J)=K
2370 NEXT J
2380 NEXT I
2390 PRINT " "
2400 PRINT "THE PRODUCT IS:"
2410 IF K<=1000 THEN GO TO 2490
2420 FOR I=1 TO M
2430 FOR J=1 TO N1
2440 PRINT C(I,J),
2450 NEXT J
2460 PRINT " "
2470 NEXT I
2480 GO TO 2590
2490 FOR I=1 TO M
2500 FOR J=1 TO N1
2510 PRINT E(I,J);
2520 IF F(I,J)<>1 THEN GO TO 2550
2530 PRINT ,
2540 GO TO 2560
2550 PRINT "/";F(I,J),
2560 NEXT J
2570 PRINT " "
2580 NEXT I
2590 PRINT " "
2600 PRINT "DO YOU WANT A FURTHER PRODUCT?  YES OR NO"
2610 INPUT G$
2620 IF G$="YES" THEN GO TO 2640
2630 GO TO 2810
2640 PRINT " "
2650 PRINT "GIVE THE DIMENSIONS OF THE NEW  B  MATRIX AS M,N"
2660 INPUT M2,N2
2670 IF N1=M2 THEN GO TO 2710
2680 PRINT " "
2690 PRINT "THE PRODUCT IS NOT DEFINED"
2700 GO TO 2640
2710 LET N=N1
2720 LET M1=M2
2730 LET N1=N2
2740 PRINT " "
2750 PRINT "DO YOU WANT TO ENTER FRACTIONS?  YES OR NO"
2760 INPUT A$
2770 LET F$="N"
2780 IF A$<>"YES" THEN GO TO 2800
2790 LET F$="B"
2800 GO TO 1660
2810 END


GIVE THE DIMENSIONS OF A AS M,N
? 3,3
GIVE THE DIMENSIONS OF B AS M,N
? 3,3
DO YOU WANT TO ENTER FRACTIONS?  YES OR NO
? NO
```

```
ENTER A BY ROWS


THE MATRIX  A  IS:
 1              1              -1
 1             -1              1
-1              1              1


ENTER B BY ROWS


THE MATRIX  B  IS:
 1              0              0
 1              1              1
 2              1              1


THE PRODUCT IS:
 0              0              0
 2              0              0
 2              2              2


DO YOU WANT A FURTHER PRODUCT?   YES OR NO
? YES

GIVE THE DIMENSIONS OF THE NEW  B  MATRIX AS M,N
? 3,3

DO YOU WANT TO ENTER FRACTIONS?   YES OR NO
?

ENTER B BY ROWS


THE MATRIX  B  IS:
 1              1              0
 1              0              1
 0              1              1


THE PRODUCT IS:
 0              0              0
 2              2              0
 4              4              4


DO YOU WANT A FURTHER PRODUCT?   YES OR NO
? NO
```

DETERMINANT

This program computes the determinant of a square matrix.

```
1000 DIM B(10,10),C(10),A$(3)
1010 DIM F(10,10)
1020 PRINT " "
1030 PRINT "ENTER THE ORDER OF THE MATRIX.";
1040 INPUT M
```

```
1050 PRINT " "
1060 PRINT "DO YOU WANT TO ENTER FRACTIONS? YES OR NO"
1070 INPUT A$
1080 PRINT " "
1090 IF A$="YES" THEN GO TO 1120
1100 PRINT "ENTER THE MATRIX BY ROWS (RETURN BETWEEN ENTRIES)"
1110 GO TO 1130
1120 PRINT "ENTER THE MATRIX OF NUMERATORS BY ROWS"
1130 FOR I=1 TO M
1140 PRINT " "
1150 PRINT "ROW";I
1160 FOR J=1 TO M
1170 INPUT B
1180 LET B(I,J)=B
1190 LET F(I,J)=1
1200 NEXT J
1210 NEXT I
1220 IF A$<>"YES" THEN GO TO 1330
1230 PRINT" "
1240 PRINT "ENTER MATRIX OF DENOMINATORS BY ROWS"
1250 FOR I=1 TO M
1260 PRINT " "
1270 PRINT "ROW";I
1280 FOR J=1 TO M
1290 INPUT F
1300 LET F(I,J)=F
1310 NEXT J
1320 NEXT I
1330 PRINT " "
1340 PRINT "THE MATRIX IS:"
1350 FOR I=1 TO M
1360 FOR J=1 TO M
1370 PRINT B(I,J);
1380 IF F(I,J)<>1 THEN GO TO 1410
1390 PRINT ,
1400 GO TO 1420
1410 PRINT "/";F(I,J),
1420 NEXT J
1430 PRINT " "
1440 NEXT I
1450 FOR I=1 TO M
1460 FOR J=1 TO M
1470 IF ABS(F(I,J))>.00001 THEN GO TO 1500
1480 PRINT "INVALID DENOMINATOR"
1490 GO TO 1230
1500 LET B(I,J)=B(I,J)/F(I,J)
1510 NEXT J
1520 NEXT I
1530 LET J=1
1540 LET D1=1
1550 FOR I=1 TO M
1560 LET K=I-1
1570 LET K=K+1
1580 IF B(K,J)<>0 THEN GO TO 1630
1590 IF K < M THEN GO TO 1570
1600 LET J=J+1
1610 IF J>M THEN GO TO 1830
```

```
1620 GO TO 1560
1630 LET D=B(K,J)
1640 LET D1=D1*D
1650 IF K>I THEN LET D1=-D1
1660 FOR J1=J TO M
1670 LET C(J1)=B(I,J1)
1680 LET B(I,J1)=B(K,J1)
1690 LET B(K,J1)=C(J1)
1700 LET B(I,J1)=B(I,J1)/D
1710 NEXT J1
1720 FOR I1=I+1 TO M
1730 IF I1=I THEN GO TO 1800
1740 LET D=B(I1,J)
1750 FOR J1=J TO M
1760 LET B(I1,J1)=B(I1,J1)-B(I,J1)*D
1770 IF ABS(B(I1,J1))>.00001 THEN GO TO 1790
1780 LET B(I1,J1)=0
1790 NEXT J1
1800 NEXT I1
1810 LET J=J+1
1820 NEXT I
1830 LET D1=D1*B(M,M)
1840 IF ABS(D1)>.00001 THEN GO TO 1860
1850 LET D1=0
1860 LET K=1
1870 LET G=ABS(D1)
1880 IF G*K-INT(G*K)<.00001 THEN GO TO 1930
1890 IF G*K-INT(G*K)>.99999 THEN GO TO 1930
1900 LET K=K+1
1910 IF K>1000 THEN GO TO 1940
1920 GO TO 1880
1930 LET D2=INT(D1*K+.00001)
1940 PRINT " "
1950 PRINT "THE DETERMINANT IS"
1960 IF K<=1000 THEN GO TO 1990
1970 PRINT D1
1980 GO TO 2020
1990 PRINT D2;
2000 IF K=1 THEN GO TO 2020
2010 PRINT "/";K
2020 PRINT " "
2030 END
```

MATRIX INVERSE

This program determines whether or not a square matrix is invertible and gives its inverse if it is.

```
1000 DIM B(10,20),C(20),A$(3)
1010 DIM E(10,10),F(10,10)
1020 PRINT " "
1030 PRINT "ENTER THE ORDER OF THE MATRIX"
1040 INPUT M
1050 PRINT " "
```

```
1060 PRINT "DO YOU WANT TO ENTER FRACTIONS?   YES OR NO"
1070 INPUT A$
1080 PRINT " "
1090 IF A$="YES" THEN GO TO 1120
1100 PRINT "ENTER THE MATRIX BY ROWS (RETURN BETWEEN ENTRIES)"
1110 GO TO 1130
1120 PRINT "ENTER MATRIX OF NUMERATORS"
1130 FOR I=1 TO M
1140 PRINT " "
1150 PRINT "ROW ";I
1160 FOR J=1 TO M
1170 INPUT B
1180 LET B(I,J)=B
1190 LET E(I,J)=B(I,J)
1200 LET F(I,J)=1
1210 NEXT J
1220 NEXT I
1230 IF A$<>"YES" THEN GO TO 1340
1240 PRINT ""
1250 PRINT "ENTER MATRIX OF DENOMINATORS BY ROWS"
1260 FOR I=1 TO M
1270 PRINT " "
1280 PRINT "ROW ";I
1290 FOR J=1 TO M
1300 INPUT F
1310 LET F(I,J)=F
1320 NEXT J
1330 NEXT I
1340 PRINT " "
1350 PRINT "THE MATRIX IS:"
1360 FOR I=1 TO M
1370 FOR J=1 TO M
1380 PRINT B(I,J);
1390 IF F(I,J)<>1 THEN GO TO 1420
1400 PRINT ,
1410 GO TO 1430
1420 PRINT "/";F(I,J),
1430 NEXT J
1440 PRINT " "
1450 NEXT I
1460 IF A$<>"YES" THEN GO TO 1550
1470 FOR I=1 TO M
1480 FOR J=1 TO M
1490 IF ABS(F(I,J))>.00001 THEN GO TO 1520
1500 PRINT "INVALID DENOMINATOR"
1510 GO TO 1240
1520 LET B(I,J)=B(I,J)/F(I,J)
1530 NEXT J
1540 NEXT I
1550 LET N=2*M
1560 FOR I=1 TO M
1570 FOR J=1 TO M
1580 LET B(I,J+M)=0
1590 NEXT J
1600 LET B(I,I+M)=1
1610 NEXT I
1620 LET J=1
```

```
1630 FOR I=1 TO M
1640 LET K=I-1
1650 LET K=K+1
1660 IF ABS(B(K,J))>.00001 THEN GO TO 1710
1670 IF K < M THEN GO TO 1650
1680 LET J=J+1
1690 IF J>N THEN GO TO 1960
1700 GO TO 1640
1710 LET D=B(K,J)
1720 FOR J1=J TO N
1730 LET C(J1)=B(I,J1)
1740 LET B(I,J1)=B(K,J1)
1750 LET B(K,J1)=C(J1)
1760 LET B(I,J1)=B(I,J1)/D
1770 NEXT J1
1780 FOR I1=1 TO M
1790 IF I1=I THEN GO TO 1860
1800 LET D=B(I1,J)
1810 FOR J1=J TO N
1820 LET B(I1,J1)=B(I1,J1)-B(I,J1)*D
1830 IF ABS(B(I1,J1))>.00001 THEN GO TO 1850
1840 LET B(I1,J1)=0
1850 NEXT J1
1860 NEXT I1
1870 LET J=J+1
1880 NEXT I
1890 LET T=0
1900 FOR J=1 TO M
1910 LET T=T+B(M,J)
1920 NEXT J
1930 IF ABS(T)>.00001 THEN GO TO 1970
1940 PRINT " "
1950 PRINT "THE MATRIX IS NOT INVERTIBLE."
1960 GO TO 2360
1970 FOR I=1 TO M
1980 FOR J=1 TO M
1990 LET B(I,J)=B(I,J+M)
2000 NEXT J
2010 NEXT I
2020 FOR I=1 TO M
2030 FOR J=1 TO M
2040 LET K=1
2050 LET G=ABS(B(I,J))
2060 IF G*K-INT(G*K)<.00001 THEN GO TO 2110
2070 IF G*K-INT(G*K)>.99999 THEN GO TO 2110
2080 LET K=K+1
2090 IF K>1000 THEN GO TO 2150
2100 GO TO 2060
2110 LET E(I,J)=INT(B(I,J)*K+.00001)
2120 LET F(I,J)=K
2130 NEXT J
2140 NEXT I
2150 PRINT " "
2160 PRINT "THE INVERSE IS:"
2170 IF K<=1000 THEN GO TO 2250
2180 FOR I=1 TO M
2190 FOR J=1 TO M
```

```
2200 PRINT B(I,J),
2210 NEXT J
2220 PRINT " "
2230 NEXT I
2240 GO TO 2350
2250 FOR I=1 TO M
2260 FOR J=1 TO M
2270 PRINT E(I,J);
2280 IF F(I,J)<>1 THEN GO TO 2310
2290 PRINT ,
2300 GO TO 2320
2310 PRINT "/";F(I,J),
2320 NEXT J
2330 PRINT " "
2340 NEXT I
2350 PRINT " "
2360 END
```

```
ENTER THE ORDER OF THE MATRIX
? 3

DO YOU WANT TO ENTER FRACTIONS?   YES OR NO
? NO

ENTER THE MATRIX BY ROWS (RETURN BETWEEN ENTRIES)

ROW  1
? 1
? 1
? 0

ROW  2
? 1
? 0
? 1

ROW  3
? 0
? 1
? 1

THE MATRIX IS:
 1                1               0
 1                0               1
 0                1               1

THE INVERSE IS:
 1 / 2            1 / 2          -1 / 2
 1 / 2           -1 / 2           1 / 2
-1 / 2            1 / 2           1 / 2
```

MATRIX POWERS

This program raises a square matrix to a sequence of positive integer powers.

```
1000 DIM A(15,15),B(15,15),A$(3)
1010 DIM C(15,15),F(15,15)
1020 DIM D(15,15)
```

```
1030 PRINT " "
1040 PRINT "ENTER THE ORDER OF   A "
1050 INPUT M
1060 PRINT "ENTER THE EXPONENT"
1070 INPUT K2
1080 LET K1=1
1090 PRINT " "
1100 PRINT "DO YOU WANT TO ENTER FRACTIONS?   YES OR NO"
1110 INPUT A$
1120 IF A$="YES" THEN GO TO 1150
1130 PRINT "ENTER THE MATRIX BY ROWS (RETURN BETWEEN ENTRIES)"
1140 GO TO 1170
1150 PRINT " "
1160 PRINT "ENTER MATRIX OF NUMERATORS BY ROWS"
1170 FOR I=1 TO M
1180 PRINT "ROW ";I
1190 FOR J=1 TO M
1200 INPUT A
1210 LET A(I,J)=A
1220 LET B(I,J)=A(I,J)
1230 LET C(I,J)=1
1240 NEXT J
1250 NEXT I
1260 IF A$<>"YES" THEN GO TO 1370
1270 PRINT " "
1280 PRINT "ENTER MATRIX OF DENOMINATORS BY ROWS"
1290 FOR I=1 TO M
1300 PRINT " "
1310 PRINT "ROW ";I
1320 FOR J=1 TO M
1330 INPUT C
1340 LET C(I,J)=C
1350 NEXT J
1360 NEXT I
1370 PRINT ""
1380 PRINT "THE MATRIX IS:"
1390 FOR I=1 TO M
1400 FOR J=1 TO M
1410 PRINT B(I,J);
1420 IF C(I,J)<>1 THEN GO TO 1450
1430 PRINT ,
1440 GO TO 1460
1450 PRINT "/";C(I,J),
1460 NEXT J
1470 PRINT " "
1480 NEXT I
1490 FOR I=1 TO M
1500 FOR J=1 TO M
1510 IF ABS(C(I,J))>.00001  THEN GO TO 1540
1520 PRINT "INVALID DENOMINATOR"
1530 GO TO 1270
1540 LET A(I,J)=A(I,J)/C(I,J)
1550 NEXT J
1560 NEXT I
1570 FOR I=1 TO M
1580 FOR J=1 TO M
1590 LET B(I,J)=A(I,J)
```

```
1600 LET C(I,J)=A(I,J)
1610 NEXT J
1620 NEXT I
1630 IF K2<>1 THEN GO TO 1670
1640 PRINT " "
1650 PRINT "THE MATRIX TO THE FIRST POWER IS ITSELF"
1660 GO TO 2160
1670 FOR I=1 TO M
1680 FOR J=1 TO M
1690 LET C(I,J)=0
1700 FOR K=1 TO M
1710 LET C(I,J)=C(I,J)+A(I,K)*B(K,J)
1720 NEXT K
1730 NEXT J
1740 NEXT I
1750 LET K1=K1+1
1760 IF K1=K2 THEN GO TO 1830
1770 FOR I=1 TO M
1780 FOR J=1 TO M
1790 LET B(I,J)=C(I,J)
1800 NEXT J
1810 NEXT I
1820 GO TO 1670
1830 FOR I=1 TO M
1840 FOR J=1 TO M
1850 LET K=1
1860 LET G =ABS(C(I,J))
1870 IF G*K-INT(G*K)<.00001 THEN GO TO 1920
1880 IF G*K-INT(G*K)>.99999 THEN GO TO 1920
1890 LET K=K+1
1900 IF K>1000 THEN GO TO 1960
1910 GO TO 1870
1920 LET D(I,J)=INT(C(I,J)*K+.00001)
1930 LET F(I,J)=K
1940 NEXT J
1950 NEXT I
1960 PRINT ""
1970 PRINT "THE MATRIX TO THE ";K2;" POWER IS:"
1980 IF K<=1000 THEN GO TO 2060
1990 FOR I=1 TO M
2000 FOR J=1 TO M
2010 PRINT C(I,J),
2020 NEXT J
2030 PRINT " "
2040 NEXT I
2050 GO TO 2160
2060 FOR I=1 TO M
2070 FOR J=1 TO M
2080 PRINT D(I,J);
2090 IF F(I,J)<>1 THEN GO TO 2120
2100 PRINT ',
2110 GO TO 2130
2120 PRINT "/";F(I,J),
2130 NEXT J
2140 PRINT " "
2150 NEXT I
2160 PRINT " "
```

```
2170 PRINT "DO YOU WANT A DIFFERENT POWER?  YES OR NO"
2180 INPUT A$
2190 IF A$="NO" THEN GO TO 2280
2200 PRINT "ENTER THE EXPONENT"
2210 INPUT K3
2220 IF K3<=K2 THEN GO TO 2250
2230 LET K2=K3
2240 GO TO 1770
2250 LET K2=K3
2260 LET K1=1
2270 GO TO 1570
2280 PRINT ""
2290 END
```

SIMPLEX ALGORITHM

This program allows a sequence of pivot operations to be performed on a matrix. Deciding which pivots to use and interpreting the results are left to the operator.

```
1000 DIM B(10,15),C(15),A$(3)
1010 DIM E(10,15),F(10,15)
1020 PRINT " "
1030 PRINT "ENTER THE DIMENSIONS OF THE MATRIX AS M,N"
1040 INPUT M,N
1050 PRINT " "
1060 PRINT "DO YOU WANT TO ENTER FRACTIONS?  YES OR NO"
1070 INPUT A$
1080 PRINT " "
1090 IF A$="YES" THEN GO TO 1120
1100 PRINT "ENTER THE MATRIX BY ROWS (RETURN BETWEEN ENTRIES)"
1110 GO TO 1130
1120 PRINT "ENTER MATRIX OF NUMERATORS BY ROWS"
1130 FOR I=1 TO M
1140 PRINT "ROW ";I
1150 FOR J=1 TO N
1160 INPUT B
1170 LET B(I,J)=B
1180 LET F(I,J)=1
1190 NEXT J
1200 NEXT I
1210 IF A$<>"YES" THEN GO TO 1320
1220 PRINT " "
1230 PRINT "ENTER MATRIX OF DENOMINATORS BY ROWS"
1240 FOR I=1 TO M
1250 PRINT " "
1260 PRINT "ROW ";I
1270 FOR J=1 TO N
1280 INPUT F
1290 LET F(I,J)=F
1300 NEXT J
1310 NEXT I
1320 PRINT " "
1330 PRINT "THE MATRIX IS:"
```

```
1340 FOR I=1 TO M
1350 FOR J=1 TO N
1360 PRINT B(I,J);
1370 IF F(I,J)<>1 THEN GO TO 1400
1380 PRINT ,
1390 GO TO 1410
1400 PRINT "/";F(I,J),
1410 NEXT J
1420 PRINT " "
1430 NEXT I
1440 FOR I=1 TO M
1450 FOR J=1 TO N
1460 IF ABS(F(I,J))>.00001 THEN GO TO 1490
1470 PRINT "INVALID DENOMINATOR"
1480 GO TO 1220
1490 LET B(I,J)=B(I,J)/F(I,J)
1500 NEXT J
1510 NEXT I
1520 PRINT " "
1530 PRINT "CHOSE THE PIVOT ROW I AND COLUMN J AS  I,J"
1540 INPUT I,J
1550 LET D=B(I,J)
1560 IF ABS(D)>.00001 THEN GO TO 1600
1570 PRINT " "
1580 PRINT "THE PIVOT ENTRY IS  0 .   CHOOSE ANOTHER."
1590 GO TO 1520
1600 FOR J1=1 TO N
1610 LET B(I,J1)=B(I,J1)/D
1620 NEXT J1
1630 FOR I1=1 TO M
1640 IF I1=I THEN GO TO 1710
1650 LET D=B(I1,J)
1660 FOR J1 = 1 TO N
1670 LET B(I1,J1)=B(I1,J1)-B(I,J1)*D
1680 IF ABS(B(I1,J1))>.00001 THEN GO TO 1700
1690 LET B(I1,J1)=0
1700 NEXT J1
1710 NEXT I1
1720 FOR I=1 TO M
1730 FOR J=1 TO N
1740 LET K=1
1750 LET G=ABS(B(I,J))
1760 IF G*K-INT(G*K)<.00001 THEN GO TO 1810
1770 IF G*K-INT(G*K)>.99999 THEN GO TO 1810
1780 LET K=K+1
1790 IF K>1000 THEN GO TO 1850
1800 GO TO 1760
1810 LET E(I,J)=INT(B(I,J)*K+.00001)
1820 LET F(I,J)=K
1830 NEXT J
1840 NEXT I
1850 PRINT " "
1860 PRINT "THE NEW MATRIX IS:"
1870 IF K<=1000 THEN GO TO 1950
1880 FOR I=1 TO M
1890 FOR J=1 TO N
1900 PRINT B(I,J),
```

```
1910 NEXT J
1920 PRINT " "
1930 NEXT I
1940 GO TO 2050
1950 FOR I=1 TO M
1960 FOR J=1 TO N
1970 PRINT E(I,J);
1980 IF F(I,J)<>1 THEN GO TO 2010
1990 PRINT ,
2000 GO TO 2020
2010 PRINT "/";F(I,J),
2020 NEXT J
2030 PRINT " "
2040 NEXT I
2050 PRINT " "
2060 PRINT "DO YOU WANT TO PIVOT AGAIN?  YES OR NO"
2070 INPUT A$
2080 IF A$="NO" THEN GO TO 2100
2090 GO TO 1520
2100 PRINT " "
2110 END
```

```
ENTER THE DIMENSIONS OF THE MATRIX AS M,N
? 2.3

DO YOU WANT TO ENTER FRACTIONS?  YES OR NO
?

ENTER THE MATRIX BY ROWS (RETURN BETWEEN ENTRIES)
ROW  1
? 1
? 0
? 2
ROW  2
? 0
? 1
? 3

THE MATRIX IS:
  1            0            2
  0            1            3

CHOSE THE PIVOT ROW I AND COLUMN J AS  I,J
? 2,3

THE NEW MATRIX IS:
  1           -2 / 3        0
  0            1 / 3        1

DO YOU WANT TO PIVOT AGAIN?  YES OR NO
?

CHOSE THE PIVOT ROW I AND COLUMN J AS  I,J
? 1,2

THE NEW MATRIX IS:
 -3 / 2        1            0
  1 / 2        0            1

DO YOU WANT TO PIVOT AGAIN?  YES OR NO
? NO
```

ANSWERS TO ODD-NUMBERED EXERCISES

SECTION 1.1 (Page 11)

1. $(-4, -8, 12)$ **3.** $(1, -1)$ **5.** $(5, 1, 1)$ **7.** $(-3, -2, 2)$

9. $(-9, -3, 1)$ **11.** $\sqrt{14}$ **13.** $\sqrt{17}$ **15.** $\sqrt{14} - \sqrt{5}$

17. $(\frac{3}{5}, -\frac{4}{5})$ **19.** $(-\frac{2}{3}, \frac{1}{3}, -\frac{2}{3})$ **21.** $(4, 2)$ **23.** $(1, -4, -7)$

25. $2\mathbf{i} - 3\mathbf{k}$ **27.** $\sqrt{14}$

33. If $r < 0$, no solutions. If $r = 0$, simply point $\mathbf{p}$. If $r > 0$:
   in $\mathbf{R}^2$, a circle centered at $\mathbf{p}$ with radius r;
   in $\mathbf{R}^3$, a sphere centered at $\mathbf{p}$ with radius r.

35. $(2/3)\mathbf{p} + (1/3)\mathbf{q}$ **37.** yes

SECTION 1.2 (Page 22)

1. 4   **3.** 4   **5.** $(11, -7, -10)$ **7.** $(-22, 14, 20)$

9. -38 **11.** obtuse   **13.** $-8/\sqrt{145}$

15. $-\frac{1}{2}$ **17.** of form $c(-4, 7, -1)$ **19.** $\sqrt{14}/2$

23. The quantity $\mathbf{v} \cdot \mathbf{w}$ is a *scalar*, and dot product is only defined for two vectors.

37. $\mathbf{u} \times \mathbf{v} = (1/168, -3/140, 1/60)$

SECTION 1.3 (Page 33)

1. $(3, 1, 2) \cdot (\mathbf{x} - (2, -1, -4)) = 0$
$3x + y + 2z = -3$

3. $(2, 1, 3) \cdot (\mathbf{x} - \mathbf{0}) = 0$
$2x + y + 3z = 0$

5. $(3, 6, 1) \cdot (\mathbf{x} - (0, 2, 0)) = 0$
$3x + 6y + z = 12$

7. $(2, 3, 0) \cdot (\mathbf{x} - (1/2, 0, 0)) = 0$
$2x + 3y = 1$

9. $(2, 1) \cdot (\mathbf{x} - (-1, 2)) = 0$

11. $(2, -1) \cdot (\mathbf{x} - (-2, 5)) = 0$

13. $\mathbf{x}(t) = (2, 1, -3) + t(1, 2, 2)$
$$\begin{aligned} x &= 2 + t \\ y &= 1 + 2t \\ z &= -3 + 2t \end{aligned}$$

15. $\mathbf{x}(t) = (2, -3, 1) + t(1, 0, 0)$
$$\begin{aligned} x &= 2 + t \\ y &= -3 \\ z &= 1 \end{aligned}$$

17. $\mathbf{x}(t) = (1 - t)(2, 0, -2) + t(1, 4, 2)$
$$\begin{aligned} x &= 2 - t \\ y &= 4t \\ z &= -2 + 4t \end{aligned}$$

19. $\mathbf{x}(t) = (2, 4, 5) + t(1, -1, -2)$
$$\begin{aligned} x &= 2 + t \\ y &= 4 - t \\ z &= 5 - 2t \end{aligned}$$

21. $\mathbf{x}(t) = (1, -1) + t(1, -3)$
$$\begin{aligned} x &= 1 + t \\ y &= -1 - 3t \end{aligned}$$

23. $(3, 1) \cdot (\mathbf{x} - (2, 3)) = 0$

25. $\sqrt{6}/2$

27. $9\sqrt{5}/5$

29. $(13/5, 1, -7/5)$

35. $2/5$

37. 0 (the planes are perpendicular)

SECTION 1.4 (Page 43)

1. $(1, 2, 5, 1)$

3. $(12, -6, 14, 40)$

5. 2

7. $6\sqrt{11}$

9. $\sqrt{31}$

11. $2\sqrt{47}$

13. $\mathbf{u}_1$ and $\mathbf{u}_3$, $\mathbf{u}_2$ and $\mathbf{u}_3$, $\mathbf{u}_2$ and $\mathbf{u}_4$, $\mathbf{u}_3$ and $\mathbf{u}_4$

17. $(2/\sqrt{31}, 1/\sqrt{31}, -1/\sqrt{31}, 0, 3/\sqrt{31}, 4/\sqrt{31})$

19. $-2x_1 - x_2 + x_4 = c$ where c is arbitrary.

21. $\mathbf{x}(t) = (1 - t)(-1, 0, 3, 2) + t(-1, 0, 4, 5)$ (two point form)
$\mathbf{x}(t) = (-1, 0, 3, 2) + t(0, 0, 1, 3)$ (point-parallel form)
$x_1 = -1, x_2 = 0, x_3 = 3 + t, x_4 = 2 + 3t$ (parametric equations)

23. $(1, -1, 1, -1, 1) \cdot (\mathbf{x} - (3, 4, 5, 6, 7)) = 0$ (point-normal form)
$x_1 - x_2 + x_3 - x_4 + x_5 = 5$ (standard form)

25. $(2, -3, 0, 1, -1, 0) \cdot (\mathbf{x} - (1, 0, 0, 0, 0, 0)) = 0$

27. product of variables

29. products of variables

35. $\mathbf{x} = (3, 1, 0, 2, 1) + r(-1, 0, 4, 0, -1) + s(-4, 1, 1, 1, 0) + t(-3, 1, 0, -1, -1)$

39. $(2, 0, 2, -2)$

REVIEW EXERCISES CHAPTER 1 (Page 46)

1. $(4, -3, 3, -3, -10)$

3. $\sqrt{34}$

5. $2\sqrt{14}$

7. $(6, -3, 2)$

9. 7

11. $(1/\sqrt{10})(1, 0, -3)$

13. $z - 3 = 0$

15. $(3, 1, -2, 1) \cdot (\mathbf{x} - (0, 0, 0, 1)) = 0$

17. $\mathbf{x} = t(1, 0, 2)$

19. $x + y + z = c, \quad c$ arbitrary

SECTION 2.1 (Page 56)

1. yes

3. no

5. no

7.
$$y + z = 6$$
$$x + y - z = 0$$
$$-x + y + z = 3$$

9. $x_1 + x_2 + x_3 - x_4 = 0$
$\quad x_1 \qquad\qquad = 1$
$\quad\quad - x_2 \quad\quad + x_4 = 3$

15. $(1 + 4s, s)$

17. $(2 - 3t, -1 - s + 2t, s, t)$

19. point

21. line

23. -1

SECTION 2.2 (Page 71)

1. $\begin{bmatrix} 2 & -3 & 1 & | & 0 \\ 1 & 0 & -2 & | & 1 \\ 0 & -4 & 1 & | & -1 \end{bmatrix}$

3. $\begin{bmatrix} 1 & -1 & 0 & | & 0 \\ 0 & 1 & -1 & | & 0 \end{bmatrix}$

5. no

7. yes

9. no

11. $(0, 2, -1)$

13. no solution

15. $(4 - 2s - 3t, s, t)$

17. $(1 + 2s, s)$

19. $(-\frac{3}{4}, -\frac{5}{4}, \frac{13}{4})$

21. $(2s, -\frac{1}{3} + \frac{5}{3}s, s)$

23. no solution

25. $(\frac{1}{2} + s, 1 + 2s - t, s, t)$

39. $k = 2$; no

41. Answers vary with computer system but are approximately those of Exercises 13–20.

SECTION 2.3 (Page 85)

1. $\begin{bmatrix} -1 & 4 & -1 \\ 5 & 0 & 5 \\ -1 & 1 & -1 \end{bmatrix}$

3. $\begin{bmatrix} -2 & -12 & 8 \\ -5 & -5 & 0 \\ 3 & 2 & -7 \end{bmatrix}$

5. $\begin{bmatrix} -1 - \lambda & 0 & 1 \\ 2 & -1 - \lambda & 3 \\ 0 & 1 & -2 - \lambda \end{bmatrix}$

7. $\begin{bmatrix} -1 & 2 & 0 \\ 0 & -1 & 1 \\ 1 & 3 & -2 \end{bmatrix}$

9. $\begin{bmatrix} -1 & -4 & 3 \\ -6 & 7 & -3 \\ 5 & 1 & 0 \end{bmatrix}$

11. $\begin{bmatrix} -1 & -6 & 5 \\ -4 & 7 & 1 \\ 3 & -3 & 0 \end{bmatrix}$

13. $\begin{bmatrix} 2 & 1 \\ -1 & 0 \\ \frac{1}{2} & -2 \end{bmatrix}$

15. $\mathbf{a}^1 = (2, 1)$
$\quad \mathbf{a}^3 = (\frac{1}{2}, -2)$
$\quad \mathbf{a}^2 = (1, 0, -2)$

17. $AB = \begin{bmatrix} 7 & -1 & 1 \\ 0 & -4 & -2 \\ 6 & -6 & 0 \end{bmatrix}$, $BA = \begin{bmatrix} -1 & 2 & 4 \\ 5 & 6 & -1 \\ 6 & 4 & -2 \end{bmatrix}$

19. Neither is defined.

21. $AB = A$, BA not defined

23. $\begin{bmatrix} -9 & 1 & 8 \\ -2 & -3 & -10 \end{bmatrix}$

25. $A^2 = \begin{bmatrix} 2 & -3 \\ -3 & 5 \end{bmatrix}$, $A^3 = \begin{bmatrix} 5 & -8 \\ -8 & 13 \end{bmatrix}$

27. $A^2 = \begin{bmatrix} 1 & 0 & 0 \\ 4 & 1 & 0 \\ 10 & 4 & 1 \end{bmatrix}$, $A^3 = \begin{bmatrix} 1 & 0 & 0 \\ 6 & 1 & 0 \\ 21 & 6 & 1 \end{bmatrix}$

35. $A^n = \begin{bmatrix} 1 & n \\ 0 & 1 \end{bmatrix}$

41. $AB = \begin{bmatrix} 1 & 1 & 2 & 2 & 3 & 3 & 4 & 4 & 5 & 5 \\ 2 & 1 & 3 & 2 & 4 & 3 & 5 & 4 & 6 & 5 \\ 3 & 1 & 4 & 2 & 5 & 3 & 6 & 4 & 7 & 5 \\ 4 & 1 & 5 & 2 & 6 & 3 & 7 & 4 & 8 & 5 \\ 5 & 1 & 6 & 2 & 7 & 3 & 8 & 4 & 9 & 5 \\ 6 & 1 & 7 & 2 & 8 & 3 & 9 & 4 & 10 & 5 \\ 7 & 1 & 8 & 2 & 9 & 3 & 10 & 4 & 11 & 5 \\ 8 & 1 & 9 & 2 & 10 & 3 & 11 & 4 & 12 & 5 \\ 9 & 1 & 10 & 2 & 11 & 3 & 12 & 4 & 13 & 5 \\ 10 & 1 & 11 & 2 & 12 & 3 & 13 & 4 & 14 & 5 \end{bmatrix}$

SECTION 2.4 (Page 96)

1. $\begin{bmatrix} 1 & -1 & 3 \\ 1 & 0 & -1 \\ -2 & 1 & 0 \end{bmatrix} \begin{bmatrix} x \\ y \\ z \end{bmatrix} = \begin{bmatrix} 1 \\ 0 \\ -1 \end{bmatrix}$

3. $\begin{bmatrix} 1 & -1 & 1 \\ -1 & 1 & 1 \end{bmatrix} \begin{bmatrix} x \\ y \\ z \end{bmatrix} = \begin{bmatrix} 0 \\ 0 \end{bmatrix}$

5. $AB = BA = I$

9. $\begin{bmatrix} \frac{1}{2} & 0 \\ \frac{3}{2} & 1 \end{bmatrix}$

11. $\begin{bmatrix} 1 & -2 & 5 \\ 0 & 1 & -2 \\ 0 & 0 & 1 \end{bmatrix}$

13. $\begin{bmatrix} 2 & 1 & -1 \\ 1 & 1 & -1 \\ 1 & 1 & 0 \end{bmatrix}$

15. not invertible

17. -6

19. $(-3, -8)$

21. $(\frac{7}{2}, -\frac{3}{2}, -\frac{3}{2})$

23. $(12, 7)$

25. $(4, -4, -3)$

27. no inverse

31. $\begin{bmatrix} 1 & \frac{1}{4} \\ 0 & \frac{1}{4} \end{bmatrix}$

33. $(1, 0, 0, 0, 0)$

SECTION 2.5 (Page 105)

1. homogeneous

3. 2

5. 3

7. (a) $ad \neq bc$
(b) $ad = bc$ but $af \neq ce$
(c) $ad = bc$ and $af = ce$

9. false

11. true

13. false　　　　　　　　**15.** infinitely many solutions

17. one solution or infinitely many solutions

21. (a) $x = 0,$　　y nonzero　　(b) x nonzero
　　(c) $x = y = 0$

REVIEW EXERCISES CHAPTER 2 (Page 110)

1. $(2s - 1, s)$　　　　　**3.** $(2s - t, s, t)$　　　**5.** $(1, 0, -1)$　　　　**7.** no solution

9. $\begin{bmatrix} 5 & -3 & -1 \\ -3 & -7 & 3 \\ -6 & 0 & 9 \end{bmatrix}$　　**11.** $\begin{bmatrix} 1 & 1 & 0 \\ 7 & 3 & -2 \\ 6 & 0 & -3 \end{bmatrix}$　　**13.** $\begin{bmatrix} 1 & 0 & -\frac{1}{3} \\ 0 & 1 & -1 \\ 0 & 0 & \frac{1}{3} \end{bmatrix}$　　**15.** does not exist

17. 3　　　　　　　　　**19.** $\begin{bmatrix} -1 & 0 & -4 \\ 0 & -1 & -12 \\ 0 & 0 & -9 \end{bmatrix}$

21. $\begin{bmatrix} 1 & -2 \\ -2 & 4 \end{bmatrix}\begin{bmatrix} x \\ y \end{bmatrix} = \begin{bmatrix} -1 \\ 2 \end{bmatrix}$, cannot be used; inverse of coefficient matrix does not exist

23. $\begin{bmatrix} 1 & -2 & 1 \\ -1 & 2 & -1 \end{bmatrix}\begin{bmatrix} x_1 \\ x_2 \\ x_3 \end{bmatrix} = \begin{bmatrix} 0 \\ 0 \end{bmatrix}$, cannot be used; not a square system

25. $\begin{bmatrix} 1 & 2 & -3 \\ 0 & -1 & 1 \\ 2 & 0 & -1 \end{bmatrix}\begin{bmatrix} u \\ v \\ w \end{bmatrix} = \begin{bmatrix} 4 \\ -1 \\ 3 \end{bmatrix}$, $(1, 0, -1)$

27. $\begin{bmatrix} 1 & 1 & 2 & -1 \\ -1 & 2 & 1 & 1 \\ -1 & 0 & -1 & 1 \end{bmatrix}\begin{bmatrix} x \\ y \\ z \\ w \end{bmatrix} = \begin{bmatrix} 1 \\ 2 \\ 3 \end{bmatrix}$, cannot be used; not a square system

29. one solution　　　　　　　　**31.** infinitely many solutions, one parameter

SECTION 3.1 (Page 121)

1.

	minor	cofactor
3, 1	-7	-7
3, 2	-1	1

3. 3　　　　　　　　**5.** -5

7. 15　　　　　　　　**9.** 0　　　　　　　　**11.** 120

13. a^3　　　　　　　**15.** $k^2(ad - bc)$　　　　**17.** $t^2 - 3t + 3$

19. -8　　　　　　　**21.** 0　　　　　　　　**27.** $\{-1, 3\}$

33. 24 cubic units

SECTION 3.2 (Page 132)

1. 2   **3.** −24   **5.** −1/720   **7.** $-x(x + 1)(2x - 1)$

9. $-(\lambda - 1)(\lambda + 1)(\lambda - 2)(\lambda + 2)$

11. $x \neq -1, 0, \frac{1}{2}; t \neq 3, (1 \pm \sqrt{33})/2; \lambda \neq \pm 1, \pm 2$

13. solution is unique   **15.** solution is not unique

17. $\det(A^T) = 4$, $\det(2A) = 32$

21. $(-1)^k \det A$, where $k = n/2$ if n is even, and $k = (n - 1)/2$ if n is odd

31. −1.94

SECTION 3.3 (Page 141)

1. $x = 1,$
$ y = -1$

3. zero determinant; Cramer's rule cannot be used

5. $x = \frac{3}{5},$ **7.** $w = 0,$
$ y = \frac{3}{2},$ $ x = 2,$
$ z = \frac{17}{10}$ $ y = 1,$
$ z = -1$

9. $\text{Adj } A = \begin{bmatrix} 4 & -2 \\ 3 & 1 \end{bmatrix}, \quad A^{-1} = \begin{bmatrix} \frac{2}{5} & -\frac{1}{5} \\ \frac{3}{10} & \frac{1}{10} \end{bmatrix}$

11. $\text{Adj } A = \begin{bmatrix} 2 & 1 & 2 \\ 0 & 1 & 2 \\ 0 & 0 & 2 \end{bmatrix}, \quad A^{-1} = \begin{bmatrix} 1 & \frac{1}{2} & 1 \\ 0 & \frac{1}{2} & 1 \\ 0 & 0 & 1 \end{bmatrix}$

13. $\text{Adj } A = \begin{bmatrix} -27 & 9 & 9 \\ 18 & -6 & -6 \\ 9 & -3 & -3 \end{bmatrix}$, no inverse exists

15. $\dfrac{1}{\det A} \begin{bmatrix} a_{22}a_{33}-a_{23}a_{32} & a_{13}a_{32}-a_{12}a_{33} & a_{12}a_{23}-a_{13}a_{22} \\ a_{23}a_{31}-a_{21}a_{33} & a_{11}a_{33}-a_{13}a_{31} & a_{13}a_{21}-a_{11}a_{23} \\ a_{21}a_{32}-a_{22}a_{31} & a_{12}a_{31}-a_{11}a_{32} & a_{11}a_{22}-a_{12}a_{21} \end{bmatrix}$

21. $w = 26.60,$
$ x = -2.16,$
$ y = 38.15,$
$ z = 13.27$

REVIEW EXERCISES CHAPTER 3 (Page 144)

1. −2   **3.** 6   **5.** $xy(x - 1)(y - 1)(y - x)$ **7.** 1

9. The inverse exists for matrices of Exercises 1, 2, and 3; not 4.

11. $\det(A^{-1}) = \det(A^T)^{-1} = \frac{1}{2}$

13. $\text{Adj } A = \begin{bmatrix} -1 & -1 & 1 \\ -1 & 1 & -1 \\ 1 & -1 & -1 \end{bmatrix}$, $\quad A^{-1} = \begin{bmatrix} \frac{1}{2} & \frac{1}{2} & -\frac{1}{2} \\ \frac{1}{2} & -\frac{1}{2} & \frac{1}{2} \\ -\frac{1}{2} & \frac{1}{2} & \frac{1}{2} \end{bmatrix}$

15. $\text{Adj } A = \begin{bmatrix} 6 & 0 & 0 \\ -3 & 3 & 0 \\ 0 & -2 & 2 \end{bmatrix}$, $\quad A^{-1} = \begin{bmatrix} 1 & 0 & 0 \\ -\frac{1}{2} & \frac{1}{2} & 0 \\ 0 & -\frac{1}{3} & \frac{1}{3} \end{bmatrix}$

17. $(\frac{1}{3}, -\frac{1}{3})$

19. $(\frac{1}{2}, \frac{1}{2}, -\frac{1}{2})$

21. zero determinant; cannot be solved by Cramer's rule

SECTION 4.1 (Page 155)

1. $\mathbf{w}_1 = 3\mathbf{u} + 2\mathbf{v}$

3. $c = -2$

5. $(0, 1, 2) = (\frac{1}{2})(1, 2, 3) + (\frac{1}{2})(-1, 0, 1)$

7. independent

9. independent

11. $(1, 1) = 1(1, 0) + 1(0, 1)$

13. independent

15. dependent, $\mathbf{v}_2 = 2\mathbf{v}_1$

17. independent

19. dependent by Theorem 2

21. independent by Theorem 3

23. dependent by Theorem 3

25. dependent from definition

35. independent

37. $\mathbf{v}_4 = \mathbf{v}_1 - (1/2)\mathbf{v}_2 + (1/2)\mathbf{v}_3$

SECTION 4.2 (Page 162)

13. does not span

15. does not span

19. $\{(1, 2, 3)\}$

21. $\{(1, 0, 1, 0), (0, 1, 0, 1)\}$

23. $\{(1, 0, 1, 1), (0, 1, -1, 1)\}$

SECTION 4.3 (Page 173)

1. not a basis

3. basis

5. not a basis; $(0, 0, 1, 1)$ is not in the subspace

7. basis

9. $\{(1, -1, 1), (2, 0, 1)\}$

11. $\{(-1, 1, 2, -1), (1, 0, -1, 1)\}$

13. $\{(1, 1, 0), (1, -1, 1), (1, 0, 0)\}$

15. $\{(1, 0, 0, 1), (0, 1, 1, 0), (1, 1, 1, 2), (0, 1, 0, 0)\}$

17. $4 > 3$, too many vectors

19. part (iii) of Theorem 5 ($m = n = 4$)

21. $\{(1, 1, 1, 0), (-2, -1, 0, 1)\}, 2$

23. $\{(1, 2, 0), (0, 0, 1)\}, 2$

25. $\{(1, 0, 1, 0), (0, -1, -2, 3)\}, 2$

27. $\{(1, -1, 3, 1), (-2, 0, -2, 0)\}, 2$

29. 1

SECTION 4.4 (Page 179)

1. row rank = column rank = 2

3. row rank = column rank = 1

5. row rank = column rank = 3

7. $\{(1, 0, 0), (0, 1, 0)\}$

9. $\{(1, -1)\}$

11. $\{(1, 0, 1, 2), (-1, 1, 1, -1), (3, 1, 0, 1)\}$

13. $\{(1, 2, 0), (0, 0, 1)\}$

15. $\{(1, 0, -1\}$

17. $\{(1, -1, 3), (0, 1, 1), (1, 1, 0)\}$

REVIEW EXERCISES CHAPTER 4 (Page 182)

1. independent

3. dependent, $(3, -1, 0, 1) = 2(1, 0, 1, 0) + (-1)(-1, 1, 2, -1)$

5. subspace

7. not a subspace

9. does span

11. basis

13. basis

15. $\{(1, 1, 0, 0), (0, 1, 1, 0), (0, 1, 0, 1)\}, 3$

17. $\{(-2, 1, 0), (-3, 0, 1)\}, 2$

19. $\{(1, 1, 1, 1, 1), (0, 1, 1, 1, 1), (0, 0, 1, 1, 1)\}, 3$

21. $\{(1, -1, 2, 0), (2, -1, 1, 1)\}, 2$

SECTION 5.1 (Page 193)

1. vector space

3. is not; not closed under addition

5. is not; not closed under multiplication by scalars

7. is not; $1\mathbf{y} = \mathbf{x} \neq \mathbf{y}$

9. vector space

11. vector space

13. subspace

15. is not; not closed under addition or multiplication by scalars

17. subspace

19. subspace

SECTION 5.2 (Page 204)

1. independent

3. dependent; $1 = (\frac{1}{2})(x + 1) + (-\frac{1}{2})(x - 1)$

5. independent

7. dependent; $1 = (1) \sin^2 x + (1) \cos^2 x$

9. independent

11. dependent; $\mathbf{v}_5 = (17/28)\mathbf{v}_1 + (17/14)\mathbf{v}_2 + (2/7)\mathbf{v}_3 + (1/4)\mathbf{v}_4$

13. generates

15. generates

17. generates

19. does not generate

21. does not generate

23. generates

25. basis

27. dependent

29. basis

31. dependent, does not generate

33. does not generate

35. dependent

37. 12

39. 2

41. 1

SECTION 5.3 (Page 210)

1. $(2, -1, 0, 3)$

3. $(-\frac{3}{2}, -\frac{7}{2})$

5. $(4, -5, -\frac{9}{2})$

7. independent

9. dependent; $(1)\begin{bmatrix} 1 & 2 \\ -1 & 0 \end{bmatrix} + (2)\begin{bmatrix} 0 & -1 \\ 1 & 1 \end{bmatrix} = \begin{bmatrix} 1 & 0 \\ 1 & 2 \end{bmatrix}$

11. dependent; $\mathbf{v}_5 = (17/28)\mathbf{v}_1 + (17/14)\mathbf{v}_2 + (2/7)\mathbf{v}_3 + (1/4)\mathbf{v}_4$

13. already a basis

15. $\left\{ \begin{bmatrix} 1 & -1 \\ 1 & 0 \end{bmatrix}, \begin{bmatrix} 1 & 1 \\ 0 & 0 \end{bmatrix}, \begin{bmatrix} 0 & 0 \\ 1 & 1 \end{bmatrix}, \begin{bmatrix} 1 & 0 \\ 0 & 0 \end{bmatrix} \right\}$

17. $\left\{ \begin{bmatrix} 1 & 0 & -1 \\ 1 & -1 & 1 \end{bmatrix}, \begin{bmatrix} 0 & 1 & 1 \\ 2 & 1 & -1 \end{bmatrix}, \begin{bmatrix} 1 & -1 & 1 \\ 1 & -1 & 1 \end{bmatrix}, \begin{bmatrix} 0 & 1 & 0 \\ 0 & 1 & 0 \end{bmatrix}, \begin{bmatrix} 1 & 0 & 0 \\ 0 & 0 & 0 \end{bmatrix}, \begin{bmatrix} 0 & 1 & 0 \\ 0 & 0 & 0 \end{bmatrix} \right\}$

19. $\{x + 1, x - 1, x^2 + 1\}$

21. $\left\{ \begin{bmatrix} 2 & 1 \\ 0 & 1 \end{bmatrix}, \begin{bmatrix} -1 & 1 \\ 0 & 0 \end{bmatrix}, \begin{bmatrix} 1 & 2 \\ 0 & 2 \end{bmatrix} \right\}$

23. $f_4 = 3f_2 + 2f_3$

SECTION 5.4 (Page 224)

1. yes

3. no; properties (b), (c), and (d) fail

5. no; property (d) fails

7. orthonormal basis

9. not orthogonal

11. orthonormal basis

13. neither orthogonal nor orthonormal

15. orthogonal and orthonormal

17. $1(1/3, -2/3, 2/3) + 2(-2/3, 1/3, 2/3) + 3(2/3, 2/3, 1/3)$

19. $\{(4\sqrt{5}/15, \sqrt{5}/3, 0, -2\sqrt{5}/15), (-2/3, 2/3, 0, 1/3), (\sqrt{5}/5, 0, 0, 2\sqrt{5}/5)\}$

21. $\{(\sqrt{2}/2, -\sqrt{2}/2, 0), (\sqrt{6}/6, \sqrt{6}/6, -\sqrt{6}/3)\}$

23. $\{(-\sqrt{2}/2, \sqrt{2}/2, 0, 0), (0, 0, 1, 0), (0, 0, 0, 1)\}$

25. $\{(1, 0, 0, 0), (0, 1, 0, 0), (0, 0, 1, 0), (0, 0, 0, 1)\}$

27. $\{1/\sqrt{2}, (\sqrt{3/2})x, (1/2)\sqrt{5/2}\,(3x^2 - 1), (1/2)\sqrt{7/2}\,(5x^3 - 3x)\}$

29. $\{1, \sqrt{2}(e^x - e + 1)/\sqrt{4e - e^2 - 3}\}$

REVIEW EXERCISES CHAPTER 5 (Page 227)

1. not a vector space; not closed under addition or multiplication by scalars

3. vector space

5. not a vector space; not closed under addition or multiplication by scalars

7. linearly independent

9. linearly dependent; $-2(x - \cos x) = -2x + 2\cos x$

11. orthogonal

13. orthogonal

15. orthogonal

17. not an inner product; if $\mathbf{p} = (x - 1)(x - 2)$, then $\mathbf{p} \neq \mathbf{0}$ but $(\mathbf{p}, \mathbf{p}) = 0$

19. not an inner product; $(A, A) = 0$ for some $A \neq 0$

21. $\{(1/\sqrt{3})(1, 1, 1, 0), (1/\sqrt{15})(1, 1, -2, 3), (1/\sqrt{35})(1, -4, 3, 3)\}$

SECTION 6.1 (Page 241)

9. linear

11. not linear

13. $n = 3, m = 2, T(\mathbf{x}_0) = (16, -3)$

15. $n = 2, m = 3, T(\mathbf{x}_0) = (-4, -3, 2)$

21. $\begin{bmatrix} -\sqrt{2}/2 & -\sqrt{2}/2 \\ \sqrt{2}/2 & -\sqrt{2}/2 \end{bmatrix}$

23. $\begin{bmatrix} 0 & 1 \\ 1 & 0 \end{bmatrix}$

SECTION 6.2 (Page 253)

1. $LK(x, y, z) = (x - y, 3x + 2y)$

3. $S^2(x, y, z) = (x, y, z)$

5. T^2 is not defined.

7. $(HG)(a, b, c) = (b + c) + (2a + b)x + (-a)x^2$

9. $(3K - 2S)(x, y, z) = (3x - 2z, 3x + y, x + 3y + 3z)$

11. $(3H)\left(\begin{bmatrix} a & b \\ c & d \end{bmatrix}\right) = (3a + 3b) + (3a - 6d)x + (3d)x^2$

13. $\begin{bmatrix} 1 & 0 & 0 \\ 1 & 1 & 0 \\ 1 & 1 & 1 \end{bmatrix}$

15. $\begin{bmatrix} 0 & 0 & 1 \\ 0 & 1 & 0 \\ 1 & 0 & 0 \end{bmatrix}$

17. $\begin{bmatrix} 1 & -1 & 0 \\ 3 & 2 & 0 \end{bmatrix}$

19. $\begin{bmatrix} 1 & 0 & 0 \\ 0 & 1 & 0 \\ 0 & 0 & 1 \end{bmatrix}$

SECTION 6.3 (Page 264)

1. $\mathbf{v}_1$ is; $\mathbf{v}_2$ is

3. $\mathbf{v}_1$ is; $\mathbf{v}_2$ is not

5. $\mathbf{v}_1$ is not; $\mathbf{v}_2 = \mathbf{0}$ is *always* in Ker (T)

7. $\mathbf{w}_1$ is not; $\mathbf{w}_2$ is

9. $\mathbf{w}_1$ is; $\mathbf{w}_2$ is

11. $\mathbf{w}_1$ is; $\mathbf{w}_2$ is not

13. nullity 2, basis for Ker $(T) - \{(-1, 1, 1, 0), (-1, 0, 0, 1)\}$, not one-to-one

15. nullity 1, basis for Ker $(T) - \{(\frac{3}{5}, -\frac{4}{5}, 1)\}$, not one-to-one

17. nullity 0, Ker $(T) = \mathbf{0}$, one-to-one

19. rank 2, basis for Im $(T) - \{(1, 1, 2), (2, 0, -4)\}$, not onto

21. rank 2, basis for Im $(T) - \{(3, 1), (1, 2)\}$, onto

23. rank 3, basis for Im $(T) - \left\{\begin{bmatrix} 0 & -1 \\ 1 & 0 \end{bmatrix}, \begin{bmatrix} 1 & 1 \\ 1 & 2 \end{bmatrix}, \begin{bmatrix} 0 & 0 \\ 0 & -1 \end{bmatrix}\right\}$, not onto

29. $T^{-1}(x, y, z) = (\frac{3}{2}x - \frac{1}{2}y - \frac{1}{2}z, x - y, -\frac{1}{2}x + \frac{1}{2}y + \frac{1}{2}z)$

31. not invertible

33. $T_{-\theta}$

35. $T_{1/c}$ if $c \neq 0$, not invertible if $c = 0$

SECTION 6.4 (Page 271)

1. $\begin{bmatrix} 0 & 1 & 0 \\ 0 & 0 & 1 \\ -1 & 0 & 0 \\ 0 & 0 & 0 \end{bmatrix}$

3. $\begin{bmatrix} 1 & 0 & 0 \\ 0 & 0 & -1 \\ 0 & 0 & -1 \\ 1 & -1 & 0 \end{bmatrix}$

5. Ker $(T) = \{\mathbf{0}\}$; one-to-one

7. Ker $(T) = \{\mathbf{0}\}$; one-to-one

9. Im $(T) = \left\{ \begin{bmatrix} a & b \\ c & 0 \end{bmatrix} \middle| a, b, c \in \mathbf{R} \right\}$

11. Im $(T) = \left\{ \begin{bmatrix} a & b \\ b & c \end{bmatrix} \middle| a, b, c \in \mathbf{R} \right\}$

13. $\begin{bmatrix} 0 & 1 & 0 & 0 & 0 \\ 0 & 0 & 2 & 0 & 0 \\ 0 & 0 & 0 & 3 & 0 \\ 0 & 0 & 0 & 0 & 4 \end{bmatrix}$

15. $\begin{bmatrix} 0 & 0 & 0 \\ 1 & 0 & 0 \\ 0 & \frac{1}{2} & 0 \\ 0 & 0 & \frac{1}{3} \end{bmatrix}$

SECTION 6.5 (Page 278)

1. $\frac{1}{4}\begin{bmatrix} 2 & 1 \\ 2 & -1 \end{bmatrix}$, $(\frac{1}{4})(1, -5)$

3. $\begin{bmatrix} 1 & 1 & -1 \\ -1 & -1 & 2 \\ -1 & 0 & 1 \end{bmatrix}$, $(1, 0, -2)$

5. $\frac{1}{5}\begin{bmatrix} 3 & -1 & 2 & -2 \\ -3 & 1 & 3 & 2 \\ 3 & -1 & 2 & 3 \\ -1 & 2 & -4 & -1 \end{bmatrix}$, $(\frac{1}{5})(-8, 3, -3, 6)$

7. $P^{-1}AP = \begin{bmatrix} 1 & 0 \\ 2 & -1 \end{bmatrix}$

9. $P^{-1}AP = \begin{bmatrix} 4 & 6 & -4 \\ -1 & -5 & 5 \\ -2 & -3 & 1 \end{bmatrix}$

11. $P^{-1}AP = \begin{bmatrix} 4 & 5 & -4 \\ -4 & -6 & 5 \\ -1 & -3 & 4 \end{bmatrix}$

13. $P^{-1}AP = \frac{1}{5}\begin{bmatrix} 14 & 12 & 8 & 15 \\ 11 & 13 & 12 & 20 \\ 24 & 17 & 3 & 15 \\ -23 & -24 & -11 & -25 \end{bmatrix}$

REVIEW EXERCISES CHAPTER 6 (Page 282)

1. Verify all properties in *general*.

3. Produce a *specific* counterexample. For example,

$$S(0, 1) + S(1, 0) = (2, 0) + (1, 0) = (3, 0) \neq (3, 1) = S((0, 1) + (1, 0)).$$

5. not defined; D yields vectors in $\mathbf{R}^3$ but F operates on vectors in $\mathbf{R}^2$

7. $E^2(x, y, z) = (x, 4x + 2y + 2z, 5x + 2y + 2z)$

9. $\begin{bmatrix} 1 & 1 \\ 1 & -2 \\ 1 & 0 \end{bmatrix}$

11. Yes. If A is the matrix of Exercise 9, then $A\mathbf{x} = \mathbf{0}$ has only the trivial solution.

13. No; $\det A = 0$, where A is the matrix that represents E.

15. $\text{Ker}\,(D) = \{\mathbf{0}\}$

17. $\mathcal{B} = \{1, x, x^2, x^3\}$, $\mathcal{B}' = \{E^{11}, E^{12}, E^{21}, E^{22}\}$

$$A = \begin{bmatrix} 1 & 0 & 1 & 0 \\ 0 & 1 & 0 & -1 \\ 0 & 0 & 2 & 0 \\ 2 & 1 & 0 & -1 \end{bmatrix}$$

19. No; $\det A \neq 0$ and $\dim\,(\mathbf{P}_3) = \dim\,(\mathbf{M}^{2,2})$

21. $\begin{bmatrix} 0 & 0 & 0 \\ 1 & 1 & 0 \\ 2 & 2 & 2 \end{bmatrix}$

SECTION 7.1 (Page 291)

1. eigenvalue: 1; eigenvectors: every $\mathbf{x} \neq \mathbf{0}$ in $\mathbf{R}^n$

3. no eigenvalues

5. eigenvalue: 1; eigenvectors: $\mathbf{x} = (0, c)$, $c \neq 0$
eigenvalue: -1; eigenvectors: $\mathbf{x} = (c, 0)$, $c \neq 0$

7. eigenvalue: 0; eigenvectors: $\mathbf{x} = -a - 2bx + ax^2 + bx^3$, a or b nonzero
eigenvalue: 1; eigenvectors: $\mathbf{x} = -a$, $a \neq 0$
eigenvalue: 2; eigenvectors: $\mathbf{x} = ax^3$, $a \neq 0$

9. $\lambda^2 - 9 = 0$; $\lambda = \pm 3$

11. $\lambda^2 - 2\lambda + 1 = 0$; $\lambda = 1$

13. $-\lambda^3 + 2\lambda = 0$; $\lambda = 0, \pm\sqrt{2}$

15. $-\lambda^3 + 3\lambda^2 = 0$; $\lambda = 0, 3$

17. $(1 - \lambda)^2(-2 - \lambda) = 0$; $\lambda = 1, -2$

19. $\lambda^4 - 4\lambda^3 = 0$; $\lambda = 0, 4$

21. $\lambda = 3$: all $(2c, c)$, c in $\mathbf{R}$; basis $\{(2, 1)\}$
$\lambda = -3$: all $(-c, c)$, c in $\mathbf{R}$; basis $\{(-1, 1)\}$

23. $\lambda = 1$: all (c, c), c in $\mathbf{R}$; basis $\{(1, 1)\}$

25. $\lambda = 0$: all $(-c, 0, c)$, c in $\mathbf{R}$; basis $\{(-1, 0, 1)\}$
$\lambda = \sqrt{2}$: all $(c, -c\sqrt{2}, c)$, c in $\mathbf{R}$; basis $\{(1, -\sqrt{2}, 1)\}$
$\lambda = -\sqrt{2}$: all $(c, c\sqrt{2}, c)$, c in $\mathbf{R}$; basis $\{(1, \sqrt{2}, 1)\}$

27. $\lambda = 0$: all $(-c - d, c, d)$, c, d in $\mathbf{R}$; basis $\{(-1, 1, 0), (-1, 0, 1)\}$
$\lambda = 3$: all (c, c, c), c in $\mathbf{R}$; basis $\{(1, 1, 1)\}$

29. $\lambda = 1$: all $(c, 0, 0)$, c in $\mathbf{R}$; basis $\{(1, 0, 0)\}$
$\lambda = -2$: all $(0, c, 0)$, c in $\mathbf{R}$; basis $\{(0, 1, 0)\}$

31. $\lambda = 0$: all $(-b - c - d, b, c, d)$, b, c, d in $\mathbf{R}$; basis $\{(-1, 1, 0, 0), (-1, 0, 1, 0), (-1, 0, 0, 1)\}$
$\lambda = 4$: all (c, c, c, c), c in $\mathbf{R}$; basis $\{(1, 1, 1, 1)\}$

33. $1, 3, 4$ **35.** $1, 3, 6, 10$

39. no eigenvalues except for $\theta = 0, \lambda = 1$ and $\theta = \pi, \lambda = -1$

SECTION 7.2 (Page 303)

1. diagonalizable; $P = \begin{bmatrix} -1 & 2 \\ 1 & 1 \end{bmatrix}$; $D = \begin{bmatrix} -3 & 0 \\ 0 & 3 \end{bmatrix}$

3. not diagonalizable

5. diagonalizable; $P = \begin{bmatrix} 1 & -1 & 1 \\ \sqrt{2} & 0 & -\sqrt{2} \\ 1 & 1 & 1 \end{bmatrix}$; $D = \begin{bmatrix} -\sqrt{2} & 0 & 0 \\ 0 & 0 & 0 \\ 0 & 0 & \sqrt{2} \end{bmatrix}$

7. diagonalizable; $P = \begin{bmatrix} -1 & -1 & 1 \\ 1 & 0 & 1 \\ 0 & 1 & 1 \end{bmatrix}$; $D = \begin{bmatrix} 0 & 0 & 0 \\ 0 & 0 & 0 \\ 0 & 0 & 3 \end{bmatrix}$

9. not diagonalizable

11. diagonalizable; $P = \begin{bmatrix} -1 & -1 & -1 & 1 \\ 1 & 0 & 0 & 1 \\ 0 & 1 & 0 & 1 \\ 0 & 0 & 1 & 1 \end{bmatrix}$; $D = \begin{bmatrix} 0 & 0 & 0 & 0 \\ 0 & 0 & 0 & 0 \\ 0 & 0 & 0 & 0 \\ 0 & 0 & 0 & 4 \end{bmatrix}$

13. orthogonal   **15.** not orthogonal

17. $\begin{bmatrix} 1/\sqrt{2} & 1/\sqrt{2} \\ 1/\sqrt{2} & -1/\sqrt{2} \end{bmatrix}$

19. $\begin{bmatrix} 1/2 & -1/\sqrt{2} & 1/2 \\ 1/\sqrt{2} & 0 & -1/\sqrt{2} \\ 1/2 & 1/\sqrt{2} & 1/2 \end{bmatrix}$

21. $\begin{bmatrix} 1/\sqrt{2} & 1/\sqrt{6} & 1/\sqrt{3} \\ -1/\sqrt{2} & 1/\sqrt{6} & 1/\sqrt{3} \\ 0 & -2/\sqrt{6} & 1/\sqrt{3} \end{bmatrix}$

23. $\begin{bmatrix} 1/\sqrt{2} & 0 & 1/\sqrt{2} & 0 \\ -1/\sqrt{2} & 0 & 1/\sqrt{2} & 0 \\ 0 & 1/\sqrt{2} & 0 & 1/\sqrt{2} \\ 0 & -1/\sqrt{2} & 0 & 1/\sqrt{2} \end{bmatrix}$

27. A is diagonalizable if $(a - d)^2 + 4bc > 0$ or if $a = d$ and $b = c = 0$. A is not diagonalizable if $(a - d)^2 + 4bc < 0$.

SECTION 7.3 (Page 313)

1. -2 **3.** 2 **5.** $(i, 0, 0)$

7. $\begin{bmatrix} i & -2i & 1 \\ -1 + i & i & 1 + 2i \\ -1 - i & 3 & -2 + 2i \end{bmatrix}$

9. $\dfrac{1}{5}\begin{bmatrix} 8 - i & 2 + i & 1 + 3i \\ 2 + i & 3 - i & -1 + 2i \\ -1 - 3i & 1 - 2i & 3 - i \end{bmatrix}$

11. $1 + 2i$

13. Neither A nor B is hermitian.

15. yes   **17.** no

19. $\{(-1, 0, 1), (0, -1, 1)\}$ **21.** $\{(1, 1)\}$

23. $\lambda_1 = -5$, $\mathbf{x}_1 = (-(3 + 4i)s, 5s)$, basis $= \{(-(3 + 4i), 5)\}$
$\lambda_2 = 5$, $\mathbf{x}_2 = ((3 + 4i)t, 5t)$, basis $= \{(3 + 4i, 5)\}$

25. $\lambda_1 = 1$, $\mathbf{x}_1 = (0, r, 0)$, basis $= \{(0, 1, 0)\}$;
$\lambda_2 = -2i$, $\mathbf{x}_2 = ((1 - i)s, 0, s)$, basis $= \{(1 - i, 0, 1)\}$
$\lambda_3 = 2i$, $\mathbf{x}_3 = (-(1 + i)t, 0, t)$, basis $= \{(-1 - i, 0, 1)\}$

27. $\lambda_1 = i$, $\mathbf{x}_1 = (s, t, 0)$, basis $= \{(1, 0, 0), (0, 1, 0)\}$

29. $\lambda_1 = -1$, $\mathbf{x}_1 = (0, ir, r)$, basis $= \{(0, i, 1)\}$
$\lambda_2 = 1$, $\mathbf{x}_2 = (s, -it, t)$, basis $= \{(1, 0, 0), (0, -i, 1)\}$

31. $P = \begin{bmatrix} -(3 + 4i)/(5\sqrt{2}) & (3 + 4i)/(5\sqrt{2}) \\ 1/\sqrt{2} & 1/\sqrt{2} \end{bmatrix}$, $D = \begin{bmatrix} -5 & 0 \\ 0 & 5 \end{bmatrix}$ (hermitian)

33. $P = \begin{bmatrix} 0 & 1 - i & 1 + i \\ 1 & 0 & 0 \\ 0 & 1 & 1 \end{bmatrix}$, $D = \begin{bmatrix} 1 & 0 & 0 \\ 0 & -2i & 0 \\ 0 & 0 & 2i \end{bmatrix}$

35. not diagonalizable

37. $P = \begin{bmatrix} 0 & 1 & 0 \\ i/\sqrt{2} & 0 & -i/\sqrt{2} \\ 1/\sqrt{2} & 0 & 1/\sqrt{2} \end{bmatrix}$, $D = \begin{bmatrix} -1 & 0 & 0 \\ 0 & 1 & 0 \\ 0 & 0 & 1 \end{bmatrix}$ (hermitian)

39. Matrices of Exercises 29 and 30 are unitary.

REVIEW EXERCISES CHAPTER 7 (Page 317)

1. $\lambda^2 - 4\lambda + 3 = 0$, $\lambda = 1, 3$

3. $(1 - \lambda)(1 - \lambda)(2 - \lambda) = 0$, $\lambda = 1, 2$

5. $(1 - \lambda)(3 - \lambda)(5 - \lambda)(7 - \lambda) = 0$, $\lambda = 1, 3, 5, 7$

7. $\lambda = 1$: $\{(-1, 1)\}$, $\lambda = 3$: $\{(1, 1)\}$

9. All are diagonalizable.

11. $\begin{bmatrix} -1 & 0 & 2 \\ 1 & -1 & 1 \\ 1 & 1 & 1 \end{bmatrix}$

13. yes

15. $\begin{bmatrix} 1/\sqrt{2} & 1/\sqrt{2} \\ -1/\sqrt{2} & 1/\sqrt{2} \end{bmatrix} \begin{bmatrix} -1/\sqrt{3} & 0 & 2/\sqrt{6} \\ 1/\sqrt{3} & -1/\sqrt{2} & 1/\sqrt{6} \\ 1/\sqrt{3} & 1/\sqrt{2} & 1/\sqrt{6} \end{bmatrix}$

17. basis

19. no; $\det A = 0$

21. 1: $\{(0, 1, 0)\}$
$(3 + \sqrt{5})/2$: $\{((-1 + \sqrt{5})i/2, 0, 1)\}$
$(3 - \sqrt{5})/2$: $\{((-1 - \sqrt{5})i/2, 0, 1)\}$

23. Each has a basis of eigenvectors (distinct eigenvalues).

25. $P = \begin{bmatrix} 0 & \alpha(-1 + \sqrt{5})i/2 & \beta(-1 - \sqrt{5})i/2 \\ 1 & 0 & 0 \\ 0 & \alpha & \beta \end{bmatrix}$, $\alpha = \dfrac{\sqrt{2}}{\sqrt{5 - \sqrt{5}}}$, $\beta = \dfrac{\sqrt{2}}{\sqrt{5 + \sqrt{5}}}$

SECTION 8.1 (Page 324)

1. $p(x) = -3 + 2x + x^2$

5. $p(x) = 1 - 2x^2 + x^3$

3. $p(x) = -1 - x^2 + x^3$

7. $p(2.5) = 0.4045$

SECTION 8.2 (Page 329)

1. $p(x) = -(4/5) + (13/10)x$

5.
$$
\begin{array}{rcl}
7b_1 & + \; 28b_3 & = -28 \\
28b_2 & + \; 196b_4 & = \;\; 0 \\
28b_1 & + \; 196b_3 & = \;\; 20 \\
196b_2 & + \; 1588b_4 & = \;\; 0
\end{array}
$$

3. $p(x) = (3.0) - (13.7)x + (7.5)x^2$

7. $p(x) = 303 - 86x$

SECTION 8.3 (Page 334)

1. current through $a: 5/6$, $b: 1/6$, $e: 1$

3. current through $a: 1.08$, $b: 0.69$, $c: 0.39$, $d: 0.80$, $e: 0.41$

5. $x_a = \frac{4}{5} - \frac{2}{5}s + \frac{2}{5}t$

$x_b = \frac{2}{3} + \frac{2}{3}s - \frac{2}{3}t$

$x_c = \qquad s$

$x_d = \frac{1}{2} \qquad - \frac{1}{2}t$

$x_e = \qquad\qquad t$

and all $x_i \geq 0$. These conditions yield the following restrictions on s and t:

$$0 \leq s$$
$$0 \leq t \leq 1$$
$$t \geq s - 2.$$

That is, (s, t) in the region:

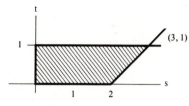

7. $x_1 = 36.7$, $x_2 = 49.3$, $x_3 = 17.9$, $x_4 = 50.7$, $x_5 = 18.8$, $x_6 = -1.4$, $x_7 = 17.3$

9. same solution as Exercise 5

SECTION 8.4 (Page 340)

1. A: $28s$, B: $10s$, C: $15s$ for $s > 0$ **3.** A: $2s$, B: s, C: s for $s > 0$

5. A: 12.51, B: 18.80, C: 9.46 **7.** A: 55.8, B: 45.2, C: 21.3

SECTION 8.5 (Page 347)

1. $t_{32} = 0.4$ **3.** $T = \begin{bmatrix} 0.2 & 0.1 & 0.2 \\ 0.5 & 0.8 & 0.4 \\ 0.3 & 0.1 & 0.4 \end{bmatrix}$

5. $(1/3, 1/3, 1/3)$ **7.** $(1/6, 17/30, 4/15)$ **9.** 0.62

11.

	W	D	B
W	0	$\frac{2}{3}$	$\frac{1}{3}$
D	$\frac{1}{2}$	0	$\frac{1}{2}$
B	$\frac{1}{3}$	$\frac{1}{3}$	$\frac{1}{3}$

$T = \begin{bmatrix} 0 & \frac{1}{2} & \frac{1}{3} \\ \frac{2}{3} & 0 & \frac{1}{3} \\ \frac{1}{3} & \frac{1}{2} & \frac{1}{3} \end{bmatrix}$

13. $(5/18, 1/3, 7/18)$ **15.** $1/81$

17. $(1/(2s + t))(s, s, t); s, t \geq 0, s + t > 0$ **21.** $(0.131, 0.689, 0.180)$

SECTION 8.6 (Page 358)

1. $(x')^2 + (y')^2 - \dfrac{(z')^2}{2} = 1$ (hyperboloid of one sheet)

3. $\dfrac{(x')^2}{4} + \dfrac{(y')^2}{24} - \dfrac{(z')^2}{6} = 0$ (cone)

5. $\dfrac{(x')^2}{18} + \dfrac{(y')^2}{36} + \dfrac{(z')^2}{36} = 1$ (ellipsoid)

7. $8(x'')^2 + 2(y'')^2 - 2(z'')^2 = 1$ (hyperboloid of one sheet)

9. $(x'')^2 - \dfrac{(z'')^2}{2} = 1$ (hyperbolic cylinder)

15. $3(x')^2 - (y')^2 + 6x' + 4y' = 0$

17. $-3(x'')^2 + (y'')^2 = 1$ (hyperbola)

SECTION 8.7 (Page 368)

1. $p(x) = 1/5$

3. $p(x) = 2(14 \ln 2 - 9) + 6(2 - 3 \ln 2)x$

5. $p(x) = -(3/35) + (6/7)x^2$

7. $p(x) = (873 \ln 2 - 603) + 6(137 - 198 \ln 2)x + 30(13 \ln 2 - 9)x^2$

9. $p(x) = -(3/35) + (6/7)x^2$

13. $p(x) = 1 + 2(\pi^2 - 6) \sin x$

11. $p(x) = 1 + 2 \sin x$

15. $p(x) = 1 + \cos x - \cos 2x$
$-1 + 2 \sin x - \sin 2x$

SECTION 8.8 (Page 380)

1. $y = c_1 e^x + c_2 e^{2x}$

5. $u = c_1 e^{-2x} + c_2 x e^{-2x}$

9. $y = e^{-x}(c_1 \cos 2x + c_2 \sin 2x)$

13. $s = c_1 e^t + c_2 e^{-t} + c_3 \cos t + c_4 \sin t$

17. $y = c_1 e^{3x} + c_2 e^{-x} - 2 \sin x + \cos x$

21. $y = -(1/4)e^{-3x} + (5/4)e^x$

25. $u = -2 - 3x + 3e^x - e^{-x}$

29. $\theta(t) = \frac{1}{4} \sin 4t$

3. $y = c_1 e^{(-1+\sqrt{2})x} + c_2 e^{(-1-\sqrt{2})x}$

7. $s = c_1 \cos (t/2) + c_2 \sin (t/2)$

11. $u = c_1 + c_2 e^{-x} + c_3 x e^{-x}$

15. $s = c_1 + c_2 t + e^t(c_3 \cos (\sqrt{3}t) + c_4 \sin (\sqrt{3}t))$

19. $u = c_1 \cos x + c_2 \sin x + e^x + e^{-x}$

23. $s = \sin t$

27. $s(t) = -2(\cos 3t - \sin 3t)$

SECTION 9.1 (Page 390)

1.

3.

5.

7.

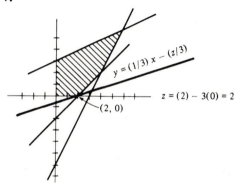

9.
$$x - y + s_1 \qquad = 2$$
$$3x - y \qquad + s_2 = 6$$
$$x, y, s_1, s_2 \geq 0$$

11. row-reduced
columns 2, 3, 5
$x_1 = 0, x_2 = 1, x_3 = 2, x_4 = 0, x_5 = 4$

13. row-reduced
columns 2, 3, 5
$x_1 = 0, x_2 = 6, x_3 = 2, x_4 = 0, x_5 = 1$
or columns 3, 4, 5
$x_1 = x_2 = 0, x_3 = 2, x_4 = 6, x_5 = 1$

15. Let x_1 be amount of savings, x_2 stocks, and x_3 property.

Standard form:
$$x_1 + x_2 + x_3 \leq 50{,}000$$
$$x_1 \qquad \leq 10{,}000$$
$$x_1 + x_2 \qquad \leq 20{,}000$$
$$x_1, x_2, x_3 \geq 0$$
$$z = 0.08x_1 + 0.15x_2 + 0.12x_3$$

Equation form:
$$x_1 + x_2 + x_3 + x_4 \qquad = 50{,}000$$
$$x_1 \qquad + x_5 \qquad = 10{,}000$$
$$x_1 + x_2 \qquad + x_6 = 20{,}000$$
$$x_1, x_2, x_3, x_4, x_5, x_6 \geq 0$$
$$z = 0.08x_1 + 0.15x_2 + 0.12x_3$$

SECTION 9.2 (Page 402)

1. bottom row:

-9	1	0	0	0	1

3.

	c^*	x_1	x_2	x_3	x_4	x_5	b
x_2	-2	3	1	2	0	0	2
x_5	2	1	0	-3	0	1	5
x_4	0	2	0	-1	1	0	8
		-5	0	-11	0	0	6

5.

	c^*	x_1	x_2	x_3	x_4	x_5	x_6	b
x_5	0	2	1	1	3	1	0	12
x_6	0	1	-3	-1	1	0	1	8
		-1	2	-1	-4	0	0	0

7. Solution of $(0, 2, 1, 0, 4)$ is maximal, since no simplex indicators are negative.

9. Solution of $(11, 0, 8, 0, 0, 0)$ is minimal, since no simplex indicators are positive.

11. $x_1 = x_2 = 0, x_3 = 1, x_4 = 9, x_5 = 8, z = 17$

13. $x_2 = 12, x_1 = x_3 = x_4 = 0, z = -24$

15. savings \$0, stocks \$20,000, property \$30,000, earnings \$6600

SECTION 9.3 (Page 408)

1.
$$2x - 3y - z + s_1 \qquad = 5$$
$$y + 2z \qquad - s_2 + a_1 \qquad = 8$$
$$x \qquad - 2z \qquad + a_2 = 3$$
$$x, y, z, s_1, s_2, a_1, a_2 \geq 0$$

3.
$$2x - y \qquad + a_1 \qquad = 7$$
$$x + 2y + s_1 \qquad = 18$$
$$-3x + 5y \qquad - s_2 \qquad + a_2 = 3$$
$$x, y, s_1, s_2, a_1, a_2 \geq 0$$

5. $x = 3, y = 8, z = 0, s_1 = 23, s_2 = 0$

7. $x_1 = 38/7, x_2 = 27/7, s_1 = 34/7, s_2 = 0$

9. No maximum exists.

11. $x = \frac{32}{5}, y = \frac{29}{5}$, minimum value: $-\frac{23}{5}$

13. $x_1 = 7, x_2 = x_3 = 0, x_4 = 12$, maximum value: 38

15. A: 50%, B: $16\frac{2}{3}$%, C: $33\frac{1}{3}$%, D: 0%; cost \$0.26$\frac{2}{3}$/lb.

INDEX